Carbohydrate Chemistry

Volume 30

A Specialist Periodical Report

Carbohydrate Chemistry

Monosaccharides, Disaccharides and Specific Oligosaccharides

Volume 30

A Review of the Literature Published during 1996

Senior Reporter
R.J. Ferrier, *Victoria University of Wellington, New Zealand*

Reporters
R. Blattner, *Industrial Research Limited, Lower Hutt, New Zealand*
K. Clinch, *Industrial Research Limited, Lower Hutt, New Zealand*
R.H. Furneaux, *Industrial Research Limited, Lower Hutt, New Zealand*
J.M. Gardiner, *UMIST, Manchester, UK*
P. B. Tyler, *Industrial Research Limited, Lower Hutt, New Zealand*
R.H. Wightman, *Heriot Watt University, Edinburgh, UK*

THE ROYAL
SOCIETY OF
CHEMISTRY
Information
Services

ISBN 0-85404-218-0
ISSN 0951-8428

Published by The Royal Society of Chemistry,
Thomas Graham House, The Science Park, Milton Road, Cambridge CB4 0WF, UK

For further information see our web site at www.rsc.org

Phototypeset by Computape (Pickering) Ltd, Pickering, North Yorkshire
Printed by Athenaeum Press Ltd, Gateshead, Tyne & Wear

Preface

Much is made these days of the increase in activity in carbohydrate chemistry and of the emergence of glycobiology as a major branch of biochemistry, and while such data as attendances at conferences and the nature of programmes at the meetings bear these out, this Specialist Periodical Report offers an alternative means of quantifying the changes. In the table, numbers of papers abstracted in this series under different chapter headings are given for volumes 10, 15, 20, 25 and 30, these exemplifying the literature published between 1976 and 1996. Since the criteria used for selecting papers to be noted here remained the same over the period (the intention being to cite every relevant publication), the data certainly confirm the view that the subject has boomed in the last 20 years. While the overall volume of publication of relevant papers has gone up almost linearly, the 1976 number of papers doubling in the following 10 years and almost tripling in 20 years, the growth has been anything but even across the subject, and the excitement associated with the new-found significances in biology is clearly reflected in the attention that has been given to the synthesis of oligosaccharides and other glycosidic compounds. The rate of growth in activity in the cyclitols is likewise extreme, reflecting the biochemical importance of the inositol phosphates. Other increasingly favoured areas are nucleosides, nitrogen-containing sugars and related compounds and sugar acids, all these groups of compounds featuring importantly in topics related to matters of significance in biochemistry and glycobiology.

The increase in use of carbohydrates for the preparation of enantiomerically pure non-carbohydrate products has been emphatically demonstrated, while the reliance on physical methods has risen in line with methological advances – notably those in X-ray diffraction analysis.

Expectations that the reviewing team would be expanded to meet the increased pressure have not been met, and the current members are thanked most sincerely for accepting significantly increasing work loads. Not only have rising numbers of papers to be abstracted, but they have to be condensed with ever-increasing severity. In the last 10 years there has been a 50% increase in the number of relevant papers to be dealt with and a 33% increase in available space. Additionally, the reviewers are now responsible for providing texts on disc and for (at least) managing the art work, the majority of which is done by Royal Society of Chemistry staff. Mr Alan Cubitt and Mrs Janet Freshwater assist with this and all liaison, and we are most appreciative of their co-operation and support.

R.J. Ferrier
July, 1998

Table *Number of papers cited in the chapters of specialist periodical reports 1976-1996*

Volume Year of Literature	10 1976	15 1981	20 1986	25 1991	30 1996
Free Sugars	47	79	87	72	75
Glycosides and Disaccharides	91	198	168	251	334
Oligosaccharides	–	–	75	109	227
Ethers and Anhydro-sugars	36	43	46	40	54
Acetals	14	32	24	7	31
Esters	74	78	78	92	87
Halogeno-sugars	20	30	40	22	29
Amino-sugars	59	57	68	71	84
Miscellaneous Nitrogen Derivatives	43	43	55	96	103
Thio and Seleno-sugars	18	21	10	25	48
Deoxy-sugars	8	15	31	36	24
Unsaturated Derivatives	14	40	36	26	47
Branched-chain Sugars	24	25	40	60	48
Aldosuloses and Other Dicarbonyl Compounds	13	13	13	9	20
Sugar Acids and Lactones	31	66	62	88	95
Inorganic Derivatives	8	23	37	35	56
Alditols and Cyclitols	17	46	64	139	184
Antibiotics	51	121	89	122	89
Nucleosides	154	151	174	300	343
NMR Spectroscopy and Conformational Features	60	63	65	89	90
Other Physical Methods	93	115	116	153	226
Separatory and Analytical Methods	53	96	77	122	23*
Synthesis of Enantiomerically Pure Non-carbohydrate Compounds	5	37	98	115	127
	933	1392	1553	2079	2444

*This figure is anomalous. See Chapter 23.

Contents

Abbreviations

The following abbreviations have been used:

Ac	acetyl
Ade	adenin-9-yl
AIBN	2,2-azobisisobutyronitrile
All	allyl
Ar	aryl
Ara	arabinose
Asp	aspartic acid
BBN	9-borabicyclo[3.3.3]nonane
Bn	benzyl
Boc	t-tutoxycarbonyl
Bu	butyl
Bz	benzoyl
CAN	ceric ammonium nitrate
Cbz	benzyloxycarbonyl
CD	circular dichroism
Cer	ceramide
CI	chemical ionization
Cp	cyclopentadienyl
Cyt	cytosin-1-yl
Dahp	3-deoxy-D-*arabino*-2-heptulosonic acid 7-phosphate
DAST	diethylaminosulfur trifluoride
DBU	1,5-diazabicyclo[5.5.0]undec-5-ene
DCC	dicyclohexylcarbodi-imide
DDQ	2,3-dichloro-5,6-dicyano-1,4-benzoquinone
DEAD	dietyl azodicarboxylate
DIBAL	di-isobutylaluminium hydride
DMAD	dimethylacetylene dicarboxylate
DMAP	4-dimethylaminopyridine
DMF	*N,N*-dimethylformamide
DMSO	dimethyl sulfoxide
Dmtr	dimethoxytrityl
e.e.	enantiomeric excess
Ee	1-ethoxyethyl
ESR	electron spin resonance
Et	ethyl
FAB	fast-atom bombardment

Fmoc	9-fluorenylmethylcarbonyl
Fru	fructose
FTIR	Fourier transform infrared
Fuc	fuctose
Gal	galactose
GalNAc	2-acetamido-2-deoxy-D-glactose
GLC	gas liquid chromatography
Glc	glucose
GlcNAc	2-acetamido-2-deoxy-D-glucose
Gly	glycine
Gua	guanin-9-yl
Hep	L-*glycero*-D-*manno*-heptose
HMPA	hexmethylphosphoric triamide
HMPT	hexamethylphosphorous triamide
HPLC	high performance liquid chromatography
IDCP	iodonium dicollidine perchlorate
Ido	idose
Im	imidazolyl
IR	infrared
Kdo	3-deoxy-D-*manno*-2-octulosonic acid
LAH	lithium aluminium hydride
LDA	lithium di-isopropylamide
Leu	leucine
LTBH	lithium triethylborohydride
Lyx	lyxose
Man	mannose
MCPBA	*m*-chloropherbenzoic acid
Me	methyl
Mem	methoxyethoxymethyl
Mmtr	monomethoxytrityl
Mom	methoxymethyl
MS	methanesulfonyl (mesyl)
NMR	nuclear magnetic resonance
NAD	nicotinamide adenine dinucleotide
NBS	*N*-bromosuccinimide
NeuNAc	*N*-acetylneuraminic acid
NIS	*N*-iodosuccinimide
NMNO	*N*-methylmorpholine-*N*-oxide
NOE	nuclear Overhauser effect
ORD	optical rotatory dispersion
PCC	pyridinium chlorochromate
PDC	pyridinium dichromate
Ph	phenyl
Phe	phenylalanine
Piv	pivaloyl
Pmb	*p*-methoxybenzyl

Pr	propyl
Pro	proline
p.t.c.	phase transfer catalysis
Py	pyridine
Rha	rhamnose
Rib	ribose
Ser	serine
SIMS	secondary-ion mass spectrometry
TASF	tris(dimethylamino)sulfonium(trimethylsilyl)difluoride
Tbdms	*t*-butyldimethylsilyl
Tbdps	*t*-butyldiphenylsilyl
Tf	trifluoromethanesulfonyl(trifyl)
Tfa	trifluoroacetyl
TFA	trifluoroacetic acid
THF	tetrahydrofuran
Thp	tetrahydropyranyl
Thr	threonine
Thy	thymin-1-yl
Tips	1,1,3,3-tetraisopropyldisilox-1,3-diyl
TLC	thin layer chromatography
Tms	trimethylsilyl
TPP	triphenylphosphine
Tps	tri-isopropylbenzenesulfonyl
Tr	triphenylmethyl(trityl)
Ts	toluene-*p*-sulfonyl(tosyl)
Ura	uracil-1-yl
UDP	uridine diphosphate
UDPG	uridine diphosphate glucose
UV	ultraviolet
Xyl	xylose

1
Introduction and General Aspects

The very unusual absence of a new volume of *Advances in Carbohydrate Chemistry and Biochemistry* limits the number of quality reviews to have been published this year, and those that have appeared tend to be referred to at the beginnings of relevant chapters. This chapter is therefore unusually brief.

Danishefsky and Roberge have summarized the Sloan-Kettering work on the use of glycals as glycosyl donors and acceptors in the synthesis of oligosaccharides of glycoconjugates.[1] The use of enzymically derived *cis*-dihydroxylated aromatics as starting materials for the iterative preparation of oligo- nor-saccharides, -inositols and pseudo-sugars has been surveyed.[2]

In the field of origin of life studies Eschenmoser and Kisakürek have discussed the potential biogenic relationship between hexose- and pentose-based nucleic acids,[3] and a paper has appeared from the same group on the base pairing between oligonucleotides comprising pentopyranosyl D- and L- nucleotides.[4]

References

1   S.J. Danishefsky and J.Y. Roberge, *Pure Appl. Chem.*, 1995, **67**, 1647.
2   T. Hudlicky, K.A. Abboud, D.A. Entwistle, R. Fan, R. Maurya, A.J. Thorpe, J. Bolonick and B. Myers, *Synthesis*, 1996, 897.
3   A. Eschenmoser and M.V. Kisakürek, *Helv. Chim. Acta*, 1996, **79**, 1249.
4   R. Krishnamurthy, S. Pitsch, M. Minton, C. Miculka, N. Windhab and A. Eschenmoser, *Angew. Chem., Int. Ed. Engl.*, 1996, **35**, 1537.

2
Free Sugars

1 Theoretical Aspects

Molecular dynamics simulation calculations aimed at providing better understanding of the relative sweetness of β-D-glucopyranose, β-D-galactopyranose, and α- and β-D-mannopyranose failed to verify the hydrophobic G site hypothesis proposed by Tinti and Nofre (*ACS Symp. Ser.*, 1991, **450**, Chapter 15).[1]

2 Synthesis

A comprehensive review (260 refs.) on the synthesis of carbohydrates from non-carbohydrate sources covers the use of benzene-derived diols and products of Sharpless asymmetric oxidation as starting materials, Dodoni's thiazole and Vogel's 'naked sugar' approaches, as well as the application of enzyme-catalysed aldol condensations.[2] The preparation of monosaccharides by enzyme-catalysed aldol condensations is also discussed in a review on recent advances in the chemoenzymic synthesis of carbohydrates and carbohydrate mimetics,[3] in parts of reviews on the formation of carbon-carbon bonds by enzymic asymmetric synthesis[4] and on carbohydrate-mediated biochemical recognition processes as potential targets for drug development,[5] as well as in connection with the introduction of three 'Aldol Reaction Kits' that provide dihydroxyacetone phosphate-dependent aldolases (27 refs.).[6] A further review deals with the synthesis of carbohydrates by application of the nitrile oxide 1,3-dipolar cycloaddition (13 refs.).[7]

A newly discovered NAD-dependent hydrogenase from celery oxidizes D-mannitol to D-mannose; several other pentitols and hexitols, especially those with the same absolute configuration at C-2 as that of D-mannitol, are oxidized to the corresponding aldoses at a slower rate.[7a]

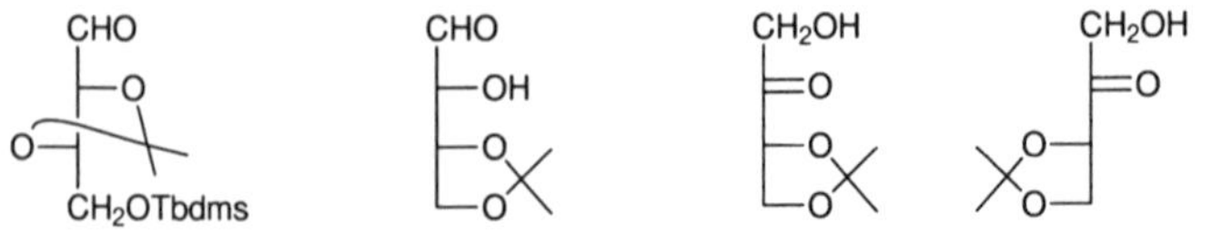

1 | **2** R = H, **3** R = Tbdms | **4** | **5**

Carbohydrate Chemistry, Volume 30

2.1   Tetroses and Pentoses – 4-*O*-*t*-Butyldimethylsilyl-2,3-*O*-isopropylidene-L-threose (**1**) has been prepared in seven efficient steps from D-xylose.[8] 3,4-*O*-Isopropylidene-D-eythrulose (**4**) has been synthesized from the known tetritol derivative **2** by primary protection as the silyl ether **3**, followed by Dess-Martin oxidation and desilylation. Compound **2** was derived from D-isoascorbic acid (see Vol. 22, p. 178, refs. 9,10). In a similar reaction sequence, the enantiomer **5** has been obtained from L-ascorbic acid.[9] The dehomologation of several di-*O*-isopropylidenehexofuranoses (*e.g.,* **6** → **7**) has been carried out in two steps without intermediate purification, by successive treatment with periodic acid in ethyl acetate, followed by sodium borohydride in ethanol.[10] Selective reduction of 3-deoxy-D-*glycero*-pentos-2-ulose (**8**) to 3-deoxy-D-*glycero*-pent-2-ose (**9**) has been achieved enzymically with aldose reductase and NADPH.[11] 4-Isopropyl-2-oxazolin-5-one (**10**) is a masked formaldehyde equivalent that is easily converted to an anion and demasked by mild acid hydrolysis. One of the three examples of its use in the synthesis of monosaccharides is shown in Scheme 1.[12]

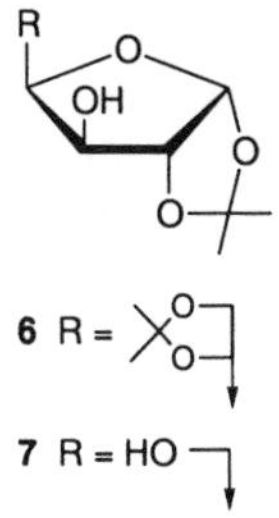

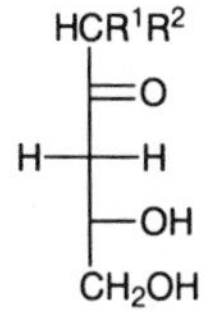

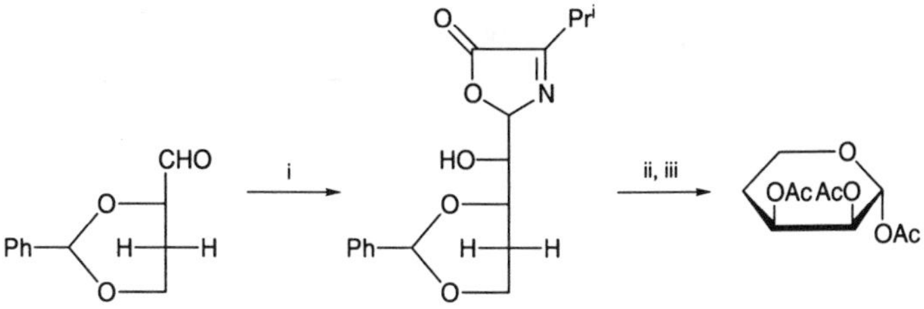

Reagents: i, **10**, Et$_3$N; ii, 5% aq. HCl; iii, Ac$_2$O, Py

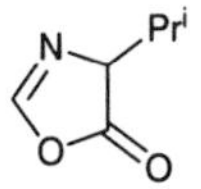

Scheme 1

The synthesis of [5-^{13}C]D-ribose from D-ribose involved diol cleavage of the original C-4–C-5 bond and formation of a new C-4–C-5 bond using a ^{13}C-enriched Wittig reagent.[13] 3,4,5,5-d_4-DL-Ribose (**12**) has been obtained from d_8-glycerol (**11**), as outlined in Scheme 2.[14]

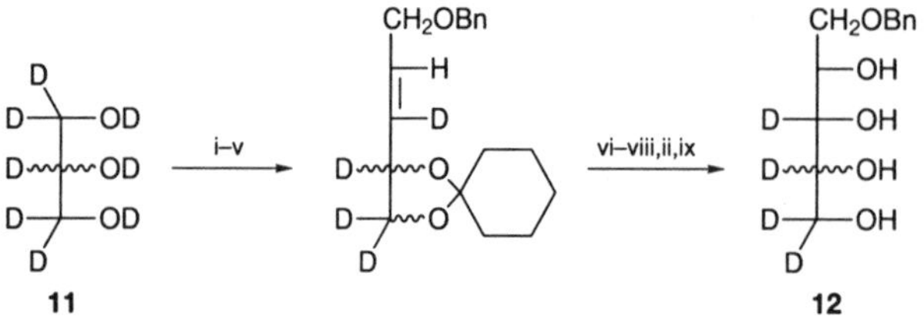

Reagents: i, cyclohexanone, $(MeO)_3CH$, H^+; ii, Swern; iii, $Ph_3P=CHCO_2Et$; iv, DiBAL
v, BnBr, NaH; vi, OsO_4, NMO; vii, 2-methoxypropene, H^+;
viii, Pd/C, H_2; ix, H_3O^+, THF

Scheme 2

2.2 Hexoses – The synthesis of hexoseptanose derivatives by ring expansion of uloses, Baeyer Villiger oxidation of inositols, or Baeyer-Fischer reaction of sugar dialdehydes has been reviewed.[15]

The methyl β-L-idoseptanoside derivative **14** was produced by oxidation/reduction $(Ru_2O\text{-}NaIO_4/NaBH_4)$ of the corresponding methyl β-D-glucoseptanoside **13**.[16] On transmetalation with $InCl_3$, δ-oxygenated allylic stannanes undergo *in situ* addition to aldehydes furnishing predominantly *anti* products. This method has been applied to the synthesis of D-altrose as shown in Scheme 3.[17] A similar route to α-L-daunosamine hydrochloride is covered in Chapter 9. The chemoenzymic synthesis of 6-substituted D-fructose analogues by use of epoxide **15** is referred to in Chapters 7, 8, 9 and 10, and that of 6-deoxy-L-sorbose in Chapter 12.

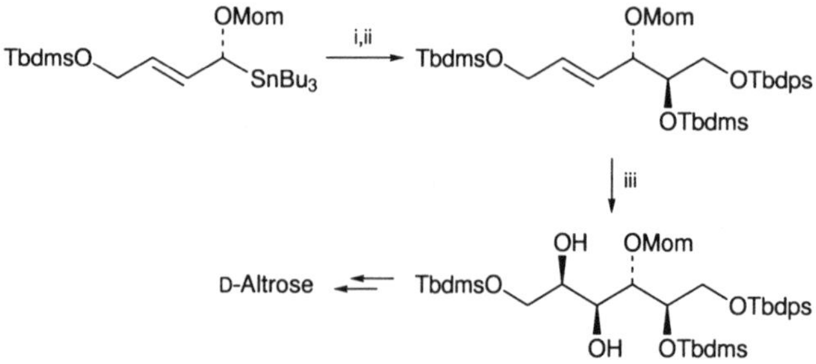

Reagents: i, TbdmsO CHO, $InCl_3$; ii, TbdmsCl, lutidine; iii, OsO_4, NMO

Scheme 3

The best conditions for producing glucose with close to 100% H-isotope labelling at C-1, with negligible formation of labelled by-products, by exchange with deuterium or tritium gas in aqueous solution have been determined.[18] Various ^{2}H-labelled D-mannoses, as well as 2,3,4,5,6-d_5- and 1,1,2,3,4,5,6,6-d_8-D-

13 R^1 = H, R^2 = OH
14 R^1 = OH, R^2 = H

mannitol, were obtained by ozonolysis/sodium borohydride reduction of the functionalized cyclohexene **16**, which was obtained in enantiomerically pure form from ^{2}H$_5$-chlorobenzene following biological hydroxylation.[19]

D-Tagatose 3-epimerase immobilized on Chitopearl beads effected the isomerization of L-sorbose and L-psicose to L-tagatose and L-fructose in 20 and 65% yield, respectively.[20] Sequential use of pyranose 2-oxidase and hydrogen over palladium permitted the one-pot chemoenzymatic conversion of D-galactose to D-tagatose in 30% yield,[21] whereas sodium borohydride reduction followed by microbial oxidation furnished D-sorbose from D-gulonolactone with high efficiency.[22] A remarkable, concurrent oxidation at C-5 and reduction at C-1 in 2,3,4,6-tetra-O-benzyl-D-galactose and -glucose on exposure to samarium(II) iodide, preoxidized with dry air in THF, offers entry into the L-sorbose and L-tagatose series. As a possible mechanism, Meerwein-Oppenauer-like oxidation/reduction *via* a transition state **17** has been proposed.[23]

On treatment with 5% sodium hydrogen carbonate for 1.5 h, ascorbigen (**18**) formed the unstable product **19** and, after longer exposure times, the 6-deoxy-L-tagatose derivative **20**.[24]

2.3 Chain-extended Sugars – The application of indium- and tin-mediated Barbier-Grignard reactions in aqueous solutions to the chain-extension of unprotected sugars has been included in two general reviews of these reac-

tions.[25,26] The two main approaches to the synthesis of the undecose system of tunicamycin, namely tail-to-tail coupling of two readily available sugar units and elongation at the non-reducing end of either D-ribose or D-galactosamine, have been reviewed (52 refs.).[27] The synthesis of a variety of higher carbon sugars by tail-to-tail coupling of simple monosaccharides is dealt with in a further review (44 refs.).[28]

The enzyme-catalyzed aldol condensation of ω-functionalized C_5- and C_6-aldehydes with Dhap (dihydroxacetone phosphate) has been used to synthesize unusual C_8- and C_9-sugars. An example is given in Scheme 4.[29] Coupling of carbohydrate carboxylic acids such as **21** and **23** with alkanoic acids by mixed

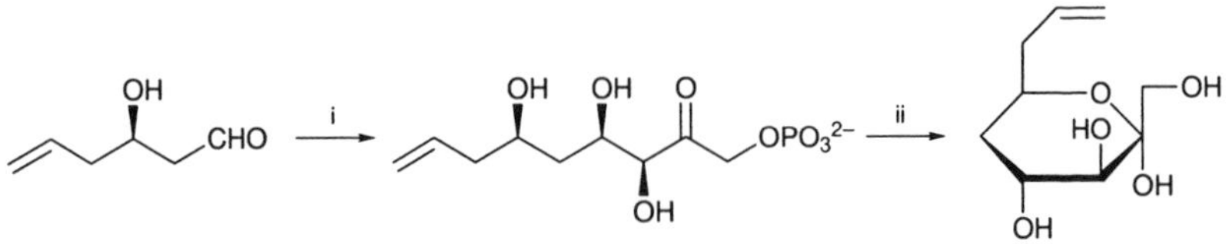

Reagents: i, Dhap, rabbit muscle aldolase; ii, phosphatase

Scheme 4

Kolbe electrolysis allowed chain-elongation at the 'reducing' or at the 'non-reducing' end, to give products such as **22** and **24**, respectively.[30] Chain-extension at the reducing as well as at the non-reducing end has also been brought about by free radical allylation of glycosyl bromides and primary bromo-deoxy sugars, respectively, with allylic sulfides or sulfones. *C*-Glycoside **25** and the branched trideoxynonuronic acid derivative **26** are typical products of this procedure.[31]

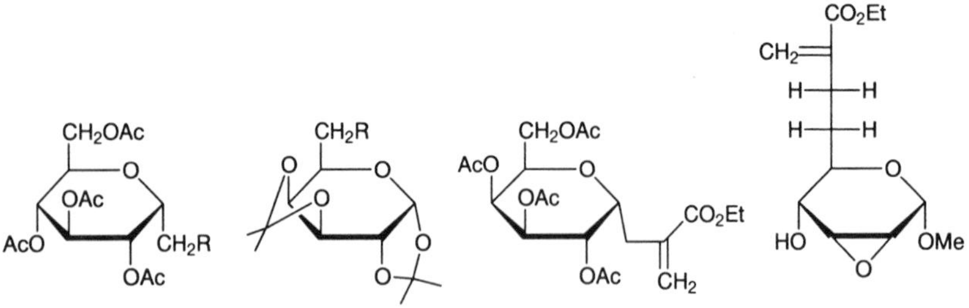

21 R = CO$_2$H
22 R = C$_5$H$_{11}$

23 R = CO$_2$H
24 R = C$_7$H$_{15}$

25

26

2.3.1. Chain-extension at the 'Non-reducing End' – Both epimers of 6-*C*-phenylglucose **28** have been synthesized from D-glucuronolactone *via* intermediates **27**.[32] Introduction of a cyclopentadiene moiety at C-6 of diacetonegalactose was effected by reaction of triflate **29** with cyclopentadienyl lithium in DMF to give **30**.[33]

Three papers describing chain-extensions by use of C-6-aldehydes have been published: the diacetonegalactose-derived dialdose **31** was extended by two carbons to give the acetamido-dideoxy-octosulose derivative **33** *via* intermediate **32** by consecutive treatment with potassium cyanide in the presence of ammonia,

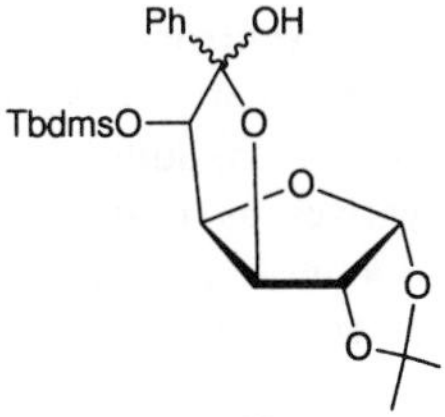

27 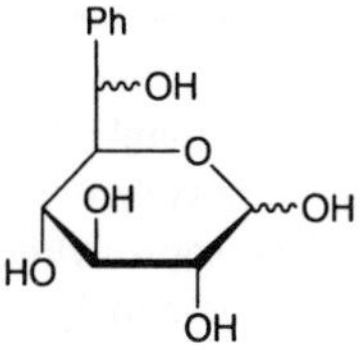 **28** 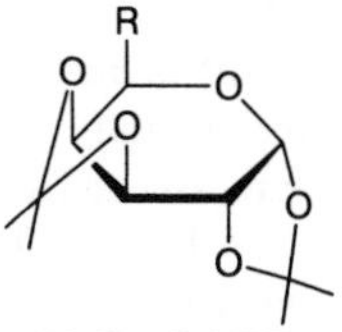 **29** R = CH₂OTf
30 R = CH₂Cp
31 R = CHO

32 R = AcHN⟶CN
33 R = AcHN⟶O=Me

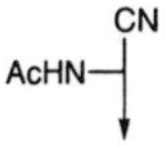

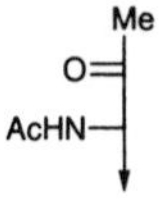

acetic anhydride in pyridine, and methylmagnesium iodide;[34] five different benzyl tetra-*O*-benzyl-L-*glycero*-α-D-*manno*-heptopyranosides with one free hydroxyl group were prepared by oxidation of suitably protected benzyl α-D-mannopyranosides at C-6 (*e.g.*, **34** → **35**), followed by reaction with benzyloxymethylmagnesium chloride and regiospecific protection/deprotection (*e.g.*, **35** → **36**).[35] Their conversion to monophosphates is covered in Chapter 7; condensation of dialdose **31** with the 5'-diazoriboside **37** gave the *C*-linked disaccharide ketose **38** as the sole product in 60% yield.[36]

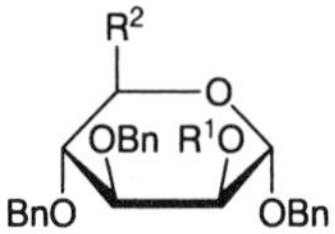

34 R¹ = All, R² = CH₂OH
35 R¹ = All, R² = CHO
36 R¹ = H, R² = ⟶OBn (CH₂OBn)

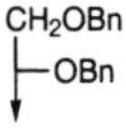

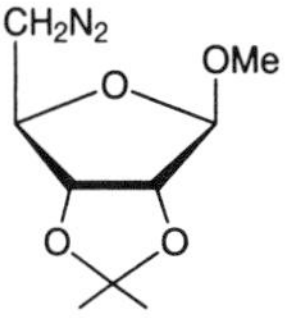

37 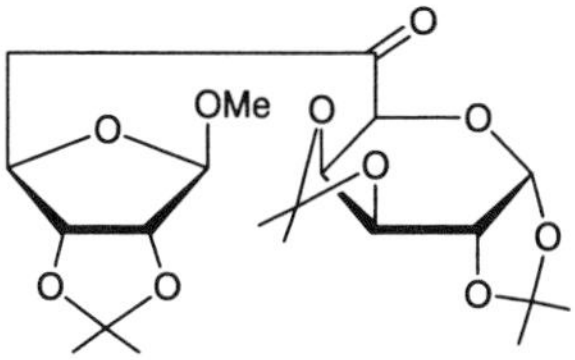 **38**

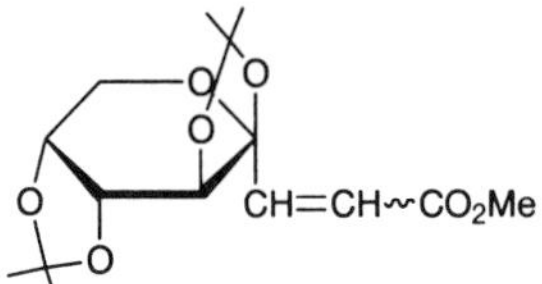

39 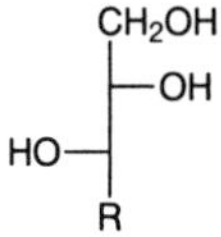 **40** 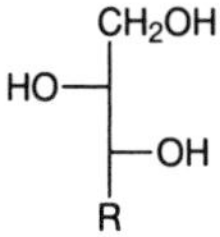 **41** 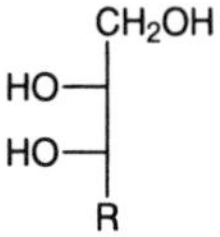 **42**

R =

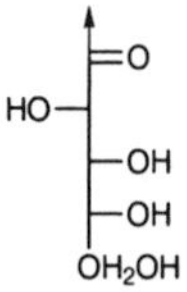

2.3.2 Chain-extension at the Reducing End – Catalytic osmylation of the D-fructose-derived (*E*)- and (*Z*)-alkenes **39** gave, after reduction with LAH, a mixture of D-*glycero*-D-*galacto*- (**40**) and D-*glycero*-D-*ido*-oct-4-ulose (**41**), and D-

glycero-D-*manno*-oct-4-ulose   (**42**),   respectively,   as   precursors   for
polyhydroxyindolizidines.[37,38] The synthesis of 2-(β-D-glycopyranosyl)-
nitroethenes and -nitroethanes from 2,6-anhydro-1-deoxy-1-nitroalditols is
covered in Chapter 3, and that of 7-deoxyheptose derivatives as rhamnose
mimics, using Kiliani ascent from 2,3-*O*-isopropylidene-L-rhamnose as the first
step, in Chapter 16. The synthesis of octono-1,4-lactones from heptoses by
Kiliani ascent is referred to in Chapter 6.

Preparations of various *spiro*-ketals have been reported: photolysis in the
presence of diacetoxyiodobenzene and iodine converted ω-hydroxyalkyl *C*-glyco-
sides into [6.5]-, [5.5]-, [5.4]-, and [4.4]-ring systems (*e.g.*, **43** → **44**).[39,40] Intramolecular Friedle-Craft reaction of 6-*O*-acetyl-3,4-di-*O*-benzoyl-2-*O*-benzyl-α-D-
glucopyranosyl chloride, followed by oxidation of the benzylic methylene group
of the cyclic ether moiety and Zemplen deacylation gave lactone **45**, which in
aqueous solution exists mainly as the hydroxy acid **46**.[41]

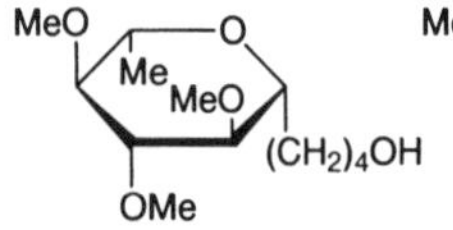

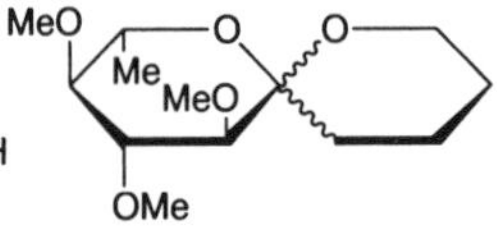

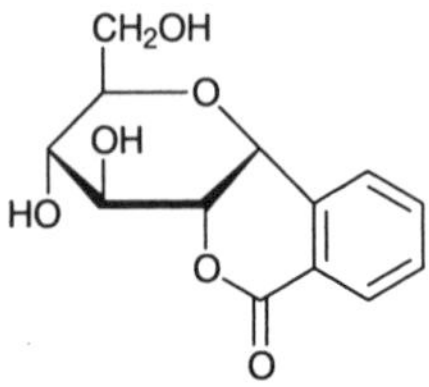

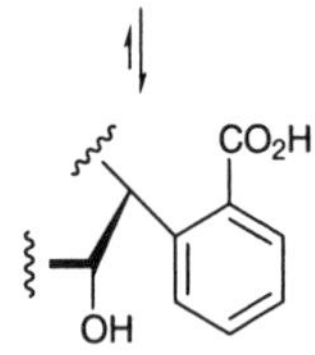

The key-steps in the synthesis of a part structure of the antifungal papulacandin are shown in Scheme 5.[42]

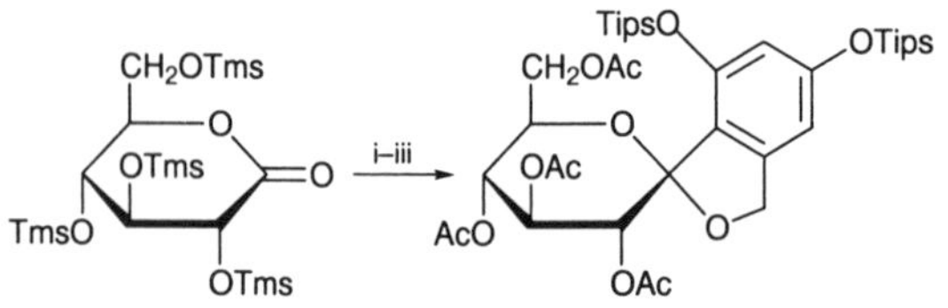

Regents: i, **47**, Bu^{t}Li, Et$_2$O; ii, Amberlite (H$^+$), MeOH; iii, Ac$_2$O, Py

Scheme 5

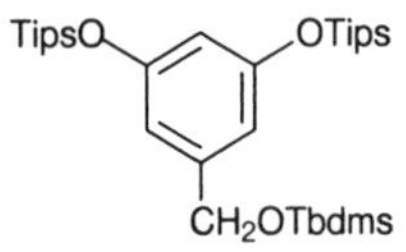

3 Physical Measurements

A 21 page review on sugar-protein recognition and sugar binding in water and non-polar solvents has appeared.[43] The melting of a β-D-glucopyranose crystal and the vitrification of the melt-fluid have been simulated by use of a molecular mechanics technique.[44] The interactions between some carbohydrates (lactose, trehalose, cellobiose) and water have been investigated by molecular mechanics and molecular dynamics methods.[45] Molecular mechanics simulations of aqueous trehalose have been carried out to investigate the role played by this sugar as protecting agent against water stress in biological systems.[46]

Reports on the crystallization of sucrose (41 pp.),[47] on the solubility of sucrose in pure water, sugar-containing water and some other solvents (24 pp.),[48] and on the rheological properties of pure and impure sucrose solutions and suspensions(28 pp.)[49] have been published.

Measurements of the enthalpies of solution of several crown ethers in aqueous D-glucose indicated that the interactions between D-glucose and the crown ethers are weak.[50] Association constants between a number of monosaccharides and pyrene have been calculated from the solubility of pyrene in aqueous solutions of these sugars at various concentrations, and the hydrophobicity of the saccharide solutions was estimated from the fluorescence spectra of the dissolved pyrene; the two properties were found to be unrelated.[51] The effects of solvent systems such as aqueous DMSO, aqueous acetonitrile or aqueous methanol, on the relative stabilities of complexes between sugars (e.g., D-ribose, D-arabinose, D-glucitol) and lanthanide cations have been studied by thin layer ligand exchange chromatography.[52]

4 Isomerization

Some methods of metal ion-catalyzed chemical and enzymic isomerization (Lobry de Bruyn-Alberda van Ekenstein rearrangement, epimerization at C-2 of aldoses, action of isomerases) of free sugars have been reviewed (136 refs.).[53]

A comparative study on the epimerization of D-glucose and D-mannose catalyzed by water-soluble organometallic complexes with nitrogen ligands showed that strong coordination of the ligand to the metal lowers the catalytic activity. Thus, none of the complexes examined was as active as ammonium heptamolybate.[54] The mutarotation of D-fructose in aqueous ethanol (1:3 - 1:9) has been investigated at temperatures between 24 and 50 °C. Increased ethanol content has been shown to favour the furanose tautomers and to retard the mutarotation of the furanose forms to β-D-fructopyranose.[55] The GlcNAc residue at the reducing end of tetrasaccharide **48**, isolated from the lyase-catalyzed degradation of heparin, epimerized to a D-mannose unit on standing in 0.25% aqueous ammonia and also under the conditions of isolation of the tetramer.[56]

Δ-UA-(1→4)-α-D-Glc$p$NAc-(1→4)-β-D-Glc$p$A-(1→4)-Glc$p$NAc

48

5 Oxidation

Platinum catalysts supported on activated charcoal, with or without promoters such as bismuth or gold, have been examined for selectivity in the air oxidation of aqueous D-glucose and D-gluconate to glucarate.[57] Palladium(II) has been found to inhibit the oxidation of aldoses by alkaline $Fe(CN_6)^{3-}$,[58] and by Ce(IV).[59] Exposure of methyl α-D-glucopyranoside in aqueous solution at 50 - 75 °C to oxygen at 15 - 20 bar in the presence of bismuth-rich ruthenium pyrochlore oxide caused glycol cleavage; oxidation of the primary hydroxyl groups occurred only to a minor extent and at a much slower rate. β-Cyclodextrin, on the other hand, reacted non-selectively, and this catalyst was therefore declared unsuitable for the controlled oxidation of starch to dicarboxylic acids.[60] Suspended nickel oxide is believed to be the reactive species in the $NiSO_4$-catalysed oxidation of sugars such as methyl α-D-glucopyranoside and β-cyclodextrin with sodium hypochlorite at pH 10. Glycol cleavage was the main reaction, accompanied to *ca.* 20% by primary oxidation.[61]

Kinetic and/or mechanistic studies on the following processes have been reported: the oxidation of several free sugars by diperiodatoargentate(III) in alkaline media;[62] the oxidation of D-glucose, D-mannose, D-fructose and L-sorbose by Mn(III) in sulfuric acid;[63] the hexacyanoferrate-catalysed oxidation of D-glucose with ammonium persulfate;[64] the 1,10-phenanthroline-catalysed oxidation of epidermic aldo- and keto-hexoses by Fe(III);[65] the oxidation of D-glucose by HOBr;[66] and the oxidation of several sugars (*e.g.*, methyl and octyl β-D-glucopyranoside, decyl β-maltoside) by catalytic $[Ru(azpy)_2(H_2O)_2]^{2+}$ and $NaBrO_3$.[67]

6 Other Aspects

A review (58 refs.) on the homogeneous catalytic hydrogenation of aldehydes and aldoses in water and in organic solvents has been published.[68] The homogeneous hydrogenation of D-glucose and D-mannose by hydrogen transfer from triethylamine/formic acid under catalysis by {Ru[tri(*m*-sulfophenyl)phosphine]} complex (RuTppts) has been compared with the heterogeneous hydrogenation by H_2 over a solid catalyst. Both methods were equally effective.[69]

D-Fructose and D-glucose gave the same products on degradation with aqueous $Ca(OH)_2$ at 100 °C. Glycosides of D-fructofuranose required temperatures of 130-140 °C and furnished similar degradation products in different proportions; glycopyranosides (methyl D-glucopyranoside, α,α'-trehalose) were stable under these conditions.[70,71] The effect of cations (Na^+, K^+, Mg^{2+}, Ca^{2+}, Al^{3+}) on the thermal degradation of concentrated aqueous sucrose solutions has been investigated.[72] In the reaction of guanosine with D-glucose, D-glucose 6-phosphate and

D-ribose in the presence of propylamine, sugar degradation products are attached to the amino group of the guanosine residue to form compounds such as **49**, several of which have been isolated in pure form.[73]

The colour reaction of hexoses with substitited phenols in hot 75% sulfuric acid have been studied.[74]

D-*glycero*-D-*ido*-2-Octulose (**50**), which constitutes 90% of the carbohydrate content of the fully hydrated leaves of the plant *Craterastigma plantagineum*. has been fully characterized by NMR analysis.[75]

References

1   E.I. Howard and J.R. Grigera, *Carbohydr. Res.*, 1996, **282**, 25.
2   T. Hudlicky, D.A. Entwistle, K.K. Pitzer and A.J. Thorpe, *Chem. Rev.*, 1996, **96**, 1195 (*Chem. Abstr.*, 1996, **124**, 290 002).
3   H.J.M. Gijsen, L.Qiao, W. Fitz and C.-H. Wong, *Chem. Rev.*, 1996, **96**, 443 (*Chem. Abstr.*, 1996, **124**, 87 497).
4   J.P. Rasor, *Spec. Chem.*, 1995, **15**, S5-S6, S9-S11, S13, (*Chem. Abstr.*, 1996, **124**, 232 893).
5   C.-H. Wong, *Pure Appl. Chem.*, 1995, **67**, 1609 (*Chem. Abstr.*, 1996, **124**, 87 494).
6   J.P. Rasor, *Chim. Oggi*, 1995, **13**, 9 (*Chem. Abstr.*, 1996, **124**, 290 001).
7   M.I. Fascio and N.B. D'Accorso, *Bol. Soc. Quim. Peru*, 1996, **62**, 1 (*Chem. Abstr.*, 1996, **125**, 248 227).
7a   J.M.H. Stoop, W.S. Chilton and D.M. Pharr, *Phytochemistry*, 1996, **43**, 1145.
8   J.Y. Kim, J.E.N. Shin and K.H. Chun, *Bull. Korean Chem. Soc.*, 1996, **17**, 478 (*Chem. Abstr.*, 1996, **125**, 114 958).
9   J. A. Marco, M. Carda, F. González, S. Rodriguez and J. Murga, *Liebigs Ann. Chem.*, 1996, 1801.
10   M. Xie, D.A. Berges and M.J. Robins, *J. Org. Chem.*, 1996, **61**, 5178.
11   J.A. Kotecha, M.S. Feather, T.J. Kubiseski and D.J. Walton, *Carbohydr. Res.*, 1996, **289**, 77.
12   A. Barco, S. Benetti, C. De Risi, G.P. Pollini, G. Spalluto and V. Zanirato, *Tetrahedron*, 1996, **52**, 4719.
13   T. Sekine, E. Kawashima and Y. Ishido, *Tetrahedron Lett.*, 1996, **37**, 7757.
14   T.J. Tolbert and J.R. Williamson, *J. Am. Chem. Soc.*, 1996, **118**, 7929.

15   Z. Pakulski, *Pol. J. Chem.*, 1996, **70**, 667 (*Chem. Abstr.*, 1996, **125**, 143 134).
16   C.J. Ng, D.C. Craig and J.D. Stevens, *Carbohydr. Res.*, 1996, **284**, 249.
17   J.A. Marshall and A.W. Garofalo, *J. Org. Chem.*, 1996, **61**, 8732.
18   M.A. Long, H. Morimoto and P.G. Williams, *J. Labelled Compd. Radiopharm.*, 1995, **36**, 1037 (*Chem. Abstr.*, 1996, **124**, 117 709).
19   T. Hudlicky, K.K. Pitzer, M.R. Stabile and A.J. Thorpe, *J. Org. Chem.*, 1996, **61**, 4151.
20   H. Itoh and K. Izumori, *J. Ferment. Bioeng.*, 1996, **81**, 351 (*Chem. Abstr.*, 1996, **125**, 86 998).
21   S. Freimund, A. Huwig, F. Giffhorn and S. Köpper, *J. Carbohydr. Chem.*, 1996, **15**, 121.
22   A. Huwig, S. Emmel and F. Giffhorn, *Carbohydr. Res.*, 1996, **281**, 183.
23   M. Adinolfi, A. Iadonisi and L. Mangoni, *Tetrahedron Lett.*, 1996, **37**, 5987.
24   M.N. Preobrazhenskaya, E.I. Lazhko, A.M. Korolev, M.L. Reznikova and I.I. Rozhkov, *Tetrahedron: Asymm.*, 1996, **7**, 461.
25   C.-J. Li, *Tetrahedron*, 1996, **52**, 5643.
26   T.H. Chan and M.B. Isaac, *Pure Appl. Chem.*, 1996, **68**, 919 (*Chem. Abstr.*, 1996, **125**, 143 177).
27   A. Zamojski and A. Banaszek, *Wiad. Chem.*, 1995, **45**, 683 (*Chem. Abstr.*, 1996, **125**, 58 855).
28   S. Jarosz, *Pol. J. Chem.*, 1996, **70**, 141 (*Chem. Abstr.*, 1996, **125**, 33 951).
29   N.-J. Kim and I.T. Kim, *Synlett*, 1996, 138.
30   M. Harenbrock, A. Matzeit and H.J. Schäfer, *Liebigs Ann. Chem.*, 1996, 55.
31   F. Pontén and G. Magnusson, *J. Org. Chem.*, 1996, **61**, 7463.
32   Y. Blériot, C.R. Veighey, K.H. Smelt, J. Cadefau, W. Stalmans, K. Bigadikke, A.L. Lane, M. Müller, D.J. Watkin and G.W.J. Fleet, *Tetrahedron: Asymm.*, 1996, **7**, 2761.
33   R. Lai and S. Martin, *Tetrahedron: Asymm.*, 1996, **7**, 2783.
34   M.A.F. Prado, R.J. Alves, A.Braga de Olivera and J. Dias de Sanzo Filho, *Synth. Commun.*, 1996, **26**, 10.
35   B. Grzescyk, O. Holst and A. Zamojski, *Carbohydr. Res.*, 1996, **290**, 1.
36   F. Sarabia-Garcia and F.J. Lopez-Herrera, *Tetrahedron*, 1996, **52**, 4757.
37   I. Izquierdo and M.T. Plaza, *J. Carbohydr. Chem.*, 1996, **15**, 303.
38   I. Izquierdo, M.T. Plaza, R. Robles and C. Rodriguez, *Tetrahedron: Asymm.*, 1996, **7**, 3593.
39   A. Martín, J.A. Salazar and E. Suárez, *J. Org. Chem.*, 1996, **61**, 3999.
40   R.L. Dorta, A. Martín, J.A. Salazar, E. Suárez and T. Prangé, *Tetrahedron Lett.*, 1996, **37**, 6021.
41   P. Verlhac, C. Leteux, L. Toupet and A. Veyrières, *Carbohydr. Res.*, 1996, **291**, 11.
42   A.G.M. Barrett, M. Pena amd J.A. Willardsen, *J. Org. Chem.*, 1996, **61**, 1082.
43   J. Haseltine and T.J. Doyle, *Org. Synth.: Theory Appl.*, 1996, **3**, 85 (*Chem. Abstr.*, 1996, **125**, 86 990).
44   E. Caffarena and J.R. Grigera, *J. Chem. Soc., Faraday Trans.*, 1996, **92**, 2285 (*Chem. Abstr.*, 1996, **125**, 196 113).
45   Y. Fukazawa, *Kenkyu Hokoku-Kanagawa-ken Sangyo Gijutsu Sogo Kenkyusho*, 1995, **1**, 5 (*Chem. Abstr.*, 1996, **125**, 168 488).
46   M.C. Donnamaria and J.R. Grigera, *Nonlinear Phenom. Complex Syst.*, 1996, **1**, 359 (*Chem. Abstr.*, 1996, **125**, 222 289).
47   G. Vaccari and G. Mantovani, *Sucrose*, 1995, 33 (*Chem. Abstr.*, 1996, **124**, 290 053).
48   Z. Bubnik and P. Kadlec, *Sucrose*, 1995, 101 (*Chem. Abstr.*, 1996, **124**, 290 054).

49 M. Mathlouthi and G. Genotelle, *Sucrose*, 1995, 126 (*Chem. Abstr.*, 1996, **124**, 290 055).

50 E.V. Parfenyuk and O.V. Kulikov, *Thermochim. Acta*, 1996, **285**, 253 (*Chem. Abstr.*, 1996, **125**, 301 345).

51 M. Shigematsu, *Carbohydr. Res.*, 1996, **292**, 165.

52 Y. Israëli, C. Lhermet, J.-P. Morel and N. Morel-Desrosiers, *Carbohydr. Res.*, 1996, **289**, 1.

53 A. De Raadt, C.W. Ekhart and A.E. Stütz, *Adv. Detailed React. Mech.*, 1995, **4**, 175 (*Chem. Abstr.*, 1996, **124**, 146 594).

54 S. Kolaric and V. Sunjic, *J. Mol. Catal. A: Chem.*, 1996, **110**, 181 (*Chem. Abstr.*, 1996, **125**, 276 303).

55 A.E. Flood, M.R. Johns and E.T. White, *Carbohydr. Res.*, 1996, **288**, 45.

56 T. Toida, I.R. Vlahof, A.E. Smith, R.E. Hileman and R.J. Linhardt, *J. Carbohydr. Chem.*, 1996, **15**, 351.

57 M. Besson, G. Fleche, P. Fuertes, P. Gallezot and F. Lahmer, *Recl. Trav. Chim. Pays-Bas*, 1996, **115**, 217.

58 B. Bajpai, A. Shukla and S.K. Upadhyay, *Int. J. Chem. Kinet.*, 1996, **28**, 413 (*Chem. Abstr.*, 1996, **125**, 58 871).

59 M. Gupta and S.K. Upadhyay, *Transition Met. Chem.*, 1996, **21**, 266 (*Chem. Abstr.*, 1996, **125**, 143 139).

60 S.J.H.F. Arts, F. van Rantwijk and R.A. Sheldon, *J. Carbohydr. Chem.*, 1996, **15**, 317.

61 E.J.M. Mombarg, A. Abbadi, F. van Rantwijk and H. van Bekkum, *J. Carbohydr. Chem.*, 1996, **15**, 513.

62 K.V. Krishna and J.P. Rao, *Transition Met. Chem.*, 1995, **20**, 344 (*Chem. Abstr.*, 1996, **124**, 30 196).

63 M.K. Ram Reddy, K. Nagy Reddy, K.C. Rajanna and P.K. Saiprakash, *Oxid. Commun.*, 1996, **19**, 381 (*Chem. Abstr.*, 1996, **125**, 329 138).

64 V.N. Kislenko, A.A. Berlin and N.V. Litovchenko. *Zh. Obshch. Khim.*, 1996, **66**, 854 (*Chem. Abstr.*, 1996, **125**, 329 140).

65 K.N. Reddy, K.R. Reddy, K.C. Rajanna and P.K. Saiprakash, *Transition Met. Chem.*, 1996, **21**, 112 (*Chem. Abstr.*, 1996, **125**, 11 233).

66 Y. Yang, B. Shen, T. Peng, G. Wang, Y. Tang and P. Zhang, *Hangzhou Daxue Xuebao, Ziran Kexueban*, 1996, **23**, 168 (*Chem. Abstr.*, 1996, **125**, 248 236).

67 A.E.M. Boelrijk, J.T. Dorst and J. Reedijk, *Recl. Trav. Chim. Pays-Bas*, 1996, **115**, 536.

68 S. Kolaric and V. Sunjic, *J. Mol. Catal. A: Chem.*, 1996, **111**, 239 (*Chem. Abstr.*, 1996, **125**, 301 322).

69 S. Kolaric and V. Sunjic, *J. Mol. Catal. A: Chem.*, 1996, **110**, 189 (*Chem. Abstr.*, 1996, **125**, 276 304).

70 B.Y. Yang and R. Montgomery, *Carbohydr. Res.*, 1996, **280**, 27.

71 B.Y. Yang and R. Montgomery, *Carbohydr. Res.*, 1996, **280**, 47

72 G. Egglestone, J.R. Vercellotti, L.A. Edye and M.A. Clarke, *J. Carbohydr. Chem.*, 1996, **15**, 81.

73 J. Nissl, S. Ochs and Th. Severin, *Carbohydr. Res.*, 1996, **289**, 55.

74 T.N. Pattabiraman and M. Mallya, *Biochem. Arch.*, 1996, **12**, 137 (*Chem. Abstr.*, 1996, **125**, 329 139).

75 O.W. Howarth, N. Pozzi, G. Vlahov and D. Bartels, *Carbohydr. Res.*, 1996, **289**, 137.

3
Glycosides and Disaccharides

1   O-Glycosides

1.1   Synthesis of Monosaccharide Glycosides – Three reviews have dealt with this important subject.[1-3] Attention will now be given to the methods which are used in glycoside synthesis and then to the types of glycosides produced.

1.1.1 Methods of synthesis of glycosides – D-Fructose treated with various alkanols in the presence of silica-alumina cracking catalysts or acidic clay affords the corresponding glycosides in the following yields: ethyl 100%, *n*-butyl 91%, octyl 60%, decyl 51%, and dodecyl 40–45%. The β-pyranosides and the two furanosides were obtained, the ratios commonly being 1:2:3, but this depends upon the reaction times.[4] A direct way of making acetylated alkyl fructofuranosides involves treatment of the sugar with the alcohols in the presence of iodine (0.05 equiv.), followed by acetylation. Similarly, D-glucose gave the furanosides in 66% yield and the polysaccharide inulin solvolysed under these conditions again to give the fructofuranosides.[5] The use of long-chain alkanes terminating in an α-diol in the synthesis of glycosides with promising surfactant properties involved conversion of the diols to the cyclic sulfates. Treatment of these esters with glucose in the presence of sodium hydride, followed by neutralization with acid and *O*-acetylation gave separable α/β mixtures of tetra-*O*-acetyl-pyranosides, the sugar being bonded to the terminal alcohol group and the penultimate hydroxyl group being esterified.[6]

Mukaiyama and colleagues have continued their investigation into ways of making specific ribofuranosides. 2,3,5-Tri-*O*-benzyl-D-ribofuranose, treated in dichloromethane with alcohols and trityl tetrakis(pentafluorophenyl)borate in catalytic amounts and excess of lithium ditrifylamide gave the furanosides, the anomeric ratios of the products being impressively 99:1 in favour of the α-isomers. One example was given of a primary sugar alcohol coupled in this way.[7] The same starting ribose ether with the same trityl agent, now in nitroethane and in the absence of the lithium salt, gave products with α/β ratios as low as 5:95.[8] Reaction of 2,3,4,6-tetra-*O*-acetyl-α- and β-D-glucose with ethyl 1,2-dibromoethyl ether in dichloromethane in the presence of DBU at −78°C gave, respectively, the α- and β- glycosides having 2-bromo-1-ethoxyethyl aglycons. In the case of both the α- and β-compounds the products were deacetylated

Carbohydrate Chemistry, Volume 30
© The Royal Society of Chemistry, 1998

and then could be resolved into the glycosides with the (R) and (S) configurations within their aglycons. The O-acetylated derivatives treated with sodium iodide and DMSO gave the corresponding 1-ethoxyethyl-2-iodo analogues; glucosidases hydrolysed the deacetylated glycosides and gave α-bromo- and α-iodoacetaldehyde which reacted with, and ultimately destroyed, the enzymic activity.[9]

Glycosyl esters remain of importance as glycosylating agents and a useful one-step procedure has been developed for the synthesis from them of glycosides protected at all positions except O-2. This involves treating, for example, penta-O-acetyl-β-D-glucopyranose with alcohols in the presence of boron trifluoride etherate in the ratios 1:4:1.5. The allyl glycoside with free hydroxyl group at C-2 was obtained in 63% yield and similar results were achieved by use of penta-O-acetyl-β-D-galactose and tetra-O-acetyl-β-D-xylose. In the case of tetra-O-acetyl-β-D-ribofuranose yields of the order of 30% were recorded.[10,11] Mukaiyama and co-workers have now applied their methoxyacetyl procedure to the synthesis of α-fucosides. 2,3,4-Tri-O-benzyl-α-L-fucosyl methoxyacetate reacts with alcohols in the presence of tin triflate to give 95% yields of glycosides with the α-configuration. The procedure is applicable to the preparation of disaccharides and with α-hydroxy-amino-acids.[12] An improved method of making glycosyl trichloroacetimidates involves the use of catalytic tetrabutylammonium bisulfate. 2,3,4,6-Tetra-O-benzyl-D-glucose afforded 96% of the α-ester.[13] Tetra-O-benzyl-D-glucopyranosyl trichloroacetimidates, whether α- or β-, reacted with alcohols in ether containing lithium perchlorate to give moderate yields of glycosides with poor anomeric selectivity. In the case of the corresponding α- or β-glycosyl phosphates lower yields were obtained, but the stereoselectivity was better, particularly for the formation of β-glycosides. The corresponding α-glycosyl bromide and chloride showed little or no reaction, but the β-fluoride gave moderate yields with some α-preference.[14] When the O-benzyl protecting groups were then replaced by acetates or pivaloates, the α- or β-trichloroacetimidates led to orthoester formation. Studies under these neutral conditions were also carried out with glycosyl dibenzyl phosphates; moderate yields of β-glycosides were produced.[15]

Fraser-Reid has written a review on his research that involved 2,3-unsaturated glycosides, and details of his synthesis of sucrose using a non-acidic glycal rearrangement to attach a 2,3-unsaturated glycosyl unit to tetra-O-acetyl β-D-fructofuranose were recorded. The paper also discusses the use of various ω-unsaturated alkyl glycosides in glycoside synthesis.[16] Reaction of compound 1 with trimethylsilylated alcohols or thiols gave the adducts 2 under either acid or base catalysis. Yields, however, were only in the 20–40% range. The reduction of the products with sodium borohydride in methanol afforded *ribo*-configurated

1   2 X = O, S   3

2-deoxyglycosides, whereas sodium borohydride/cerium chloride in methanol gave the *arabino*-epimers.[17] Many of the glycosides and particularly disaccharides noted in this Chapter have been made using thioglycosides. Phenylselenyl glycosides give *O*-glycosides under photoirradiation (see Section 2 of this Chapter). *S*-Ethyl 1,3,4,6-tetra-*O*-benzyl-2-thio-fructofuranoside, activated with DMTST or *N*-iodosuccinimide, gave excellent yields of fructofuranosyl glycosides and disaccharides, the α-anomers predominating whether α- or β-glycosylating agents were used.[18] Related work by the same authors used Ogawa's intramolecular approach which involved incorporating at O-3 of the 1-thiofructofuranoside a *p*-methoxybenzyl ether. This functional group in the presence of DDQ and alcohol gave a benzal acetal on the thioglycoside which, on activation with DMTST, delivered the alkoxy aglycon group from the β-direction. In this way compounds containing β-D-fructofuranosyl residues were obtained in close to 80% yield.[19] Stereoselective syntheses of α-D-galactofuranosides were developed using the *O*-benzyl protected *S*-ethylthio galactofuranoside as donor and *N*-bromoimides and *N*-bromoamides as promoters.[20]

Usual activation of the Kahne phenyl sulfoxide glycosylation method is by use of triflic anhydride, trimethylsilyl triflate, or triflic acid and methyl propiolate. This, however, has been found to be unsatisfactory in cases of glycosylating agents which do not have 2-deoxy groups. It has now been reported that triethyl phosphite with catalytic amounts of triflic acid cause the reaction to give disaccharide products in variable yields.[21]

Considerable work continues to be done with glycosyl halides. Tetra-*O*-benzyl-β-D-glucosyl fluoride reacts with simple alcohols in the presence of ytterbium triflate in acetonitrile to give glycosides with high β- to α-ratios. When ether is used as solvent α-selectivity is observed. The use of a sugar primary alcohol trimethylsilyl ether in the presence of lanthanum perchlorate afforded 94% of disaccharide, the α/β ratio being 3:1.[22] Tetra-*O*-benzyl-α-D-glucopyranosyl fluoride with L-menthol trimethylsilyl ether in the presence of dimethylgallium chloride in dichloromethane plus various additives gave good yields of glycosides with the β-compounds predominating, their ratios depending upon the nature of the additives. Dicyclopentadienylzirconium chloride gave an α/β ratio of 1:5.[23] Some glycosylations using 2,3,4-tri-*O*-chlorosulfonyl-β-L-fucopyranosyl chloride with silver salts as promoters, in contrast to previous reports, have resulted in the β-glycosides. The corresponding β-D-xylopyranosyl chloride afforded mainly α-compounds which makes the rationalization of the results difficult.[24] Reaction of acetobromoglucose with the zinc salts of alcohols or thiols gives, as expected, mainly β-products.[25] 'Normal' acetohalogen sugars with alcohols in the presence of iodine (1.5 mol. equiv.) and DDQ (0.75 mol. equiv.) afford good yields of, again, the normal glycosides, that is, for example, the β-galactosides and α-mannosides.[26] 2,3,4-Tris-(trimethylsilyl) α-L-fucopyranosyl iodide, generated *in situ* using trimethylsilyl iodide, on treatment with alcohols in the presence of a base, gave the α-fucopyranosides (including disaccharides). No further activator is required, and the reaction proceeds with simple alcohols and with sugar secondary alcohols.[27]

Isopropenyl and other substituted vinyl glycosides are increasing in significance as glycosylating agents. The isopropenyl compounds have been described as being stable to storage, activatable by electrophiles, and suitable as precursors of glycosides in the D-*gluco-* and D-*galacto-* series. Under conditions that retard the formation of glycosyl carbocations (weak electrophiles, non-polar solvents, electron withdrawing O-protecting groups) additions to the vinyl ether double bond compete with glycosylation processes.[28] A full paper from the preliminary *Tetrahedron Lett.*, 1994, **35**, 3593 gives details of the isomerization of the aglycon of tetrabenzyl β-D-glucopyranosyl glycoside of but-1-en-3-ol. Rearrangement with a rhodium salt affords the but-2-en-2-ol glycoside which gives good yields of disaccharides with primary or secondary sugar alcohols in the presence of trimethylsilyl triflate. The β-anomeric configuration is appreciably favoured.[29] The same but-2-en-2-ol glycosides, treated with dibenzyl phosphate in the presence of NIS and trimethylsilyl triflate as activators, afford the corresponding glycosyl phosphates.[30] Boons and colleagues have reported that a range of 1-substituted prop-1-en-1-ol glycosides can be made from the corresponding 1-substituted allyl compounds by use of tris(triphenylphosphine)rhodium chloride/butyl lithium.[31] Furanosyl pent-4-enyl glycosides in the D-glucose, D-galactose and D-mannose series have been used in the preparation of corresponding furanosides, including a 1-glyceryl derivative.[32]

The glycosidation reaction which depends upon the condensation of sugar 1,2-O-cyanoalkylidene derivatives and tritylated acceptors continues to attract the attention of Kotchetkov and colleagues. The rate-determining step of the condensation has been shown to be the reaction between the donor and the triphenylmethyl cation.[33] The stereochemical outcome of the reaction is dependent upon the concentration of trityl perchlorate used as catalyst. A new mechanism for 1,2-*cis*-glycoside formation by this method was proposed.[34] The nature of the O-protecting groups used in the acceptor also has an effect on the anomeric ratio produced.[35]

Vasella and co-workers have published further on their glycosylidene glycosidation procedure and have carried out important studies on the effect of hydrogen bonding on glycosylations occurring on compounds **3**. They found that with X = F, 59% of the β-glycoside formed was at position b, whereas when X was hydrogen 38% reaction occurs at this position. The sites of glycosylation depend upon the extent of intramolecular hydrogen bonding, this bonding favouring the hydroxyl groups involved.[36]

Extensive interest in the use of enzymes in glycoside synthesis continues, and two reviews have been produced on the subject.[37,38] Studies have been reported on the reaction between glucose and alcohols catalysed by β-glucosidase,[39] and a specific synthesis of n-octyl β-D-glucopyranoside has been reported. The procedures involved a heterogeneous system with high glucose concentrations and yielded up to 58% product.[40] (Z)-But-2-ene-diol has been converted enzymically to the mono-β-glucopyranoside which was acetylated. Chemical epoxidation then gave a 1:1 mixture of the diastereoisomers and a specific lipase allowed the production of the tetraacetylated glucoside of the 4-hydroxy-(2R)(3S)-epoxide in 90% yield. This therefore affords a method of making the glucoside of the enantiomerically pure epoxide.[41] Enzymic glucosylation of racemic 1-hydro-xyethylbenzene can be effected with enantioselectivity according to the enzyme that is used. An example is given of an enzyme which reacts with the (*S*)-alcohol and another that reacts with the (*R*).[42] Allyl β-D-galactopyranoside was obtained in 38% yield as the only product starting from lactose and allyl alcohol using a streptococcal β-galactosidase.[43] Using the same enzyme the same workers produced 2-fluoroethyl β-D-galactopyranoside in yields up to 55%.[44] A non-ionic lipid coated β-D-galactosidase has been used to catalyse transgalactosylation from *p*-nitrophenyl β-D-galactopyranoside to alcohols in diisopropyl ether. Various primary, secondary and tertiary alkanols and 5-phenylpentanol were used.[45] Other papers have been published on the enzymic production of alkyl β-D-xylopyranosides,[46] and methyl fructoside made using β-D-fructofuranosidase with sucrose as a source and methanol as substrate.[47]

1.1.2 Classes of glycosides – A review in Chinese on the subject of synthesis of β-D-mannopyranosides has appeared.[48] A newer procedure involves the coupling of alcohols in the presence of triflic anhydride and a hindered base with the sulfoxides derived from *S*-ethyl 3-*O*-benzyl-4,6-*O*-benzylidene-2-*O*-t-butyl-dimethylsilyl-1-thio-α-D-mannopyranoside. In the case of primary alcohols, yields were greater than 80% and selectivities greater than 10:1 in favour of β-anomers. With secondary alcohols, selectivity was much lower.[49] Transmannosylation with an *Aspergillus niger* enzyme using mannobiose as source allowed the preparation of alkyl β-D-mannopyranosides in yields up to 81% for the methyl compound. Yields dropped with increase in alkyl chain length, but the octyl compound was made satisfactorily.[50]

In the area of amino-sugars, D-glucosamine pentaacetate, used with 2 equivalents of acceptor and camphorsulfonic acid as catalyst with azeotropic removal of acetic acid, gave good yields of β-glycosides. The 4-pentenyl glycoside was made with 86% efficiency.[51] A comparison of the *N*-phthalimido and *N*-2,2,2-trichloroethoxycarbonyl (Troc)*N*-protecting groups for glucosamine and galactosamine glycosyl donors, including the glycosyl acetate, bromide and *S*-ethyl glycosides, showed that normally the Troc-protected donors gave β-glycosides in higher yields than the phthalimido compounds. The trichloroethoxycarbonyl (Troc) group can be removed using zinc.[52] T. Ogawa and colleagues have used the 4,5-dichlorophthaloyl group for amino protection in the case of an *S*-methyl β-thioglycoside of glucosamine which yielded β-thioglycosides. Ethylenediamine or hydrazine in methanol can be used to remove it.[53] 3,4,6-Tri-*O*-benzoyl-2-benzoyl-oximino-2-deoxy-α-D-*lyxo*-hexopyranosyl bromide has been used in the synthesis of α-D-talosaminides; the α-linked L-serine glycoside and disaccharides were described.[54] Butyl 2-acetamido-4,6-*O*-benzylidene-2-deoxy-3-*O*-lactyl-β-D-gluco-pyranoside was coupled with a dipeptide via the carboxylic acid group to give butyl 2-acetamido-2-deoxy-muramoyl-L-alanyl-D-isoglutamine.[55]

The glycosylating L-iduronic acid derivatives **4** and **5** proved to be equally effective as donors and were much better than the corresponding *S*-ethyl or *S*-phenyl thioglycosides or glycosyl fluorides.[56] The uronosides **6** and **7**, which are anionic surfactants derived from glucose, are inhibitors of HIV.[57]

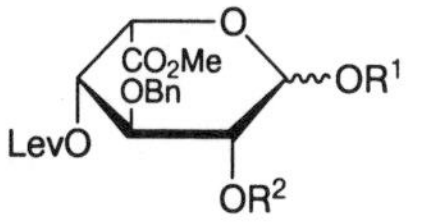

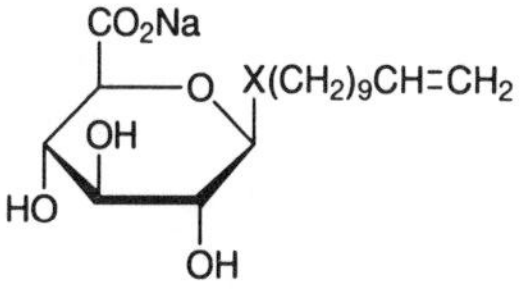

4 $R^1 = OC(=NH)CCl_3$, $R^2 = Ac$
5 $R^1 = OCH_2CH_2CH_2CH=CH_2$, $R^2 = Bz$

6 X = O
7 X = S

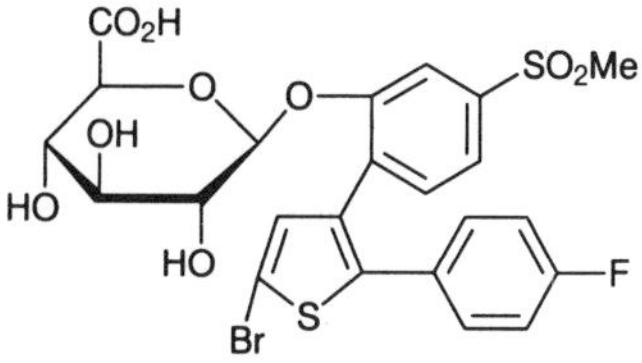

8

In the area of aryl glycosides a novel electrochemical method of synthesizing such compounds by cathodic reduction of glycosyl halides has been applied to phenols and polyhydroxyphenols and gave encouraging yields and product stereoselectivity.[58] A set of *p*-substituted phenyl β-D-glucopyranosides were produced from the corresponding phenols and penta-*O*-acetyl-β-D-glucopyranose using boron trifluoride as catalyst,[59] and phenyl β-D-glucopyranosides carrying variously substituted *p*-amino groups were examined as potential anti-HIV agents.[60] α-D-Mannose has been linked to several substituted biphenyls to give glycomimetic selectin inhibitors,[61] and glucuronide **8** has been made by the trichloroacetimidate method as a drug metabolite.[62] Other specific compounds to have been prepared are the mono-β-D-glucopyranoside of 9,10-dihydroxyphenan-threne,[63] and a range of β-D-glucopyranosides of various substituted tropo-lones.[64] The thiazole derivative **9** was used in the preparation of ulosonic acid-derived 'calixsugars' of the type **10**.[65]

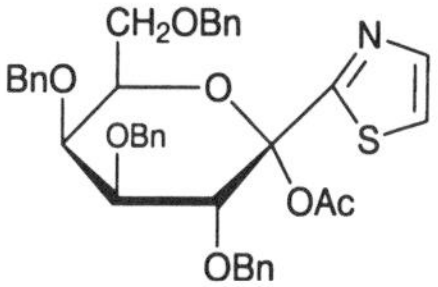

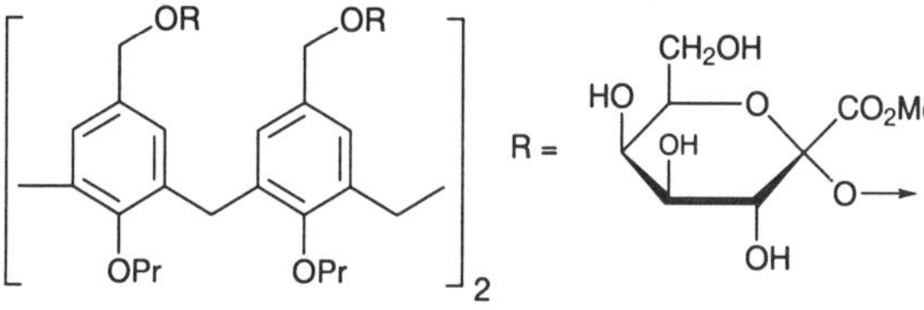

9   **10**

Addition to various glycals of pentane-2,4-dione-3-thione in pyridine gave hetero Diels-Alder adducts of the type **11**.[66] On methylenation these compounds gave the dienes, e.g. **12**, which ring-opened on treatment with alcohols in the presence of triflic acid to give 2-thio-compounds which, on reductive desulfuriza-tion, afforded 55-95% yield of β-2-deoxy-glycosides including disaccharides (Scheme 1).[67]

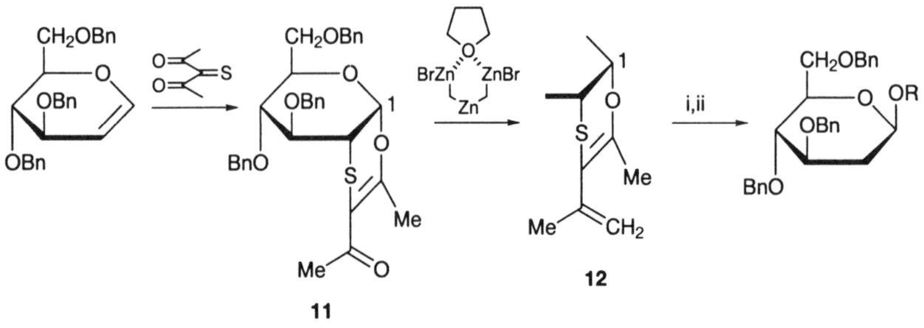

Reagents: i, ROH, TfOH; ii, Raney Ni

Scheme 1

Considerable attention is being turned to synthetic compounds with more than one monosaccharide substituent, and branching structures of the dendrimer type are becoming popular. Reactions of 'pentaerythritol tetrabromide' with β-D-glucopyranosyloxyalkanoic acid cesium salts have been used in the synthesis of compounds with structure **13**.[68] Tri-α-D-mannosylated clusters were made as photoaffinity ligands for mannose-binding proteins by glycosylating temporarily mono-protected pentaerythritol. Linkages were direct or via spacer groups including one carrying a diazirine unit.[69] Dendrimers of the type **14** and more complex related materials containing up to 18 terminal sugars have been described,[70] and related compounds containing α-D-mannopyranosyl units are of interest because of their potential bioactivity linked to their likeness to high-mannose glycoproteins. Compound **15**,[71] as well as related materials akin to the 'monomeric' parts of structure **15**, terminating in a carboxylic acid on the unglycosylated aromatic ring, inhibit the binding of concanavalin A to yeast mannan.[72] By use of tetra-*O*-acetyl-α-D-mannopyranosyl isothiocyanate the unit **16** has been produced and combined in pairs and in triplets to give mannose-terminated dendrimer compounds.[73] Solid phase coupling of *O*-protected 5-hydroxypentyl α-D-mannopyranoside has led to compound **17** in work related to antisense nucleic acid segments.[74]

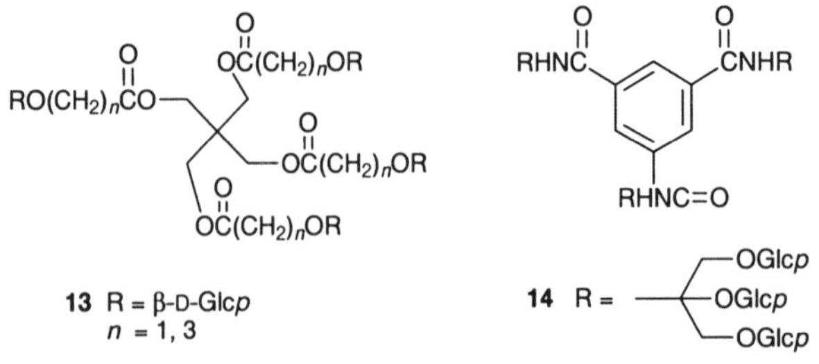

13 R = β-D-Glc*p*
      n = 1, 3

14 R = ⟨ —OGlc*p*
              —OGlc*p*
              —OGlc*p*

Descotes and co-workers have reported further examples of 'bolaforms', for example **18**, produced by metathesis coupling of the aglycons of corresponding glycosides having terminal alkene units, the coupling being catalysed by a

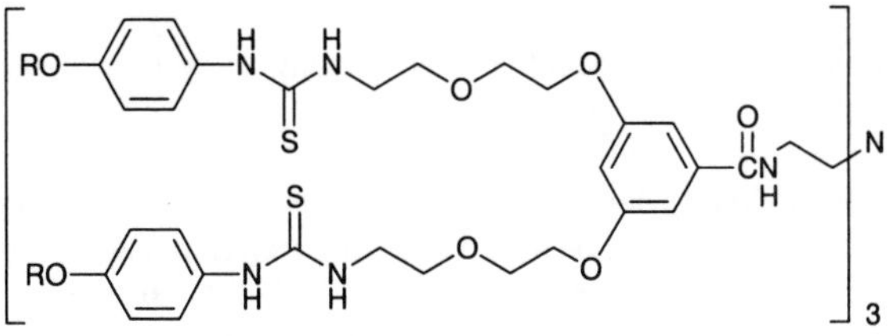

15 R = α-D-Manp

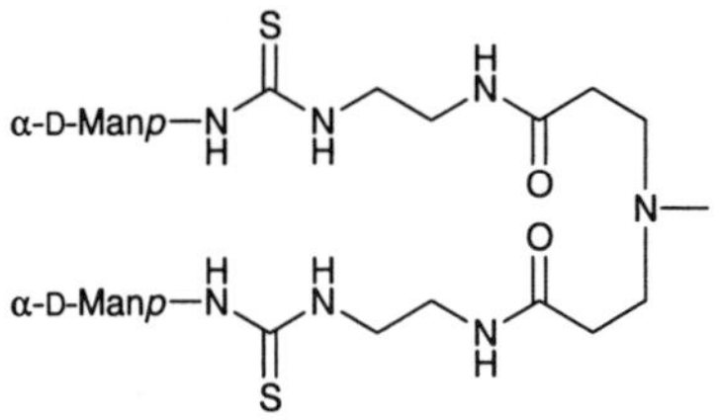

16

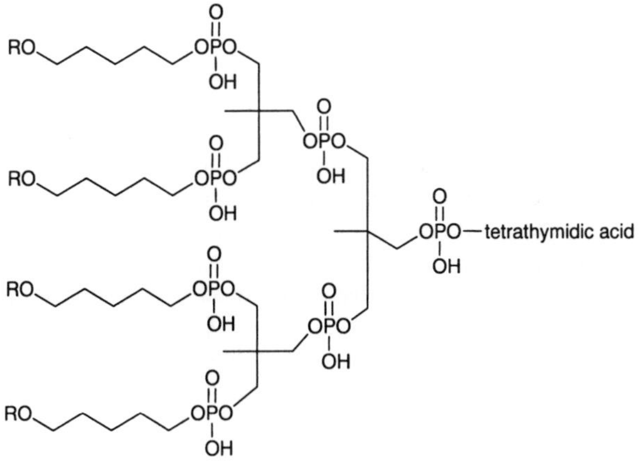

17 R = α-D-Manp

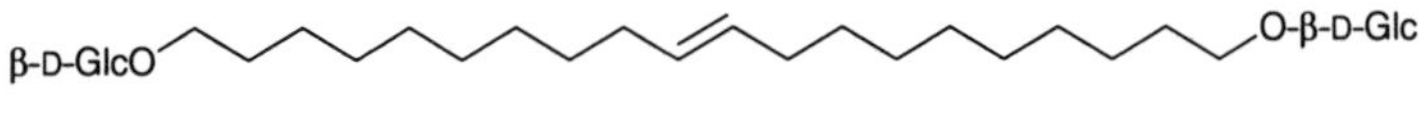

18

tungsten complex.[75] 'Ring opening metathesis polymerisation' allowed the preparation of 'neoglycoproteins' based on vinyl-linked tetrahydrofuran 3,4-dicarboxylic acids esterified to D-glucose or D-mannose units by way of two-carbon spacers. The products (both *O*- and *C*-glycosides) were tested for their specifity in binding with concanavalin A.[76]

Reference to the synthesis of glycosides of 3,6-dideoxyhexoses, the 3-deoxy groups being introduced by reductive ring opening of cyclic sulfates, is made in Chapter 12.

1.2   Synthesis of Glycosylated Natural Products and Their Analogues – Very appreciable interest remains in glycosides having functionalized acyclic aglycons. Dimeric coupling of derivatives of compound **19** led to **20**, the core portion of cycloviracin B$_1$, which is a recently discovered anti-herpes compound.[77]

19 R = Piv   **20**

Several analogues of 3-*O*-β-D-glucopyranosyl-1-*O*-hexadecyl-2-*O*-methyl-*sn*-glycerol with different substituents at C-2 of the glycosyl unit were prepared as analogues of the anti-tumour glyceride compound edelfosine.[78,79] Specifically deuterium-labeled diastereoisomers of dipalmitoyl-1-β-D-glucopyranosyl-1-thiol-*sn*-glycerol have been reported.[80] New surfactants having glycosyl units at O-1 and O-3 and long-chain alkyl substituents at O-2 have been described,[81] as have analogues with a *C*-alkyl chain rather than an *O*-alkyl chain at C-2 of a propane-1,3-diol unit.[82] 2-*O*-β-D-Glucopyranosylglycerol carrying a range of long-chain acyl substituents at O-1 is described in Chapter 7. Tatsuta and Yasuda have reported the synthesis of caloporoside, **21**, which is based on β-D-manno-pyranosyl-(1→5)-mannitol and is a phosphorylase C inhibitor.[83] The deacetylated compound has also been described, it being an inhibitor of the GABA$_A$ receptor ion channel.[84] The simple β-D-mannopyranoside having the sugar linked to the non-carbohydrate acyclic feature of the molecule has also been prepared.[85]

Various cerebrosides to have been reported are the β-D-glucopyranosyl compound which is the major component of *Penicillium* cerebrosides,[86] an α-galactosylated cerebroside, the ceramide asymmetric parts of which were made from lyxose, which is described in Chapter 24, and the β-galactosyl compound **22** which shows anti-viral activity.[87]

The synthesis of glycoconjugates by chemical and enzymic methods has been the subject of a review in Japanese.[88] D-Glucosamine has been β-linked to the hydroxyl position of serine and threonine using *N*-dithiasuccinoyl protection of the amino group introduced by treatment with $(EtOCS)_2S$.[89] α-Fucosylation of these amino acids has been effected by use of the sugar tetraacetate, Fmoc amino protection and boron trifluoride catalysis,[90] and α-fucosylthreonine has been peptide-linked to hydroxythreonine and then glutaric acid to give a siaLe^x mimic.[91] *N*-Acetylgalactosaminyl-substituted threonine has been obtained by

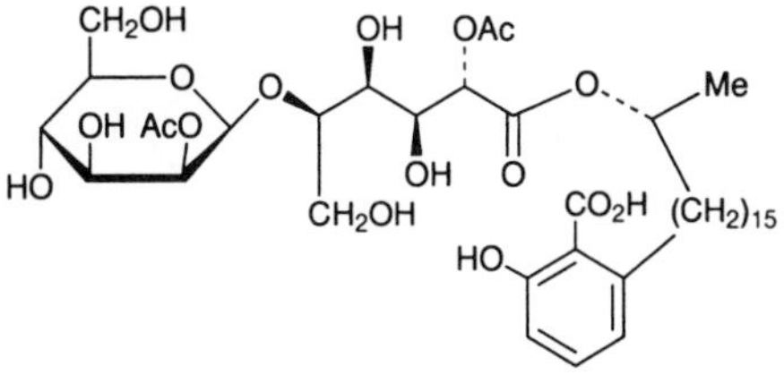

21

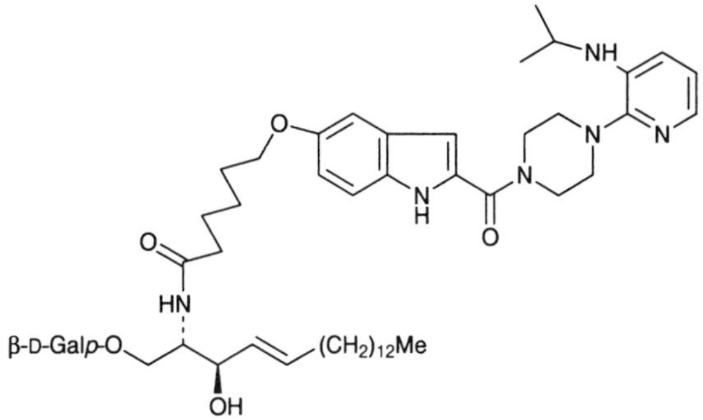

22

lipase hydrolysis of its (methoxyethoxy)ethyl ester.[92] In the field of glycosyl peptides compound **23** has been prepared to represent a new class of glycopepti-domimetic compounds and is based on an N-substituted oligo-glycine.[93] Other peptide compounds to have been prepared are a β-D-glucopyranosyl serine which was amide-linked to a nitrogen-containing chromophore, the product being used to study the influence of glycosylation on elementary structural units involved in protein folding.[94] β-D-N-Acetylglucosamine linked to serine has been incorpo-rated into a phosphorylated hexapeptide representing glycosylated fragments of the large sub-unit of mammalian RNA polymerase II.[95] The galactosyl tripeptide derivative **24**, during de-O-acetylation, underwent epimerization and β-elimina-tion to give **25**.[96] Various L-fucosylated threonine-containing oligopeptides and related compounds have been synthesized as liposome-like mimics for siaLe^x. Several were found to be good inhibitors of SLe^x-binding to E-selectin.[97,98] A variety of oxime-linked glycopeptides were prepared using the hydroxylamine derivatives **26** (R = H) and free sugars to give glycosylated compounds of structure **26** (R = glycosyl). Monosaccharides, glucobioses and glucotrioses were used in the work.[99]

In the area of glycosylated cyclitols and related compounds, the glucosaminyl-allosamizoline **27** has been made largely by the Danishefsky approach,[100] and 2,6-di-O-(α-D-mannopyranosyl)-D-myo-inositol derivatives **28** have been re-ported.[101] The glycosylated episulfide **29** has been described and was prepared by glycosylation of an alcohol derived ultimately from D-glucose.[102] A related compound which inhibited binding of E-selectin to siaLe^x is *trans*-cyclohexane-

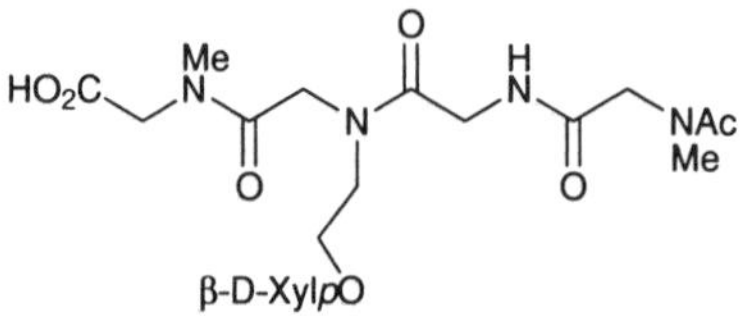

23

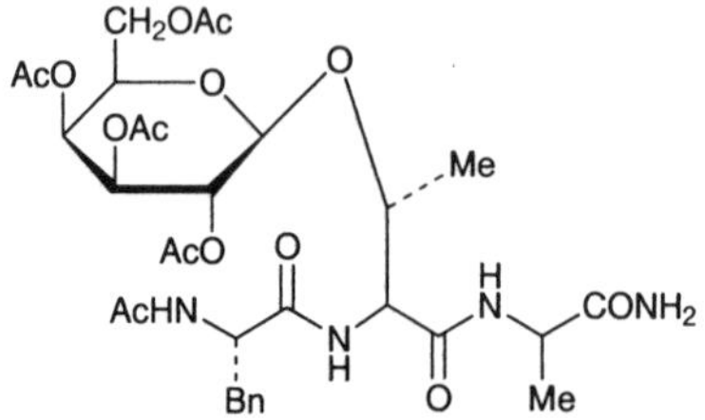

24

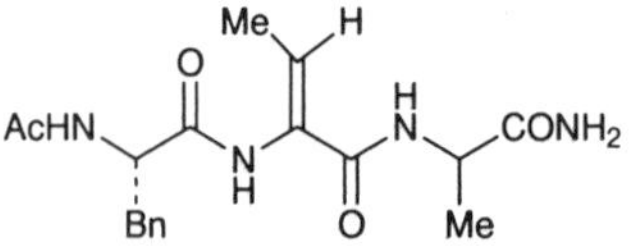

25

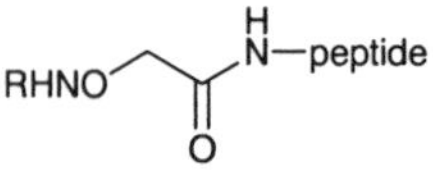

26

1,2-diol carrying an α-L-fucosyl unit and a 4-benzyloxy-3-(3-carboxyethyl)phenyl as hydroxyl substituents.[103] Reaction of tri-*O*-acetyl-2-phthalimido-D-glucal with a monohydroxy deoxyinositol phosphate ester in the presence of boron trifluoride gave a 2,3-unsaturated inositol phosphate for use in the synthesis of anchor compound analogues.[104]

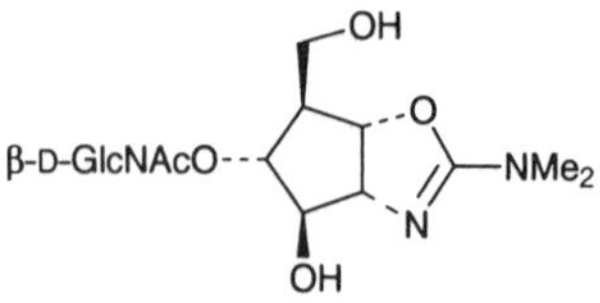

27

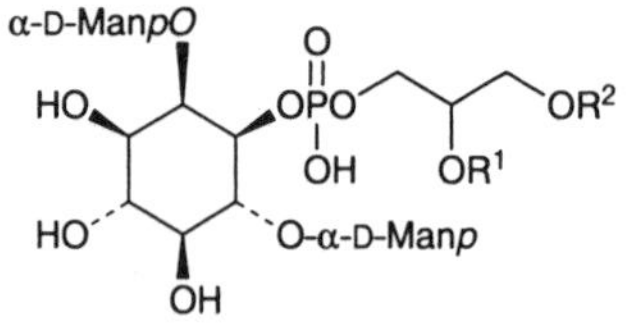

28 R¹, R² = long-chain acyl

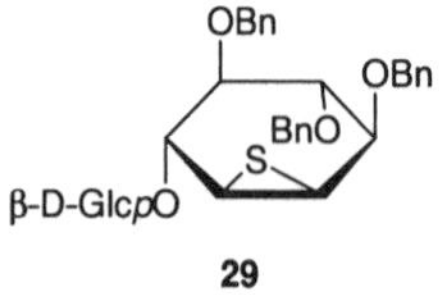

29

In the field of steroidal glycosides several syntheses have been reported: that of some analogues of blattellastanoside A, the steroid aggregation pheromone of the German cockroach,[105] a tri-*O*-acetyl-α-L-rhamnopyranosyl derivative of strophanthidin which showed, on deacetylation, cardiotonic activity,[106] and the β-D-glucopyranoside of 3-α-(2′-hydroxyethoxy)-cholest-5-ene.[107] Glycosides to have been made from triterpenes are: β-glucosides of oleanolic and ursolic acid,[108] 2-deoxy-α-D-glucosides of glycyrrhetic acid and allobetulin (from triacetyl D-glucal),[109] rhamnosides of 20,24-epoxycycloartane-16β,25-diols,[110] 14 synthetic mono- and disaccharide bufalyl glycosides with anti-viral and cytotoxic activities,[111] and β-D-glucopyranosides of 20(S)-protopanaxidiol obtained by the alkaline hydrolysis of ginsenosides.[112] Rhamnosylated derivatives of cyclosiversiosides A and H have also been described.[113]

A variety of ring A fluorinated anthracyclines, for example 30, were synthesized either by glycosylation of fluorinated aglycons or by modification of the aglycons of daunorubicin or idarubicin, and their antitumor properties were evaluated.[114] In related work, 4-demethoxy-6,7-dideoxydaunomycinone analogues, for example 31, have been synthesized.[115]

30

31 R = Complex, chiral
acyl group

Quercetin 7-*O*-glucoside has been prepared by an electrochemical method involving the use of lithium perchlorate in acetonitrile as electrolyte,[116] two papers have described the preparation of glucosides of isoflavones,[117,118] the influence of the glycosyl group on the chemical reactivity of 3-glycosylated flavylium ions has been investigated,[119] and selective chemical glucosylation of 4,7-dihydroxycoumarin has been described.[120]

Interest remains in glycosides of *N*-heterocyclic compounds. 4'- And 5'-*O*-(α-D-glucopyranosyl)pyridoxine were prepared in 1:1 ratio by use of an α-glucosidase.[121] In the area of bicyclic nitrogen heterocycles, compounds 32, benzoxazinoid glycosides of *Gramineae*, were prepared by use of a trichloroacetimidate,[122] and the biopterinyl glucoside 33 was made by enzymic methods in an investigation of the biosynthesis of tetrahydrolimipterin which was isolated from *Chlorobium limicola*.[123] Several iridoid glycosides, for example 8-epi-loganin, have been used to make pyridine monoterpene alkaloids, for example 34, following deglucosylation with β-glucosidase.[124] A series of 2-hydroxy-1,2-dihydroacronycine glycosides, for example 35 (with R = 2,6-dideoxy-α-L-*arabino*-hexopyranose),[125] have been made and their effects on the inhibition of L1210 cells were evaluated.

Several aryl glycosides which are closely related to natural products have been made, for example β-D-glucopyranosides of various mono-lignol derivatives of the type 36 which were easily produced from the relevant hydroxybenzaldehydes condensed with monomethyl malonate.[126] Several 4'-deoxyphlorizin glycosides, for example 37, were prepared, the illustrated compound being a promising inhibitor of the sodium-glucose co-transporter and hence a potential anti-diabetic drug.[127] The glucuronide metabolite 38 of an antilipidemic drug was prepared using an acylated glycosyl bromide.[128]

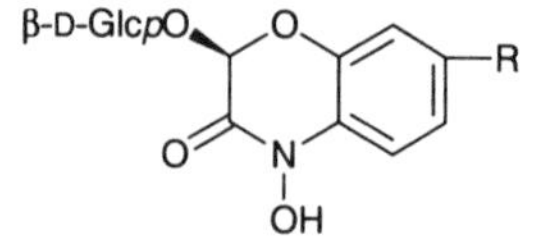

32 R = H, OMe

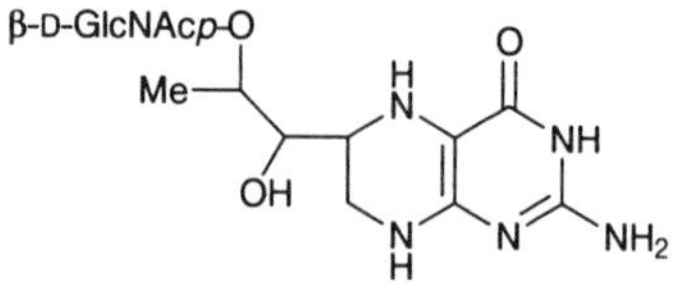

33

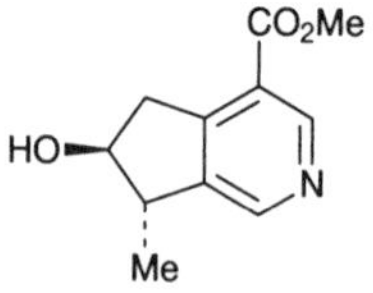

34

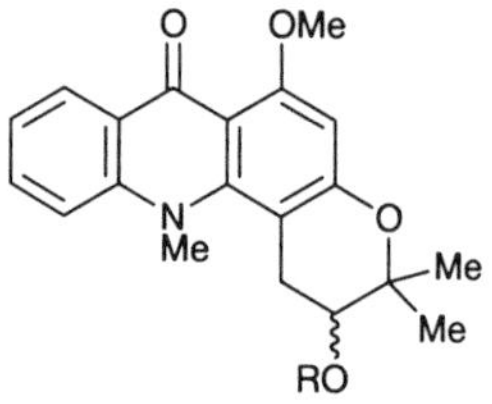

35

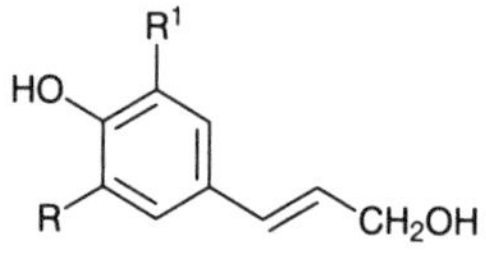

36

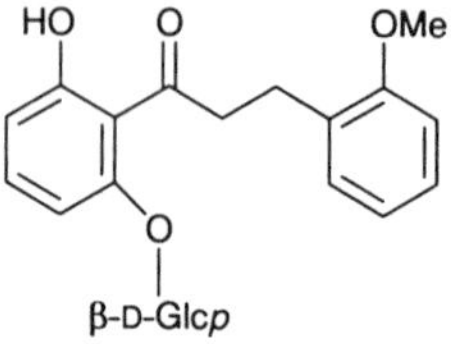

37

β-D-Glucopyranosides of *p*-hydroxyacetophenone, *o*-hydroxycinnamic acid and fluorescein (i.e. **39**) were made in good yields from the aglycon alcohols using high concentration plant cell cultures.[129] Several other benzene-ring-containing glycosides have been described. While the caesium salt of phenylpyruvic acid, on treatment with acetobromoglucose, led to the glycosyl ester, the methyl ester of the acid reacted under basic conditions to give the enol glycoside (Scheme 2). In both cases the (*Z*)-isomers were initially formed as illustrated, and both isomerized to the (*E*)-forms on photolysis in methanol. Compound **40** is an acetate of a natural product isolated from the leaves of *Aspalathus linearis*.[130] The unusual glycosylating agent **41**, used with the corresponding alcohol and boron trifluoride etherate as catalyst, afforded a route to etoposide A **42**.[131] Various derivatives of this compound having acyl groups – some containing *N*-heterocyclic moieties – on the phenolic hydroxyl group have been reported.[132]

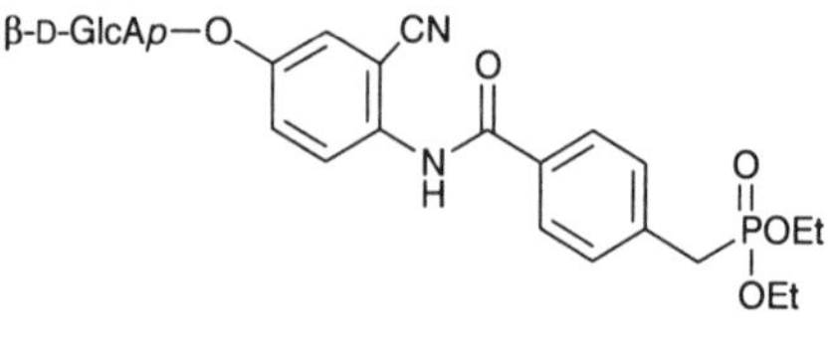

38

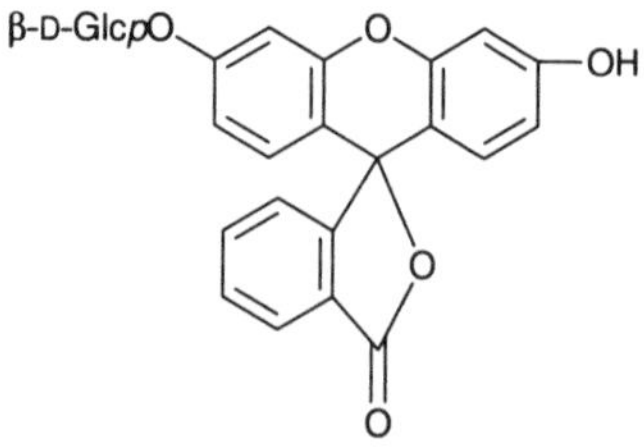

39

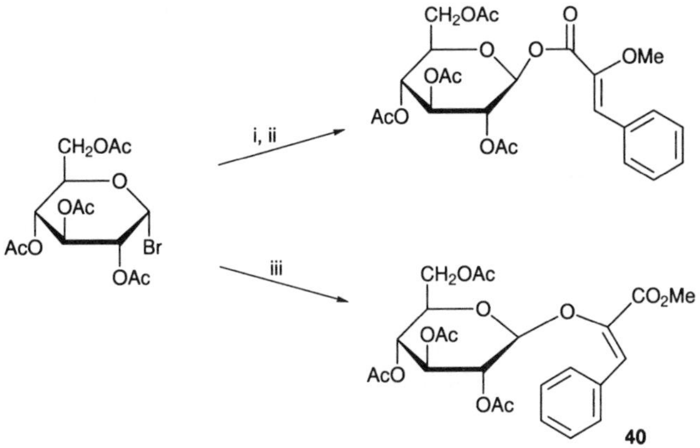

Reagents: i, PhCH=C(OH)CO$_2^-$Cs$^+$; ii, CH$_2$N$_2$; iii, PhCH=C(OH)CO$_2$Me, NaH

Scheme 2

1.3 O-Glycosides Isolated from Natural Products – Although, as always, many new naturally-occurring glycosides have been characterized, this year few involving novel features in the carbohydrate component have been described. Exceptions are the observation that extraction of frozen fresh leaves of *Cerbera manghas* yields β-D-glucopyranosides of cyclo-pentano-terpenoids, while similar treatment of dried leaves gives β-D-allopyranosides. The explanation has not been found.[133] The fused-ring glycoside ester **43** was obtained following acetylation of a product isolated from the rhizome of *Curculigo capitulata*. It is not clear how many of the acetyl groups occur in the natural product.[134] The new chromone glycoside 3-O-α-L-rhamnopyranosyl-3,5,7-trihydroxychromone isolated from *Smilax glabra*, has been structurally characterized by 2D INADEQUATE NMR and molecular modelling procedures.[135]

1.4 Synthesis of Disaccharides and Their Derivatives – A review in Japanese has described convenient syntheses of di- and trisaccharide units which are involved in molecular recognition events. In the main it relates to enzymic methods.[136]

In the field of non-reducing disaccharides coupling of sugars protected at all hydroxyl groups except those at the anomeric centres by use of trimethylsilyl triflate and molecular sieves causes dehydration to give non-reducing disaccharides, the products being very dependent upon the sieve used. Davison SP 7-8461 gave best results and favoured α-linked products. Thus 2,3,4,6-tetra-O-benzyl-D-glucose gave the α,α-, α,β- and β,β-trehaloses in 76% overall yield, the ratios being 2.8:1.0:0.1. With 2,3,5-tri-O-benzyl-D-arabinose only the α,α-furanose compound was produced. Likewise 4,6-di-O-acetyl-2,3-dideoxy-D-*erythro*-hex-2-

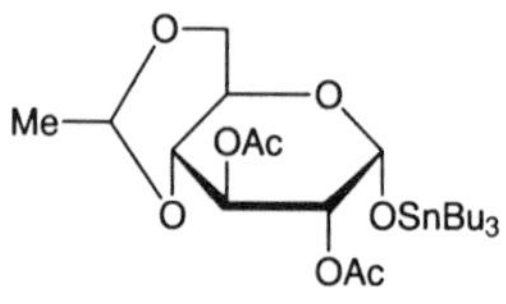

41

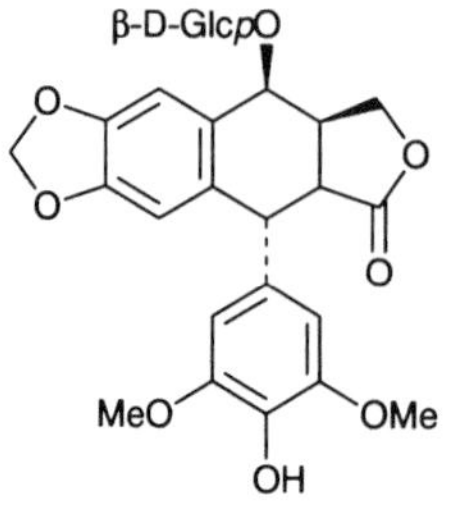

β-D-GlcpO

42

CH₂OAc

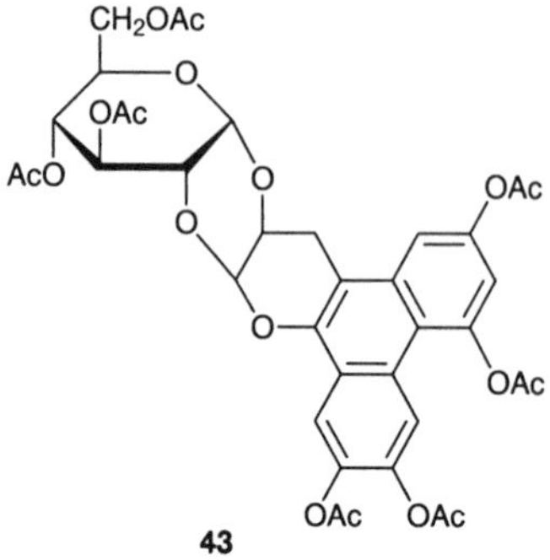

43

enose gave the α,α-disaccharide exclusively.[137] 3-*O*-Acetyl-2,4,6-tri-*O*-benzyl-1-*O*-trimethylsilyl-β-D-galactose coupled with tetra-*O*-benzyl-α-D-mannopyranosyl fluoride with boron trifluoride etherate as activator gave a fully substituted β-D-galactosyl α-D-mannopyranoside which was used to prepare the 3-*O*-glycolyl ether with the substituent in the galactosyl moiety.[138]

Enzymic methods were used to link α-D-glucopyranose to the anomeric positions of α-D-lyxopyranose and β-D-xylulose.[139] Various dichloro-dideoxy sucroses and products derived by selective nucleophilic displacement of single chlorine atoms are referred to in Chapter 8. Trehalose derivatives carrying long-chain branched fatty acid residues are referred to in Chapter 19.

As far as reducing disaccharides are concerned, many of the synthetic procedures referred to at the beginning of this chapter are relevant, but the literature contains examples of such syntheses using newer procedures. A novel intramolecular approach relies upon the removal of carbon dioxide from diglycosyl carbonates. Several examples have been described including that from the disaccharide analogue **44** which, on treatment with trimethylsilyl triflate, afforded 67% of mixed glucobiose anomers, the maltose and cellobiose derivatives being produced in the ratio of 42:58. In the case of the preparation of the 1,6-linked analogues the yield was increased to 85%, and the α/β ratio was 32:68.[140]

A systematic investigation of the use of glycosyl acetates with trimethylsilyl triflate as activator in disaccharide synthesis has been conducted.[141]

The cationic heterocyclic aminals **45** are novel glycosylating agents. Compound **45** (R = H, X = Cl), treated with allyl 4,6-*O*-benzylidene-α-D-glucopyranoside, afforded 90% of the α-(1→2)-linked kojibiose derivative.[142] Tetra-*O*-benzyl-α-D-

CH₂OBn CH₂OBn CH₂OAc

44   **45**

glucopyranosyl diphenylphosphate in acetonitrile containing lithium perchlorate and lithium iodide gave the β-glycosyl iodide as intermediate which, with methyl 2,3,4-tri-*O*-benzyl-α-D-glucopyranoside, afforded 54% yield of the 1,6-linked glucobioses, the α/β ratio being 2:1. In the absence of the iodide the glycosyl phosphate itself gave 83% yield, the α/β ratio being 1:3.[143] DCC-coupling of the maltose glycoside **46** gave 28% of the cyclic dimer, a so-called 'glycophane'.[144] Reaction of methyl 2,3,4,6-tetra-*O*-methyl-α-D-glucopyranoside with 1,5-anhydro-2,3,4-tri-*O*-methyl-D-glucitol and the 2,3,6-trimethyl isomer in the presence of trimethylsilyl triflate gave 1,6- and 1,4-linked products, respectively, indicating that reductive cleavage studies of permethylated polysaccharides using this reagent together with triethylsilane can give rise to *trans*-glycosylation with coupling between the released anhydroalditol and reducing sugar derivatives.[145] The previously described *O*-acylated-*O*-maltosyl tyrosines (*J. Chem. Soc., Perkin Trans. 1*, 1993, 2119) have been incorporated, using solid phase peptide synthesis procedures, into a long-chain peptide related to glycogenin.[146] Tetra-*O*-benzyl-D-glucopyranosyl fluoride, coupled with 2,3,4-tri-*O*-acetyl-D-glucopyranosyl fluoride in the presence of the activating dicyclopentadienylhafnium dichloride and silver chloride, gave the 1,6-linked products in α/β ratio 2:1, indicating that the disarming concept relating to the acetate function at C-2 of the latter fluoride was in operation.[147] Different dicaproyl esters of gentiobiose, representing non-peptide antagonists for atrial natriuretic peptide receptor, have been made.[148] 1,3-Dipolar addition of peracetyl β-cellobiosyl azide to acetylene dicarboxylates gave *N*-cellobiosyltriazoles, and similar derivatives were made from lactose and melibiose.[149] β-Cellobiose having acetylenic *C*-substituents at positions 1 and 4′ have been reported.[150] (See also the *C*-glycoside Section 3.1)

Immobilised glucoamylase in the presence of organic solvents can be used to produce e.g. maltose, isomaltose and panose.[150a]

Solid phase procedures have been used to link α-D-Glc-(1→2)-β-D-Gal via a spacer arm to a 13-amino-acid peptide to give an analogue of the immunodominant T cell epitope of type II collagen.[151] Likewise solid phase coupling gave access to β-D-Glc-(1→6)-D-Gal.[152] Enzymic procedures were used to link glucose to O-4 of 6-deoxy-D-glucose, D-mannosamine and D-mannose with β-anomeric linkages,[153] to L-rhamnose via an α-1,4-linkage, the source of the glucose being a cyclodextrin,[154] to D-arabinose by way of an α-1,3-linkage,[155] and methyl β-D-arabinofuranoside with an α-1,5-bond.[156]

Trichloroacetimidate technology was used to link a glucose moiety to a D-fucoside *en route* to compound **47** which is a derivative of the macrolactone disaccharide segment of tricolorin A a Mexican plant product with inhibitory

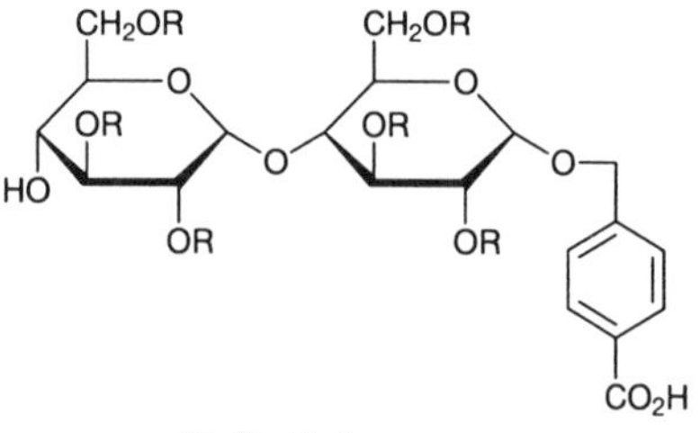

46 R = Pmb

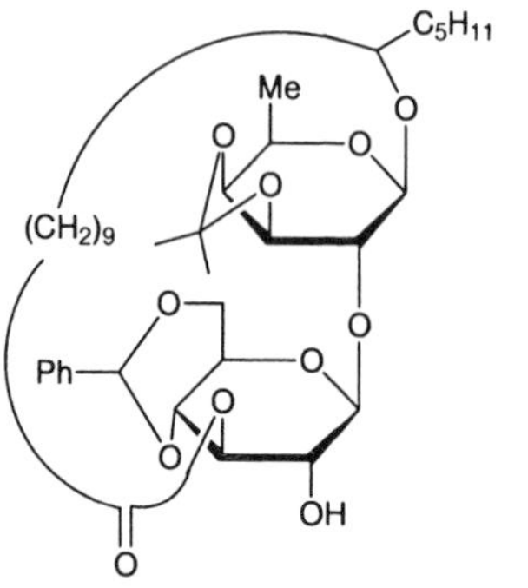

47

properties on growth of other plants.[157] Two enzymes were used to assist in the synthesis of β-D-Glcp-(1→3)-D-GalNH$_2$.[158]

Hepta-*O*-acetylmaltal, on treatment with a cynoethylcopper derivative in the presence of boron trifluoride was converted to compound **48** in 67% yield. This novel reaction is referred to further in Chapter 13.[159]

Considerable interest continues in the synthesis of galactose-containing disaccharides, almost all for reasons associated with their biological significance. 6-Hydroxyhexyl β-D-galactopyranoside has been made in 48% yield from the free sugar by an enzymic method, and the product was then used to prepare α-D-Galp-(1→4)-β-D-Gal as the 6-hydroxyhexyl glycoside by chemical coupling.[160] α-D-Galp-(1→6)-D-Gal β-linked to O-1 of glycerol suppresses the inhibitory effect of the 6-sulfate of glyceryl α-D-glucopyranoside on yeast α-glucosidase.[161] The T-antigen β-D-Galp-(1→3)-D-GalNAc continues to excite interest and the glycosylating agent **49** has been described as an excellent donor for α-linking the disaccharide to serine or threonine.[162] Three reports have appeared on the synthesis of the disaccharide α-linked to amino acids and peptides.[163–165]

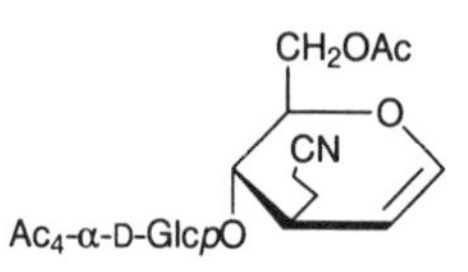

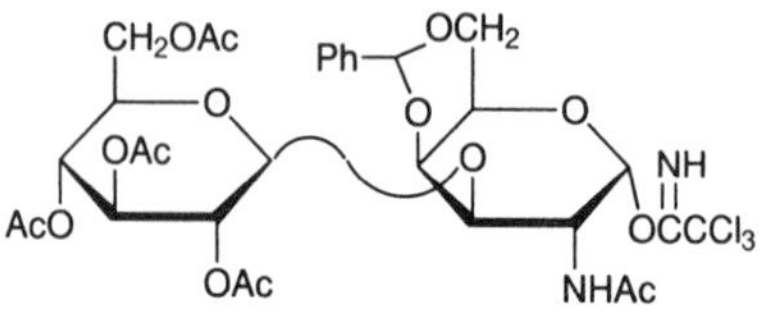

48   **49**

Enzymic methods appear to be particularly useful for the synthesis of *N*-acetyllactosamine: a new three enzyme reaction sequence uses sucrose synthase, UDP-glucose 4′-epimerase and β-(1→4)-galactosyl transferase and gives the product with *in situ* regeneration of UDP-glucose and UDP-galactose. A repetitive batch technique gave yields as high as 57%.[166] Two groups have used lactose as donor and the β-galactosidase of *Bacillus circulans*, and in the course of the work have also transferred galactose to other monosaccharides.[167,168] Related work used various β-galactosidases, *p*-nitrophenyl β-D-galactopyranoside as donor and *N*-acetylgalactosamine was also used as an acceptor. The β-(1→4)-product was mainly obtained with *N*-acetylglucosamine, but with the galactosamine (1→3)-, (1→4)- and (1→6)-products were also formed.[169] *S*-Ethyl 6-*O*-benzyl-1-thio-glucosaminide and *N*-phthalimidoglucosamine *S*-ethyl thioglycoside were used as acceptors in related enzymic preparations.[170]

Random chemical galactosylation using the trichloroacetimidate method of a β-glycoside of *N*-acetylglucosamine gave all possible disaccharides without the 6-linked products dominating, suggesting that useful libraries could be made by this procedure.[171]

S-Ethyl 1-thio-β-D-glucopyranoside as acceptor in further enzymic work has led to yields in the 30-40% range of the lactose thioglycoside together with small proportions of the (1→6)-linked isomer. In the course of the work galactose was also transferred to the *S*-ethyl α-thioglucoside, the corresponding β-thiogalactoside, β-thio-*N*-acetylglucosaminide and β-thioxyloside.[172] Lactoses labelled at C-1, C-2 and C-6 of the glucose moieties have also been made enzymically.[173]

3-*O*-Acetyl-2,2′,6,6′-tetra-*O*-benzoyl-3′,4′-*O*-isopropylidene-α-D-lactosyl trichloroacetimidate has been advocated as a lactosyl donor for higher saccharide synthetic work.[174] A lactosyl ceramide containing a hydroxymethylene group as the inter-sugar unit has been described.[175] β-2-Aminoethyl lactoside carrying a sulfate ester at O-3 of the galactose moiety was coupled via the amino group to the carboxylic acid group of peptides to give selectin-targeted glycopeptides.[176] In related work lactose was converted into the spacer-linked diglycine derivative **50** and this was then polymerized to give a set of flexible poly-*N*-linked lactosylglycines.[177] The biotin-containing lactose derivative bearing a chromogenic diazine, **51**, which is a photoprobe for GM$_3$ synthase, has been described.[178]

50

51

Lactosides and 1-thiolactosides with long chain amide-containing aglycons which mimic lactosylceramides have been made as substrates for *endo*-glycosyl ceramidases,[179] and related lactosylamines carrying *N*-acetyl and *N*-alkyl (octyl, decyl) substituents, which were made in two steps from lactose, have been made as a new set of surfactants.[180]

Reaction of fully etherified 1,6-anhydrolactose with Lewis acids, e.g. antimony pentachloride, caused polymerization to give a 'comb-shaped' product based on an α-1,6-linked glucose backbone with β-D-galactosyl substituents.[181]

By use of galactofuranosyl pentabenzoate β-D-Gal*f*-(1→4)-D-GlcNAc, a trypanosomal glycoprotein component, has been made.[182] The isomeric β-D-Gal*p*-(1→n)-D-Xyl has been produced enzymically; the 2-, 3- and 4- substitution ratios being 8.6:1.4:1. The synthesis could be carried out on a multigram scale using lactose or *o*-nitrophenyl β-D-galactopyranoside as donors.[183] Glycolipids derived from lactonamide and containing two long chain alkyl groups with three amino acid residues between the hydrophilic and hydrophobic parts have been made.[184]

The synthesis of mannopyranosides seems to be particularly suitable for the application of intramolecular glycosylation methods, and Stork's group has continued to contribute in this area. Connection of an *S*-phenyl thiomannopyranoside sulfoxide, *O*-protected at all positions except C-2, separately to monohydroxy carbohydrates by way of dimethylsilyl bridges, and activation of the sulfoxides using triflic anhydride, has given good access to a range of 2-, 3- or 6-hydroxy-connected mannosyl disaccharides. In the case of (1→4)-linked compounds lower yields were obtained, and in the course of this part of the work compound **52** was made. When treated with triflic anhydride, instead of giving the expected 4-linked disaccharide, this bridged product afforded an unusual tricyclic compound, **53**, in 82% yield. This, treated with tetrabutylammonium fluoride, afforded the 6-*O*-β-D-mannopyranosyl glucopyranoside having free hydroxy groups at C-4 and C-2'.[185] In related work a phthalic diester bridge was used to connect O-6 of *S*-phenyl tri-*O*-methyl-1-thio-α-D-mannopyranoside with O-2 of methyl-α-D-glucopyranoside silylated at O-6, and activation with *N*-iodosuccinimide caused coupling of the unprotected O-3 to the mannosyl anomeric centre. Following acetylation, the α-1,3-linked disaccharide derivative **54** was obtained. A further case involved phthaloyl-linking of the O-6 groups of the same mannothioside and methyl 2-*O*-benzoyl-α-D-glucopyranoside, and the result was the preparation of the α-(1→4)-linked mannosyl glucoside in 72% yield. The sites of the spacer linkages and the reaction temperatures affected the selectivities of the reactions.[186]

A straightforward Königs-Knorr glycosylation reaction to produce α-D-Man*p*-(1→2)-α-D-Man was deemed to be less expensive and less time-consuming than the preparation of this compound by enzymic methods.[187] A new specific 1,2-α-mannosidase isolated from *Aspergillus phoenicis*, however, could change this situation. With its use yields of 20% of the disaccharide, 8% of the corresponding trisaccharide and 3% of the tetrasaccharide were reported.[188] Significantly, β-selective mannosylation was achieved, by use of 2,3,4,6-tetra-*O*-benzyl-α-D-mannosyl fluoride, of carbohydrate primary alcohols in the presence of tin triflate and lanthanum perchlorate. A Man-(1→6)-Glc*p*OMe derivative was obtained in 97% yield with the α/β ratio 26:74, and similar selectivities were achieved on forming 1,6-linkages with mannose and galactose acceptors. When secondary alcohols were used, however, the reaction favoured α-products slightly.[189]

52

53

54

O-Substituted *S*-ethyl 1-thio-β-D-fructofuranoside derivatives activated with IDCP led predominantly to α-linked disaccharides, the 6-substituted galactose being produced in high yield with exclusively α-linking.[190]

In the area of amino-sugar disaccharides deprotection of tetrachlorophthalimido groups has been effected by Merrifield resins carrying alkylamino substituents, and the procedure has been applied to making a derivative of methyl 6-*O*-(2-amino-2-deoxy-β-D-glucopyranosyl)-α-D-glucopyranoside.[191] The glycosyl donors **55** and **56** containing trichloroethoxycarbonyl-protected amino groups, on activation with triflic acid and NIS and with trimethylsilyl triflate respectively, gave quantitative yields of the β-linked disaccharide with a primary carbohydrate alcohol, but the efficiency of the condensation with secondary alcohols was reduced, particularly with **55**. The advantages of this particular *N*-protecting group are that it increases the glycosyl donor activity by forming oxazolinium intermediates instead of the less reactive oxazolines, and it is readily removed under non-basic conditions. Compounds **55** and **56** are easily prepared in good yields from D-glucosamine via a glycosyl acetate or 1-hydroxy compound respectively.[192] Chitobiose α-peracetate, treated with alcohols in the presence of trimethylsilyl triflate, reacts as the oxazoline to give β-linked glycosidic products. With simple alcohols yields greater than 90% glycosides were obtained; with diacetonegalactose the yield was 80%.[193] Chitobiose and chitotriose have been β-glycosidically linked to various exoglycosidase inhibitors and screened for inhibition of chitinase activity.[194] The chitobiosylamine derivatives **57** and **58** have been synthesized and tested for their inhibition of chitinase.[195] The action of a bacterial chitinase on a *Fusarium* cell wall component gave β-D-GlcNH$_2$-(1→2)-GlcNAc as main product.[196]

Transferases have been used in the preparation of β-D-GlcNAc-(1→6)-D-GalNAc and the corresponding dimer of GalNAc[197] and related higher saccharides (See Chapter 4). Chemical methods were employed to make β-D-GlcNAc-(1→3)-α-D-GalNAc-*O*-Thr as a building block for core units of mucin glycoproteins.[198] The 4-amino-4,6-dideoxy-sugar derivatives **59**, which represent the terminus of the *O*-polysaccharide of *Vibrio cholerae* with spacer-linked aglycons, have been described.[199] In the area of sulfated amino-sugar disaccharides, β-D-

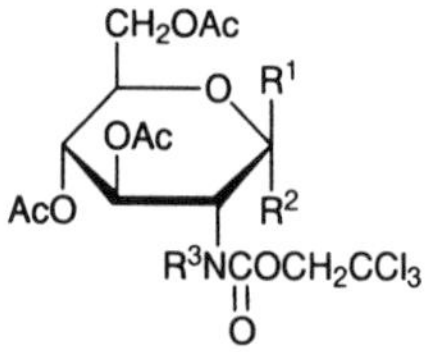

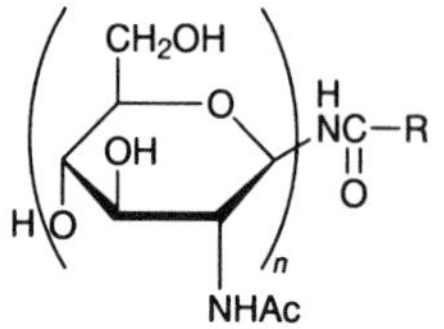

55 R^1 = SEt, R^2 = H, R^3 = Ac
56 R^1 = H, R^2 = OC(NH)CCl$_3$, R^3 = H

57 R =

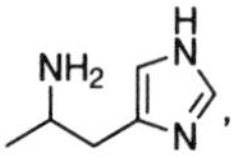

58 R =

59

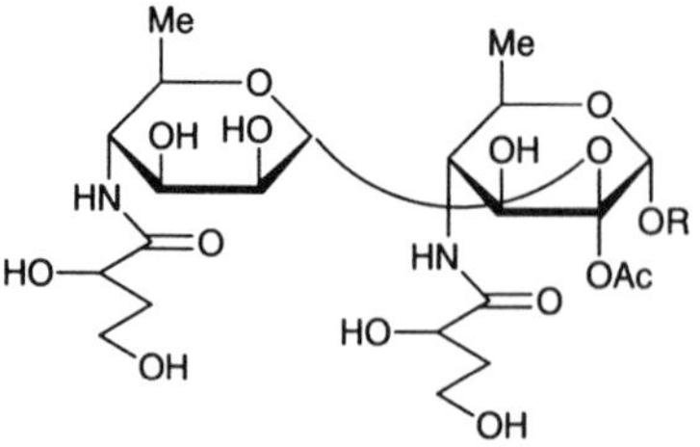

R = (CH$_2$)$_5$CO$_2$H,
(CH$_2$)$_5$CONH$_2$,
(CH$_2$)$_2$S(CH$_2$)$_2$CO$_2$H,
(CH$_2$)$_2$S(CH$_2$)$_2$CONH$_2$

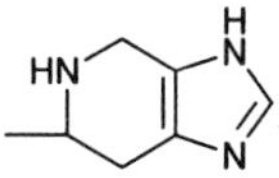

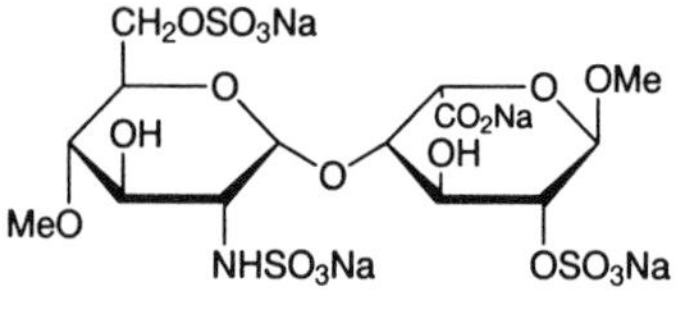

60

GlcNAc-(1→3)-β-D-FucOMe, carrying a sulfate at O-3 of the glucosamine residue, has been made in connection with work on the species-specific aggregation of dissociated cells of *Microciona prolifera*,[200] and the heparin-related disaccharide **60** was made for the study of heparin/platelet binding.[201] A set of compounds e.g. **61**, with varying long-chain alkyl substituents, has been made as bioactive tritium-labelled analogues of Lipid A.[202] A derivative of β-D-GlcNAc-(1→2)-β-D-GlcA has been made as an advanced synthetic intermediate *en route* to a moenomycin A analogue.[203]

A disaccharide derivative consisting of daunosamine α-(1→4)-linked to L-oliose has been coupled to anthracyclins to give anti-tumour products **62**.[204] In the area of monodeoxy-sugar glycosides, α-L-Fuc-(1→2)-β-D-GalpOMe and the corresponding methyl α-glycoside have been described.[205] 2,4-Di-*O*-methyl-L-rhamnose, in the α-1,3-linked dimeric form, has been found as one of the carbohydrate components of Japanese sea hare glycosides together with the 3-carbamate of the same monosaccharide.[206] 3,6-Dideoxy-D-*arabino*-hexose

61 $R^1 = CH_2\overset{\displaystyle OCO(CH_2)_{12}Me}{\underset{\displaystyle |}{CH}}(CH_2)_{10}Me$ $R^2 = CH_2\overset{\displaystyle OCO(CH_2)_{10}Me}{\underset{\displaystyle |}{CH}}(CH_2)_{10}Me$

$R^3 = CH_2\overset{\displaystyle OBz}{\underset{\displaystyle |}{CH}}(CH_2)_{10}Me$ $R^4 = CH_2\overset{\displaystyle OH}{\underset{\displaystyle |}{CH}}(CH_2)_{10}Me$

62 R = H, OMe

(tyvelose) has been α-1,3-coupled to glycosides of GalNAc to give models to assist in the determination of the anomeric configuration of a tyvelose-containing epitope of a glycan.[207] Tyvelose and abequose have been α-(1→3)-linked to allyl α-D-mannopyranoside in studies related to *Salmonella* antigens.[208]

In further studies of heparin-related disaccharides β-D-GlcpA-(1→4)-GlcNH$_2$ has been made carrying a sulfate ester at either O-2 or O-3 of the glucuronic acid residue.[209] Benzyl and methyl 2,3,4-tri-*O*-benzyl-D-glucuronate have been coupled with alcohols in the presence of *p*-nitrobenzenesulfonyl chloride and silver triflate to give high yields of glycosides with the α-anomers predominating and the reaction was used to make α-D-GlcA-(1→3)-L-Araf compounds.[210] β-D-GlcpA-(1→6)-β-D-Gal has been linked to glycyrrhetic acid to give a product which was studied for cytoprotective activities.[211] α-D-Glc-(1→2)-L-Rha was found to be particularly active in causing hypocotyl elongation during a study of 20 analogues of the 2,3-unsaturated disaccharide acid lepidomide **63**.[212] In the field of ulosonic acid disaccharides compound **64** was made from a 1-*C*-furanylmannose thioglycoside which was coupled to diacetonegalactose prior to oxidation of the furan ring. Barton decarboxylation then gave the β-D-Man-(1→6)-α-D-Gal derivative in 45% yield with β/α ratio 25:1.[213]

A 2-thio-neuraminic acid glycosyl xanthate gives good yields of α-linked disaccharides when used in conjunction with phenylselenyl triflate.[214]

A suitable α-Neu5Ac-(2→8)-α-Neu5Ac donor is the glycosyl fluoride of the fully acetylated methyl diester carrying phenylthio groups at C-3 of both rings.[215] The synthesis of the *bis*-neuraminic acid 1,2:8′,9′-lactam **65** has been reported.[516]

63   **64**

65

Several disaccharides terminating in pentoses have been reported. Tri-*O*-benzyl-D-ribofuranosyl xanthates have been coupled to various sugars in good yield to give mainly β-linked disaccharides,[217] and some 2-deoxyribonucleosides have been β-D-ribofuranosylated at O-3 by use of a tri-*O*-benzoyl-β-D-ribofuranosyl acetate using tin tetrachloride as promoter.[218] α-L-Ara-(1→6)-β-D-Glc has been made as the *p*-(2-nitroethyl)phenyl glycoside,[219] and the same disaccharide, as the geranyl glycoside, has been isolated as the aroma precursor from leaves of a green tea.[220] β-D-Xyl-(1→6)-β-D-Glc*p* has been prepared during the synthesis of neohancoside A and B which are monoterpene glycosides found in a Mongolian plant known to have antitumour activity.[221] The *p*-hydroxyacetophenone glycoside of this xylosylglucose occurs in *Aster batangensis* and has been synthesized chemically.[222] 1,2-Anhydro-3,4-di-*O*-benzyl-β-D-lyxo- and -α-L-ribopyranose have been used to make disaccharides containing α-D-lyxose or β-L-ribose,[223] and the same workers have used 1,2-anhydro-3,5-di-*O*-benzyl-β-D-lyxose to obtain α-D-Lyx*f*-(1→6)-D-Gal in 85% yield. They also described other 1,2-anhydrofuranose derivatives.[224]

Chapter 19 contains reference to other disaccharides, some with branched-chain sugars.

1.5 Disaccharides with Anomalous Linking or Containing Modified Rings – The search for glycosidase inhibitors is producing a range of compounds with major modifications, for example the hydroxylamino derivative **66**.[225] *N*-Acetylneuraminic acid has been glycosidically linked to 5-deoxycytidine extended by two methylene groups to give a carbon-linked CMP-Neu*N*Ac analogue,[226] and the same sugar acid and α-L-fucopyranose have been glycosidically bonded to hexane-1,6-diol to give a sialyl Le[x] analogue in which the Glc*N*Ac and the D-Gal units are replaced by an alkyl spacer.[227] α-Galactosylation of a 4-thiogalactose derivative led to an *S*-linked galactobiose made for conformational studies of protein binding.[228] Other *S*-linked disaccharides are referred to in Chapter 11. *S*-Linked α-D-Neu*N*Ac-(1→4)-D-Gal has been made as a potential rotavirus inhibitor[229] and a selenium-linked isomaltose analogue has been made via a glucosyl selenide salt.[230]

66

67 R = H, CH$_2$OH
R^1 = H, NHBoc

A carbamaltose with the ring oxygen atom in the reducing moiety replaced by methylene has been produced together with analogues having the interunit oxygen atom replaced by nitrogen or sulfur.[231] Related carbadisaccharides with carbocyclic 'non-reducing' groups to have been described are the analogues of α-D-Man*p*-(1→4)-D-Man, α-D-Glc*p*-(1→4)-D-Man, and β-D-Glc*N*Ac-(1→4)-D-Man.[232] The fucosylated 2-aminocyclohexanol derivatives **67** are sialyl Le[x] mimics, and can be considered to be derivatives of a carbadisaccharide.[233] See Chapters 4 and 18 for references to related compounds, and also earlier in the present Chapter for mention of some glycosylinositol derivatives.

Disaccharide analogues with nitrogen in the ring to have been described are deoxynojirimicin having β-D-galactose substituted at one of each of the four hydroxy groups. The galactosidase used for the glycosylation yielded 6%, 20%, 26% and 7% of the disaccharides linked through the 2-, 3-, 4- and 6-hydroxyl groups respectively (carbohydrate number).[234] Danishefsky's group has reported details of its synthesis of allosamidin, and in the course of the work described the allosamine-containing disaccharide analogue **68**.[235] Fleet's group has described compound **69** which was isolated from fruits of *Nicandra physalodes*.[236]

The α-1,6-linked glucobiose with sulfur in place of the ring oxygen atom in the non-reducing moiety has been made from gentiobiose by an ingenious method involving thiolysis of the octaacetate to give an intermediate with an acyclic non-reducing moiety in the form of the hemithioacetal.[237]

1.6 Reactions, Complexation and Other Features of *O*-Glycosides – A short review has appeared on recent work directed towards the clarification of the

68   **69**

mechanisms of action of glycoside hydrolases which act with retention of configuration.[238] The enzymic cleavage of *o*-nitrophenyl β-D-galactopyranoside proceeds in frozen aqueous solution.[239] The cellulase-catalysed cleavage of cellobiose is not enhanced by microwave radiation, but neither is the activity diminished.[240] Hydroxyl groups at C-2, C-3 and C-4 in the substrates are necessary for the action of jack bean and almond α-mannosidases, but that at C-6 is not important.[241]

Fraser-Reid's group continues to study the Br⁺ activation of *n*-pentenyl glycosides and has reported on measured and predicted trends in the activation process. Solvation energies obtained by a method developed by Still *et al.* gave activation energies in good agreement with observed trends.[242] They have also reported on the reactions of ω-alkenyl glycosides in aqueous *N*-bromosuccinimide and found the pentenyl compounds particularly subject to hydrolysis, while allyl, butenyl and hexenyl tended to give bromohydrins formed by addition processes.[243] Dibromination has been used to protect the double bond of pentenyl glycosides during other reactions, and new methods for restoring the double bonds use samarium diiodide, which involves a one-electron transfer process, or sodium iodide in refluxing butenone which is a nucleophilic procedure. The latter, in particular, was highly efficient.[244]

Hydrolysis studies on sucrose, leucrose and isomaltulose with Y-zeolites indicate that the pores of the catalyst can include a disaccharide and one or more water molecules. The processes are accelerated by dealumination of the catalysts.[245]

A study has been made of hydrogen abstraction by chemically produced radicals of the α-D-mannoside type **70**. H-1 abstraction leads to β-mannosides, whereas H-2 abstraction competes and leads to 2-ulosides, and *O*-protecting groups determine the relative significance of the two processes. An analogous photochemical reaction permits α- to β-mannoside conversions, α-D-Man*p*-(1→6)-D-Glc being converted to the β-D-Man-(1→6)-D-Glc.[246]

Unprotected methyl glycosides have been converted in a one-pot process to acetylated glycosyl fluorides by initial treatment with HF followed by addition of acetic anhydride. The process is intended as a gas chromatographic analytical tool with conversions in excess of 95%.[247] Efforts to convert *O*-protected derivatives of α-D-Rha-(1→2)-α-D-Rha-SEt to the (methoxycarbonyl)undecyl (or other) glycosides with a radical cation reagent (BrC₆H₄)N⁺HSbCl₆⁻ failed, but afforded the corresponding monosaccharide glycosides.[248]

Methyl β-D-galactopyranoside, treated with deuterated water in the presence of Raney nickel as catalyst and ultrasonic radiation, gives rapid C-deuteration at positions 2-, 3- and 4- without inversion of configuration. Relative rates of the deuteration at these positions were very dependent on solvent, but reaction at C-4 was most favoured in the three oxygenated solvents used.[249] A report on the methanolysis of cyclohexyl 2-*N*-acetyl-2-deoxy-*N*-methylamino-tri-*O*-methyl-D-glucopyranosides is reported in Chapter 9. *p*-Nitrophenyl glycosides can be cleaved with hydrogen and palladium catalysts, and with ammonium ceric nitrate in acetonitrile/water. The processes are almost quantitative.[250]

The metathesis coupling of the alkene groups of the corresponding pentenyl 6-*O*-pentenoyl-β-D-glucopyranoside using the tungsten complex W(OAr)₂-Cl₄.PbBu₄, gave lactone **71**. Several related examples were reported, yields being in excess of 50%.[251]

70   **71**

72

The paraquat cation **72** binds phenyl-β-D-glucopyranoside more strongly than the α-anomer in aqueous solution. Computation study results concurred with this finding.[252,253] Binding of glycosides with inorganic species is treated in Chapter 17. The association of *O*- and *C*-glycosides with concanavalin A has been measured using fluorescence anisotropy. The free energies of binding for both types of glycosides were similar indicating that the recognition components were also similar.[254]

The self-diffusion coefficients for methyl α- and β-D-glucopyranosides in aqueous solutions have been measured in the concentration ranges 10-70% over the temperature range 250–424 K. At concentrations greater than 40% the mobility of the water and glucoside molecules is significantly lower in the solutions of the α-anomer.[255]

2   S-, Se- and Te-Glycosides

S-Ethyl 1-thio-glycoside tetraacetates of β-D-galactose, β-D-glucose and 2-deoxy-2-phthalimido-β-D-glucose uncontaminated by α-isomers are available in almost quantitative yield from the peracetates by use of ethanethiol and boron trifluoride etherate at –78 °C in chloroform/ether (3:2). The reactions, however, require 6–7 days for completion, and in the D-mannose and L-rhamnose series give 80:20 α/β mixtures.[256] A new approach to the preparation of S-phenyl thioglycosides involves converting peracetates to *p*-methoxyphenyl glycosides. These were then converted cleanly to the glycosyl bromides or chlorides by the use of zinc halide and acetyl halide in dichloromethane. From the halides the S-phenyl β-thioglycosides were made using thiophenol and boron trifluoride etherate. This method was applied in the glucose, N-phthalimidoglucose, N-phthalimidogalactose and N-Troc-glucosamine and lactose series.[257]

Activation of O-substituted S-methyl or S-phenyl thioglycosides as glycosyl donors with iodosobenzene and various acids for the synthesis of di- and higher saccharides led preferentially to α- or β-linkages according to the conditions used. α-Linked compounds were favoured with: 2-O-benzyl and 6-O-(2,2,2-trichloro-ethoxycarbonyl) groups on the donor, tin(II) or tin(IV) chloride, bismuth trichloride or antimony chloride all with silver perchlorate as activators and benzene as solvent. For β-linked compounds the following were required: benzoate ester group at O-2 and triflic acid, triflic anhydride, trimethylsilyl triflate, tin ditriflate (which was best), or ytterbium triflate as catalyst. Good yields and stereoselectivities were achieved.[258] O-Benzylated S-methyl 1-thio-β-D-galactopyranoside has been activated by use of iodine. Linked with primary sugar alcohols it gave disaccharides with yields of 90% and α/β ratios of about 1.5:1, whereas with secondary alcohols 80% yields and α/β ratios of about 2:1 were observed.[259]

Coupling of S-ethyl 3,6-di-O-benzyl-4-O-methyl-1-thio-α-D-mannopyranoside carrying different substituents at O-2 with methyl 4-O-acetyl-2-O-methyl-α-L-fucopyranoside and its D-enantiomer occurred with a small double stereodiffer-entiation effect for the 2-O-methyl- and -O-benzyl donors. α-Selectivity was slightly lower for the D,L combination compared with the D,D combination.[260] This stereochemical effect was further investigated by coupling of S-ethyl 2-O-acetyl-3-O-benzyl-4-O-methyl-1-thio-α-D-rhamnopyranoside and its enantiomer with a chloroacetyl rather than acetyl at O-2, with methyl 4-O-acetyl-2-O-methyl-α-L-fucopyranoside. While the L,D combination gave the disaccharide in the ratio 2:1, the L,L afforded only an α-linked disaccharide.[261] These important results highlight the significance of double stereodifferentiation in disaccharide synthesis.

S-2-Chloroethyl 1-thio-β-D-galactopyranoside and the corresponding glucoside are suitable glycosylating agents for the preparation of glycoconjugates.[262]

The attempted coupling of an S-phenyl 2-trichloracetylamino-2-deoxy-4,6-O-isopropylidene-1-thio-β-D-galactopyranoside with the glycosylating trichloroace-timidate of a methyl-D-glucuronate derivative, with trimethylsilyl triflate as activator, did not lead to disaccharide synthesis as expected. Instead the S-phenyl group of the galactose derivative exchanged and the product obtained in 80%

yield was the *S*-phenyl β-thioglycoside of methyl glucuronate. This is a most surprising result and the surprise is increased by the fact that the coupling of the trichloroacetimidate with an *S*-phenyl β-thioglucoside carrying a free hydroxyl at C-3 gave a disaccharide in 85% yield.[263] Compound **73** was produced by treatment of pentaacetyl β-D-glucosamine with Lawesson's reagent followed by deacetylation and is a potent competitive inhibitor of jack bean *N*-acetyl-hexosaminidase.[264]

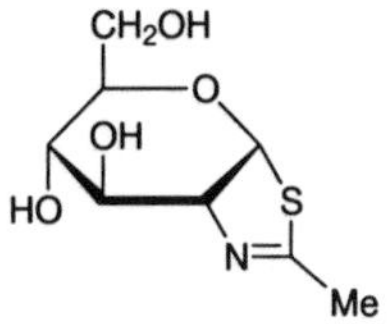

73

Four molecules of *N*-acetylneuraminic acid linked through sulfur to an amide spacer and hence to tetrahydroxy-*p*-*t*-butylcalix[4]arene gives a product that binds tightly to wheat germ agglutenin.[265] The same acid was linked glyco-sidically through sulfur to a dimethylmercury group to give a heavy atom product for use in protein sialic acid complex crystallography.[266] The *bis*-macrocyclic bola-amphiphile **74** forms pores in vesicle membranes that selectively channel metal ions with the following efficiency $Cs^+>K^+>Na^+>Cl^-$.[267] The dendritic α-thiosialoside **75** has been made, and related compounds having up to 16 sialic acid units were described.[268]

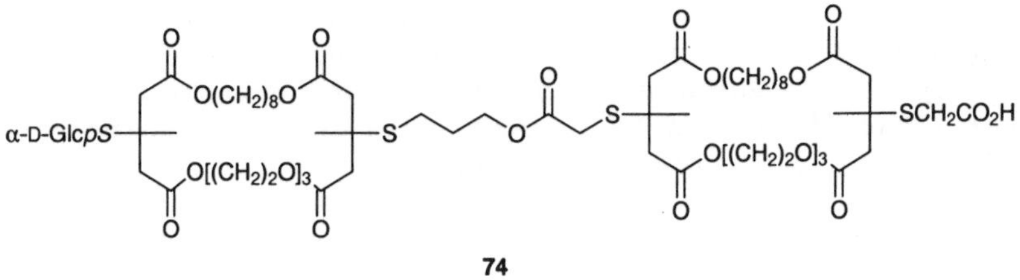

74

Tetra-*O*-acetyl-1-thio-β-D-glucopyranose has been linked through sulfur to benzyl groups carrying diphenylphosphino and dicyclohexylphosphino groups on the *ortho* positions to give P,S ligands and their palladium complexes.[269] A thioglycoside derived from 2-bromo-3-alkyljuglone has been reported to have cytostatic activity.[270] A very efficient and selective oxidation of thioglycosides to corresponding sulfoxides involves the use of hydrogen peroxide, acetic anhydride in dichloromethane in the presence of silica which accelerates the reaction. With 2-azido-2-deoxy- and 2-deoxy-2-*N*-phthalimido sugars, however, yields were somewhat lower.[271] Anomerization of *S*-alkyl tetra-*O*-benzyl-β-D-gluco-pyranosides occurs very rapidly in the presence of IDCP, giving α/β ratios of 3:2

for the *S*-methyl and *S*-ethyl compounds, but 1:7 for the *S*-isopropyl. For the tertiary butyl analogue there was no reaction. The process occurs by inter-molecular exchange, at least in the case of the *S*-methyl compound.[272]

In the area of glucosinolate chemistry, sinigrin **76**, has been made by methods suitable for a ten gram reaction, starting with 4-nitrobut-1-ene and 1-thio-glucose 2,3,4,6-tetra-*O*-acetate,[273] and a similar approach was adopted in the preparation of the methylthioalkyl analogue **77**.[274]

75 R = α-NeuNAc

76 R = allyl
77 R = (CH₂)ₙSMe

In the areas of phenylselenyl- and phenylteluryl-glycosides, *Se*-phenyl 2,3,4,6-tetra-*O*-acetyl-1-selenyl-β-D-glucopyranoside, treated with diphenyldiselenide and tributyltin hydride, gave the product of direct reductive cleavage of the carbon-selenium bond and 1,3,4,6-tetra-*O*-acetyl-2-deoxy-α-D-glucopyranoside which is the product of acetyl migration in the initially formed glycosyl radical. The reaction was studied in some detail.[275] *Se*-Phenyl tetra-*O*-methyl-1-seleno-β-D-glucopyranoside irradiated with alcohols in the presence of aromatic sensitisers gives *O*-glycosides in about 70% yields with α/β ratios near 1:3. Coupling with 1,2:3,4-di-*O*-isopropylidene-α-D-galactose was effected.[276] Glycosyl bromides and diaryl-ditelurides in the presence of sodium borohydride were used to make *Te*-aryl *O*-substituted 1-teluryl-β-D-glucopyranosides almost quantitatively. Phenyl and *p*-tolyl substituents were involved together with acetyl, benzyl and benzoyl *O*-substituents.[277]

3   *C*-Glycosides

3.1   Pyranoid Compounds – β-D-Glucosyl-, β-D-*N*-acetylglucosaminyl-, β-D-mannosyl- and β-D-galactosyl-nitromethane were ozonolysed to give the corresponding *C*-formyl compounds which, on treatment with nitromethane, gave products which were converted to the α-D-hexosyl 2-nitroethanes.[278] From the 2-acetamido-2-deoxy-β-D-glucopyranosylnitromethane the corresponding alde-hyde, carboxylic acid and 2-hydroxy-1-nitroethyl compounds were produced by the same research group.[279] Peracetylated glucosaminyl chloride, treated with allyltributyltin and a radical initiator, gave the corresponding allyl α-C-glycosides in 66% yield, α/β ratio 10:1, but when the N-acetyl group was replaced by N-phthalimido, 71% yield was obtained and the α/β ratio was reversed to 1:10.[280] Allyl phenyl sulfide and the corresponding sulfone treated with glycosyl bromides in the presence of tributyltin radicals offer alternative ways of making C-allyl glycosides, and in an extension of this work glycosyl radical addition to ethyl 2-(phenylthiomethyl)acrylate afforded the α-C-glycoside with a 2-ethoxy-carbonylallyl aglycon.[281] In related studies glycosyl radicals have been added to acylsulfonyl oxime ethers, for example $PhSO_2CH(=NOBn)$ to give C-glycosides with C-acyl substituents in the form of the N-benzyl oximes.[282] 2-Acetamido-2,4,6-tri-O-benzyl-2-deoxy-α-D-galactopyranosyl pyridylsulfone, on treatment with samarium diiodide and cyclohexanone, gave the 1-hydroxycyclohexyl C-glycosides with the α/β ratio 10:1. Other carbonyl compounds reacted similarly, and much poorer α/β selectivity was observed when the β-linked sulfones were used.[283]

Glycal derivatives continue to offer considerable scope for the preparation of C-glycosides. Various acetylated glycals have been treated with allyltrimethyl-silane in the presence of montmorillonite to give excellent yields of 2,3-unsatu-rated C-allyl glycosides with the α/β ratios being about 4:1 or, with galactal derivatives, much greater.[284] Analogously, acetylated glycals treated with diethyl-zinc afforded 2,3-unsaturated C-ethyl compounds with α-selectivity which, in the glucal series, was not particularly high, but only the α-product was obtained from tri-O-acetyl-D-galactal.[285] Likewise, tri-O-acetylglucal treated with the zincalkyl **78** in the presence of boron trifluoride etherate afforded the corresponding unsaturated C-glycosides with α/β ratio 9:1. The major compound was converted into the α-C-mannoside with the 2-aminobutan-4-oyl aglycon.[286]

Photobromination of 2,3,4,6-tetra-O-acetyl-β-D-galactosyl carboxamide gave the C-glycosidic glycosyl bromide which, on treatment with zinc in the presence of a base, underwent elimination and led to D-galactal carrying the carboxamide group at C-1. From this, various D-galactal C-glycosides were prepared as potential inhibitors of β-D-galactosidase.[287]

Photocyclisation of the 2,3-unsaturated α-C-glycoside having a 3-oxo-2-methyl-but-1-yl aglycon afforded the mixed isomers **79** to offer a new way of making functionalized cyclopentanes.[288] In related work, Se-phenyl 2-allylamino-3,4,6-tri-O-benzyl-2-deoxy-1-selenyl-D-glucopyranoside, on treatment with tri-butyltin hydride/triethylboron in a stereoselective process, afforded the bicyclic **80**.[289] The exo-1-methylene derivatives produced from tetra-O-benzyl-D-glucono-lactone or -galactonolactone, on reaction with diethyl malonate radical, gave β-C-glycosides having the diethyl malonylmethyl aglycon, and similar work was carried out in the furanoid series.[290] Compounds with exo-difluoromethylene groups at C-1 undergo addition when treated with diethyl phosphite and di-t-butyl peroxide to give glycosyldifluoromethyl phosphonates.[291] The diacetylene **81**, treated with butyllithium followed by sodium methoxide, underwent 1,2-Wittig rearrangement to give the isomeric dialkyne **82**. Likewise the isomer of the starting material with inverted stereochemistry at C-1 gave the anomerically

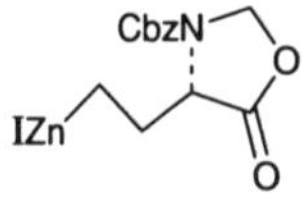

78 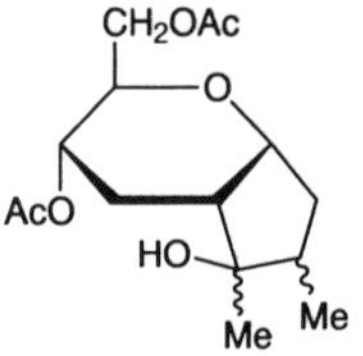   **79** 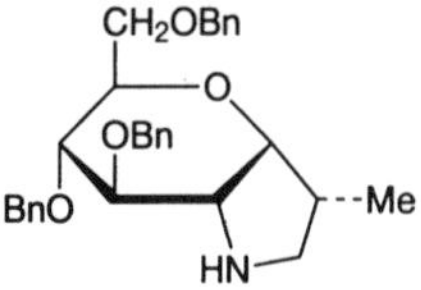   **80**

inverted isomer of **82**.[292] The 1,4-diacetylene **83** has been obtained by reaction of the corresponding 4-alkynyl-4-deoxy-β-1,6-anhydride with 1-diethylchlorosilyl-2-trimethylsilyl-ethyne followed by butyllithium and aluminium trichloride,[293] and the corresponding terminally substituted diacetylene derived from cellobiose has been reported.[294]

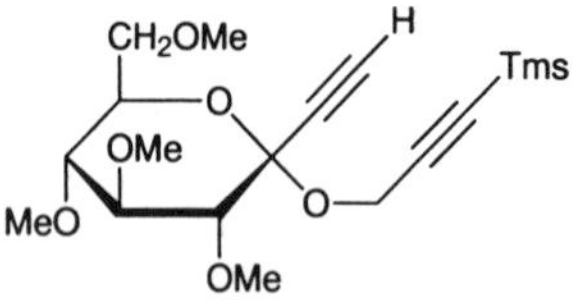

81 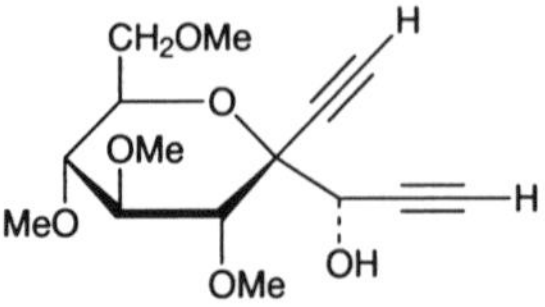   **82** 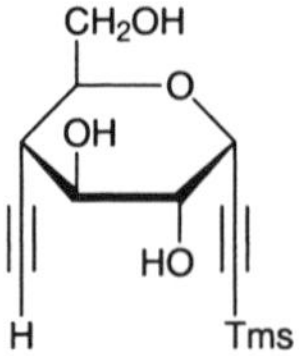   **83**

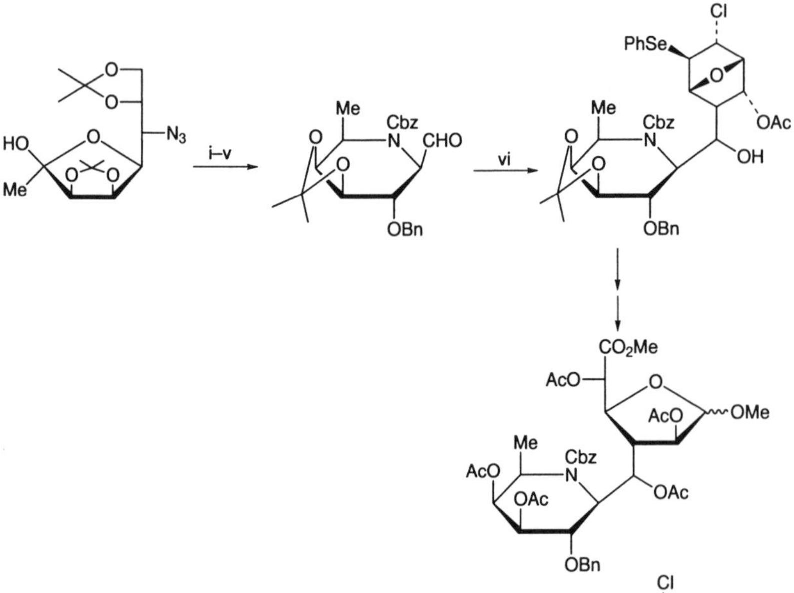

Reagents: i, H_2, Pd/C; ii, BnBr; iii, CbzCl; iv, AcOH, H_2O; v, $NaIO_4$; vi,

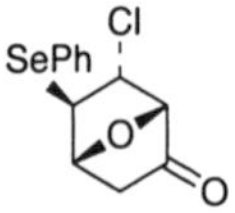

Scheme 3

Samarium iodide has been used to link perbenzylated deoxynojirimicin having an iodomethyl group at the 'anomeric centre' to diisopropylidene-D-*galacto*-dialdose and produce a disaccharide analogue with the two rings joined by way of a hydroxyethyl linkage.[295] A related D-*fuco*-nojorimicin having the anomeric centre linked by a 1-carbon bridge to C-3 of a hexuronofuranoside has been produced by Vogel's laboratory in the manner outlined in Scheme 3,[296] and the related compound 84 has also been described by Vogel and colleagues, both units having been derived from a 7-oxa-2,2,1-bicycloheptane system.[297]

84   85

Considerable attention has been given to other *C*-linked disaccharide compounds, a further study having been conducted on the dimerization which occurs when glycal derivatives are treated with Lewis acids. For example, compound 85 is obtainable in 77% yield from triacetyl-D-galactal on treatment with acetyl perchlorate and the corresponding di-*O*-acetylxylal-derived dimer was described for the first time. The report of a cyclopentane derivative which was produced similarly from tribenzylglucal is now known to be in error.[298] During studies of free radical reactions of tetraacetylglucosyl bromide and *p*-methoxyphenol several products were reported including that of radical dimerization of the glycosyl moiety. A reinvestigation of the electrochemical reaction of the same glycosyl bromide also gave the three C-1–C-1-linked glycosyl dimers in 70% overall yield.[299]

The α-D-Man-(1→2)-α-D-Man analogues with a methylene and a hydroxy-methylene linkage have been made by coupling an appropriate glycosyl pyridyl sulfone with a 2-deoxy-2-*C*-formyl-α-D-mannoside derivative with samarium diiodide used as catalyst.[300] Sinaÿ and colleagues continue with their intramolecular radical approach to *C*-linked disaccharides and have generated the analogue 86 in 43% yield along with smaller proportions of the two possible diastereoisomers. Initially the two moieties were linked by way of a silyl bridge between O-2 of a phenylselenyl glucoside and O-3 of a 5-deoxy-pent-4-enose compound.[301] The *C*-linked lactose analogue having a hydroxyl group in the two possible orientations on the bridging methylene have been examined by NMR and theoretical methods, and while one diastereomer is conformationally like lactose in solution, the other is quite different.[302] A ceramide derived from them is described in reference 175. Further work has revealed for the first time that *C*-linked saccharides (notably *C*-linked lactose) may bind to proteins in different conformations from those of the natural ligands.[303]

(Trifluoromethyl)trimethylsilane together with tetrabutylammonium triphenyltin difluoride generate difluoromethylacyl nucleophiles from acyltrimethyl-silanes and, in this way, triacetylglucal may be converted into compounds **87** by use of benzoyltrimethylsilane. By application of this methodology the disac-charide analogues **88** were produced as indicated in Scheme 4.[304]

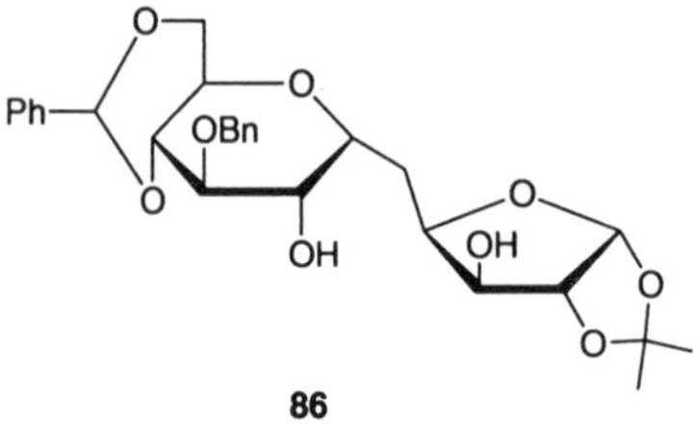

86 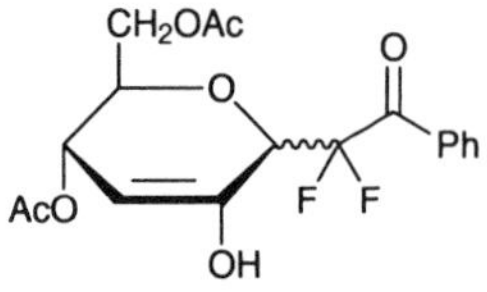   **87**

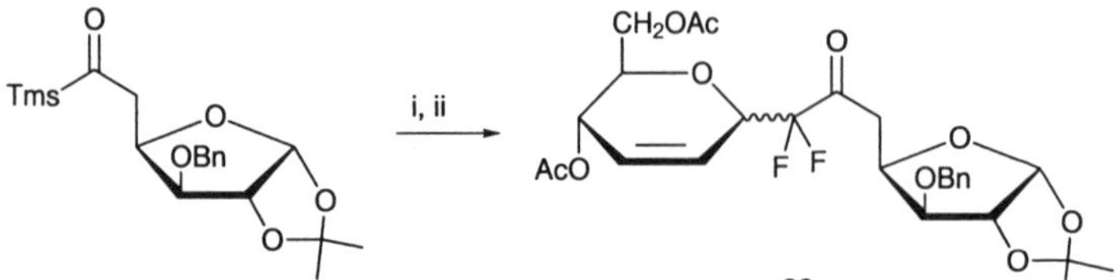

88

Reagents: i, CF$_3$Tms, Bu$_4$NPh$_3$SnF$_2^-$(cat.); ii, tri-O-acetyl-D-glucal, BF$_3$·Et$_2$O

Scheme 4

In the main, disaccharide analogues of this type are made by addition reactions to carbonyl compounds. A Wittig condensation between a phosphorane derived from 1,2:3,4-diisopropylidene-D-*galacto*-dialdose and a 3,4:6,7-diisopropylidene-2,5-anhydroheptose has given access to a methylene-linked β-D-Man*f*-(1→6)-D-Gal analogue. In the course of the same work, the methylene-linked analogue of α-D-Gal*p*-(1→6)-D-Gal was described.[305] In related work a 6-deoxy-6-nitro-α-D-glucopyranoside was coupled with a 2,6-anhydroheptose ether to give access to the methylene analogue of β-D-Gal-(1→6)-D-Glc,[306] and samarium diiodide-induced coupling between tetra-O-benzyl-α-D-mannopyranosyl chloride and a derivative of a 6-deoxy-D-*gluco*-heptodialdose gave the hydroxymethylene-linked analogues of α-D-Man-(1→6)-D-Glc.[307] Compound **89**, which can be considered to be a ketosyl *C*-linked disaccharide, was obtained following the coupling of a D-glucose-based hex-6-ynose with a non-4-ulose derivative obtained from gluconolactone following ring opening with allylmagnesium bromide.[308]

Appreciable interest has been shown in compounds having sugars *C*-linked to amino acids and hence peptides. 3,4,6-Tri-O-acetyl-2-deoxy-2-phthalimido-β-D-glucosyl cyanide has been used to link *N*-acetylglucosamine by way of a methylene group to asparagine and hence a pentapeptide which was then enzymically converted into a high-mannose glycopeptide containing a methylene inter-unit link.[309] Closely related work describes *N*-acetylglucosamine linked by

89

90 $n = 0, 1, 2$

way of a carbonyl group to the *N*-terminus of various peptides.[310] Compounds **90** represent a set of so-called 'carbopeptoids' which are *C*-glycosides of glucosamine peptide-linked together.[311] In their work on mimetics of the tetrasaccharide siaLe^x, Wong and colleagues have reported on α-*C*-fucosides linked by way of two carbon atoms to amino groups which were linked to peptides, these being produced from a 2-aminoethyl *C*-fucoside derivative.[312,313] Related studies have led to the D-mannose *C*-glycosides **91** and **92** amongst others.[314] Compound **93** was developed in studies of ligands for cell surface carbohydrate receptors together with a library containing 96 related compounds produced in a combinatorial approach for screening purposes.[315]

Aryl *C*-glycosides are still of interest, mainly in relationship to their presence in natural products, and Schmidt and colleagues have developed a new approach starting from an *O*-substituted *aldehydo*-D-arabinose and an aryl Wittig reagent. The derived alkenes were then electrophilically cyclized to give aryl 2-deoxy-2-iodo and hence 2-deoxy-β-*C*-glucopyranosides.[316] They have also used their trichloroacetimidate method and activated benzene derivatives to prepare *C*-aryl glycosides of glucosamine.[317] Intramolecular Friedel-Crafts reaction of 2-*O*-benzyl-α-D-glucosyl chloride derivatives catalysed by silver tetrafluoroborate afforded the *C*-glycoside **94** in 74% yield. From it the *o*-carboxyphenyl α-*C*-glucoside was produced.[318] The coupling of *O*-substituted 2,6-dideoxy-D-*arabino*-hexopyranosyl acetates with β-naphthol in the presence of silver perchlorate and bis-cyclopentadienylhafnium chloride gave high yields of α-linked compounds, for example **95**. This methyl ether did not undergo anomerization under the conditions of its synthesis, whereas the corresponding diacetate did. Related work was carried out on further activated naphthalene derivatives.[319] In work related to the aryl *C*-glycosides chrysomycins further *C*-naphthyl glycosides were produced via the racemic **96** which itself was made from a racemic substituted

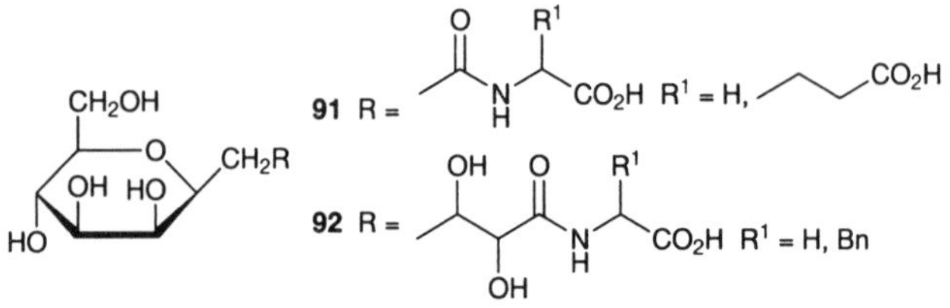

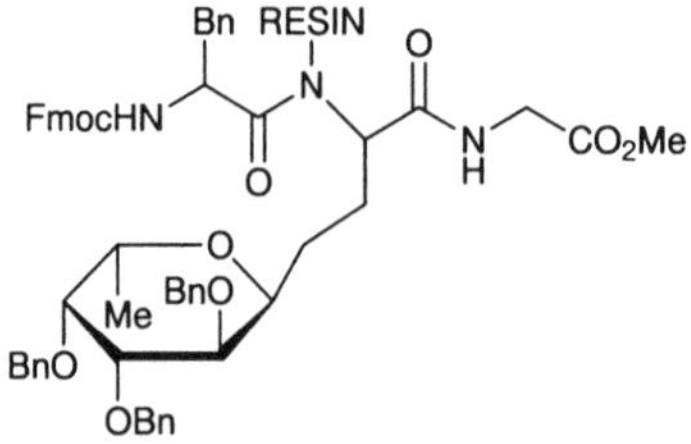

91 R =   (R¹ = H, ...CO₂H structure)

92 R =   (R¹ = H, Bn structure)

93

furanone derivative.[320] Further reference is made to naphthyl *C*-glycosides in Chapter 19. An improved synthesis of the D-glucuronic acid *C*-benzyl glycoside **97**, which has cancer preventative properties, has used sodium hypochlorite and Tempo as oxidising reagents to introduce the acid function.[321]

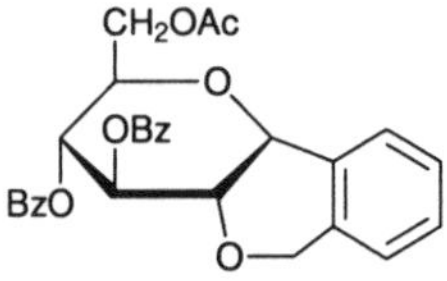

94

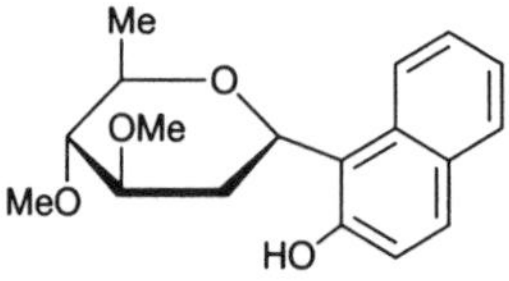

95

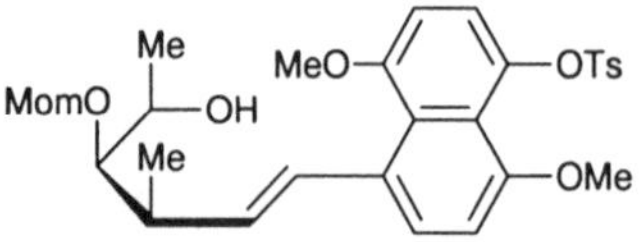

96

Compounds having an oxygen-linked and a carbon-linked substituent at the anomeric centre can be used to give *C*-glycosides. For example compounds **98** have been derived by lithiumalkyl addition to the appropriate 2-azido-2-deoxy-aldonic acid lactone, and reductive cleavage using triethylsilane and boron trifluoride at low temperature gave the *C*-glycoside **99**. This offers a route to 2-acetamido-2-deoxy-β-D-*C*-glucopyranosides.[322] Other compounds with both *C*- and *O*-linked substituents at the anomeric centre are considered to be extended chain compounds and are referred to in Chapter 2.

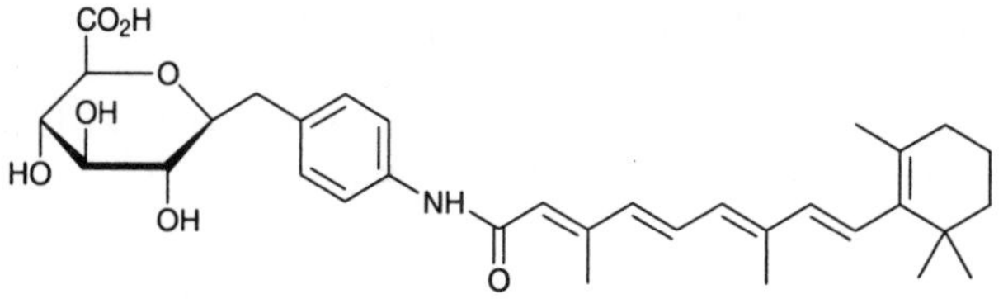

97

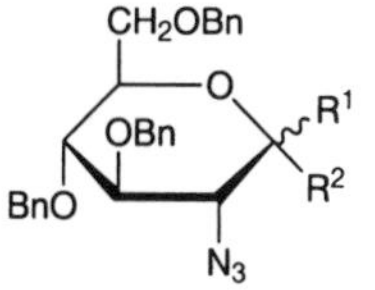

98 R^1 = Me, Bn, Ph; R^2 = OH
99 R^1 =Me, Bn, Ph; R^2 = H

3.2 Furanoid Compounds – A report from a symposium on developing strategies for the stereoselective synthesis of compounds, including monocyclic *C*-glycosides, by catalytic cyclization reactions included reference to the anti-tumour agent goniofufurone and the central subunit **100** of the antibiotic efrotomycin.[323]

Several further compounds with simple 'aglycons' have been described, and Fleet and his group who have continued their work on 2-triflates of aldonolactones have reported synthetic routes to compound **101** and its C-2 epimer starting from the 2-triflate of 3,5:6,7-di-*O*-isopropylidene-D-*glycero*-D-*gulo*-heptono-lactone.[324] The thiocarbamate *C*-glycoside **102** was produced from the corresponding glycosyl acetate in high yield and with excellent β-selectivity by use of 1-(*t*-butyldimethylsilyloxy)vinyl benzyl ether (a ketene acetal) with the unusual

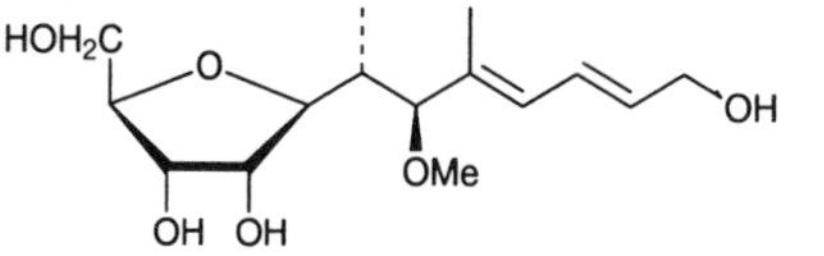

100

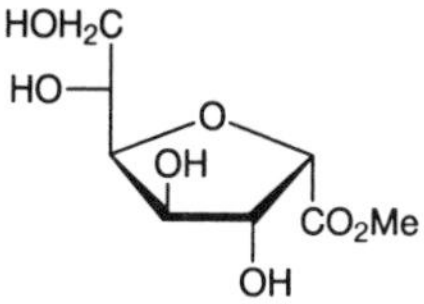

101

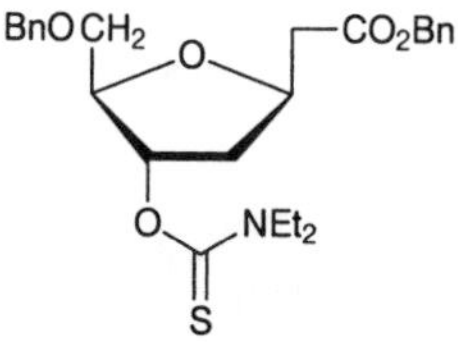

102

promoter tris(triflyloxy)silyl chloride.[325] Compound **103**, made from 6-*O*-piva-loyl-D-galactal by electrophilic addition of O-4 at the anomeric centre, was ring-opened nucleophilically using allyltrimethylsilane and trimethylsilyl triflate to give the *C*-allyl glycoside **104** in 89% yield with α/β-selectivity 1:9.[326]

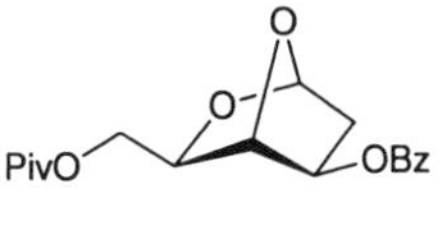

103 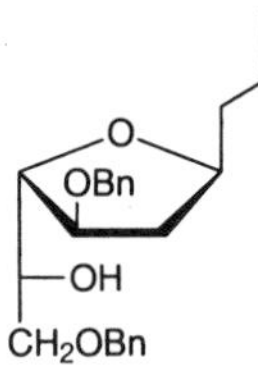   **104**

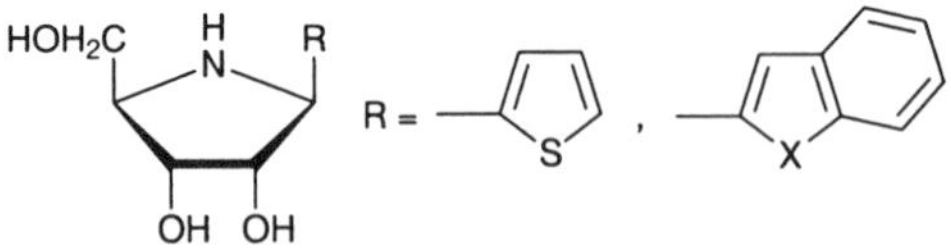

105 X = O, N

From an *O*-substituted D-ribono-γ-lactone, treated with aryllithiums, the corresponding *C*-phenyl[327] and *p*-(trifluoroacetylamino)phenyl[328] β-D-ribofura-nosides were produced and then incorporated into oligo-ribonucleotides following the introduction of a phosphate precursor at O-3. A set of β-*C*-aryl 2-deoxy-2-ribofuranosyl glycosides having substituted phenyl, naphthyl, 9-phenanthrenyl and 1-pyrenyl groups were produced using the corresponding

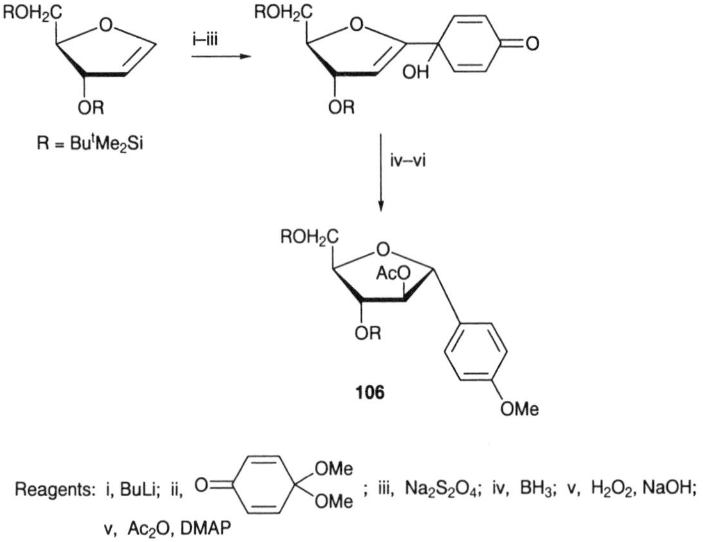

R = Bu^{t}Me$_2$Si

106

Reagents: i, BuLi; ii, [structure] ; iii, Na$_2$S$_2$O$_4$; iv, BH$_3$; v, H$_2$O$_2$, NaOH;

v, Ac$_2$O, DMAP

Scheme 5

glycosyl chloride and diaryl cadmiums.[329] In the area of substituted-phenyl *C*-glycosides, compound **106** was synthesized using an umpolung approach from a furanoid glycal (Scheme 5).[330]

Starting from an *O*-substituted ribofuranose derivative, the '*C*-azanucleosides' **105** were synthesized using aryllithium reagents followed by introduction of an amino group and ring closure,[331] and further β-D-ribofuranosyl *C*-glycosides have been made incorporating imidazole[332] and various indoles, pyrroles and pyrazoles.[333] An *O*-substituted 2-deoxy-D-ribofuranosyl cyanide has been elaborated to the corresponding methylene-linked thiamine derivative.[334] See Chapter 20, Section 10 for other nucleosides containing *C*-linkages. Chapter 11 contains reference to a set of α-linked *C*-glycosides of 2-deoxy-ribofuranose having the ring oxygen atoms replaced by sulfur.

References

1   H. Waldmann, *Org. Synth. Highlights II*, 1995, 289 (*Chem. Abstr.*, 1996, **125**, 168 460).

2   A. Cheriti and A. Kessat, *Bull. Union Physiciens*, 1996, **90**, 107 (*Chem. Abstr.*, 1996, **125**, 196 106).

3   G. Russo and L. Panza, *Trends Org. Chem.*, 1993, **4**, 191 (*Chem. Abstr.*, 1996, **125**, 58 864).

4   A. T. J. W. de Goede, M. P. J. van Deurzen, I. G. van der Leij, A. M. van der Heijden, J. M. A. Baas, F. van Rantwijk and H. van Bekkum, *J. Carbohydr. Chem.*, 1996, **15**, 331.

5   C. G. J. Verhart, C. T. M. Fransen, B. Zwannenberg and G. J. F. Chittenden, *Recl. Trav. Chim. Pays-Bas*, 1996, **115**, 133.

6   W. Klotz and R. R. Schmidt, *Synthesis*, 1996, 687.

7   H. Uchiro and T. Mukaiyama, *Chem. Lett.*, 1996, 271.

8   H. Uchiro and T. Mukaiyama, *Chem. Lett.*, 1996, 79.

9   H. Ebrahim, D. J. Evans, J. Lehmann and L. Ziser, *Carbohydr. Res.*, 1996, **286**, 189.

10   M.-Z. Liu H.-N. Fan, Z.-W. Guo and Y.-Z. Hui, *Carbohydr. Res.*, 1996, **290**, 233.

11   M.-Z. Liu, H.-N. Fan, Z.-W. Guo and Y.-Z. Hui, *Chin J. Chem.*, 1996, **14**, 190 (*Chem. Abstr.*, 1996, **125**, 33 959).

12   T. Mukaiyama, K. Takeuchi, S. Higuchi and H. Uchiro, *Chem. Lett.*, 1996, 1123.

13   V. J. Patil, *Tetrahedron Lett.*, 1996, **37**, 1481.

14   G. Böhm and H. Waldmann, *Liebigs Ann. Chem.*, 1996, 613.

15   G. Böhm and H. Waldmann, *Liebigs Ann. Chem.*, 1996, 621.

16   B. Fraser-Reid, *Acc. Chem. Res.*, 1996, **29**, 57 (*Chem. Abstr.*, 1996, **124**, 176 622).

17   K. Michael and H. Kessler, *Tetrahedron Lett.*, 1996, **37**, 3453.

18   C. Krog-Jensen and S. Oscarson, *J. Org. Chem.*, 1996, **61**, 1234.

19   C. Krog-Jensen and S. Oscarson, *J. Org. Chem.*, 1996, **61**, 4512.

20   K. Osumi, M. Enomoto and H. Sigimura, *Carbohydr. Lett.*, 1996, **2**, 35 (*Chem. Abstr.*, 1996, **125**, 196 140).

21   I. Alonso, N. Khiar and M. Martn-Lomas, *Tetrahedron Lett.*, 1996, **37**, 1477.

22   W.-S. Kim, S. Hosono, H. Sasai and M. Shibasaki, *Heterocycles*, 1996, **42**, 795.

23   K. Koide, M. Ohno and S. Kobayashi, *Synthesis*, 1996, 1173.

24   R. E. Hubbard, J. G. Montana, P. V. Murphy and R. J. K. Taylor, *Carbohydr. Res.*, 1996, **287**, 247.

25   K. N. Gurudutt, L. J. M. Rao, S. Rao and S. Srinivas, *Carbohydr. Res.*, 1996, **285**, 159.

26   K. P. R. Kartha, M. Aloui and R. A. Field, *Tetrahedron Lett.*, 1996, **37**, 8807.

27   T. Uchiyama and O. Hindsgaul, *Synlett.*, 1996, 499.

28   H. K. Chenault, A. Castro, L. F. Chafin and J. Yang, *J. Org. Chem.*, 1996, **61**, 5024.

29   G.-J. Boons and S. Isles, *J. Org. Chem.*, 1996, **61**, 4262.

30   G.-J. Boons, A. Burton and P. Wyatt, *Synlett.*, 1996, 310.

31   G.-J. Boons, A. Burton and S. Isles, *J. Chem. Soc., Chem. Commun.*, 1996, 141.

32   R. Velty, T. Benvegnu and D. Plusquellec, *Synlett.*, 1996, 817.

33   P. I. Kitov, Y. E. Tsvetkov, L. V. Backinowsky and N. K. Kochetkov, *Izv. Akad. Nauk, Ser. Khim.*, 1993, 1485 (*Chem. Abstr.*, 1996, **125**, 196 120).

34   P. I. Kitov, Y. E. Tsvetkov, L. V. Backinowsky and N. K. Kochetkov, *Izv. Akad. Nauk, Ser. Khim.*, 1995, 1158 (*Chem. Abstr.*, 1996, **124**, 56 445).

35   P. I. Kitov, Y. E. Tsvetkov, L. V. Backinowsky and N. K. Kochetkov, *Mendeleev Commun.*, 1995, 176 (*Chem. Abstr.*, 1996, **124**, 117 750).

36   A. Zapata, B. Bernet and A. Vasella, *Helv. Chim. Acta*, 1996, **79**, 1169.

37   H. J. M. Gijsen, L. Qiao, W. Fitz and C.-H. Wong, *Chem. Rev.*, 1996, **96**, 443.

38   H. Waldmann, *Org. Synth. Highlights II*, 1995, 157 (*Chem. Abstr.*, 1996, **125**, 168 459).

39   Y. Kimura, C. Panintrarux, S. Adachi and R. Matsuno, *Ann. N. Y. Acad. Sci.*, 1995, **750**, 312 (*Chem. Abstr.*, 1996, **124**, 87 529).

40   C. Panintrarux, S. Adachi and R. Matsuno, *J. Mol. Catal. B: Enzym.*, 1996, **1**, 165 (*Chem. Abstr.*, 1996, **125**, 114 966).

41   A. Trincone and E. Pagnotta, *Tetrahedron: Asymm.*, 1996, **7**, 2773.

42   Y. S. Lee, D. I. Ito, K. Koya, S. Izumi and T. Hirata, *Chem. Lett.*, 1996, 719.

43   D. E. Stevenson and R. H. Furneaux, *Carbohydr. Res.*, 1996, **284**, 279.

44   D. E. Stevenson and R. H. Furneaux, *Enzyme Microb. Technol.*, 1996, **18**, 513 (*Chem. Abstr.*, 1996, **125**, 34 028).

45   Y. Okahata and T. Mori, *J. Chem. Soc., Perkin Trans. 1*, 1996, 2861.

46   P. Drouet, M. Zhang, M.-D. Legoy, *Ann. N. Y. Acad. Sci.*, 1995, **750**, 306 (*Chem. Abstr.*, 1996, **124**, 87 528).

47   M. Rodriguez, A. Gomez, F.Gonzalez, E. Barzana and A. Lopez-Munguia, *Appl. Biochem. Biotechnol.*, 1996, **59**, 163 (*Chem. Abstr.*, 1996, **125**, 58 886).

48   Y. Du, *Huanjing Huaxue*, 1996, **15**, 168 (*Chem. Abstr.*, 1996, **124**, 343 812).

49   D. Crich and S. Sun, *J. Org. Chem.*, 1996, **61**, 4506.

50   H. Itoh and L. Y. Kamiyami, *J. Ferment. Bioeng.*, 1995, **80**, 510 (*Chem. Abstr.*, 1996, **124**, 176 692).

51   C. G. Sowell, M. T. Livesay and D. A. Johnson, *Tetrahedron Lett.*, 1996, **37**, 609.

52   U. Ellervik and G. Magnusson, *Carbohydr. Res.*, 1996, **280**, 251.

53   H. Shimizu, Y. Ito, Y. Matsuzaki, H. Iijima and T. Ogawa, *Biosci. Biotech. Biochem*, 1996, **60**, 73.

54   E. Kaji, Y. Osa, K. Takahashi and S. Zen, *Chem. Pharm. Bull.*, 1996, **44**, 15.

55   V. O. Kuryanov, A. Y. Zemlyakov and V. Y. Chirva, *Ukr. Khim. Zh.*, 1994, **60**, 858 (*Chem. Abstr.*, 1996, **124**, 117 730).

56   C. Tabeur, F. Machetto, J.-M. Mallet, P. Duchaussoy, M. Petitou and P. Sinay, *Carbohydr. Res.*, 1996, **281**, 253.

57   A. Leydet, C. Jeantet-Segonds, P. Barthélémy, B. Boyer and J. P. Roque, *Recl. Trav. Chim. Pays-Bas*, 1996, **115**, 421.

58   M. Benedetto, G. Miglierini, P. R. Mussini, F. Pelizzoni, S. Rondinini and G. Sello, *Carbohydr. Lett.*, 1995, **1**, 321 (*Chem. Abstr.*, 1996, **124**, 290 028).

59   E. Smits, J. B. F. N. Engberts, R. M. Kellogg and H. A. van Doren, *J. Chem. Soc.,*
     Perkin Trans. 1, 1996, 2873.
60   J. C. Briggs and A. H. Haines, *Carbohydr. Res.*, 1996, **282**, 293.
61   B. Dupré, H.Bui, I. L. Scott, R. V. Market, K. M. Keller, P. J. Beck and T. P.
     Kogan, *Bioorg. Med. Chem. Lett.*, 1996, **6**, 569.
62   S. Nakamura, M. Kondo, K. Goto, M. Nakamura, Y. Tsuda and K. Shishido,
     Heterocycles, 1996, **43**, 2747.
63   Z. W. Guo and Y. Z. Hui, *Chin. Chem. Lett.*, 1996, **7**, 423 (*Chem. Abstr.*, 1996, **125**,
     114 994).
64   Z.-W. Li, Z.-H. Li, Z.-T. Jin and K. Imafuku, *Biosci. Biotech. Biochem*, 1996, **60**,
     2095.
65   A. Marra, A. Dondoni and F. Sansone, *J. Org. Chem.*, 1996, **61**, 5155.
66   C. Capozzi, A. Dios, R. W. Franck, A. Geer, C. Marzabadi, S. Menichetti, C. Nativi
     and M. Tamarez, *Angew. Chem. Int. Ed. Engl.*, 1996, **35**, 777.
67   C. H. Marzabadi and R. W. Franck, *J. Chem. Soc., Chem. Commun.*, 1996, 2651.
68   W. H. Binder and W. Schmid, *Monatsh. Chem.*, 1995, **126**, 923 (*Chem. Abstr.*, 1996,
     124, 146 648).
69   J. Lehmann and U. P. Weitzel, *Carbohydr. Res.*, 1996, **294**, 65.
70   P. R. Ashton, S. E. Boyd, C. L. Brown, N. Jayaraman, S. A. Nepogodiev and J. F.
     Stoddart, *Angew. Chem. Int. Ed. Engl.*, 1996, **35**, 1115.
71   D. Pagé, S. Aravind and R. Roy, *J. Chem. Soc., Chem. Commun.*, 1996, 1913.
72   D. Pagé and R. Roy, *Bioorg. Med. Chem. Lett.*, 1996, **6**, 1765.
73   T. K. Lindhorst and C. Kieburg, *Angew. Chem. Int. Ed. Engl.*, 1996, **35**, 1953.
74   E. R. Wijsman, D. Filippov, A. R. P. M. Valentijn, G. A. van der Marel and J. H.
     van Boom, *Recl. Trav. Chim. Pays-Bas*, 1996, **115**, 397.
75   J. Ramza, G. Descotes, J.-M. Basset and A. Much, *J. Carbohydr. Chem.*, 1996, **15**, 125.
76   K. H. Mortell, R. V. Weatherman and L. L. Kiessling, *J. Am. Chem. Soc.*, 1996, **118**,
     2297.
77   S. Valarde, J. Urbina and M. R. Pena, *J. Org. Chem.*, 1996, **61**, 9541.
78   R. K. Erukulla, X. Zhou, P. Samadder, G. Arthur and R. Bittman, *J. Med. Chem.*,
     1996, **39**, 1545.
79   J. A. Marino-Albernas, R. Bittman, A. Peters and E. Mayhew, *J. Med. Chem.*, 1996,
     39, 3241.
80   S. I. Dubovskaya, M. V. Anikin, V. V. Chupin and G. A. Serebrennikova, *Bioorg.*
     Khim., 1994, **20**, 1242 (*Chem. Abstr.*, 1996, **124**, 56 428).
81   A. Terjung, K. H. Jung and R. R. Schmidt, *Liebigs Ann. Chem.*, 1996, 1313.
82   R. R. Schmidt and K. Jankowski, *Liebigs Ann. Chem.*, 1996, 867.
83   K. Tatsuta and S. Yasuda, *J. Antibiotics*, 1996, **49**, 713.
84   K. Tatsuta and S. Yasuda, *Tetrahedron Lett.*, 1996, **37**, 2453.
85   A. Fürstner and I. Konetzki, *Tetrahedron Lett.*, 1996, **37**, 15071.
86   K. Mori and K. Uenishi, *Liebigs Ann. Chem.*, 1996, 1.
87   I. Islam, R. R. Hinshaw, K.-T. Chong, A. Kato, R. T. Borchardt and J. F. Fisher,
     Bioorg. Chem., 1995, **23**, 499 (*Chem. Abstr.*, 1996, **124**, 202 849).
88   S. Nishimura, *Kobunshi*, 1996, **45**, 539 (*Chem. Abstr.*, 1996, **125**, 168 462).
89   K. J. Jensen, P. R. Hansen, D. Venugopal and G. Barany, *J. Am. Chem. Soc.*, 1996,
     118, 3148.
90   M. Elofsson, S. Roy, L. A. Salvador and J. Kihlberg, *Tetrahedron Lett.*, 1996, **37**,
     7645.
91   S.-H. Wu, M. Shimazaki, C.-C. Lin, L. Qiao, W. J. Moree, G. Weitz-Schmidt and
     C.-H. Wong, *Angew. Chem. Int. Ed. Engl.*, 1996, **35**, 88.

92   J. Eberling, P. Braun, D. Kowalczyk, M. Schultz and H. Kunz, *J. Org. Chem.*, 1996, **61**, 2638.

93   J. M. Kim and R. Roy, *Carbohydr. Lett.*, 1995, **1**, 465 (*Chem. Abstr.*, 1996, **124**, 317 640).

94   W. Steffan, M. Schutkowski and G. Fischer, *J. Chem. Soc., Chem. Commun.*, 1996, 313.

95   T. Pohl and H. Waldmann, *Angew. Chem. Int. Ed. Engl.*, 1996, **35**, 1720.

96   P. Sjölin, M. Elofsson and J. Kihlberg, *J. Org. Chem.*, 1996, **61**, 560.

97   C. C. Lin, M. Shimazaki, M.-P. Heck, S. Aoki, R. Wang, T. Kimura, H. Ritzèn, S. Takayama, S.-H. Wu, G. Weitz-Schmidt and C.-H. Wong, *J. Am. Chem. Soc.*, 1996, **118**, 6826.

98   C.-C. Lin, T. Kimura, S.-H. Wu, G. Weitz-Schmidt and C.-H. Wong, *Bioorg. Med. Chem. Lett.*, 1996, **6**, 2755.

99   S. E. Cervigni, P. Dumy and M. Mutter, *Angew. Chem. Int. Ed. Engl.*, 1996, **35**, 1230.

100   W. D. Shrader and B. Imperiali, *Tetrahedron Lett.*, 1996, **37**, 599.

101   Y. Watanabe, T. Yamamoto and S. Ozaki, *J. Org. Chem.*, 1996, **61**, 14.

102   P. Letellier, A. El Meslouti, D. Beaupere and R. Uzan, *Synthesis*, 1996, 1435.

103   A. Liu, K. Dillon, R. M. Campbell, D. C. Cox and D. M. Huryn, *Tetrahedron Lett.*, 1996, **37**, 3785.

104   M. M. Silva, J. Cleophax, A. A. Benicio, M. V. Almeida, J.-M. Delaumeny, A. S. Machado and S. D. Gero, *Synlett.*, 1996, 764.

105   K. Mori, T. Nakayama and M. Sakuma, *Bioorg. Med. Chem.*, 1996, **4**, 401 (*Chem. Abstr.*, 1996, **125**, 58 876).

106   Z. A. Khushbaktova, V. N. Syrov and N. Sh. Pal'yants, *Khim.-Farm. Zh.*, 1995, **29**, 22 (*Chem. Abstr.*, 1996, **124**, 117 733).

107   D. Todorova, A. Ivanova and Ts. Milkova, *Dokl. Bulg. Akad. Nauk, 1994*, **47**, 41 (*Chem. Abstr.*, 1996, **124**, 117 716).

108   A. Cheriti and G. Balansard, *Nat. Prod. Lett.*, 1995, **7**, 47 (*Chem. Abstr.*, 1996, **125**, 276 323).

109   L. A. Baltina, O. B. Flekhter and E. Vasiljieva, *Mendeleev Commun.*, 1996, 63 (*Chem. Abstr.*, 1996, **125**, 33 957).

110   M. I. Isaev, *Khim. Prir. Soedin*, 1993, 830 (*Chem. Abstr.*, 1996, **124**, 87 511).

111   M. Takechi, C. Uno and Y. Tanaka, *Phytochemistry* 1996, **41**, 125.

112   E. H. Chang, N. G. Je and K. S. Im, *Yakhak Hoechi*, 1996, **40**, 163 (*Chem. Abstr.*, 1996, **125**, 58 879).

113   N. Sh. Pal'yants, R. U. Umarova, M. B. Gorovits and N. K. Abubakirov, *Khim. Prir. Soedin*, 1993, 621 (*Chem. Abstr.*, 1996, **124**, 56 439).

114   F. Pasqui, F. Canfarini, A. Giolitti, A. Guidi, V. Pestellini and F. Arcamone, *Tetrahedron*, 1996, **52**, 185.

115   Z. Dienes and P. Vogel, *J. Org. Chem.*, 1996, **61**, 6958.

116   J. F. R. Patricia, T. Jasintha, P. Nagarajan, N. S. Subramanian and N. Sulochana, *Bull. Electrochem.*, 1995, **11**, 324 (*Chem. Abstr.*, 1996, **124**, 9 132).

117   V. G. Pivovarenko and V. P. Khilya, *Khim. Prir. Soedin*, 1993, 220 (*Chem. Abstr.*, 1996, **124**, 56 432).

118   V. P. Khilya, A. Aitmambetov, V. G. Pivovarenko, D. M. Zakharik and Y. L. Shvachko, *Khim. Prir. Soedin*, 1994, 347 (*Chem. Abstr.*, 1996, **124**, 56 441).

119   H. Elhajiji, O. Dangles and R. Brouillard, *Colloq.-Inst. Natl. Rech. Agron.*, 1995, **69**, 193 (*Chem. Abstr.*, 1996, **124**, 87 519).

120 V. F. Traven, L. S. Krasavina, E. Y. Dmitrieva and E. A. Carberry, *Heterocycl. Commun.*, 1996, **2**, 309 (*Chem. Abstr.*, 1996, **125**, 301 357).
121 Y. Suzuki, Y. Doi, K. Uchida and H. Tsuge, *Oyo Toshitsu Kagaku*, 1996, **43**, 369 (*Chem. Abstr.*, 1996, **125**, 301 365).
122 M. Kluge and D. Sicker, *Tetrahedron Lett.*, 1996, **37**, 10389.
123 D.-M. Kang, S. J. Kim and J. Yim, *Pteridines*, 1995, **6**, 93 (*Chem. Abstr.*, 1996, **124**, 290 077).
124 S. M. Frederiksen and F. R. Stermitz, *J. Nat. Prod.*, 1996, **59**, 41 (*Chem. Abstr.*, 1996, **124**, 290 020).
125 S. Mitaku, A.-L. Skaltsounis, F. Tillequin, M. Koch, Y. Rolland, A. Pierre and G. Atassi, *Pharm. Res.*, 1996, **13**, 939 (*Chem. Abstr.*, 1996, **125**, 143 147).
126 N. Terashima, S. A. Ralph and L. L. Landucci, *Holzforshung*, 1996, **50**, 151 (*Chem. Abstr.*, 1996, **125**, 33 966).
127 K. Tsujihara, M. Hongu, K. Saito, M. Inamasu, K. Arakawa, A. Oku and M. Matsumoto, *Chem. Pharm. Bull.*, 1996, **44**, 1174.
128 K. Goto, S. Nakamura, Y. Morioka, M. Kondo, S. Naito and K. Tsutsumi, *Chem. Pharm. Bull.*, 1996, **44**, 547.
129 C. Schroeder, R. Lauterbach and J. Stöckigt, *Tetrahedron*, 1996, **52**, 925.
130 C. Marais, J. A. Steenkamp and D. Ferreira, *J. Chem. Soc., Perkin Trans. 1*, 1996, 2915.
131 K. Vogel, J. Sterling, Y. Herzig and A. Nudelman, *Tetrahedron*, 1996, **52**, 3049.
132 X. Tian, Z.-Q. Yan, Y.-Z. Chen and X.-Q. Mu, *Chem. Res. Chin. Univ.*, 1995, **11**, 79 (*Chem. Abstr.*, 1996, **124**, 9 114).
133 F. Abe and T. Yamauchi, *Chem. Pharm. Bull.*, 1996, **44**, 1797.
134 S.-S. Lee, W.-L. Chang and C.-H. Chen, *Tetrahedron Lett.*, 1996, **37**, 4405.
135 Z. Li, D. Li, N. L. Owen, Z. Len, Z. Cao and Y. Yi, *Magn. Reson. Chem.*, 1996, **34**, 512 (*Chem. Abstr.*, 1996, **125**, 114 970).
136 T. Usui and T. Murata, *Yuki Gosei Kagaku Kyokaishi*, 1996, **54**, 607 (*Chem. Abstr.*, 1996, **125**, 114 956).
137 G. H. Posner and D. S. Bull, *Tetrahedron Lett.*, 1996, **37**, 6279.
138 K. Hiruma, T. Kajimoto, G. Weitz-Schmidt, I. Ollmann and C.-H. Wong, *J. Am. Chem. Soc.*, 1996, **118**, 9265.
139 M. Grothus, A. Steigel, M.-R. Kula and L. Elling, *Carbohydr. Lett.*, 1994, **1**, 83 (*Chem. Abstr.*, 1996, **125**, 114 973).
140 T. Iimora, T. Shibazaki and S. Ikegami, *Tetrahedron Lett.*, 1996, **37**, 2267.
141 N. E. Nifant'ev, E. A. Khatuntseva, A. S. Shashkov and K. Bock, *Carbohydr. Lett.*, 1996, **1**, 399 (*Chem. Abstr.*, 1996, **124**, 343 848).
142 Z.-J. Li, L.-B. Wang and Z.-T. Huang, *Carbohydr. Res.*, 1996, **295**, 77.
143 U. Schmid and H. Waldmann, *Tetrahedron Lett.*, 1996, **37**, 3837.
144 J. C. Morales and S. Penadés, *Tetrahedron Lett.*, 1996, **37**, 5011.
145 C. K. Lee and E. J. Kim, *Carbohydr. Res.*, 1996, **280**, 59.
146 A. M. Jansson, K. J. Jensen, M. Meldal, J. Lomako, W. M. Lomako, C. E. Olsen and K. Bock, *J. Chem. Soc., Perkin Trans. 1*, 1996, 1001.
147 M. I. Barrena, R. Echarri and S. Castillón, *Synlett.*, 1996, 675.
148 Y. Qiu, Y. Nakahara and T. Ogawa, *Biosci. Biotech. Biochem*, 1996, **60**, 986.
149 C. Pető, G. Batta, Z. Györgydeák and F. Sztaricskai, *J. Carbohydr. Chem.*, 1996, **15**, 465.
150 A. Ernst and A. Vasella, *Helv. Chim. Acta*, 1996, **79**, 1279.
150a H. El-Sayed and E. Laszlo, *Acta Aliment.*, 1994, **23**, 359 (*Chem. Abstr.*, 1996, **124**, 9 151).

151   J. Broddefalk, K.-E. Bergquist and J. Kihlberg, *Tetrahedron Lett.*, 1996, **37**, 3011.
152   M. Adinolfi, G. Barone, L. de Napoli, A. Iadonisi and G. Piccialli, *Tetrahedron Lett.*, 1996, **37**, 5007.
153   M. A. Tariq and K. Hayashi, *Biochem. Biophys. Res. Commun.*, 1995, **214**, 568 (*Chem. Abstr.*, 1996, **124**, 56 492).
154   T. Kometani, Y. Terada, T. Nishimura, T. Nakae, H. Takii and S. Okada, *Biosci. Biotech. Biochem*, 1996, **60**, 1176.
155   S. Kitahata and H. Murakami, *Kagaku to Kogyo*, 1995, **69**, 372 (*Chem. Abstr.*, 1996, **124**, 87 507).
156   H. Kakinuma, H. Yuasa and H. Hashimoto, *Carbohydr. Res.*, 1996, **284**, 61.
157   D. P. Larson and C. H. Heathcock, *J. Org. Chem.*, 1996, **61**, 5208.
158   K.-F. Hsiao, S.-T. Chen, K.-T. Wang and S.-H. Wu, *Synlett.*, 1996, 966.
159   S. N. Thorn and T. Gallagher, *Synlett.*, 1996, 857.
160   G. Vic, J. J. Hastings and D. H. G. Crout, *Tetrahedron: Asymm.*, 1996, **7**, 1973.
161   H. Kurihara, S. Tada, K. Takahashi and M. Hatano, *Biosci. Biotech. Biochem*, 1996, **60**, 932.
162   D. Qiu, S. S. Gandhi and R. R. Koganty, *Tetrahedron Lett.*, 1996, **37**, 595.
163   T. Tsuda and S.-I. Nishimura, *J. Chem. Soc., Chem. Commun.*, 1996, 2779.
164   S. Hanessian, D. Qiu, H. Prabhanjan, G. V. Reddy and B. Lou, *Can. J. Chem.*, 1996, **74**, 1738.
165   T. Vuljanic, K.-E. Bergquist, H. Clausen, S. Roy and J. Kihlberg, *Tetrahedron*, 1996, **52**, 7983.
166   A. Zervosen and L. Elling, *J. Am. Chem. Soc.*, 1996, **118**, 1836.
167   T. Usui, S. Marimoto, Y. Hayakawa, M. Kawaguchi, T. Murata, Y. Matahira and Y. Nishida, *Carbohydr. Res.*, 1996, **285**, 29.
168   A. Vetere and S. Paoletti, *Biochem. Biophys. Res. Commun.*, 1996, **219**, 6 (*Chem. Abstr.*, 1996, **124**, 261 535).
169   J.-H. Yoon and K. Ajisaka, *Carbohydr. Res.*, 1996, **292**, 153.
170   K. G. Nilsson and A. Eliasson, *Biotechnol. Lett.*, 1995, **17**, 717 (*Chem. Abstr.*, 1996, **124**, 30 180).
171   Y. Ding, J. Labbe, O. Kanie and O. Hindsgaul, *Bioorg. Med. Chem.*, 1996, **4**, 683 (*Chem. Abstr.*, 1996, **125**, 114 984).
172   G. Vic, J. J. Hastings, O. W. Howarth and D. H. G. Crout, *Tetrahedron: Asymm.*, 1996, **7**, 709.
173   D. Monti, E. Giouse, S. Riva and L. Panza, *Gazz. Chim. Ital.*, 1996, **126**, 303 (*Chem. Abstr.*, 1996, **125**, 114 978).
174   R. W. Bassily, *Spectrosc. Lett.*, 1996, **29**, 497 (*Chem. Abstr.*, 1996, **125**, 11 258).
175   H. Dietrich, C. Regele-Mayer and R. R. Schmidt, *Carbohydr. Lett.*, 1994, **1**, 115 (*Chem. Abstr.*, 1996, **125**, 114 974).
176   J. Brüning and L. L. Kiessling, *Tetrahedron Lett.*, 1996, **37**, 2907.
177   R. Roy and U. K. Saha, *J. Chem. Soc., Chem. Commun.*, 1996, 201.
178   Y. Hatanaka, M. Hashimoto, K. I.-P. J. Hidari, Y. Sanai, Y. Tezuka, Y. Nagai and Y. Kanaoka, *Chem. Pharm. Bull.*, 1996, **44**, 1111.
179   Y. Mirua, T. Arai and T. Yamagata, *Carbohydr. Res.*, 1996, **289**, 193.
180   F. Costes, M. El Ghoul, M. Bon, I. Rico-Latte, *Langmuir*, 1995, **11**, 3644 (*Chem. Abstr.*, 1996, **124**, 117 797).
181   T. Yoshida, Y. Yasuda, K. Hattori and T. Uryu, *Macromol. Rapid Commun.*, 1995, **16**, 881 (*Chem. Abstr.*, 1996, **124**, 176 705).
182   C. Gallo-Rodriguez, O. Varela and R. M. de Lederkremer, *J. Org. Chem.*, 1996, **61**, 1886.

183   J. J. Aragón, F. J. Cañada, A. Fernández-Mayoralas, R. López, M. Martín-Lomas and D. Villanueva, *Carbohydr. Res.*, 1996, **290**, 209.
184   Z. Z. Zhang, K. Fukunaga, Y. Sugimura, K. Nakao and T. Shimizu, *Chin. Chem. Lett.*, 1996, **7**, 119 (*Chem. Abstr.*, 1996, **124**, 343 931).
185   G. Stork and J. J. La Clair, *J. Am. Chem. Soc.*, 1996, **118**, 247.
186   S. Valverde, A. M. Gómez, J. C. López and B. Herradón, *Tetrahedron Lett.*, 1996, **37**, 1105.
187   Z. Szurmai and L. Jánossy, *Carbohydr. Res.*, 1996, **296**, 279.
188   S. Suwasono and R. A. Rastall, *Biotechnol. Lett.*, 1996, **18**, 851 (*Chem. Abstr.*, 1996, **125**, 196 142).
189   W.-S. Kim, H. Sasai and M. Shibasaki, *Tetrahedron Lett.*, 1996, **37**, 7797.
190   Y.-L. Li and Y.-L. Wu, *Tetrahedron Lett.*, 1996, **37**, 7413.
191   P. Stangier and O. Hindsgaul, *Synlett.*, 1996, 179.
192   W. Dullenkopf, J. H. C. Castro-Palomino, L. Manzoni and R. R. Schmidt, *Carbohydr. Res.*, 1996, **296**, 135.
193   B. Yu, Q. Ouyang, C. Li and Y. Hui, *J. Carbohydr. Chem.*, 1996, **15**, 297.
194   M. G. Peter, J. P. Ley, S. Petersen, M. H. M. G. Schumacher-Wandersleb, K.-D. Spindler-Barth, M. Spindler-Barth, A. Turberg and M. Londershausen, *Chitin World [Proc. Intl. Conf. Chitin Chitosan] 6th*, 1994, 359 (*Chem. Abstr.*, 1996, **125**, 58 928).
195   J. P. Ley and M. G. Peter, *J. Carbohydr. Chem.*, 1996, **15**, 51.
196   T. Fukamizo, Y. Honda, H. Toyoda, S. Ouchi and S. Goto, *Biosci. Biotech. Biochem*, 1996, **60**, 1705.
197   S. Singh, M. Scigelova, G. Vic and D. H. G. Crout, *J. Chem. Soc., Perkin Trans. 1*, 1996, 1921.
198   E. Meinjohanns, M. Meldal, A. Schleyer, H. Paulsen and K. Bock, *J. Chem. Soc., Perkin Trans. 1*, 1996, 985.
199   Y. Ogawa, P.-s. Lei and P. Kováč, *Carbohydr. Res.*, 1996, **288**, 85.
200   Z.-W. Guo, S.-J. Deng and Y.-Z. Hui, *J. Carbohydr. Chem.*, 1996, **15**, 965.
201   Y. Suda, K. Bird, T. Shiyama, S. Koshida, D. Marques, K. Fukase, M. Sobel and S. Kusomoto, *Tetrahedron Lett.*, 1996, **37**, 1053.
202   K. Fukase, I. Kinoshita, Y. Suda, Y. Aoki, W.-C. Liu, M. Oikawa, M. Kurosawa, U. Zähringer, U. Saydel, E. Th. Rietschel, S. Kusumoto, *Synlett.*, 1996, 252.
203   D. Weigelt, R. Kraehmer and P. Welzel, *Tetrahedron Lett.*, 1996, **37**, 367.
204   F. Animati, F. Arcamone, M. Berettoni, A. Cipollone, M. Franciotti and P. Lombardi, *J. Chem. Soc., Perkin Trans. 1*, 1996, 1327.
205   D. K. Watt, D. J. Brasch, D. S. Larsen, L. D. Melton and J. Simpson, *Carbohydr. Res.*, 1996, **285**, 1.
206   H. Sone, H. Kigoshi and K. Yamada, *J. Org. Chem.*, 1996, **61**, 8956.
207   M. A. Probert, J. Zhang and D. R. Bundle, *Carbohydr. Res.*, 1996, **296**, 149.
208   K. Zegelaar-Jaarsveld, S. C. van der Plas, G. A. van der Marel and J. H. van Boom, *J. Carbohydr. Chem.*, 1996, **15**, 665.
209   N. Razi, J. Kreuger, L. Lay, G. Russo, L. Panza, B. Lindahl and U. Lindahl, *Glycobiology*, 1995, **5**, 807 (*Chem. Abstr.*, 1996, **124**, 232 952).
210   S. Koto, T. Miura, M. Hirooka, A. Tomaru, M. Iida, M. Kanemitsu, K. Takenaka, S. Masuzawa, S. Miyaji, N. Kuroyanagi, M. Yagashita, S. Zen, K. Yago and F. Tomonaga, *Bull. Chem. Soc. Jpn.*, 1996, **69**, 3247.
211   S. Saito, S. Nagase, M. Kawase and Y. Nagamura, *Eur. J. Med. Chem.*, 1996, **31**, 557 (*Chem. Abstr.*, 1996, **125**, 276 387).
212   K. Yamada, T. Anai, S. Kosemura, S. Yamamura and K. Hasegawa, *Phytochemistry* 1996, **41**, 671.

213   D. Crich, J.-T. Hwang and H. Yuan, *J. Org. Chem.*, 1996, **61**, 6189.
214   V. Martichonok and G. M. Whitesides, *J. Org. Chem.*, 1996, **61**, 1702.
215   T. Kondo, T. Tomoo, H. Abe, M. Isobe and T. Goto, *J. Carbohydr. Chem.*, 1996, **15**, 857.
216   T. Ercégovic and G. Magnusson, *J. Org. Chem.*, 1996, **61**, 179.
217   J. Bogusiak and W. Szeja, *Carbohydr. Res.*, 1996, **295**, 235.
218   S. N. Mikhailov, E. De Clercq and P. Herdewijn, *Nucleosides Nucleotides*, 1996, **15**, 1323.
219   M. Yoshikawa, S. Yoshizumi, T. Murakami, H. Matsuda, J. Yamahara and N. Murakami, *Chem. Pharm. Bull.*, 1996, **44**, 492.
220   M. Nishikitani, K. Kubota, A. Kobayashi and F. Sugawara, *Biosci. Biotech. Biochem*, 1996, **60**, 929.
221   Y. Konda, T. Toida, E. Kaji, K. Takeda and Y. Harigaya, *Tetrahedron Lett.*, 1996, **37**, 4015.
222   Y. Shao, Y. L. Li and B. N. Zhou, *Phytochemistry* 1996, **41**, 1593.
223   G. Yang and F. Kong, *Carbohydr. Lett.*, 1994, **1**, 137 (*Chem. Abstr.*, 1996, **125**, 87 025).
224   Y. Du and F. Kong, *J. Carbohydr. Chem.*, 1996, **15**, 797.
225   L. Sun, P. Li, D. W. Landry and K. Zhao, *Tetrahedron Lett.*, 1996, **37**, 1547.
226   Y. Hatanaka, M. Hashimoto, K. I.-P. Jwa Hidari, Y. Sanai, Y. Nagai and Y. Kanaoka, *Heterocycles*, 1996, **43**, 531.
227   G. Dekany, K. Wright, P. Ward and I. Toth, *J. Carbohydr. Chem.*, 1996, **15**, 383.
228   U. Nilsson, R. Johnsson and G. Magnusson, *Chem. Eur. J.*, 1996, **2**, 295.
229   M. J. Kiefel, B. Beisner, S. Bennett, I. D. Holmes and M. von Itzstein, *J. Med. Chem.*, 1996, **39**, 1314.
230   S. Czernecki and D. Randriamandimby, *J. Carbohydr. Chem.*, 1996, **15**, 183.
231   H. Tsunoda, S. Sasaki, T. Furuya and S. Ogawa, *Liebigs Ann. Chem.*, 1996, 159.
232   S. Ogawa, K. Hirai, M. Ohno, T. Furuya, S. Sasaki and H. Tsunoda, *Liebigs Ann. Chem.*, 1996, 673.
233   R. Wang and C.-H. Wong, *Tetrahedron Lett.*, 1996, **37**, 5427.
234   M. Kojima, T. Seto, Y. Kyotani, A. Ogawa, S. Kitazawa, K. Mori, S. Maruo, T. Ohgi and Y. Ezure, *Biosci. Biotech. Biochem*, 1996, **60**, 694.
235   D. A. Griffith and S. J. Danishefsky, *J. Am. Chem. Soc.*, 1996, **118**, 9526.
236   R. C. Griffiths, A. A. Watson, H. Kizu, N. Asano, H. J. Sharp, M. G. Jones, M. R. Wormald, G. W. J. Fleet and R. J. Nash, *Tetrahedron Lett.*, 1996, **37**, 3207.
237   H. Hashimoto, M. Kawanishi and H. Yuasa, *Chem. Eur. J.*, 1996, **2**, 556.
238   A. E. Stütz, *Angew. Chem. Int. Ed. Engl.*, 1996, **35**, 1926.
239   K. Kimura, *Mol. Cryst. Liq. Cryst. Sci. Technol., Sect. A*, 1996, **277**, 467 (*Chem. Abstr.*, 1996, **125**, 33 958).
240   K. G. Kabza, J. E. Gestwicki, J. L. McGrath and H. M. Petrassi, *J. Org. Chem.*, 1996, **61**, 9599.
241   T. Nishio, Y. Miyake, H. Tsuji, W. Hakamata, K. Kadokura and T. Oku, *Biosci. Biotech. Biochem*, 1996, **60**, 2038.
242   C. W. Andrews, R. Rodebaugh and B. Fraser-Reid, *J. Org. Chem.*, 1996, **61**, 5280.
243   R. Rodebaugh and B. Fraser-Reid, *Tetrahedron*, 1996, **52**, 7663.
244   J. R. Merritt, J. S. Debenham and B. Fraser-Reid, *J. Carbohydr. Chem.*, 1996, **15**, 65.
245   C. Buttersack and D. Laketic, *Stud. Surf. Sci. Catal.*, 1995, **98**, 190 (*Chem. Abstr.*, 1996, **124**, 202 802).
246   D. Crich, S. Sun and J. Brunckova, *J. Org. Chem.*, 1996, **61**, 605.
247   B. A. Bergamaschi and J. I. Hedges, *Carbohydr. Res.*, 1996, **280**, 345.

248   W. Zou and W. A. Szarek, *J. Chem. Soc., Chem. Commun.*, 1996, 1195.

249   E. A. Cioffi, *Tetrahedron Lett.*, 1996, **37**, 6231.

250   K. Fukase, T. Yasukochi, Y. Nakai and S. Kusumoto, *Tetrahedron Lett.*, 1996, **37**, 3343.

251   G. Descotes, J. Ramza, J.-M. Basset, S. Pagano, E. Gentil and J. Banoub, *Tetrahedron Lett.*, 1996, **37**, 10903.

252   S. A. Staley and B. D. Smith, *Tetrahedron Lett.*, 1996, **37**, 283.

253   M. A. Lipton, *Tetrahedron Lett.*, 1996, **37**, 287.

254   R. V. Weatherman and L. L. Kiessling, *J. Org. Chem.*, 1996, **61**, 534.

255   A. Heinrich-Schramm, C. Buttersack and H.-D. Lüdemann, *Carbohydr. Res.*, 1996, **293**, 205.

256   S. K. Das and N. Roy, *Carbohydr. Res.*, 1996, **296**, 275.

257   Z. Zhang and G. Magnusson, *Carbohydr. Res.*, 1996, **295**, 41.

258   K. Fukase, I. Kinoshita, T. Kanoh, Y. Nakai, A. Hasuoka and S. Kusumoto, *Tetrahedron*, 1996, **52**, 3897.

259   K. P. R. Kartha, M. Aloui and R. A. Field, *Tetrahedron Lett.*, 1996, **37**, 5175.

260   K. Zegelaar-Jaarsveld, H. I. Duynstee, G. A. van der Marel and J. H. van Boom, *Tetrahedron*, 1996, **52**, 3575.

261   K. Zegelaar-Jaarsveld, S. A. W. Smits, N. C. R. van Straten, G. A. van der Marel and J. H. van Boom, *Tetrahedron*, 1996, **52**, 3593.

262   M. Ticha, M. Cerny and T. Trnka, *Glycoconjugate J.*, 1996, **13**, 681 (*Chem. Abstr.*, 1996, **125**, 276 320).

263   F. Belot and J.-C. Jacquinet, *Carbohydr. Res.*, 1996, **290**, 79.

264   S. Knapp, D. Vocadlo, Z. Gao, B. Kirk, J. Lou and S. G. Withers, *J. Am. Chem. Soc.*, 1996, **118**, 6804.

265   S. J. Meunier and R. Roy, *Tetrahedron Lett.*, 1996, **37**, 5469.

266   M. J. Kiefel and M. von Itzstein, *Tetrahedron Lett.*, 1996, **37**, 7307.

267   T. M. Fyles, D. Loock, W. F. van Straaten-Nijenhuis and X. Zhou, *J. Org. Chem.*, 1996, **61**, 8866.

268   D. Zanini and R. Roy, *J. Org. Chem.*, 1996, **61**, 7348.

269   P. Barbaro, A. Currao, J. Herrmann, R. Nesper, P. S. Pregosin and R. Salzmann, *Organometallics*, 1996, **15**, 1879 (*Chem. Abstr.*, 1996, **124**, 290 012).

270   S. G. Polonik, A. M. Tolkach, E. B. Shentsova and N. I. Uvarova, *Khim.-Farm. Zh.*, 1995, **29**, 9 (*Chem. Abstr.*, 1996, **124**, 202 774).

271   R.Kakarla, R. G. Dulina, N. T. Hatzenbuhler, Y. W. Hui and M. J. Sofia, *J. Org. Chem.*, 1996, **61**, 8347.

272   G.-J. Boons and T. Stauch, *Synlett.*, 1996, 906.

273   W. Abramski and M. Chmielewski, *J. Carbohydr. Chem.*, 1996, **15**, 109.

274   M. Mavratzotis, V. Dourtoglou, C. Lorin and P. Rollin, *Tetrahedron Lett.*, 1996, **37**, 5699.

275   D. Crich, X.-Y. Jiao, Q. Yao and J. S. Harwood, *J. Org. Chem.*, 1996, **61**, 2368.

276   T. Furuta, K. Takeuchi and M. Iwamura, *J. Chem. Soc., Chem. Commun.*, 1996, 157.

277   S. Yamago, K. Kokubo, S. Masuda and J.-i. Yoshida, *Synlett.*, 1996, 929.

278   M. Petrušová, J. N. BeMiller, A. Krihova and L. Petruš, *Carbohydr. Res.*, 1996, **295**, 57.

279   M. Petrušová, J. N. BeMiller and L. Petruš, *Tetrahedron Lett.*, 1996, **37**, 2341.

280   B. A. Roe, C. G. Boojamra, J. L. Griggs and C. R. Bertozzi, *J. Org. Chem.*, 1996, **61**, 6442.

281   F. Pontén and G. Magnusson, *J. Org. Chem.*, 1996, **61**, 7463.

282   S. Kim, I. Y. Lee, J.-Y. Yoon and D. H. Oh, *J. Am. Chem. Soc.*, 1996, **118**, 5138.

283   D. Urban, T. Skydstrup, C. Riche, A. Chiaroni and J.-M. Beau, *J. Chem. Soc., Chem. Commun.*, 1996, 1883.

284   K. Toshima, N. Miyamoto, G. Matsuo, M. Nakata and S. Matsumura, *J. Chem. Soc., Chem. Commun.*, 1996, 1379.

285   S. N. Thorn and T. Gallagher, *Synlett.*, 1996, 185.

286   B. J. Dorgan and R. F. W. Jackson, *Synlett.*, 1996, 859.

287   L. Kiss and L. Somsák, *Carbohydr. Res.*, 1996, **291**, 43.

288   J. Cassy and S. Ibhi, *Carbohydr. Res.*, 1996, **291**, 189.

289   S. Cernecki, E. Ayadi and J. Xie, *Tetrahedron Lett.*, 1996, **37**, 9193.

290   L. Cipolla, L. Liguori, F. Nicotra, G. Torri and E. Vismara, *J. Chem. Soc., Chem. Commun.*, 1996, 1253.

291   T. F. Herpin, J. S. Houlton, W. B. Motherwell, B. P. Roberts and J.-M. Weibel, *J. Chem. Soc., Chem. Commun.*, 1996, 613.

292   K. Tomooka, H. Yamamoto and T. Nakai, *J. Am. Chem. Soc.*, 1996, **118**, 3317.

293   R. Bürli and A. Vasella, *Helv. Chim. Acta*, 1996, **79**, 1159.

294   A. Ernst and A. Vasella, *Helv. Chim. Acta*, 1996, **79**, 1279.

295   O. R. Martin, L. Liu and F. Yang, *Tetrahedron Lett.*, 1996, **37**, 1991.

296   A. Baudat and P. Vogel, *Tetrahedron Lett.*, 1996, **37**, 483.

297   E. Frérot, C. Marquis and P. Vogel, *Tetrahedron Lett.*, 1996, **37**, 2023.

298   A. L. J. Byerley, A. M. Kenwright and P. G. Steel, *Tetrahedron Lett.*, 1996, **37**, 9093.

299   A. Alberti, M. A. Della Bona, D. Macciantelli, F. Pelizzoni, G. Sello, G. Torri and E. Vismara, *Tetrahedron*, 1996, **52**, 10241.

300   O. Jarreton, T. Skydstrup and J.-M. Beau, *J. Chem. Soc., Chem. Commun.*, 1996, 1661.

301   A. J. Fairbanks, E. Perrin and P. Sinaÿ, *Synlett.*, 1996, 679.

302   J.-F. Espinosa, H. Dietrich, M. Martín-Lomas, R. R. Schmidt and J. Jiménez-Barbero, *Tetrahedron Lett.*, 1996, **37**, 1467.

303   J. F. Espinosa, F. J. Cañada, J. L. Asenio, H. Dietrich, M. Martín-Lomas, R. R. Schmidt and J. Jiménez-Barbero, *Angew. Chem. Int. Ed. Engl.*, 1996, **35**, 303.

304   T. Brigaud, O. Lefebvre, R. Plantier-Royon and C. Portella, *Tetrahedron Lett.*, 1996, **37**, 6115.

305   A. Dondoni, A. Boscarato and H. Zuurmond, *Tetrahedron Lett.*, 1996, **37**, 7587.

306   W. R. Kobertz, C. R. Bertozzi and M. D. Bednarski, *J. Org. Chem.*, 1996, **61**, 1894.

307   S.-C. Hung and C.-H. Wong, *Tetrahedron Lett.*, 1996, **37**, 4903.

308   H. Streicher, A. Geyer and R. R. Schmidt, *Chem. Eur. J.*, 1996, **2**, 502.

309   L.-X. Wang, J.-Q. Fan and Y. C. Lee, *Tetrahedron Lett.*, 1996, **37**, 1975.

310   M. Hoffman, F. Burkhart, G. Hessler and H. Kessler, *Helv. Chim. Acta*, 1996, **79**, 1519.

311   Y. Suhara, J. E. K. Hildreth and Y. Ichikawa, *Tetrahedron Lett.*, 1996, **37**, 1575.

312   T. J. Woltering, G. Weitz-Schmidt and C.-H. Wong, *Tetrahedron Lett.*, 1996, **37**, 9033.

313   T. Uchiyama, T. J. Waltering, W. Wong, C.-C. Lin, T. Kajimoto, M. Takebaashi, G. Weitz-Schmidt, T. Asakura, M. Noda and C.-H. Wong, *Bioorg. Med. Chem.*, 1996, **4**, 1149 (*Chem. Abstr.*, 1996, **125**, 248 275).

314   T. G. Marron, T. J. Woltering, G. Weitz-Schmidt and C.-H. Wong, *Tetrahedron Lett.*, 1996, **37**, 9037.

315   D. L. Sutherlin, T. M. Stark, R. Hughes and R. W. Armstrong, *J. Org. Chem.*, 1996, **61**, 8350.

316   A. T. Khan, W. Ahmed and R. R. Schmidt, *Carbohydr. Res.*, 1996, **280**, 277.

317   J. C. Castro-Palomino and R. R. Schmidt, *Liebigs Ann. Chem.*, 1996, 1623.

318 P. Verlhac, C. Leteux, L. Toupet andd A. Veyrières, *Carbohydr. Res.*, 1996, **291**, 11.

319 T. Hosoya, Y. Ohashi, T. Matsumoto and K. Suzuki, *Tetrahedron Lett.*, 1996, **37**, 663.

320 D. J. Hart, G. H. Merriman and D. G. J. Young, *Tetrahedron Lett.*, 1996, **37**, 14437.

321 M. F. Wong, K. L. Weiss and R. W. Curley, *J. Carbohydr. Chem.*, 1996, **15**, 763.

322 E. Ayadi, S. Czernecki and J. Xie, *J. Chem. Soc., Chem. Commun.*, 1996, 347.

323 D. Craig, A. H. Payne, M. W. Pennington and J. P. Tierney, *Electron. Conf. Trends. Org. Chem.*, 1995 (Pub. 1996), Paper 26 (*Chem. Abstr.*, 1996, **125**, 11 311).

324 C. J. F. Bichard, T. W. Brandstetter, J. C. Estevez, G. W. J. Fleet, D. J. Hughes and J. R. Wheatley, *J. Chem. Soc., Perkin Trans. 1*, 1996, 2151.

325 T. Mukaiyama, H. Uchiro, N. Hirano and T. Ishikawa, *Chem. Lett.*, 1996, 629.

326 V. Jaouen, A. Jégou and A. Veyrières, *Synlett.*, 1996, 1218.

327 J. Matulic-Adamic, L. Beigelman, S. Portmann, M. Egli and N. Usman, *J. Org. Chem.*, 1996, **61**, 3909.

328 J. Matulic-Adamic and L. Beigelman, *Tetrahedron Lett.*, 1996, **37**, 6973.

329 R. X.-F. Ren, N. C. Chaudhuri, P. L. Paris, S. Rumney and E. T. Kool, *J. Am. Chem. Soc.*, 1996, **118**, 7671.

330 K. A. Parker and D.-S. Su, *J. Org. Chem.*, 1996, **61**, 2191.

331 M. Yokoyama, T. Akiba, Y. Ochiai, A. Momotake and H. Togo, *J. Org. Chem.*, 1996, **61**, 6079.

332 S. Harusawa, Y. Murai, H. Moriyama, T. Imazu, H. Ohishi, R. Yoneda and T. Kurihara, *J. Org. Chem.*, 1996, **61**, 4405.

333 M. Yokoyama, M. Nomura, H. Togo and H. Seki, *J. Chem. Soc., Perkin Trans. 1*, 1996, 2145.

334 J. H. Boal, A. Wilk, C. L. Scremin, G. N. Gray, L. R. Phillips and S. L. Beaucage, *J. Org. Chem.*, 1996, **61**, 8617.

4
Oligosaccharides

1 General

As previously, this chapter deals with specific tri- and higher oligosaccharides. Most references relate to their synthesis by chemical, enzymic or chemicoenzymic methods. Chemical features of the cyclodextrins are noted separately; their complexing features have not been surveyed. With the increasing use of enzymic synthetic methods and combinatorial procedures more examples are appearing in the literature of oligosaccharide mixtures which are difficult to classify. Several reviews have emphasised the importance of oligosaccharides biologically and also their synthesis.[1-10] They incorporate surveys of synthetic methods, including the combinatorial approach.[1]

A review has come from Danishefsky's group on the use of glycals in the synthesis of oligosaccharides and glycoconjugates.[11] Wong and coworkers have presented two extensive reviews on combined chemical and enzymic synthetic methods,[12,13] a specific review has appeared (in Korean) on the enzymic synthesis of fructooligosaccharides,[14] and a further (in Japanese) has dealt with transglycosylation enzymes which lead to routes to di- and trisaccharides.[15] A symposium report (in Japanese) describes the novel application of catalytic antibodies in oligosaccharide synthesis.[16]

2 Trisaccharides

Compounds in sections 2.1–2.3 are now categorized according to their non-reducing end sugars.

2.1 Linear Homotrisaccharides – In the series of glucose trimers α-D-Glc-(1→2)-α-D-Glc-(1→3)-α-D-Glc-O(CH$_2$)$_8$CO$_2$Me has been made for use in an assay of α-glucosidase I activity,[17] and the α-(1→6)-linked trimer has been made by polymer-supported solution synthetic methods.[18] Fraser-Reid has described a pentenyl glycoside of the α-(1→2)-linked mannose trimer for use in the synthesis of glycoprotein membrane high mannose compounds.[19] In related work the α-(1→6)-linked mannose trimer was prepared as its O-octyl- and S-octyl glycosides in work also related to core structures of anchoring compounds.[20] The unusual β-(1→4)-linked mannotriose has been isolated from the hydrolysate of a seed

Carbohydrate Chemistry, Volume 30
© The Royal Society of Chemistry, 1998

polysaccharide.[21] The α-(1→2)-linked rhamnose trimer joined to spacer groups and hence for joining to proteins has been described,[22] and chitotriose linked through nitrogen to various histidine-like molecules is reported in Chapter 10.

2.2 Linear Heterotrisaccharides – The glucosyl mannobiose β-D-Glc-(1→6)-α-D-Man-(1→6)-α-D-Man has been synthesized and linked to peptides as models for the phytoalexin elicitor glycoprotein.[23] Maltose has been linked to a mannose by chemical methods to give α-D-Glc*p*-(1→4)-α-D-Glc*p*-(1→4)-D-Man.[24] Enzymic procedures were used to make a set of trisaccharides including cellobiosyl fructose and gentiobiosyl fructose.[25]

Considerable interest continues in trisaccharides terminating in D-galactose. α-D-Gal-(1→6)-β-D-Gal-(1→4)-D-Glc and α-D-Gal-(1→3)-β-D-Gal-(1→4)-D-Glc have been produced as their ethylthio glycosides using enzymic methods,[26] and similar procedures were used to make β-D-Gal-(1→4)-β-D-Gal-(1→4)-D-GlcNAc glycosides or thioglycosides.[27] Similarly β-D-Gal-(1→3)-β-D-Gal-(1→4)-β-D-Xyl was produced as a serine or coumarin glycoside,[28] and α-D-Gal-(1→4)-β-D-Gal-(1→4)-β-D-Glc was made as a glycoside with a bis-sulfone carrying long-chain alkyl groups together with analogues having 2-, 3- or 6-deoxy units in place of the glucose moiety as neoglycolipids.[29] β-D-Gal-(1→4)-β-D-Glc-(1→6)-D-Glc was made as its 4,6-pyruvyl acetal in relationship to work on *Klebsiella* polysaccharides.[30] In related work on bacterial antigens β-D-Gal-(1→4)-α-D-Glc-(1→4)-D-Glc was synthesized.[31] Galactose-terminating trimers containing amino sugars to have been produced in connection with immunological work are α-D-Gal-(1→3)-β-D-GlcNAc-(1→2)-D-Man,[32] β-D-Gal-(1→4)-β-D-GlcNAc-(1→2)-α-D-Man and several analogues with modifications in the galactose moiety,[33] and β-D-Gal-(1→4)-β-D-GlcNAc-(1→6)-D-GalNAc.[34] From Boons' group have come descriptions of trisaccharide libraries with D-galactose and L-fucose terminal groups introduced by means of but-2-en-2-yl ethers as glycosylating agents.[35]

Mannose-terminating compounds made in connection with glycoprotein work have the sugar either α-[36] or β-[37,38] linked through O-4 of chitobiose. The first of these reports exemplifies the use of polymer-supported reactions. Chemical and enzymic methods were used in the production of the latter trimer.

Julibroside, which is extracted from the silktree and used in Chinese medicine, contains a complex saponin having β-D-Xyl-(1→2)-α-L-Ara-(1→6)-D-Glc as one of the oligosaccharide components. A further component is a branched-chain tetrasaccharide comprising two units of glucose, one of L-rhamnose and one of xylofuranose.[39]

Several trisaccharides to have been prepared terminate in amino-sugars. These include a compound having glucosamine carrying a long-chain *N*-acyl group β-(1→4)-linked to cellobiose,[40] and several compounds with *N*-acetylglucosamine linked to lactose have been produced enzymically.[41] The mannosamine-terminating compound β-D-ManNAc-(1→4)-α-D-Glc-(1→3)-L-Rha, which is a repeating unit of a *Streptococcus pneumoniae* capsular polysaccharide, has been made by use of a 2-oximinoglycosyl bromide donor.[42]

Compounds terminating in sugar acids continue to be of interest, α-D-GlcA-

$(1\rightarrow3)$-α-L-Ara$f$-$(1\rightarrow3)$-D-Xyl having been prepared by a direct glycosylating reaction involving methyl 2,3,4-tri-*O*-benzyl glucuronate used together with *p*-nitrobenzenesulfonyl chloride and silver triflate. With secondary sugar alcohols yields of about 80% were achieved with α/β ratios about 5:1.[43] Main attention has been on compounds terminating in sialic acids, in particular synthetic studies associated with the putting together of gangliosides GM_3 and GM_4 have focussed on the sialylated lactose α-D-NeuNAc-$(2\rightarrow3)$-β-D-Gal-$(1\rightarrow4)$-D-Glc. It has been produced by use of a 3-thio-thioglycoside of neuraminic acid on a gram scale as the 2-trimethylsilylethyl glycoside.[44] In other studies the *S*-ethyl glycoside[45] and glycosyl trichloroacetimidate[46] were made for the purpose of coupling the trimer to ceramides and long-chain alcohols. It has been also described as a ceramide glycoside carrying various alternatives to the acetylamino group.[47] The *p*-nitrophenyl glycoside was used for conversion to the *p*-acrylamido analogue followed by copolymerization with acrylamide to give water soluble polymers which contained the GM_3 trisaccharide and retained antigenic activity.[48] An analogue of the same GM_3 ganglioside having a 3,8-dideoxyneuraminic acid unit was prepared in a study of structure/activity relationships.[49] The closely related trisaccharide α-D-NeuNH$_2$-$(2\rightarrow6)$-β-D-Gal-$(1\rightarrow4)$-β-D-Glc carrying an *N*-glycolyl substituent was made as its 2-trimethylsilylethyl glycoside.[50] A range of *N*-acyllactosamines have been tested as α-$(2\rightarrow3)$-sialyltransferase substrates and many were tolerated thus giving access to α-D-NeuNAc-$(2\rightarrow3)$-β-D-Gal-$(1\rightarrow4)$-D-GlcNH$_2$ derivatives.[51]

Work on bacterial polysaccharides has led to the preparation of α-L-Fuc-$(1\rightarrow2)$-α-L-Fuc-$(1\rightarrow3)$-β-D-GalNAc and the isomer with the non-reducing β-L-fucosyl-linkage at the terminal position,[52] and to that of α-L-Fuc-$(1\rightarrow2)$-α-D-Gal$(1\rightarrow3)$-α-D-Glc-OMe and several analogues including some with β-L-fucosyl-linkages and modifications at C-2 in the glucose unit.[53] Danishefsky's group have used their glycal assembly methods in solid phase syntheses of α-L-Fuc-$(1\rightarrow2)$-β-D-Gal-$(1\rightarrow4)$-D-Glc.[54] β-D-Qui-$(1\rightarrow3)$-α-L-Rha-$(1\rightarrow3)$-L-Rha has been synthesized with 4-*O*-lactoyl and 3-*O*-methyl substitution on the terminal quinovose unit and 4-*O*-methyl on the central rhamnose during work on the glycopeptidolipids from a *Mycobacterium*.[55] α-L-Rha-$(1\rightarrow4)$-β-D-Glc-$(1\rightarrow6)$-D-Glc has been prepared synthetically and is a common trisaccharide component of saponins isolated from various Chinese medicinal plants.[56]

Several more complex trisaccharides involving modified sugars have been prepared: for example **1**, which is an aureolic acid-related compound[57] made from a 2-phenylthioglycosyl trichloroacetimidate derived from the glycal **2** by solid phase technology,[58] and the C-3 trisaccharide **3** of antibiotic PI-080, the unsaturated unit of which was coupled using the relevant unsaturated glycosyl tetrazole derivative.[59]

2.3   Branched Homotrisaccharides – β-D-Glc-$(1\rightarrow2)$-[β-D-Glc-$(1\rightarrow3)$]-α-D-Glc and the epimer β-D-Glc-$(1\rightarrow2)$-[β-D-Glc-$(1\rightarrow3)$]-α-D-Man have been prepared as α-methyl glycosides for conformational studies by NMR and Monte Carlo simulation methods.[60]

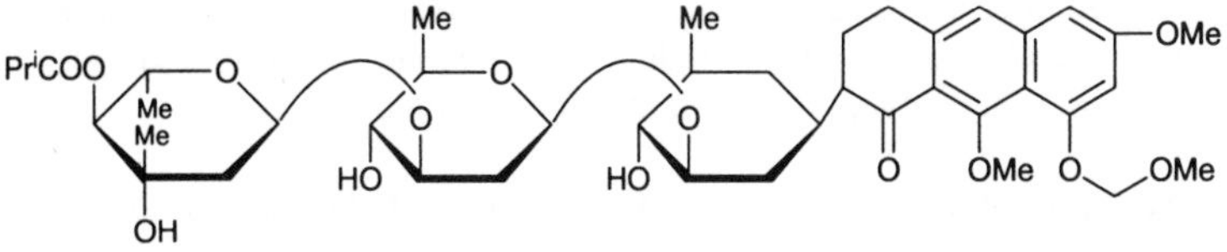

1

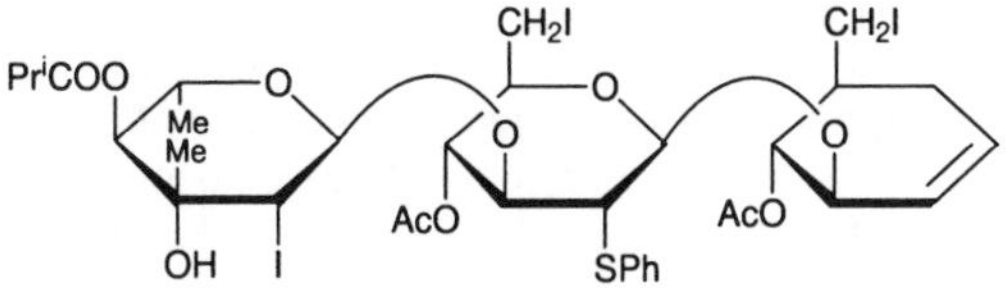

2

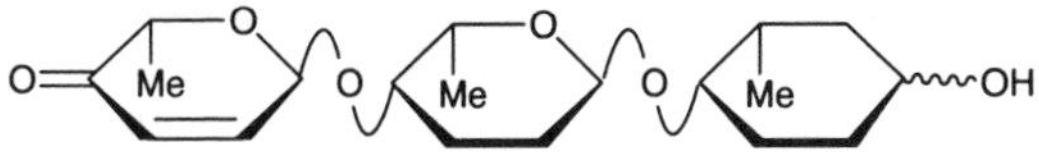

3

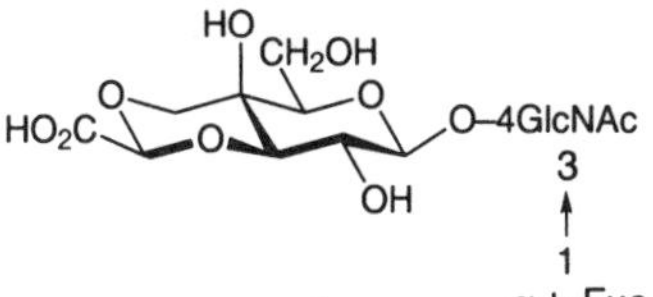

4   α-L-Fuc

2.4   Branched Heterotrisaccharides – Compounds in this section are categorized according to their reducing end sugars.

β-D-Gal-(1→3)-[α-L-Fuc-(1→4)]-D-Glc was converted to the phosphate at O-3 of the galactose moiety by protecting all the hydroxyl groups except those at C-2, C-3 and C-4 of this unit and using dibutyltin oxide followed by dibenzylphosphorochloridate.[61] The related siaLe^x trisaccharide β-D-Gal-(1→4)-[α-L-Fuc-(1→3)]-D-Glc, with a sulfate ester group at O-3 of the galactose unit, was coupled through a bridging link to an acrylamide polymer and showed inhibition of both L and E selectins,[62] and in related work the trisaccharide having a glycolyl ether group at O-3 of the galactose, and in the glycosylamine form, was *N*-coupled to a long chain unsaturated alkyl group to give a compound which formed liposomes with P-selectin inhibitory activity.[63]

Considerable work continues on analogues of the above trisaccharides having glucosamine as the reducing moiety. The Le^x compound α-D-Gal-(1→4)-[α-L-Fuc-(1→3)]-D-GlcNAc and the Le^a isomer with the galactose and fucose interchanged have been made using chemical procedures depending upon glycosyl sulfoxides for coupling,[64] and the same compounds and several analogues have been produced by the trichloroacetimidate coupling method.[65] Other workers have used fucosyl transferases to bond the fucose moiety in related work.[66] Compound **4** is a siaLe^x mimic, but is inactive in E-selectin binding.[67]

Glycosylation of methyl 2-*O*-benzoyl-α-L-rhamnopyranoside has led to a set of branched trisaccharides having β-D-gluco-, α-D-manno-, α-L-rhamno-, and β-D-fructo-pyranosyl substituents at O-3 and O-4 of rhamnose.[68]

2.5   Analogues of Trisaccharides and Compounds with Anomalous Linking – In studies of conformationally constrained analogues of siaLe[x], L-fucose and D-galactose, the latter carrying a glycoyl ether group at O-3, were coupled to cyclohexane-1,2-*trans*-diol and catechol. The former product was active biologically as the natural trisaccharide, whereas the aromatic analogue was inactive.[69,70] Related compounds based on both enantiomers of cyclohexane-1,2-*trans*-diol carrying α-L-fucose and 4-benzyloxy-3-(2-carboxyethyl)phenyl substituents showed inhibition of binding of E-selectin to siaLe[x] bonded to polyacrylamide.[71] Hexane 1,6-diol carrying *N*-acetylneuraminic and α-L-fucose substituents can also be considered to be a siaLe[x] model compound.[72]

The chitinase inhibitor allosamidine is a trisaccharide analogue with two *N*-acetylallosamine units bonded to a highly functionalized cyclopentano-oxazoline unit – allosamizoline. Analogues have now been described with one *N*-acetylallosamine unit, with the disaccharide bonded to an alternative hydroxyl of the cyclopentane, and thirdly with the oxazoline ring inverted.[73]

Two recently synthesized oligosaccharide-containing forms of calicheomycin γ$_1^1$ bind strongly to certain duplex DNA sequences.[74] Danishefsky has published a long paper on the details of his total synthesis of this compound and of dynemicin A, including the carbohydrate components.[75]

In the area of trisaccharides with anomalous linking, the (1→3)- and (1→6)-linked glucotrioses with sulfur replacing oxygen as the interunit atoms, have been reported. In the same work the *S*-linked (1→3)-, (1→6) trimer was also described,[76] and β-D-Gal-(1→4)-[α-L-Fuc-(1→3)]-D-GlcNAc with the fucose having sulfur as the ring hetero-atom has also been prepared.[77] In work related to the trisaccharides for H type I blood group determinants analogues of α-L-Fuc-(1→2)-α-D-Man-(1→3)-D-GlcNAc were made ingeniously by linking the first and the third units together with a highly functionalized acyclic bridge which was ring closed to give the central unit. A range of analogues was produced, the notable feature being that the sugars were *C*-linked and not *O*-linked.[78]

3   Tetrasaccharides
Compounds of this set and higher oligosaccharides are classified according to whether they have linear or branched structures and then by the nature of the sugars at the reducing termini.

3.1   Linear Homotetrasaccharides – Cellulose oligomers up to the 20-mer have been assembled from a key tetrasaccharide glycoside which was converted into the glycosyl trichloroacetimidate and separately to a derivative with O-4 of the non-reducing end deprotected.[79] Several specifically deoxygenated derivatives of β-maltosyl-(1→4)-α,α-trehalose have been specifically prepared, sulfated and tested for smooth muscle cell antiproliferative activity.[80-82] Iterative methods have been applied to make the α-(1→2)-D-mannose tetramer[18] and di- to tetra-oligomers of 4-amino-4,6-dideoxy-α-D-mannose carrying a 3-deoxy-L-*glycero*-tetronic acid substituent on nitrogen of each unit were prepared as the terminal unit of the polysaccharide of *Vibrio cholerae*,[83] and the α-(1→4)-L-rhamnotetraose was prepared as a long-chain alkyl glycoside which is the natural resin glycoside merremoside i.[84] N-Acetylchitotetraose may be produced, together with smaller fragments, on enzymic hydrolysis of chitin.[85]

3.2   Linear Heterotetrasaccharides – The total synthesis of ganglioside GD$_3$, α-NeuAc-(2→8)-α-NeuAc-(2→3)-β-D-Gal-(1→4)-β-D-Glc-1-Cer, has been completed, the sialic acid units being introduced using a 3-*S*-phenyl thioglycosyl fluoride,[86,87] and the 9-*O*-acetyl derivative of this compound, having the acetyl group on the terminal sialic acid unit, which is associated with melanoma, has also been described, the acetylation being conducted enzymically.[88]

α-L-Fuc-(1→3)-β-D-GlcNAc-(1→3)-β-D-Gal-(1→4)-D-Glc has been synthesized as a versatile tetrasaccharide for the preparation of glycosphingolipids and sulfated oligosaccharides related to siaLe^x.[89] The N-acetyllactosaminyl lactose glycoside β-D-Gal-(1→4)-α-D-GlcNAc-(1→6)-β-D-Gal-(1→4)-β-D-Glc, and the isomer with β-linking between the disaccharide units, have been studied as metastatic inhibitors.[90] Koenigs-Knorr glycosylation of sucrose has resulted in the preparation of tri- and tetrasaccharides.[91]

α-L-Rha-(1→2)-α-L-Rha-(1→2)-β-D-Glc-(1→2)-β-D-Gal, having a long-chain alkyl aglycon terminating in a carboxylic acid, which is lactone-linked to position 3 of the glucose unit, is a tricolorin phytogrowth inhibitor of *Convolvulaceae*. It and four related macrocyclic lactones were investigated by NMR and mass spectral procedures.[92] The Le^x and Le^a analogue, β-Gal-(1→4)-β-GlcNAc-(1→3)-β-Gal-(1→4)-GlcNAc-β-OBn, which is an N-acetyllactosamine dimer, was prepared with a sulfate at position 3 of the terminal galactose unit,[93] and ΔUA-(1→4)-α-D-GlcNAc-(1→4)-β-D-GlcA-(1→4)-D-GlcNAc, in which the terminal uronic acid unit is the 4,5-unsaturated compound derived from glucuronic acid, was isolated from a hydrolysate of heparin, and was found to epimerize at the reducing moiety under the conditions of oligosaccharide fractionation, that is in dilute ammonia solution. The glucosamine/ mannosamine equilibrium ratio was 6:4.[94]

Two synthetic studies have been concerned with fragments of the cell wall phenolic glycosides of a *Mycobacterium*. The relevant compounds consist of key *p*-2-(aminoethyl)phenyl glycosides of the α-(1→3)-linked fucotriose carrying simple substituents such as methyl or acetyl at positions 2 and 4 of the fucose units. In addition, 4-*O*-methyl-α-D-mannose was bonded to position 3 of the non-reducing fucose moiety. The issue of double stereodifferentiation during the coupling was examined.[95] This same phenomenon was pursued in the second study which involved the introduction of a 2,6-dideoxy-4-*O*-methyl-α-D-*arabino*-hexose unit.[96] A convergent procedure was used in the production of the 5-aminopentyl glycoside of β-D-Glc-(1→3)-β-D-GlcNAc-(1→3)-α-D-Gal-(1→4)-β-L-Rha, carrying a 4,6-carboxyethylidene acetal on the glucosamine unit, which is the repeating unit of the immunodominant polysaccharide of *Streptococcus pneumoniae* type 27.[97] Studies of novel steroidal saponin oligosaccharides of a

Leptadenia species have yielded the cymarose (2,6-dideoxy-3-*O*-methyl-D-*ribo*-hexose) and digitoxose (2,6-dideoxy-D-*ribo*-hexose)-containing tetrasaccharide β-D-Cym-(1→4)-3-OMe-α-D-Gal-(1→4)-β-D-Dig-(1→4)-D-Cym.[98] β-D-GlcA-(1→3)-β-D-Gal-(1→3)-β-D-Gal-(1→4)-β-D-Xyl-Ser is a tetramer of the linkage region of proteoglycans and has been synthesized.[99]

3.3   Branched Homotetrasaccharides.– The only member of this set to have been noted is the synthetic product having glucose β-(1→3)-linked in a trimer with β-D-glucose branch-linked at C-6 of the central moiety. All the interunit linkages were via sulfur rather than oxygen.[100]

3.4   Branched Heterotetrasaccharides – The D-glucose-terminating siaLe^x analogue α-D-NeuNAc-(2→3)-β-D-Gal-(1→4)-[α-L-Fuc-(1→3)]-D-Glc has been described as well as the related compound with a sulfate ester replacing the neuraminic acid,[101] and a closely similar compound having the L-fucose branch attached to O-2 of the galactose instead of the glucose moiety, and a related compound having neuraminic acid 3-linked to the same galactose unit instead of fucose have been made in very high yields by use of a sialyl transferase.[102] Other work has produced α-Neu5Ac-(2→3)-β-D-Gal*p*-(1→3)-[α-L-Fuc-(1→2)]-1-deoxy-D-Glc with a 1,5-anhydroglucitol terminal unit, and the isomers having its di- and monosaccharide substituents interchanged, which are other epitope analogues.[103] The glucose-terminating tetramer α-L-Rha-(1→4)-β-D-Glc-(1→6)-[β-D-Glc-(1→2)]-D-Glc is a component of a new triterpene saponin as is β-D-Glc-(1→3)-α-L-Rha-(1→2)-[β-D-Api*f*-(1→3)]-D-Fuc.[104]

β-D-GalNAc-(1→3)-α-D-Gal-(1→3)-[α-L-Fuc-(1→2)]-β-D-Gal, synthesized as the octyl glycoside, was the latest blood group type determinant to have been prepared.[105] The further siaLe^x analogues β-D-Gal-(1→4)-[α-L-Fuc-(1→3)]-β-D-GlcNAc-(1→3)-β-D-Gal-*O*-Cer with the sialic acid replaced by a galactosyl unit bearing anionic substituents, for example phosphate or sulfate at O-3, have been made,[106] and the *n*-pentenyl glycoside method has been used to prepare the tetrasaccharide α-D-Man-(1→2)-α-D-Man-(1→2)-[β-D-Gal-(1→4)]-α-D-Man of a *Leishmania* lipophosphoglycan as the 4,5-dibromopentyl glycoside.[107]

Extensive interest continues in the synthesis of the siaLe^a and siaLe^x which have *N*-acetylglucosamine as reducing function. A new synthesis of siaLe^x, α-Neu5Ac-(2→3)-β-D-Gal-(1→4)-[α-L-Fuc-(1→3)]-β-D-GlcNAc, has been described.[108] The analogue with a deoxy-group at the anomeric centre of the reducing moiety was 20 times more potent an inhibitor than the parent tetramer towards P- and L-selectin binding. This report also described the 1-deoxy-analogue which had a hydroxyl group in place of the *N*-acetylamino at C-2 and the 1,2-dideoxy compound.[109] A fully protected derivative of the siaLe^x tetra-saccharide having, however, a free amino group, was made to enable the preparation of various 1-*O*- and 2-*N*-substituted analogues.[110] The parent tetra-saccharide has also been described with glycosylamine linking to peptides and glycosidic linking to various amide-containing aglycons.[111] A cyclic peptide carrying three *N*-linked units of the tetrasaccharide inhibits adhesion of HL$_{60}$

cells only slightly better than does the relevant monomeric unit.[112] Reference 108 also describes the preparation of the tetramer of siaLe^a, and a further paper describes the preparation of this compound and analogues with a sulfate ester replacing the sialic acid.[113]

Both small and large lipophilic acyl groups on nitrogen are well tolerated in sialyl-lactosamine when the derivatives were used as acceptors for a cloned fucosyl transferase. In this way many siaLe^x analogues have been made.[114] α-NeuAc-(2→3)-β-D-Gal-(1→3)-[α-NeuAc-(2→6)]-α-D-GalNAc-L-Serine, having a lactone bridge between O-4 of the galactose moiety and the carboxyl group of the adjacent sialic acid unit, was prepared as a building block for sialoglycopeptide synthesis.[115]

The tetramer β-D-GlcNH$_2$-(1→4)-β-D-GlcNAc-(1→4)-[α-L-Fuc-(1→6)]-D-GlcNAc carrying a long-chain unsaturated N-acyl substituent on the non-reducing end moiety, which is a signalling compound that triggers the formation of root nodules in legumes, has been synthesized.[116]

3.5   Analogues of Tetrasaccharides and Compounds with Anomalous Linkings – The tetramer **5** (R = OH), which is a hydroxy derivative of β-acarbose [**5** (R = H)] has been made by the coupling of a 1-epivalienamine derivative, 1,6:3,4-dianhydro-2-O-benzyl-D-galactose and benzyl 2,3,6,2′,3′,6′-hexa-O-benzyl-β-cellobioside.[117] Compounds α-NeuNAc-(2→3)-β-D-Gal-(1→X)-[α-L-Fuc-(1→X)]-R, where X = 2 or 3 and R = the N-acetylglucosamine analogue **6**, are 2-acetamido-deoxynojirimicin-containing analogues of siaLe^x and siaLe^a.[118]

Tetrasaccharide analogues having the sugar units S- rather than O-linked to have been described are N-acetylchitotetraose,[119] and an analogue of siaLe^x made as the S-heptyl thioglycoside.[120] See reference 100 for a similar branched glucotetraose analogue. By linking up appropriate disaccharide units Vasella and co-workers have been able to produce the acetylene-linked tetramer compound **7**.[121] The phosphate-linked tetramer **8** and a similar hexamer have been synthesized as fragments of the end of the phosphoglycan of a *Leishamania* lipophosphoglycan by linking disaccharide phosphate components.[122]

4   Pentasaccharides

4.1   Linear Homopentasaccharides – The novel and potentially very useful compound **9** has been used in the iterative synthesis of di- to penta-saccharides of *Mycobacterium tuberculosis* polysaccharide II, which are α-(1→2)-linked glucose-based compounds, the pentamer being made as a pentyl α-D-glycoside having an N-acylhydrazide function at the terminus of the aglycon.[123] Sulfated alkyl β-(1→3)-linked glucooligosaccharides, including the pentamer, have potent inhibitory effects on HIV infection, and while the pentaoside showed cytotoxicity, higher oligosaccharides were less toxic.[124] An alternatively α- and β-(1→4)-linked glucopentaose has been made synthetically as a simple heparan sulfate model,[125] and oligosaccharides of this series including the pentamer were chemically

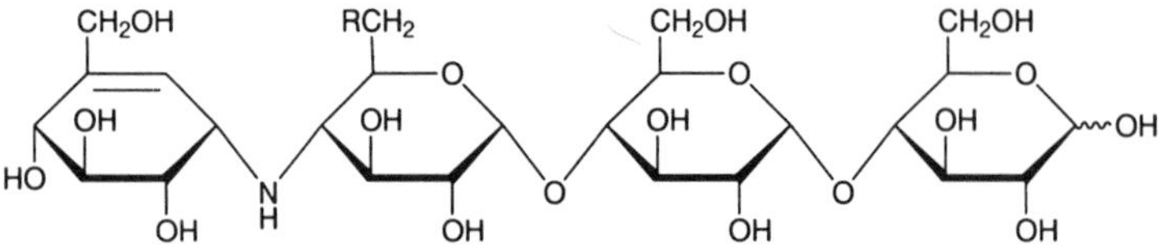

5

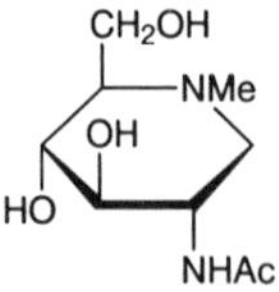

6

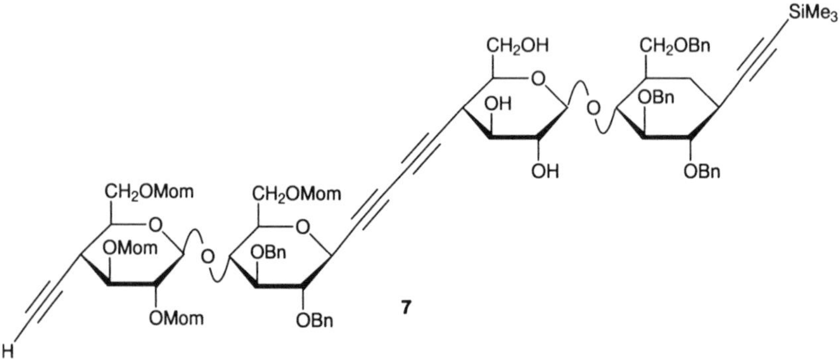

7

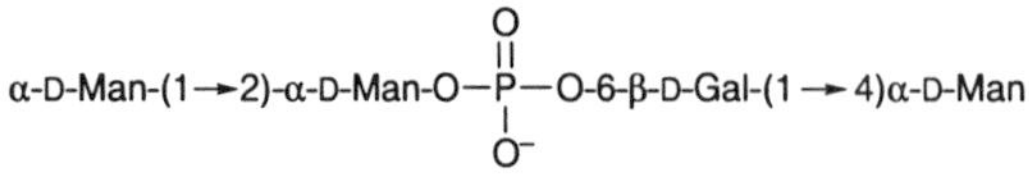

$$\alpha\text{-D-Man-}(1\rightarrow2)\text{-}\alpha\text{-D-Man-O}-\overset{\displaystyle O}{\underset{\displaystyle O^-}{P}}-O\text{-}6\text{-}\beta\text{-D-Gal-}(1\rightarrow4)\alpha\text{-D-Man}$$

8

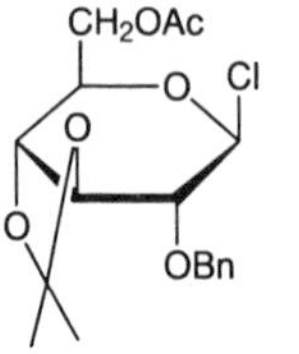

9

sulfated and tested for antiproliferative effect on smooth muscle cells in comparison with heparin. The oligosaccharides were deemed to be heparan model compounds and were biologically more active than the sulfated heparin compounds.[126]

Solid phase technology and use of 6-*O*-acetyl-2,3,4-tri-*O*-benzyl-α-D-glucosyl trichloroacetimidate led to (1→6)-linked pentasaccharides.[127]

Chitopentaose was fully deacetylated by use of a chitin deacylase,[128] and the trityl/1,2-cyanoethylidene condensation reaction was applied to prepare a linear β-(1→5)-linked D-galactofuranan, and when the reaction was run at low concentration intramolecular condensation was observed and cyclic oligosaccharides produced.[129]

4.2 Linear Heteropentasaccharides – Two reports have been published on the putative pentasaccharide binding unit of heparan sulfate to fibroblast growth factor 2, *i.e.* α-L-IdoA-(1→4)-α-D-GlcNH₂-(1→4)-α-L-IdoA-(1→4)-α-D-GlcNH₂-(1→4)-α-L-IdoA, carrying sulfate groups at the two amino functions and at O-2 of the reducing terminal acid group.[130,131] In related work α-D-Glc-(1→4)-β-D-Glc-(1→4)-α-D-GlcA-(1→4)-α-L-IdoA-(1→4)-α-D-Glc-O-Me, carrying sulfate groups at O-1, O-2 and O-3 of the reducing and central units and O-6 of the non-reducing unit, and methyl ethers at all other hydroxyl positions except O-4 of the non-reducing terminus which was linked by way of a long spacer arm to various oligosaccharides, was produced and the compounds were evaluated for anti-thrombin and anti-factor Xa activity.[132] Petitou and coworkers have described computer modelling of the antithrombin III-heparin-thrombin complex which has led to the synthesis of novel glycoconjugates, whose factor Xa and thrombin inhibitory activities can be controlled to allow access to anticoagulants with unprecedented characteristics.[133]

Two examples of β-D-GlcA-(1→3)-β-D-Gal-(1→4)-β-D-GlcNAc-(1→3)-β-D-Gal-(1→4)-β-D-Glc-O-Cer varying in the ceramide component with sulfate ester groups on O-3 of the glucuronic acid moiety have been prepared as glucuronyl paraglobosides.[134]

4.3 Branched Pentasaccharides – A chemical procedure involving a 2 + 2 coupling followed by fucosylation has given a practical synthesis of α-D-NeuNAc-(2→3)-β-D-Gal-(1→4)-[α-L-Fuc-(1→3)]-β-D-GlcNAc-(1→3)-D-Gal having the siaLe^x tetrasaccharide β-(1→3)-linked to galactose.[135] An analogue, with the acetyl group of the glucosamine unit replaced by long-chain aromatic-containing polyamide substituents, has been incorporated into liposomes for examination as inhibitor of E-selectin mediated cell adhesion,[136] and in closely parallel work combined chemical and enzymic methods were used to produce the same pentasaccharide having benzoyl and β-naphthoyl substituents on the nitrogen atom.[137] A review has appeared on the 3′-sulfated Le^a pentasaccharide which is the most potent E-selectin ligand reported so far.[138]

The synthesis of α-NeuNAc-(2→3)-[β-D-Gal-(1→3)-β-D-GalNAc-(1→4)]-β-D-Gal-(1→4)-β-D-Glc-O-Cer, which is the ganglioside GM₁, has been reported by Schmidt's group,[139] and the lactosyl-*N*-acetyllactosaminyl galactose β-D-Gal-(1→4)-β-D-Glc-(1→6)-[β-D-Gal-(1→4)]-β-D-GlcNAc-(1→3)-β-D-GalOMe, having a lactyl ether bonded to O-3 of the branching galactose unit, has similarly been prepared. It is a compound which has been implicated in Group B *Streptococcal* infections.[140] Pentasaccharide β-D-Glc-(1→3)-β-D-Glc-(1→4)-[β-D-Xyl-(1→2)]-β-D-Qui-(1→2)-D-Xyl, carrying a methyl ether group at O-3 of the terminal glucose unit and sulfate esters at O-4 and O-6 of the reducing xylose unit and the two glucose units respectively, has been isolated from a marine saponin.[141]

5   Hexasaccharides

As has become customary in these volumes, an abbreviated method is now used for representing higher saccharides. Sugars will be numbered as follows, and linkages will be indicated in the usual way:

1 D-Glc*p*	**2** D-Man*p*	**3** D-Gal*p*
4 D-Glc*p*NAc	**5** D-Gal*p*NAc	**6** Neu*p*Ac
7 L-Rha*p*	**8** L-Fuc*p*	**9** D-Xyl*p*
10 D-Glc*p*NH$_2$	**11** D-Glc*p*A	**12** D-Qui (6-deoxy-D-glucopyranose)
13 L-*Glycero*-D-*manno*-heptose	**14** L-Ara*f*	**15** Kdn

5.1   Linear Hexasaccharides – The β-1,6-glucohexaose carrying 3-*O*-caproyl substituents on units 2, 4 and 6 has been made during studies of the lipooligo-saccharide HS-142-1, which is a novel non-peptide antagonist for the atrial natriuretic peptide receptor.[142,143] Different enzymes were used to cause partial hydrolysis of chitosan, and methods were developed for producing mixtures containing 72% of the hexamer with smaller amounts of the pentamer and heptamer.[144] The hexamer comprising six 1,4-linked units of 2,6-di-*O*-benzyl-3-*O*-methyl-α-D-mannose has been made during work on a *micobacterium* poly-saccharide,[145] and similar work on a further bacterial product led to the preparation of a hexamer comprising α-(1→2)-linked 4-(3-deoxy-L-*glycero*-tetro-namido)-4,6-dideoxy-α-D-mannopyranose, which is an *N*-acylated perosamine.[146] A maltohexaose derivative produced by opening *O*-methylated α-cyclodextrin is reported in reference 186.

A European and a Japanese group have completed the preparation of sialylgalactosylgloboside **10**.[147,148] The latter workers also made the isomer

(α6) 2 → 3 (β3) 1→ 3 (β5) 1→ 3 (α3) 1→ 4 (β3) 1→ 4 (β1)–Cer   **10**

having the sialic acid (2→6)-linked,[149] and a further one with the sialic acid (2→6)-linked and also galactose in place of *N*-acetylgalactosamine.[149] Hexasac-charide **11**, which corresponds to a fragment of the porcine zona pellucida

SO$_3$Na
6|
(β3) 1→ 4 (β4) 1→3 (β3) 1→ 4 (β4) 1→ 3 (β3) 1 → 3 (α5)–O–(CH$_2$)$_3$NH$_2$ **11**

glycoprotein that coats the oocyte and is able to inhibit the porcine sperm-oocyte interaction, was prepared,[150] and the synthesis of compound **12**, which is the

(α8) 1→ 2 (β3) 1→ 3 (β5) 1→ 3 (α3) 1→ 4 (β3) 1→ 4 (3)   **12**

hexasaccharide moiety of Globo H antigen, has been reported.[151] Using their glycal technology Danishefsky's group has put together a very similar antigen which is a substance very heavily expressed on human breast tumours. The compound produced was a fully-substituted derivative of **13** with a sulfonamido

(α8) 1→ 2 (β3) 1→ 3 (α5) 1→ 3 (α3) 1→ 4 (β3) 1→ 4–D-glucal   **13**

group in place of the *N*-acetylamino group of the galactosamine unit and a glycal in the reducing terminal position.[152]

5.2 Branched Hexasaccharides – Compounds of this set with glucose at the reducing end to have been prepared are siaLe^x ganglioside compounds **14**, with sulfate ester groups separately at the 6-positions of residues 3 and 4,[153] **15**, which

(α6) 2→ 3 (β3) 1→4 (β4) 1→ 3 (β3) 1→ 4 (β3) 1 → 3 (α5)–O–Cer   **14**

with branch:
```
        (α8)
         1
         ↓
         6
```

is a hexasaccharide related to a *Streptococcus* capsular polysaccharide component and has an *O*-lactyl substituent at O-3 of the branching galactose moiety,[154] and compound **16**, a fragment of a further streptococcal polymer.[155]

15
```
        (β3)
         1
         ↓
         4
(β3) 1→ 4 (β1) 1→6 (β4) 1→ 3 (β3) 1→ 4 (β1)–O⌒⌒NH₂
```

16   (α6) 2→ 3 (β3) 1→4 (β4) 1→ 3 (β3) 1→ 4 (1)
```
                    (β1)
                     1
                     ↓
                     4
```

Mannose-terminating compounds to have been described are **17**, which is the Le^x determinant linked to a linear mannotriose,[156] and **18** with two lactosamine

17
```
        (β8)
         1
         ↓
         3
(β3) 1→ 4 (β4) 1→6 (α2) 1→ 6 (β2) 1→ 6 (2)
        (β4)
         1
         ↓
         3
(α7) 1→ 2 (α7) 1→3 (α7) 1→ 2 (α7)   19
        (β4)
         1
         ↓
         3
```

18   (β3) 1→ 4 (β4) 1→2 (α2) 1→ 6 (α2)
```
        (β3) 1→ 4 (β4)
                   1
                   ↓
                   6
```

units linked to a mannobiose.[157,158] The haemolytic streptococcal group A cell-wall polysaccharide unit **19** has also been prepared.[159]

6   Heptasaccharides

Pokeweed lectin-B was determined by spectroscopic methods alone to have structure **20**.[160] A maltoheptaose derivative, produced by opening O-methylated β-cyclodextrin, is reported in reference 186.

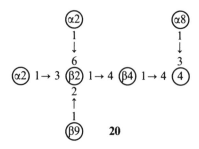

7   Octasaccharides

7.1   Linear Octasaccharides – A convergent procedure based on trichloroaceti-midate coupling has been used to prepare cellooctaose.[161] A malto-octaose derivative, produced by opening O-methylated γ-cyclodextrin, is reported in reference 186. Solid phase methods were employed to make the β-1,3-tetramer of lactosamine,[162] and an iterative approach was used in the preparation of the tetramer of β-D-GlcNAc-(1→4)-β-D-GlcA-(1→3) giving an octamer section of hyaluronic acid.[163]

7.2   Branched Octasaccharides – Syntheses have been recorded of compounds **21** and **22**, which are the octamers of the lipopolysaccharide of *Moraxella* *catarrhalis*[164] and the VIM-2 ganglioside isolated from human myelogenous leukemia cells.[165] Compound **23** is the octasaccharide of the core fucosylated

biantennary glycoprotein of serum and it has been chemically synthesized,[166] and **24**, the novel β-galactofuranose-containing high mannose oligosaccharide of an

enzyme from a bacterium, has been identified by spectroscopic methods. Several variations of this octamer were also reported.[167]

8   Higher Saccharides

Compounds **25** and **26** were used as the key reactants in the synthesis of the branched mannononaose **27**, the acetal ring system of the former disarming the phenylselenyl group to allow specific reaction of **26** as glycosylating agent.[168]

Oligomerization of the sulfur-linked cellobiosyl fluoride **28** by use of a cellulase gave a set of products with even numbers of glucose residues of general structure **29**, lower members of which, up to the decaose, were separated. Higher members of the series were obtained as a mixture.[169] The branched chain decasaccharide **30**, with and without C-13 labelling in the acetyl group of the sialic acid units,

25   **26**   **28**

29

(α6) 2→ 3 (β3)   (α6) 2→ 3 (β3)
                 1   1
                 ↓   ↓
                 4   4
(β1) 1→ 6 (β4) 1→ 3 (β3) 1→ 4 (β1) 1→ 6 (β4) 1→ 3 (1)   **30**

and an analogue with propanoyl groups instead of acetyl groups on these units, have been made.[170]

In the undecaose set, compounds **31** and **32** have been prepared by chemical

(α2) 1→ 2 (α2)
                 1
                 ↓
                 6
(α2) 1→ 2 (α2) 1→ 3 (α2)
                          1
                          ↓
                          6
(α2) 1→ 2 (α2) 1→ 2 (α2) 1→ 3 (β2) 1→ 4 (β4) 1→ 4 (4)   **31**

(α6) 2→ 3 (β3) 1→ 4 (α4) 1→ 2 (α2)
                                      1
                                      ↓
                                      6
(α6) 2→ 3 (β3) 1→ 4 (α4) 1→ 2 (α2) 1→ 3 (β2) 1→ 4 (β4) 1→ 4 (β4) (4) **32**

synthesis, some couplings occurring with remarkable efficiencies. In the case of the latter oligomer, which was prepared in protected form, difficulties were found with deprotection.[171] Enzymic transglycosylation of compound **31** allowed the generation of a glycoside with an aglycon having a spacer group linked to an acryloyl group which was copolymerized with acrylamide. The polymeric product was much more efficient as an inhibitor of glycoprotein binding than was **31** itself.[172]

By careful control of reaction conditions and reactant proportions, 2-deoxy-maltooligosaccharides have been obtained in yields of about 45% starting from D-glucal with a phosphorylase, inorganic phosphate and a primer such as maltotetraose. The chain lengths were dependent upon reaction times, with the degree of polymerization being about 12 after 3-4 h and 20 after one day.[173]

The tridecacompound **33**, which contains two siaLe^x tetrasaccharide units was put together relatively readily by enzymic methods. The reducing end of the initial oligosaccharide required for the synthesis was linked to asparagine.[174]

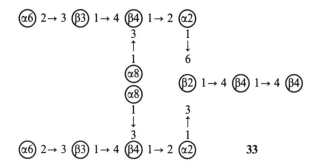

Mass spectrometric methods are now able to assist with specific characterization of relatively long-chain oligomers, and the MALDI-TOF technique has been used in the investigation of a cell wall xylan linked by β-(1→3)-bridges and with chain length of 25. Such analysis can be carried out on impure samples.[175]

Increasingly, enzymes are being used in the synthesis of oligosaccharides, and in this type of work normally ranges of products of different chain lengths are obtained. Oligosaccharides to have been synthesized in this way are: α-1,2-linked glucooligomers,[176] and β-1,4-linked *N*-acetylglucosamine compounds, that is, artificial chitin. Interestingly, the substrate for this reaction was an oxazoline derived from chitobiose.[177] In analogous, but purely chemical synthetic work, oligomers of DP~20 corresponding to fragments of cellulose have been generated from 3,6-di-*O*-benzyl-α-D-glucopyranose 1,2,4-orthopivaloate by oligomerisation using trityl tetrafluoroborate which gave β-(1→4)-linked 2-*O*-pivaloyl products.[178]

9   Cyclodextrins

This report concentrates almost exclusively on the chemistry of the cyclodextrins themselves and disregards the large amount of material that has been reported on the properties of these compounds as binding hosts, the properties of the complexes they form, and their effects as reaction catalysts.

A novel enzymic synthesis of cyclodextrins uses a specific enzyme from *Bacillus macerans* and employs aqueous organic solvent mixtures. No product was obtained in water, but in a mixture of water and cyclohexane, 44% of β-cyclodextrin was obtained. Extraordinarily, acyclic alkanes used with water did not permit cyclodextrin formation, but benzene, xylene and chloroform, ethanol, butanol and pentanol did permit such synthesis.[179] Cycloinulohexaose (otherwise α-cyclofructin) can be produced from inulin enzymically and has been shown to be a useful cryoprotective agent during freezing and freeze drying of liposomes, having a higher collapse temperature than does trehalose.[180] The electrostatic and lipophilicity potential profiles of the cyclofructin have been compared with those of the cyclodextrins. The former has no central cavity, but is well suited to the regioselective complexing and incorporation of metal ions.[181] Chemical methods have also been used to make cyclodextrin compounds. Tin tetrachloride-catalysed oligomerization of 1,2,2′,3,3′,4′,6-hexaacetyl-β-maltose and 6′-*O*-tritylmaltose heptaacetate has afforded the acetylated hexa-, octa- and decamers containing alternative α-(1→4)- and β-(1→6)-linkages.[182] Cyclic compounds derived from heterodisaccharides have also been reported, trityl perchlorate catalysing the oligomerization of compound **34** to give the 6- and 8-sugar-containing cyclic products based on α-L-Rha-(1→4)-α-D-Man-(1→4).[183]

34

35

Statistical analyses based on crystallographic data have been conducted on the cyclodextrins and the results were used to compile contact surfaces/molecular lipophilicity patterns.[184] Pyrolyses of cyclodextrins have been studied by use of a Curie-point pyrolyser.[185]

Considerable attention has been given to various cyclodextrin ethers. Thiolysis of fully methylated cyclodextrins, using phenylthiotrimethylsilane and zinc bromide in dichloroethane, caused cleavage of a single glycosidic bond and from the α-cyclodextrin ether 28% of the fully substituted corresponding hexaose *S*-phenyl thioglycoside was isolated following benzoylation of the liberated hydroxyl group on the terminal sugar unit. Similar results were obtained with α- and γ-cyclodextrin ethers.[186]

Treatment of the per-2,6-di-*t*-butylmethylsilyl ethers with sodium hydride and methyl iodide or allyl bromide in THF results surprisingly in alkylation at O-2, illustrating that the 2-silyl ether migrated to O-3 prior to ether formation. Following desilylation the products are per-2-*O*-alkyl cyclodextrins.[187] A range of specific β-cyclodextrin ethers, for example the per-2,3-di-*O*-methyl, per-2,3-di-*O*-methyl-6-*O*-*t*-butyldimethylsilyl- and 6-*t*-butyldimethylsilyl-β-cyclodextrins, have been used in studies of their immobilization on gold surfaces.[188] From mono-6-*O*-*t*-butyldimethylsilyl-β-cyclodextrin, the derivative carrying methyl ether groups at all positions except one primary position was obtained after desilylation.[189] NMR. and X-ray diffraction studies on methyl ethers of cyclodextrins are reported in Chapters 21 and 22 respectively.

Mono-2-hydroxypropyl β-cyclodextrin has been made directly in 28% yield using epoxypropane in alkaline conditions,[190] and glycolyl ether groups have been introduced at O-6 and separately at O-3, to give products with negatively charged species to examine the possibility that they would link electrostatically with ammonium group-containing cyclodextrins to give heterodimers in aqueous solution.[191]

Heptakis 6-*O*-*t*-butyldimethylsilyl-β-cyclodextrin, on reaction with 4-chloromethyl-*N*-methyl-2-nitroaniline, affords the mono-3-substituted benzyl ether.[192] β-Cyclodextrin carrying a 2-(naphthylmethyl) group at O-6 exists with the aromatic rings within the cavity to an extent which is very temperature dependent. Temperature can therefore be used to control the degree of complexing of the aromatic system with a fluorescent naphthalene compound.[193,194] Heptakis-[2,3-di-*O*-acetyl-6-deoxy-6-iodo]-β-cyclodextrin treated with 6-methoxycarbonyl-2-naphthol allowed access to the compound having naphthoic acid substituents at all of the primary positions, and this forms a very stable 1:1 complex with a merocyanine laser dye which is a mimic of the antenna function in photosynthesis and shows promise as a photochemical microreactor.[195] Mono-[6-*O*-(8-quinolyl)]-β-cyclodextrin has been reported, and the stabilities of inclusion complexes with amino acid 'guests' have been described.[196]

Selective substitution at the primary positions on two adjacent rings in cyclodextrins is now possible following the finding that 1,3-di-(chlorosulfonyl)-4,6-dimethoxybenzene forms a cyclic disulfonate with these hydroxyl groups on heptakis-2,3-dimethyl β-cyclodextrin. Subsequent straightforward reactions led to 6a,6b-di-*p*-allyloxyphenyl-β-cyclodextrin permethyl ether.[197]

An abbreviated route to heptakis 2,3-anhydro-D-*manno*-β-cyclodextrin has been reported,[198] and likewise a modified route to the per 3,6-anhydro-compound via the hepta-6-iodo derivative has been reported.[199] A paper which reports the mono-2,3-*allo*-anhydride and the mono-3,6-anhydride of β-cyclodextrin also reports the 2,3'-anhydro derivative with the anhydro ring linking two adjacent sugar units.[200] Other Japanese workers have reported the derivative **35** containing a benzylidene acetal ring also linking the same 2,3'-hydroxyl positions of the heptakis-6-*O*-pivaloate of β-cyclodextrin. Clearly this work opens many opportunities for making further selectively substituted compounds, and the authors prepared a nonadecyl-*O*-benzyl ether of β-cyclodextrin. By reductively ring-opening the benzylidene acetal, on the other hand, they produced a mono 2-substituted ether in good yield. The analogous 2,3′-acetals were reported of α- and γ-cyclodextrin.[201]

A bis-cyclic sulfonate spanning the primary positions of adjacent glucose units of β-cyclodextrin was referred to above, and a diphenyl-4,4′-disulfonyl reagent has been shown to link the primary positions of the A- and D-related glucose units. This, interestingly, was cleaved with taka-amylase A and gave an acyclic heptaglucose with a bridging link between sugars 2 and 5.[202] In related work 9,10-dicyano-2,6-(dichlorosulfonyl)anthracene was used as a bidentate capping agent with β-cyclodextrin. In the course of this work 9,10-dicyano-2-methyl-anthracene was linked through the methyl group to O-6 of β-cyclodextrin.[203] A large excess of toluene *p*-sulfinyl chloride was used to fully protect β-cyclodextrin with *p*-tolylsulfinyl groups and the product was used as a chiral stationary phase for chromatography.[204] β-Cyclodextrin has also been capped by use of diols linked through carbonate ester groups. They were made by use of carbonyldi-imidazole as carbonylating agent,[205] and the same cyclodextrin condensed with 7 molar equivalents of boric acid with removal of water gives a product which, it is believed, could involve esterification of pairs of secondary hydroxyl groups. It catalyses the hydroxylation of alkenes with *t*-butyl peroxide.[206]

Aminodeoxy derivatives of cyclodextrins seem increasingly to attract attention, and as an indication of their significance heptakis-6-amino-6-deoxy-β-cyclodex-trin catalyses hydrogen/duterium exchange in active hydrogen substances.[207] The diamino compound, having amino groups at the 6-positions of adjacent glucose rings, has been made following azide displacement of the relevant known disulfonate. X-ray analysis of the product was carried out.[208] The *N*-2-aminoethyl derivative of the mono-6-amino compound was studied as a host for anilino-naphthalene sulfonates,[209] and the diethylenetriamine analogue with an extended chain has also been reported.[210] The *N*-aminoethyl derivative mentioned above has been amide-coupled to the diketopiperazine dimer of aspartic acid to give a derivative with the side-chain **36**, which exists with the diketopiperazine held on the 6-surface of the cyclodextrin ring.[211] Related work has incorporated the *N*-acetyltyrosinyl and *N*-acetyltryptophanyl groups **37** and **38** at the nitrogen of 6-

36

37

38

amino-6-deoxy-cyclodextrin, and while the former is strongly self-included with the aromatic ring within the cyclodextrin cavity, the indole ring of the latter is only weakly self-included. These products were then examined as hosts for a range of compounds including some monoterpenes.[212] Closely related studies have used diastereoisomers prepared by attaching L- and D-tyrosine and N-formyl L- and D-phenylglycine.[213] Stoddart and co-workers have reported compounds carrying an amino-acid group at C-6 of each of the sugar rings of β-cyclodextrin, notably those having phenylalanine amide-coupled to the 6-amino-6-deoxy compound and cysteine S-coupled to C-6.[214] Other compounds to have been linked by amide coupling are 6-amino-6-deoxy-β-cyclodextrin and 6-amino-6-deoxy-β-glucose the amino groups of which were connected by way of a thiono group, the monosaccharide being in the form of the β-glycosylamine amide-bonded to glycine.[215] Novel urea coupling between two β-cyclodextrin molecules has been effected by treatment of the mono-6-azido-6-deoxy derivative with triphenylphosphine and carbon dioxide.[216]

Mono-6-anilino-6-deoxy-β-cyclodextrin has been described,[217] and the analogue with o-phenylenediamine groups N-bonded to C-6 of the A- and C-rings of α-cyclodextrin has allowed the preparation of the interesting compound **39** by condensation with the appropriate p-cresol dialdehyde. The product was extremely unsymmetrical with all six sugar rings being in different magnetic environments.[218] Opening of the epoxide ring of the mono-2,3-anhydro-allo-β-cyclodextrin with imidazole gave predominantly the 3-deoxy-3-imidazolyl D-glucose compound, and the manno-epoxide also afforded the product of attack at C-3 and gave mainly the 3-deoxy-3-imidazolyl-D-altro-compound.[219] The same workers then produced bisimidazolyl derivatives from the corresponding A,C-manno-diepoxides.[220]

Enzymic condensation, by use of cyclodextrin glycosyltransferases, of α-D-maltosyl fluoride, which has sulfur in place of the interunit oxygen atom, gave cyclic oligomers having 8-, 10- and 12-sugar units in 16%, 14% and 7% yield respectively. These analogues, therefore, had sulfur replacing oxygen in alternate interunit bonds.[221] Mono-6-thio-β-cyclodextrin and long chain sulfur-substituted derivatives have been made in connection with their immobilization on gold surfaces[188] and coupling of 6-deoxy-6-iodo-β-cyclodextrin with sodium sulfide has given a dimer with a sulfur linkage between the C-6 carbon atoms. In the same work coupling was effected between rings by use of epichlorohydrin.[222] Heptakis-[6-S-(2,3-dihydroxypropyl)-6-thio]-β-cyclodextrin and several related compounds have been prepared as well as corresponding sulfones. Some products have notably improved water solubility.[223] Compound **40** is a monothio-linked derivative of β-cyclodextrin having unique photochromic behaviour in exhibiting normal and reverse photochromism. In aqueous solution it existed with the aromatic ring system within the cyclodextrin cavity, but in the dark the molecule was 'open'.[224]

Per-6-bromo-6-deoxy derivatives of α-, β- and γ-cyclodextrin have been produced using triphenylphosphine and bromine in DMF, the yields being almost quantitative. Nucelophilic displacements then led to per-cyano-, azido- and sulfur-linked compounds in excellent yields. It has therefore become easy to

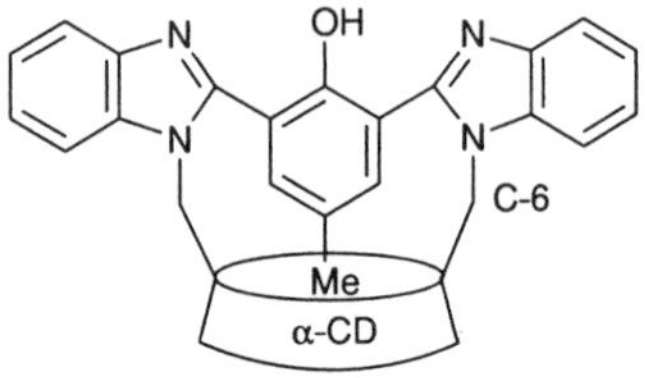

39 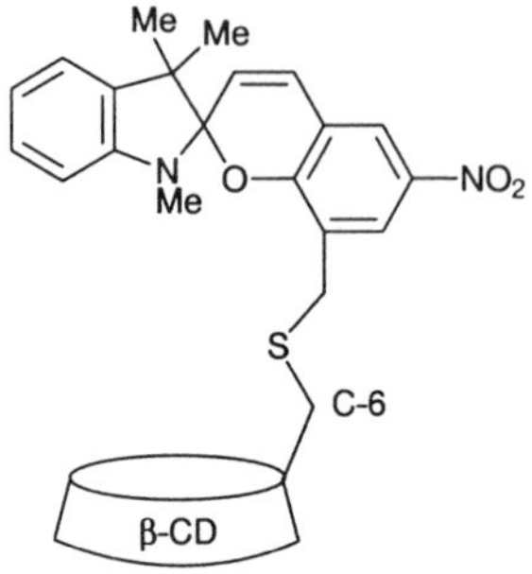   40

make per-primary-substituted cyclodextrin derivatives.[225] Maltotriosyl fluoride
having a 6-deoxy-6-iodo substituent on the non-reducing terminal unit underwent
enzymic cyclodimerization to give the $6^A,6^D$-dideoxy-$6^A,6^D$-diiodo-α-cyclodex-
trin.[226]

α-Cyclodextrin carrying an α-D-glucopyranosyl group on C-6 was enzymically
glycosylated to give each of the positional isomers of the monogalactosyl
derivatives having the new unit at the 2- or the 6-positions.[227]

References

1   Y. Ding, O. Kanie, J. Labbe, M. M. Palcic, B. Ernst and O. Hindsgaul, *Adv. Exp.
     Med. Biol.*, 1995, **376**, 261 (*Chem. Abstr.*, 1996, **125**, 11 262).
2   T. Kajimoto and Y. Ida, *Kobunshi*, 1996, **45**, 534 (*Chem. Abstr.*, 1996, **125**, 168 461).
3   G.-J. Boons, *Drug Discovery Today*, 1996, **1**, 331 (*Chem. Abstr.*, 1996, **125**, 276 299).
4   P. J. Garegg, *Med. Res. Rev.*, 1996, **16**, 345 (*Chem. Abstr.*, 1996, **125**, 143 136).
5   G.-J. Boons, *Tetrahedron*, 1996, **52**, 1095.
6   G.-J. Boons, *Contemp. Org. Synth.*, 1996, **3**, 173 (*Chem. Abstr.*, 1996, **125**, 143 135).
7   S. Hashimoto and M. Nakajima, *Farumashia*, 1996, **32**, 1375 (*Chem. Abstr.*, 1996,
     125, 301 326).
8   G. Russo and L. Panza, *Trends Org. Chem.*, 1993, **4**, 191 (*Chem. Abstr.*, 1996, **125**,
     58 864).
9   F. Baressi and O. Hindsgaul, *Mod. Synth. Methods*, 1995, **7**, 281 (*Chem. Abstr.*,
     1996, **124**, 146 589).
10   D. M. Whitfield and S. P. Douglas, *Glycoconjugate J.*, 1996, **13**, 5 (*Chem. Abstr.*,
     1996, **124**, 290 000).
11   S. J. Danishefsky and M. T. Bilodeau, *Angew. Chem. Int. Ed. Engl.*, 1996, **35**, 1381.
12   C. H. Wong, *Pure Appl. Chem.*, 1995, **67**, 1609 (*Chem. Abstr.*, 1996, **124**, 87 494).
13   H. J. M. Gijsen, L. Qiao, W. Fitz and C.-H. Wong, *Chem. Rev.*, 1996, **96**, 443
     (*Chem. Abstr.*, 1996, **124**, 87 497).
14   J. W. Yun and S. K. Song, *Kongop Hwahak*, 1994, **5**, 561 (*Chem. Abstr.*, 1996, **124**, 9
     101).
15   T. Usui and T. Murata, *Yuki Gosei Kagaku Kyokaishi*, 1996, **54**, 607 (*Chem. Abstr.*,
     1996, **125**, 114 956).

16   Y. Iwabuchi, Y. Sugihara, S. Fukuyama and I. Rujii, *Tennen Yuki Kagobutsu Toronkai Koen Yoshishu*, 1995, 150 (*Chem. Abstr.*, 1996, **124**, 232 923).

17   C. H. Scaman, O. Hindsgaul, M. M. Palcic and O. P. Srivastava, *Carbohydr. Res.*, 1996, **296**, 203.

18   L. Jiang, R. C. Hartley and T.-H. Chan, *J. Chem. Soc., Chem. Commun.*, 1996, 2193.

19   C. Roberts, C. L. May and B. Fraser-Reid, *Carbohydr. Lett.*, 1994, **1**, 89 (*Chem. Abstr.*, 1996, **125**, 87 023).

20   T. Ziegler, R. Dettmann, M. Duszenko and V. Kolb, *Carbohydr. Res.*, 1996, **295**, 7.

21   R. B. Singh, *Asian J. Chem.*, 1995, **7**, 475 (*Chem. Abstr.*, 1996, **124**, 9 171).

22   J. C. Castro Palomino, M. Hernandez Rensoli and V. Verez Bencomo, *J. Carbohydr. Chem.*, 1996, **15**, 137.

23   T. Takeda, T. Kanemitsu, N. Shimizu, Y. Ogihara and M. Matsubara, *Carbohydr. Res.*, 1996, **283**, 81.

24   A. K. Misra and N. Roy, *Indian J. Chem., Sect. B: Org. Chem. Incl. Med. Chem.*, 1995, **34B**, 865 (*Chem. Abstr.*, 1996, **124**, 176 681).

25   J. Biton, J.-M. Michel, D. Le Beller, V. Pelenc, F. Paul, P. F. Monsan and G. Gellf, *Ann. N. Y. Acad. Sci.*, 1995, **750**, 321 (*Chem. Abstr.*, 1996, **124**, 117 734).

26   G. Vic, M. Scigelova, J. J. Hastings, O. W. Howarth and D. H. G. Crout, *J. Chem. Soc., Chem. Commun.*, 1996, 1473.

27   S. Takayama, M. Shimazaki, L. Qiao and C.-H. Wong, *Bioorg. Med. Chem. Lett.*, 1996, **6**, 1123.

28   K. Fukase, T. Yasakochi, Y. Suda, M. Yoshida and S. Kusumoto, *Tetrahedron Lett.*, 1996, **37**, 6763.

29   Z. Zhang and G. Magnusson, *J. Org. Chem.*, 1996, **61**, 2383.

30   T. Ziegler and G. Schuele, *J. Prakt. Chem/Chem.-Ztg.*, 1996, **338**, 238 (*Chem. Abstr.*, 1996, **125**, 33 969).

31   A. K. Choudhury and N. Roy, *Synlett.*, 1996, 3937.

32   A. K. Misra, S. Basu and N. Roy, *Synth. Commun.*, 1996, **26**, 2857.

33   J. A. L. M. van Dorst, C. J. van Iteusden, A. F. Voskamp, J. P. Kamerling and J. F. G. Vliegenthart, *Carbohydr. Res.*, 1996, **291**, 63.

34   S. Singh, M. Scigelova, G. Vic and D. H. G. Crout, *J. Chem. Soc., Perkin Trans. 1*, 1996, 1921.

35   G.-J. Boons, B. Heskamp and F. Hout, *Angew. Chem. Int. Ed. Engl.*, 1996, **35**, 2845.

36   Z.-G. Wang, S. P. Douglas and J. J. Krepinsky, *Tetrahedron Lett.*, 1996, **37**, 6985.

37   I. Matsuo, M. Isomura, R. Walton and K. Ajisaka, *Tetrahedron Lett.*, 1996, **37**, 8795.

38   S. Singh, M. Scigelova and D. H. G. Crout, *J. Chem. Soc., Chem. Commun.*, 1996, 993.

39   L. Ma, G. Tu, S. Chen, R. Zhang, L. Lai, X. Xu and Y. Tang, *Carbohydr. Res.*, 1996, **281**, 35.

40   I. Robina, E. López-Barba and J. Fuentes, *Tetrahedron*, 1996, **52**, 10771

41   Y. Matahira, A. Tashiro, T. Sato, H. Kawagishi and T. Usui, *Glycoconjugate J.*, 1995, **12**, 664 (*Chem. Abstr.*, 1996, **124**, 87 614).

42   E. Kaji, Y. Osa, M. Tanaike, Y. Hosokawa, H. Takayanagi and A. Takada, *Chem. Pharm. Bull.*, 1996, **44**, 437.

43   S. Koto, T. Miura, M. Hirooka, A. Tomaru, M. Iida, M. Kanemitsu, K. Takenaka, S. Masuzawa, S. Miyaji, N. Kuroyanagi, M. Yagashita, S. Zen, K. Yago and F. Tomonaga, *Bull. Chem. Soc. Jpn.*, 1996, **69**, 3247.

44   V. Martichonok and G. M. Whitesides, *J. Am. Chem. Soc.*, 1996, **118**, 8187.

45   T. Tomoo, T. Kondo, H. Abe, S. Tsukamoto, M. Isobe and T. Goto, *Carbohydr. Res.*, 1996, **284**, 207.

46   A. Hasegawa, N. Suzuki, H. Ishida and M. Kiso, *J. Carbohydr. Chem.*, 1996, **15**, 623.
47   A. Hasegawa, N. Suzuki, F. Kozawa, H. Ishida and M. Kiso, *J. Carbohydr. Chem.*, 1996, **15**, 639.
48   S. Cao and R. Roy, *Tetrahedron Lett.*, 1996, **37**, 3421.
49   M. Yoshida, H. Ishida, M. Kiso and A. Hasegawa, *Carbohydr. Res.*, 1996, **280**, 331.
50   T. Sugata and R. Higuchi, *Tetrahedron Lett.*, 1996, **37**, 2613.
51   G. Baisch, R. Öhrlein, M. Streiff and B. Ernst, *Bioorg. Med. Chem. Lett.*, 1996, **6**, 755.
52   V. P. Kamath and O. Hindsgaul, *Carbohydr. Res.*, 1996, **280**, 323.
53   J. Ø. Duus, N. Nifant'ev, A. S. Shashkov, E. A. Khatuntseva and K. Bock, *Carbohydr. Res.*, 1996, **288**, 25.
54   S. J. Danishefsky, J. T. Randolph and J. Y. Roberge, *Polym. Prepr. (Am. Chem. Soc., Div. Polym. Chem.)*, 1994, **35**, 977 (*Chem. Abstr.*, 1996, **124**, 9 152).
55   K. Zegelaar-Jaarsfeld, S. C. van der Plas, G. A. van der Marel and J. H. van Boom, *J. Carbohydr. Chem.*, 1996, **15**, 591.
56   M.-Z. Liu, Z.-W. Guo and Y.-Z. Hui, *Carbohydr. Lett.*, 1995, **1**, 387 (*Chem. Abstr.*, 1996, **124**, 290 050).
57   W. R. Roush, K. Briner, B. S. Kesler, M. Murphy and D. J. Gustin, *J. Org. Chem.*, 1996, **61**, 6098.
58   J. A. Hunt and W. R. Roush, *J. Am. Chem. Soc.*, 1996, **118**, 9998.
59   A. Sobti, K. Kim and G. A. Sulikowski, *J. Org. Chem.*, 1996, **61**, 6.
60   P.-E. Jansson, A. Kjellberg, T. Rundlöf and G. Widmalm, *J. Chem. Soc., Perkin Trans. 2*, 1996, 33.
61   D. D. Manning, C. R. Bertozzi, S. D. Rosen and L. L. Kiessling, *Tetrahedron Lett.*, 1996, **37**, 1953.
62   R. Roy, W. K. C. Park, O. P. Srivastava and C. Foxall, *Bioorg. Med. Chem. Lett.*, 1996, **6**, 1399.
63   W. Spevak, C. Foxall, D. H. Charych, F. Dasgupta and J. O. Nagy, *J. Med. Chem.*, 1996, **39**, 1018.
64   L. Yan and D. Kahne, *J. Am. Chem. Soc.*, 1996, **118**, 9239.
65   N. Imazaki, H. Koike, H. Miyauchi and M. Hayashi, *Bioorg. Med. Chem. Lett.*, 1996, **6**, 2043.
66   G. Baisch, R. Öhrlein and A. Katopodis, *Bioorg. Med. Chem. Lett.*, 1996, **6**, 2953.
67   G. Thoma, F. Schwarzenbach and R. O. Duthaler, *J. Org. Chem.*, 1996, **61**, 514.
68   E. A. Khatuntseva, A. S. Shashkov and N. E. Nifant'ev, *Bioorg. Khim.*, 1996, **22**, 376 (*Chem. Abstr.*, 1996, **125**, 301 370).
69   J. C. Prodger, M. J. Bamford, M. I. Bird, P. M. Gore, D. S. Holmes, R. Priest and V. Saez, *Bioorg. Med. Chem.*, 1996, **4**, 793 (*Chem. Abstr.*, 1996, **125**, 222 303).
70   M. J. Bamford, M. Bird, P. M. Gore, D. S. Holmes, R. Priest, J. C. Prodger and V. Saez, *Bioorg. Med. Chem. Lett.*, 1996, **6**, 239.
71   A. Liu, K. Dillon, R. M. Campbell, D. C. Cox and D. M. Huryn, *Tetrahedron Lett.*, 1996, **37**, 3785.
72   G. Dekany, K. Wright, P. Ward and I. Toth, *J. Carbohydr. Chem.*, 1996, **15**, 383.
73   R. Blattner, R. H. Furneaux and G. P. Lynch, *Carbohydr. Res.*, 1996, **294**, 29.
74   K. C. Nicolaou, B. M. Smith, K. Ajito, H. Komatsu, L. Gomez-Paloma and Y. Tor, *J. Am. Chem. Soc.*, 1996, **118**, 2303.
75   S. J. Danishefsky and M. D. Shair, *J. Org. Chem.*, 1996, **61**, 16.
76   M.-O. Contour-Galcera, J.-M. Guillot, C. Ortiz-Mellet, F. Pfleiger-Carrara, J. Defaye and J. Gelas, *Carbohydr. Res.*, 1996, **281**, 99.

77   M. Izumi, O. Tsuruta, H. Hashimoto and S. Yazawa, *Tetrahedron Lett.*, 1996, **37**, 1809.
78   D. P. Sutherlin and R. W. Armstrong, *J. Am. Chem. Soc.*, 1996, **118**, 9802.
79   T. Nishimura and F. Nakatsubo, *Tetrahedron Lett.*, 1996, **37**, 9215.
80   H. P. Wessel, N. Iberg, M. Trumtel and M.-C. Viaud, *Bioorg. Med. Chem. Lett.*, 1996, **6**, 27.
81   H. P. Wessel, M. Trumtel and R. Minder, *J. Carbohydr. Chem.*, 1996, **15**, 523.
82   H. P. Wessel, M.-C. Viaud and M. Trumtel, *J. Carbohydr. Chem.*, 1996, **15**, 769.
83   P. Lei, Y. Ogawa and P. Kovac, *Carbohydr. Res.*, 1996, **281**, 47.
84   I. Kitagawa, N. I. Baek, K. Kawashima, Y. Yokokawa, M. Yoshikawa, K. Ohashi and H. Shibuya, *Chem. Pharm. Bull.*, 1996, **44**, 1680.
85   S.-i. Aiba, *Chitin Chitosan, [Proc., Asia-Pac. Chitin Chitosan Symp.]*, 1994 (Pub. 1995), 119 (*Chem. Abstr.*, 1996, **124**, 117 749).
86   T. Kondo, T. Tomoo, H. Abe, M. Isobe and T. Goto, *Chem. Lett.*, 1996, 337.
87   T. Kondo, T. Tomoo, H. Abe, M. Isobe and T. Goto, *J. Carbohydr. Chem.*, 1996, **15**, 857.
88   S. Takayama, P. O. Livingston and C.-H. Wong, *Tetrahedron Lett.*, 1996, **37**, 9271.
89   U. S. Chowdhury, *Tetrahedron*, 1996, **52**, 12775.
90   X. X. Zhu, P. Y. Ding and M. S. Cai, *Carbohydr. Res.*, 1996, **296**, 229.
91   J. Ning, C. Li and S. Wang, *Beijing Daxue Xuebao, Ziran Kexueban*, 1995, **31**, 684 (*Chem. Abstr.*, 1996, **125**, 11 253).
92   M. Bah and R. Pereda-Miranda, *Tetrahedron*, 1996, **52**, 13063.
93   G. V. Reddy, R. K. Jain, R. D. Locke and K. L. Matta, *Carbohydr. Res.*, 1996, **280**, 261.
94   T. Toida, I. R. Vlahof, A. E. Smith, R. E. Hileman and R. J. Linhardt, *J. Carbohydr. Chem.*, 1996, **15**, 351.
95   K. Zegelaar-Jaarsveld, H. I. Duynstee, G. A. van der Marel and J. H. van Boom, *Tetrahedron*, 1996, **52**, 3575.
96   K. Zegelaar-Jaarsveld, S. A. W. Smits, N. C. R. van Straten, G. A. van der Marel and J. H. van Boom, *Tetrahedron*, 1996, **52**, 3593.
97   G. Schüle and T. Ziegler, *Tetrahedron*, 1996, **52**, 2925.
98   S. Srivastav, D. Deepak and A. Khare, *Trends Carbohydr. Chem., Carbohydr. Conf., 9th*, 1993 (Pub. 1995), 61 (*Chem. Abstr.*, 1996, **124**, 343 854).
99   J. Tamura, K. W. Neumann and T. Ogawa, *Liebigs Ann. Chem.*, 1996, 1239.
100 M.-O. Contour-Galcera, C. Ortiz-Mellet and J. Defaye, *Carbohydr. Res.*, 1996, **281**, 119.
101 Y. Wada, T. Saito, N. Matsuda, H. Ohmoto, K. Yoshino, M. Ohashi, H. Kondo, H. Ishida, M. Kiso and A. Hasegawa, *J. Med. Chem.*, 1996, **39**, 2055.
102 Y. Kajihara, T. Yamamoto, H. Nagae, M. Nakashizuka, T. Sakakibara and I. Terada, *J. Org. Chem.*, 1996, **61**, 8632.
103 M. Yoshida, T. Suzuki, H. Ishida, M. Kiso and A. Hasegawa, *J. Carbohydr. Chem.*, 1996, **15**, 147.
104 M. Freder, G. Reznicek, J. Jurinitsch, W. Kubelka, W. Schmidt, M. Schubert-Zsilavecz, E. Hastinger and J. Reiner, *Helv. Chim. Acta*, 1996, **79**, 385.
105 O. Kanie, Y. Ito and T. Ogawa, *Tetrahedron Lett.*, 1996, **37**, 4551.
106 M. Yoshida, Y. Kawakami, H. Ishida, M. Kiso and A. Hasegawa, *J. Carbohydr. Chem.*, 1996, **15**, 399.
107 A. Arasappan and B. Fraser-Reid, *J. Org. Chem.*, 1996, **61**, 2401.
108 N. Nifant'ev, Y. E. Tsvetkov, A. S. Shashkov, L. O. Kononov, V. M. Menshov, A. B. Tuzikov and N. V. Bovin, *J. Carbohydr. Chem.*, 1996, **15**, 939.

109 H. Ohmoto, K. Nakamura, T. Inoue, N. Kondo, Y. Inoue, K. Yoshino, H. Kondo, H. Ishida, M. Kiso and A. Hasegawa, *J. Med. Chem.*, 1996, **39**, 1339.

110 M. Hayashi, M. Tanaka, M. Itoh and H. Miyauchi, *J. Org. Chem.*, 1996, **61**, 2938.

111 U. Sprengard, H. Kunz, C. Hüls, W. Schmidt, D. Seiffge and G. Kretzschmar, *Bioorg. Med. Chem. Lett.*, 1996, **6**, 509.

112 U. Sprengard, M. Schudok, W. Schmidt, G. Kretzschmar and H. Kunz, *Angew. Chem., Int. Ed. Engl.*, 1996, **35**, 321.

113 R. K. Jain, R. Vig, R. D. Locke, A. Mohammad and K. L. Malta, *J. Chem. Soc., Chem. Commun.*, 1996, 65.

114 G. Baisch, R. Öhrlein, A. Katopodis and B. Ernst, *Bioorg. Med. Chem. Lett.*, 1996, **6**, 759.

115 Y. Nakahara, H. Iijima and T. Ogawa, *Carbohydr. Lett.*, 1994, **1**, 99 (*Chem. Abstr.*, 1996, **125**, 115 015).

116 J. S. Debenham, R. Rodebaugh and B. Fraser-Reid, *J. Org. Chem.*, 1996, **61**, 6478.

117 J. C. McAuliffe, R. V. Stick and B. A. Stone, *Tetrahedron Lett.*, 1996, **37**, 2479.

118 M. Kiso, H. Furui, H. Ishida and A. Hasegawa, *J. Carbohydr. Chem.*, 1996, **15**, 1.

119 L.-X. Wang and Y. C. Lee, *J. Chem. Soc., Perkin Trans. 1*, 1996, 581.

120 T. Eisele, A. Toepfer, G. Kretzshmar and R. R. Schmidt, *Tetrahedron Lett.*, 1996, **37**, 1389.

121 A. Ernst, L. Gobbi and A. Vasella, *Tetrahedron Lett.*, 1996, **37**, 7959.

122 A. V. Nikolaev, T. J. Rutherford, M. A. J. Ferguson and J. S. Brimacombe, *J. Chem. Soc., Perkin Trans. 1*, 1996, 1559.

123 E. P. Dubois, J. B. Robbins and V. Pozsgay, *Bioorg. Med. Chem. Lett.*, 1996, **6**, 1387.

124 K. Katsuraya, K. Inazawa, H. Nakashima and T. Uryu, *Front. Biomed. Biotechnol.*, 1996, **3**, 195 (*Chem. Abstr.*, 1996, **125**, 222 308).

125 H. P. Wessel, R. Minder and G. Englert, *J. Carbohydr. Chem.*, 1996, **15**, 201.

126 H. P. Wessel and N. Iberg, *Bioorg. Med. Chem. Lett.*, 1996, **6**, 427.

127 J. Rademann and R. R. Schmidt, *Tetrahedron Lett.*, 1996, **37**, 3989.

128 K. Tokuyasu, M. Ohnishi-Kameyama and K. Hayashi, *Biosci. Biotech. Biochem*, 1996, **60**, 1598.

129 S. A. Nepogod'ev, L. V. Backinowsky and N. K. Kochetkov, *Izv. Akad. Nauk. Ser. Khim.*, 1993, 1480 (*Chem. Abstr.*, 1996, **125**, 196 135).

130 J. Kovensky, P. Duchaussoy, M. Petitou an P. Sinay, *Tetrahedron: Asymm.*, 1996, **7**, 3119.

131 J. Kovensky, P. Duchaussoy, M. Petitou and P. Sinay, *Carbohydr. Lett.*, 1996, **2**, 73 (*Chem. Abstr.*, 1996, **125**, 222 302).

132 P. Westerduin, J. E. M. Bastén, M. A. Broekhoven, V. de Kimpe, W. H. A. Kuijpers and C. A. A. van Boekel, *Angew. Chem. Int. Ed. Engl.*, 1996, **35**, 331.

133 P. D. J. Grootenhuis, P. Westerduin, D. Meuleman, M. Petitou and C. A. A. van Boeckel, *Nat. Struct. Biol.*, 1995, **2**, 736 (*Chem. Abstr.*, 1996, **124**, 56 502).

134 Y. Isogai, T. Kawasa, H. Ishida, M. Kiso and A. Hasegawa, *J. Carbohydr. Chem.*, 1996, **15**, 1001.

135 T. Kiyoi, Y. Nakai, H. Kondo, H. Ishida, M. Kiso and A. Hasegawa, *Bioorg. Med. Chem.*, 1996, **4**, 1167 (*Chem. Abstr.*, 1996, **125**, 329 206).

136 S. A. DeFrees, L. Phillips, L. Guo and S. Zalipsky, *J. Am. Chem. Soc.*, 1996, **118**, 6101.

137 J. Y. Ramphal, M. Hiroshige, B. Lou, J.-J. Gaudino, M. Hayashi, S. M. Chen, L. C. Chiang, F. C. A. Gaeta and S. A. DeFrees, *J. Med. Chem.*, 1996, **39**, 1357.

138 S. David, *Chemtracts: Org. Chem*, 1995, **8**, 166 (*Chem. Abstr.*, 1996, **124**, 56 416).

139   U. Greilich, R. Brescello, K.-H. Jung and R. R. Schmidt, *Liebigs Ann. Chem.*, 1996, 663.

140   W. Zou and H. J. Jennings, *J. Carbohydr. Chem.*, 1996, **15**, 257.

141   O. A. Drozdova, S. A. Avilov, A. I. Kalinovskii, V. A. Stonik, Y. M. Mil'grom and Y. V. Rashkes, *Khim. Prir. Soedin*, 1993, 369 (*Chem. Abstr.*, 1996, **124**, 56 437).

142   Y. Qiu, Y. Nakahara and T. Ogawa, *Biosci. Biotechnol., Biochem.*, 1996, **60**, 1308 (*Chem. Abstr.*, 1996, **125**, 301 367).

143   Y. Qiu, Y. Nakahara and T. Ogawa, *Biosci. Biotech. Biochem*, 1996, **60**, 1308.

144   S. Aiba and E. Muraki, *Adv. Chitin Sci.*, 1996, **1**, 192 (*Chem. Abstr.*, 1996, **125**, 33 973).

145   W. Liao and D. Lu, *Carbohydr. Res.*, 1996, **296**, 171.

146   Y. Ogawa, P.-s. Lei and P. Kováč, *Carbohydr. Res.*, 1996, **293**, 173.

147   J. M. Lassaletta, K. Carlsson, P. J. Garegg and R. R. Schmidt, *J. Org. Chem.*, 1996, **61**, 6873.

148   H. Ishida, R. Miyawaki, M. Kiso and A. Hasegawa, *Carbohydr. Res.*, 1996, **284**, 179.

149   H. Ishida, R. Miyawaki, M. Kiso and A. Hasegawa, *J. Carbohydr. Chem.*, 1996, **15**, 163.

150   N. M. Spijker, C. A. Keuning, M. Hooglugt, G. H. Veeneman and C. A. A. van Boeckel, *Tetrahedron*, 1996, **52**, 5945.

151   J. M. Lassaletta and R. R. Schmidt, *Liebigs Ann. Chem.*, 1996, 1417.

152   T. K. Park, I. J. Kim, S. Hu, M. t. Bilodeau, J. T. Randolph, O. Kwon and S. J. Danishefsky, *J. Am. Chem. Soc.*, 1996, **118**, 11488.

153   S. Kemba, H. Ishida, M. Kiso and A. Hasegawa, *Carbohydr. Res.*, 1996, **285**, C1.

154   W. Zou and H. J. Jennings, *J. Carbohydr. Chem.*, 1996, **15**, 279.

155   W. Zou and H. J. Jennings, *J. Carbohydr. Chem.*, 1996, **15**, 925.

156   R. K. Jain and K. L. Matta, *Carbohydr. Res.*, 1996, **282**, 101.

157   X.-X. Zhu, P. Y. Ding and M.-S. Cai, *Tetrahedron: Asymm.*, 1996, **7**, 2833.

158   X. X. Zhu, P. Y. Ding and M. S. Cai, *Tetrahedron Lett.*, 1996, **37**, 8549.

159   F.-I. Auzanneau, F. Forooghian and B. M. Pinto, *Carbohydr. Res.*, 1996, **291**, 21.

160   Y. Kimura, K.-i. Yamaguchi and G. Funatsu, *Biosci. Biotech. Biochem*, 1996, **60**, 537.

161   T. Nishimura and F. Nakatsubo, *Carbohydr. Res.*, 1996, **294**, 53.

162   H. Shimizu, Y. Ito, O. Kanie and T. Ogawa, *Bioorg. Med. Chem. Lett.*, 1996, **6**, 2841.

163   G. Blatter and J.-C. Jacquinet, *Carbohydr. Res.*, 1996, **288**, 109.

164   K. Ekelöf and S. Oscarson, *J. Org. Chem.*, 1996, **61**, 7711.

165   T. Ehara, A. Kameyama, Y. Yamada, H. Ishida, M. Kiso and A. Hasegawa, *Carbohydr. Res.*, 1996, **281**, 237.

166   J. Seifert and C. Unverzagt, *Tetrahedron Lett.*, 1996, **37**, 6527.

167   M. Ohta, S. Emi, H. Iwamoto, J. Hirose, K. Hiromi, H. Itoh, M. Shin, S. Murao and F. Matsuura, *Biosci. Biotech. Biochem*, 1996, **60**, 1123.

168   P. Grice, S. V. Ley, J. Pietruszka and H. W. M. Priepke, *Angew. Chem. Int. Ed. Engl.*, 1996, **35**, 197.

169   V. Moreau and H. Driguez, *J. Chem. Soc., Perkin Trans. 1*, 1996, 525.

170   W. Zou, J.-B. Brisson, Q.-L. Yang, M. van der Zwan and H. J. Jennings, *Carbohydr. Res.*, 1996, **295**, 209.

171   Y. Nakahara, S. Shibayama, Y. Nakahara and T. Ogawa, *Carbohydr. Res.*, 1996, **280**, 67.

172   J.-Q. Fan, M. S. Quesenberry, K. Takegawa, S. Iwahara, A. Kondo, I. Kato and Y. C. Lee, *J. Biol. Chem.*, 1995, **270**, 17730 (*Chem. Abstr.*, 1996, **124**, 9 206).

173   B. Evers and J. Thiem, *Starch/Staerke*, 1995, **47**, 434 (*Chem. Abstr.*, 1996, **124**, 146 636).
174   C.-H. Lin, M. Shimazaki, C.-H. Wong, M. Koketsu, R. L. Juneja and M. Kim, *Bioorg. Med. Chem.*, 1995, **3**, 1625 (*Chem. Abstr.*, 1996, **124**, 202 793).
175   T. Yamagaki, M. Maeda, K. Kanazawa, Y. Ishizuka and H. Nakanishi, *Biosci. Biotech. Biochem*, 1996, **60**, 1222.
176   M. Quirasco, A. Lopez-Munguia, V. Pelenc, M. Remaud, F. Paul and P. Monsan, *Ann. N. Y. Acad. Sci.*, 1995, **750**, 317 (*Chem. Abstr.*, 1996, **124**, 87 557).
177   S. Kobayashi, T. Kiyosada and S.-i. Shoda, *J. Am. Chem. Soc.*, 1996, **118**, 13113.
178   F. Nakatsubo, H. Kamitakahara and M. Hori, *J. Am. Chem. Soc.*, 1996, **118**, 1677.
179   T. Morita, N. Yoshida and I. Karube, *Appl. Biochem. Biotechnol.*, 1996, **56**, 311 (*Chem. Abstr.*, 1996, **125**, 11 252).
180   K. Ozaki and M. Hayashi, *Chem. Pharm. Bull.*, 1996, **44**, 2116.
181   S. Immel and F. W. Lichtenthaler, *Liebigs Ann. Chem.*, 1996, 39.
182   H. Driguez and J.-P. Utille, *Carbohydr. Lett.*, 1994, **1**, 125 (*Chem. Abstr.*, 1996, **125**, 87 024).
183   R. R. Ashton, C. L. Brown, S. Menzer, S. A. Nepogodiev, J. F. Stoddart and D. J. Williams, *Angew. Chem. Int. Ed. Engl.*, 1996, **35**, 580.
184   F. W. Lichtenthaler and S. Immel, *Liebigs Ann. Chem.*, 1996, 27.
185   M. Wada, S. Fujishige, S. Uchino and N. Oguri, *Kobunshi Ronbunshu*, 1996, **53**, 20 (*Chem. Abstr.*, 1996, **124**, 232 924).
186   N. Sakairi and H. Kuzuhara, *Carbohydr. Res.*, 1996, **280**, 139.
187   D. Icheln, B. Gehrcke, Y. Piprek, P. Mischnick, W. A. König, M. A. Dessoy and A. F. Morel, *Carbohydr. Res.*, 1996, **280**, 237.
188   G. Nelles, M. Weisser, R. Back, P. Wohlfart, G. Wenz and S. Mittler-Neher, *J. Am. Chem. Soc.*, 1996, **118**, 5039.
189   Z. Chen, J. S. Bradshaw and M. L. Lee, *Tetrahedron Lett.*, 1996, **37**, 6831.
190   A.-Y. Hao, L.-H. Tong, F.-S. Zhang and D.-S. Jin, *Hecheng Huaxue*, 1995, **3**, 369 (*Chem. Abstr.*, 1996, **124**, 290 063).
191   F. Guillo, B. Hamelin, L. Jullien, J. Canceill, J.-M. Lehn, L. De Robertis and H. Driguez, *Bull. Soc. Chim. Fr.*, 1995, **132**, 857 (*Chem. Abstr.*, 1996, **124**, 146 620).
192   S. Tran, P. Forgo and V. T. D'Souza, *Tetrahedron Lett.*, 1996, **37**, 8309.
193   S. R. McAlpine and M. A. Garcia-Garibay, *J. Am. Chem. Soc.*, 1996, **118**, 2750.
194   S. R. McAlpine and M. A. Garcia-Garibay, *J. Org. Chem.*, 1996, **61**, 8307.
195   L. Jullien, J. Canceill, B. Valeur, E. Bardez, J.-P. Lefèvre, J.-M. Lehn, V. Marchi-Artzner and R. Pansu, *J. Am. Chem. Soc.*, 1996, **118**, 5432.
196   Y. Liu, Y. M. Zhang and Y. T. Chen, *Chin. Chem. Lett.*, 1996, **7**, 523 (*Chem. Abstr.*, 1996, **125**, 222 283).
197   Z. Chen, J. S. Bradshaw, G. Yi, D. Pyo, D. R. Black, S. S. Zimmerman, M. L. Lee, W. Tong and V. T. D'Souza, *J. Org. Chem.*, 1996, **61**, 8949.
198   A. R. Khan, L. Barton and V. T. D'Souza, *J. Org. Chem.*, 1996, **61**, 8301.
199   P. R. Aston, G. Gattuso, R. Königer, J. F. Stoddart and D. J. Williams, *J. Org. Chem.*, 1996, **61**, 9553.
200   K. Fujita, Y. Okabe, K. Ohta, H. Yamamura, T. Tahara, Y. Nogami and T. Koga, *Tetrahedron Lett.*, 1996, **37**, 1825.
201   N. Sakairi, N. Nishi and S. Tokura, *Carbohydr. Res.*, 1996, **291**, 53.
202   K. Fujita, M. Mizouchi, A. Kiyooka, K. Koga and K. Ohta, *Tetrahedron Lett.*, 1996, **37**, 4035.
203   B. K. Hubbard, L. A. Beilstein, C. H. Heath and C. J. Abelt, *J. Chem. Soc., Perkin Trans. 2*, 1996, 1005.

204   C. Roussel, C. Popescu and L. Fabre, *Carbohydr. Res.*, 1996, **282**, 307.
205   F. Trotta, G. Moraglio, O. Zerbinati and A. Nonnato, *J. Inclusion Phenom. Mol. Recognit. Chem.*, 1995 (Pub. 1996), **23**, 269 (*Chem. Abstr.*, 1996, **125**, 222 280).
206   S. Bhat and S. Chandrasekaran, *Tetrahedron Lett.*, 1996, **37**, 3581.
207   W. H. Binder and F. M. Menger, *Tetrahedron Lett.*, 1996, **37**, 8963.
208   B. Di Blasio, S. Galdiero, M. Saviano, C. Pedone, E. Benedetti, E. Rizzarelli, S. Pedotti, G. Vecchio and W. A. Gibbons, *Carbohydr. Res.*, 1996, **282**, 41.
209   N. Ito, N. Yoshida and K. Ichakawa, *J. Chem. Soc., Perkin Trans. 2*, 1996, 965.
210   V. Cucinotta, F. D'Alessandro, G. Impellizzeri, G. Maccarrone, E. Rizzarelli and G. Vecchio, *J. Chem. Soc., Perkin Trans. 2*, 1996, 1785.
211   G. Impellizzeri, G. Pappalardo, E. Rizzarelli and C. Tringali, *J. Chem. Soc., Perkin Trans. 2*, 1996, 1435.
212   M. Eddaoudi, H. Parrot-Lopez, S. Frizon de Lamotte, D. Ficheaux, P. Prognon and A.W. Coleman, *J. Chem. Soc., Perkin Trans. 2*, 1996, 1711.
213   K. Takahashi and Furushoh, *Polym. J. (Tokyo)*, 1996, **28**, 458 (*Chem. Abstr.*, 1996, **125**, 58 894).
214   P. R. Ashton, R. Königer, J. F. Stoddart, D. Alker and V. D. Harding, *J. Org. Chem.*, 1996, **61**, 903.
215   J. M. García Fernández, C. O. Mellet, S. Maciejewski and J. Defaye, *J. Chem. Soc., Chem. Commun.*, 1996, 2741.
216   F. Sallas, J. Kovács, I. Pintér, L. Jicsinszky and A. Marsura, *Tetrahedron Lett.*, 1996, **37**, 4011.
217   Y. Liu, Y. M. Zhang and Y. T. Chen, *Chin. Chem. Lett.*, 1996, **7**, 207 (*Chem. Abstr.*, 1996, **124**, 343 861).
218   D.-Q. Yuan, K. Koga, M. Yamaguchi and K. Fujita, *J. Chem. Soc., Chem. Commun.*, 1996, 1943.
219   D.-Q. Yuan, K. Ohta and K. Fujita, *J. Chem. Soc., Chem. Commun.*, 1996, 821.
220   W.-H. Chen, D.-Q. Yuan and K. Fujita, *Tetrahedron Lett.*, 1996, **37**, 7561.
221   L. Bornaghi, J.-P. Utille, D. Penninga, A. K. Schmidt, L. Dijkhuizen, G. E. Schulz and H. Driguez, *J. Chem. Soc., Chem. Commun.*, 1996, 2541.
222   R. Breslow and B. Zhang, *J. Am. Chem. Soc.*, 1996, **118**, 8495.
223   H. H. Baer and F. Santoyo-Gonzalez, *Carbohydr. Res.*, 1996, **280**, 315.
224   F. Hamada, K. Hoshi, Y. Higuchi, K. Murai, Y. Akagami and A. Ueno, *J. Chem. Soc., Perkin Trans. 2*, 1996, 2567.
225   B. I. Gorin, R. J. Riopelle and G. R. J. Thatcher, *Tetrahedron Lett.*, 1996, **37**, 4647.
226   C. Apparu, S. Cottaz, C. Bosso and H. Driguez, *Carbohydr. Lett.*, 1995, **1**, 349 (*Chem. Abstr.*, 1996, **124**, 290 049).
227   Y. Okada, K. Koizumi and S. Kitahata, *Carbohydr. Res.*, 1996, **287**, 213.

5
Ethers and Anhydro-sugars

1 Ethers

1.1 Methyl Ethers – The 3-*O*-methyl-β-D-allopyranosyl- and the corresponding 6-deoxy- sugar units have been encountered as constituents of oxypregnane-oligosaccharide saponins from *Stephanotis lutchuensis*.[1] All the mono-*O*-methyl ethers of methyl α-isomaltoside have been prepared in order to study their relative binding affinity to glucoamylase.[2]

1.2 Other Alkyl and Aryl Ethers – A number of 3-*O*-aminoalkyl-1,2-*O*-isopro-pylidene-α-D-gluco-or xylo-furanoses have been synthesized and evaluated for antiviral activity.[3] Some 6-*O*-alkyl-α-D-glycopyranosides have been prepared by displacement of 6-*O*-tosylates with alkoxides,[4] while the regioselectivity of 3- or 5-*O*-ether formation on 1,2-*O*-isopropylidene-4,6-di-*O*-benzyl-*myo*-inositol has been shown to depend on the nature of the *O*-alkylating agent.[5] A facile synthesis of methyl 2-or 3-*O*-allyl-5-*O*-benzyl-β-D-ribofuranoside has been described.[6]

The synthesis of peptidomimetics based on a D-glucose or D-allose scaffold featured 2,3-di-*O*-alkyl-6-*O*-aryl-4-*O*-carboxymethyl-pyranose compounds,[7] while a muramic acid analogue has been used in the solid phase synthesis of peptido-glycan monomers for the generation of a combinatorial library.[8]

O-Cyanomethyl ethers have been shown to be versatile intermediates for making compounds bearing *O*-(2-aminoethyl), *O*-carboxymethyl, *O*-carboxami-domethyl and *O*-formamidinemethyl ether groups, useful in the study of sugar-protein interactions.[9] The 'DATE' protecting group, *O*-1,1-dianisyl-2,2,2-tri-chloroethyl ether, has been used as a protecting group for the 2'-OH group in ribonucleotide synthesis. Cleavage is effected using lithium cobalt(I) phthalocy-anine which induces reductive fragmentation.[10] *O*-Tributylstannylmethyl ethers have been lithiated (BuLi) and treated with electrophiles to give a range of *O*-alkyl and -substituted alkyl ethers.[11]

The synthesis of four diastereomeric lignin-carbohydrate complex model compounds **1** has been described,[12] while an amphiphilic porphyrin substituted with both 6-*O*-galactopyranosyl and cholesteryloxy units has been synthesized.[13]

A study on the selectivity of stannylene acetal-mediated alkylation of methyl 4,6-*O*-benzylidene-α-D-glucopyranoside has concluded that increased bulk of the alkyl groups on tin and a non-polar or no co-solvent increase the proportion of *O*-2 mono-ethers formed.[14] The tin-mediated allylation and benzylation of 1,2-*O*-

Carbohydrate Chemistry, Volume 30
© The Royal Society of Chemistry, 1998

1

isopropylidene-*myo*-inositol is discussed in Chapter 18. A rather complicated synthesis of allyl 2,6-di-*O*-benzyl-α-D-glucopyranoside involved an inversion of configuration at C-4 of allyl 2,6-di-*O*-benzyl-3-*O*-(4-methoxybenzyl)-α-D-galacto-pyranoside.[15] A general procedure for the phase transfer catalysed *O*-allylation of carbohydrate derivatives has employed allyl bromide in dichloromethane with 50% aqueous sodium hydroxide and tetrabutylammonium bromide as catalyst.[16] A lengthy synthesis of 3-*O*-allyl-L-fucopyranose from D-arabinose has been reported.[17] The *O*-benzylation of some carbohydrate primary-secondary vicinal diols with benzyl chloride in KOH/DMSO has been reported to give predomi-nantly the secondary mono-*O*-benzyl ether product in some cases.[18]

The use of DDQ in dichloromethane/water for the cleavage of primary allyl and benzyl ethers has been reported,[19] and anhydrous ferric chloride in dichloro-methane debenzylated sugar benzyl ethers, without affecting acetals, benzoates, phthalimides and glycosidic linkages.[20] 4-Methoxy- and 3,4-dimethoxy-benzyl ethers have been deprotected with catalyic DDQ and ferric chloride.[21] The combination of Ti(O^{i}Pr)$_4$ and BuMgCl has been used to cleave allyl ethers to the corresponding alcohols.[22]

Variable levels of asymmetric induction has been reported in the epoxidation of 3- and 5-*O*-alkenylglycofuranoses using MCPBA;[23] the [2+2]-cycloaddition of tosyl isocyanate to but-1-enyl ethers is covered in Chapter 10. The reductive cleavage of 4,6-*O*-(α-methylbenzylidine) acetals of α-D-glucopyranoside with LiAlH$_4$/AlCl$_3$ has afforded single regio- and stereo-selective isomers of the 4-*O*-or 6-*O*-α-methylbenzyl ethers depending on the stereochemistry of the starting acetal centre.[24]

A number of 15- and 18-membered ring mono-aza crown ethers incorporating ether bonded glucose or galactose residues have been synthesized and their complexing abilities with Li$^+$, Na$^+$, K$^+$ and NH$_4^+$ cations were measured.[25]

1.3 Silyl Ethers – Tin-mediated *tert*-butyldimethylsilylation of some methyl glycosides and D-glucal afforded exclusively the products of silylation at the primary hydroxy groups.[26] Silylation of alcohols (including hindered and tertiary alcohols) has been effected using *bis-N,O-(tert*-butyldimethylsilyl)acetamide in the presence of catalytic amounts of Bu$_4$NF.[27] Partial silylation of methyl and allyl α-D-mannopyranoside gave variable mixtures of 2,6- and 3,6-disilyl ethers (TbdmsCl, imidazole). The 3,6-disilyl ether was isomerized to the 2,6-isomer with imidazole in DMF, but with BuLi in THF it was converted into the 4,6-isomer.[28] Selective silylation of pyranose 1,2-diols afforded the anomeric *O*-silyl ethers. Subsequent benzylation (NaH, DMF, BnBr) was preceded by 1→2 silyl group migration so that the benzyl 2-*O*-silyl glycosides were obtained.[29]

Both 1:1 and 1:2 complexes of Bu₄NF and BF₃.OEt₂ have been assessed for their unique properties for the selective removal of different silyl ethers.[30] The Tips silyl group can be removed with K10 clay in aqueous methanol.[31]

2   Intramolecular Ethers (Anhydro-sugars)

2.1   Oxirans – The use of pyranose epoxides in the preparation of sugar amino acids and peptides, aminodeoxy, halodeoxy, branched-chain, cyclopropanated and aziridino sugars has been reviewed.[32]

A number of 1,2-anhydro-per-*O*-benzyl-aldo-furanose and -pyranose compounds, including 1,2-anhydro-3,5,6-tri-*O*-benzyl-α-L-gulofuranose, 1,2-anhydro-3,5-di-*O*-benzyl-β-D-arabinofuranose,[33] 1,2-anhydro-3,5,6-tri-*O*-benzyl-β-D-mannofuranose, 1,2-anhydro-3,5-di-*O*-benzyl-α-D-xylofuranose,[34] 1,2-anhydro-3,4,6-tri-*O*-benzyl-β-D-altropyranose and -α-D-allopyranose[35] and 1,2-anhydro-3,4-di-*O*-benzyl-6-deoxy-β-L-talopyranose have been prepared by standard methods.[36]

Treatment of 1,2-anhydro-3,4,6-tri-*O*-benzyl-α-D-glucopyranose with various hydride/Lewis acid systems led to 1,2-diol products, whereas with LiAlH₄ in refluxing ether the 1,5-anhydro-glucitol tri-ether was obtained.[37] Polymerisation of 1,2-anhydro-3,4-di-*O*-benzyl-β-L-rhamnopyranose (Tf₂O, CH₂Cl₂) gave high yields of α-(1→2)-linked rhamnopyranan.[38] Stereoselective glycosidation of 1,2:3,4-di-*O*-isopropylidene-α-D-galactopyranose with 1,2-anhydro-aldo-pyranose or -furanose derivatives was effected without added catalyst to give 1,2-*trans*-glycosides.[34,39]

Benzyl 2,3-anhydro-α-D-ribopyranoside was opened by nucleophiles exclusively at *C*-3 to give 3-substituted-3-deoxy-D-xylose derivatives,[40] whereas benzyl 3,4-anhydro-β-D-ribopyranoside, can be regioselectively opened at C-3 or C-4 with various nucleophiles (CN, SPL, N₃, OMe, Br, I) depending on the choice of metal counterion.[41]

Some 2,3-epoxyamides of aldonic acids have been prepared by reaction of acyclic aldose derivatives with sulfur ylids (Scheme 1).[42] *Syn*-epoxidation of some racemic 2-benzyloxy-4-alkenamides by way of iodohydrin formation and subsequent base treatment has afforded epoxides such as **2**.[43] Epoxidation of hex-2-enopyranosyl phosphonates has been effected with H₂O₂/sodium tungstate.[44]

2.2   Other Anhydrides – The readily available diol **3** has been converted into the 1,6-anhydro-derivatives **4** and **5** (camphorsulfonic acid, toluene, reflux) in a 7:1 ratio and the galactopyranoside diol **6** afforded a 1:1 ratio of the corresponding 1,6-anhydro-compounds under the same conditions.[45] An acid-catalysed rearrangement of D-galactal has afforded the unsaturated 1,6-anhydro-sugar **7**.[46] The Sharpless asymmetric dihydroxylation applied to the terminal alkene in 2-vinylfuran, followed by further oxidation reactions, has led selectively to both

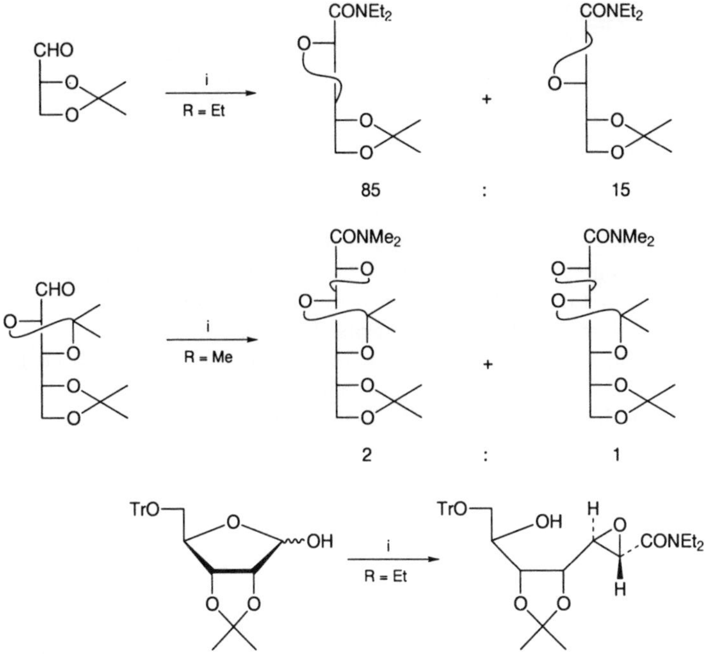

Reagents: i, Me₂S=CHCONR₂, CHCl₃

Scheme 1

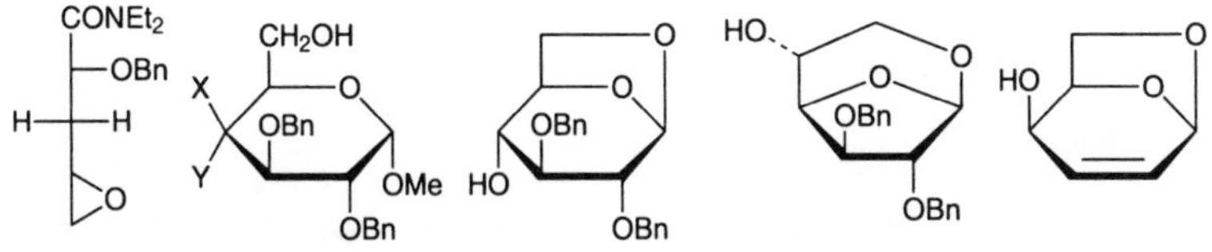

(±) **2**   **3** X = H, Y = OH   **4**   **5**   **7**
            6 X = OH, Y = H

enantiomers of levoglucosenone.[47] The synthesis of (+)-levoglucosenone and its 5-*C*-hydroxymethyl analogue is mentioned in Chapter 13.

Basic solvolysis (NaOH, MeOH) of the 5,6-thiocarbonates **8** gave rise to the 3,6-anhydride **9** by attack of O-3 on C-6.[48]

Polymerisation of 1,4-anhydro-2,3-di-*O*-benzyl-α-D-ribopyranose (BF₃.OEt₂) has afforded (1→5)-linked α-D-ribofuranose polymers.[49] The effects of 2-, 3- and 6-*O*-pivalate and benzyl substituents on the polymerisation of 1,4-anhydro-α-D-glucopyranose derivatives have been examined and it was concluded that 2-*O*-pivaloyl and 3-*O*-benzyl groups were indispensable for the formation of stereo-regular (1→5)-β-D-glucofuranose polymers.[50] 1,6-Anhydro-4-*O*-benzyl-2,3-di-*O*-dodecyl-β-D-mannopyranose has been polymerized under acidic conditions to

8 X, Y = O, S   **9**

give a stereoregular $(1 \rightarrow 6)$-α-D-mannopyranose polymer,[51] and the copolymerization of 1,6-anhydro-2,4-di-*O*-benzyl-3-*O*-1'-(methoxycarbonyl)ethyl-β-D-glucopyranose with 1,6-anhydro-2,3,4-tri-*O*-benzyl-β-D-glucopyranose (PF$_5$, CH$_2$Cl$_2$) has led to polymers with various contents of carboxyl groups.[52]

The acetimidate **10** in the presence of BF$_3$.OEt$_2$ has produced the 1,2':2,1'-dianhydride **11** with loss of the 2'-*O*-benzyl group.[53] Treatment of 1-*erythro*-2-pentulose with pyridinium poly(HF) under thermodynamic conditions gave the previously unknown β-L-1,2':2,4'-di-furanose-dianhydride as the major product.[54]

10   **11**

References

1   K. Yoshikawa, N. Okada, Y. Kann and S. Arihara, *Chem. Pharm. Bull.*, 1996, **44**, 1790.

2   R.U. Lemieux, U. Spohr, M. Bach, D.R. Cameron, T.P. Frandsen, B.B. Stoffer, B. Svensson and M.M. Palcic, *Can. J. Chem.*, 1996, **74**, 319.

3   R.P. Tripathi, V. Singh, A.R. Khan, A.P. Bhanduri, G. Saxena and K. Chandra, *Indian J. Chem.*, *Sect. B: Org. Chem. Incl. Med. Chem.*, 1995, **34B**, 791 (*Chem. Abstr.*, 1996, **124**, 9 111).

4   F. Wijnbergen, H. Regeling, B. Zwanenburg, G.J.F. Chittenden and N. Rehnberg, *Carbohydr. Res.*, 1996, **280**, 151.

5   C.-Y. Yuan, H.-X. Zhai and S.-S. Li, *Chin. J. Chem.*, 1996, **14**, 271 (*Chem. Abstr.*, 1996, **125**, 276 363).

6   T. Desai, J. Gigg and R. Gigg, *Carbohydr. Res.*, 1996, **280**, 209.

7   T. Le Diguarher, A. Boudon, C. Elwell, D.E. Paterson and D.C. Billington, *Biorg. Med. Chem. Lett.*, 1996, **6**, 1983.

8   T.Y. Chan, A. Chen, N. Allanson, R. Chen, D. Liu and M.J. Sofia, *Tetrahedron Lett.*, 1996, **37**, 8097.

9   C. Malet and O. Hindsgaul, *J. Org. Chem.*, 1996, **61**, 4649.
10 R. Klösel, S. König, S. Lehnhoff and R.M. Karl, *Tetrahedron*, 1996, **52**, 1493.
11 H.C. Hansen, O. Hindsgaul and M. Bols, *Tetrahedron Lett.*, 1996, **37**, 4211.
12 T. Tokimatsu, T. Umezawa and M. Shimada, *Holzforschung*, 1996, **50**, 156 (*Chem. Abstr.*, 1996, **125**, 33 967).
13 H.K. Hombrecher, C. Schell and J. Thiem, *Bioorg. Med. Chem. Lett.*, 1996, **6**, 1199.
14 H. Qin and T.B. Grindley, *J. Carbohydr. Chem.*, 1996, **15**, 95.
15 T. Desai, J. Gigg and R. Gigg, *Aust. J. Chem.*, 1996, **49**, 305.
16 A. Bhattacharjee, P. Chattopadhyay, A.P. Kundu, R. Mukhopadhyay and A. Bhattacharjya, *Indian J. Chem. Sect. B: Org. Chem. Incl. Med. Chem.*, 1996, **35B**, 69 (*Chem. Abstr.*, 1996, **124**, 176 657).
17 M. Izumi, O. Tsuruta and H. Hashimoto, *Carbohydr. Res.*, 1996, **280**, 287.
18 E.A. El'Perina, M.I. Struchkova, M.I. Serkebaev and E.P. Serebryakov, *Izv. Akad. Nauk, Ser. Khim.*, 1993, 776 (*Chem. Abstr.*, 1996, **124**, 117 701).
19 J.S. Yadav, S. Chandrasekhar, G. Sumithra and R. Kache, *Tetrahedron Lett.*, 1996, **37**, 6603.
20 R. Rodebaugh, J.S. Debenham and B. Fraser-Reid, *Tetrahedron Lett.*, 1996, **37**, 5477.
21 S. Chandrasekhar, G. Sumithra and J.S. Yadav, *Tetrahedron Lett.*, 1996, **37**, 1645.
22 J. Lee and J.K. Cha, *Tetrahedron Lett.*, 1996, **37**, 3663.
23 A.R. Khan, R.P. Tripathi and A.P. Bhaduri, *Indian J. Chem.*, *Sect. B: Org. Chem. Incl. Med. Chem.*, 1996, **35B**, 405. (*Chem. Abstr.*, 1996, **125**, 11 246).
24 J. Hajko, G. Szabovik, J. Kerekgyarto, M. Kajtar and A. Liptak, *Aust. J. Chem.*, 1996, **49**, 357.
25 P. Bako and L. Toke, *J. Inclusion Phenom. Mol. Recognit. Chem.*, 1995, **23**, 195 (*Chem. Abstr.*, 1996, **125**, 168 477).
26 M.W. Bredenkamp, *S. Afr. J. Chem.*, 1995, **48**, 154 (*Chem. Abstr.*, 1996, **125**, 222 275).
27 D.A. Johnson and L.M. Taubner, *Tetrahedron Lett.*, 1996, **37**, 605.
28 S. Arias-Perez and J. Santos, *Tetrahedron*, 1996, **52**, 10785.
29 J.M. Lassaletta, M. Meichle, S. Weiler and R.R. Schmidt, *J. Carbohydr. Chem.*, 1996, **15**, 241.
30 S.-i. Kawahara, T. Wada and M. Sekine, *Tetrahedron Lett.*, 1996, **37**, 509.
31 J.-i. Asakura, M.J. Robins, Y. Asaka and T.H. Kim, *J. Org. Chem.*, 1996, **61**, 9026.
32 W. Voelter, K.M. Khan and M.S. Shekhani, *Pure Appl. Chem.*, 1996, **68**, 1347 (*Chem. Abstr.*, 1996, **125**, 222 262).
33 X. Ding and F. Kong, *Carbohydr. Res.*, 1996, **286**, 161.
34 Y. Du and F. Kong, *J. Carbohydr. Chem.*, 1996, **15**, 797.
35 Y. Du, W. Mao and F. Kong, *Carbohydr. Res.*, 1996, **282**, 315.
36 Z. Gan, F. Kong, G. Xu and Y. Wang, *Huanjing Huaxue*, 1996, **15**, 174 (*Chem. Abstr.*, 1996, **125**, 11 230).
37 T.M. Flaherty and J. Gervay, *Tetrahedron Lett.*, 1996, **37**, 961.
38 Q. Chen and F. Kong, *Carbohydr. Res.*, 1996, **283**, 207.
39 G. Yang and F. Kong, *Carbohydr. Lett.*, 1994, **1**, 137 (*Chem. Abstr.*, 1996, **125**, 87 025).
40 P.K. Vasudeva and M. Nagarajan, *Tetrahedron*, 1996, **52**, 5607.
41 P.K. Vasudeva and M. Nagarajan, *Tetrahedron*, 1996, **52**, 1747.
42 F.J. López-Herrera, M.S. Pino-Gonzáles, F. Sarabia-Garca, A. Heras-López, J.J. Ortega-Alcántara and M.G. Pedraza-Cabrián, *Tetrahedron: Asymm.*, 1996, **7**, 2065.
43 P.E. Maligres, S.A. Weissman, V. Upadhyay, S.J. Cianciosi, R.A. Reamer, R.M.

Purick, J. Sager, K. Rossen, K.K. Eng, D. Askin, R.P. Volante and P.J. Reider, *Tetrahedron*, 1996, **52**, 3327.

44   D.-X. Liem, I.D. Jenkins, B.W. Skelton and A.H. White, *Aust. J. Chem.*, 1996, **49**, 371.

45   S. Caron, A.I. McDonald and C.H. Heathcock, *Carbohydr. Res.*, 1996, **281**, 179.

46   R. Haeckel, G. Lauer and F. Oberdorfer, *Synlett*, 1996, 21.

47   T. Taniguchi, K. Nakamura and K. Ogasawara, *Synlett*, 1996, 971.

48   Y. Tsuda and K. Shibayama, *Chem. Pharm. Bull.*, 1996, **44**, 1476.

49   Y. Choi, T. Yoshida, T. Mimura, Y. Kaneko, H. Nakashima, N. Yamamoto and T. Uryu, *Carbohydr. Res.*, 1996, **282**, 113.

50   H. Kamitakakara and F. Nakatsuto, *Macromolecules*, 1996, **29**, 1119 (*Chem. Abstr.*, 1996, **124**, 202 812).

51   K.-I. Kanno, W. Kinoshita, K. Kobayashi and K. Hatanaka, *Polym. J.*, 1995, **27**, 911 (*Chem. Abstr.*, 1996, **124**, 87 565).

52   K. Hatanaka, T. Minamisawa and K.-I. Kanno, *Polym. J.*, 1995, **27**, 1016 (*Chem. Abstr.*, 1996, **124**, 87 570).

53   E.P. Dubois, A. Neszmélyi, H. Lotter and V. Pozsgay, *Tetrahedron Lett.*, 1996, **37**, 3627.

54   J.M. Garcia Fernandez, R.-R. Schnelle and J. Defaye, *Aust. J. Chem.*, 1996, **49**, 319.

6
Acetals

1 Acyclic Acetals

On treatment with $Ac_2O/HOAc/H_2SO_4$, methyl 2-acetamido-2-deoxy-β-d-gluco-pyranoside (1) is converted into the acyclic acetal-like species 4, although the α-anomer 2 undergoes acetolysis to form, as expected, the 1-acetate 3 under these conditions.[1]

2 Ethylidene, Isopropylidene, Cyclohexylidene and Benzylidene Acetals

Trichloroethylidene acetals are readily dechlorinated by tributyltin hydride, thus providing convenient access to ethylidene acetals with *endo*-H configuration.[2]

A number of new acetonation conditions have been examined. Use of acetone in the presence of K 10 clay gave the 2,3-*O*-isopropylidene derivatives of d-ribofuranose and l-rhamnopyranose and the expected di-*O*-isopropylidene derivatives of d-xylose, d-glucose, d-galactose and d-mannose in excellent yields.[3] Good to excellent yields of the expected acetonides of methyl α-d-manno-, α-d-gluco-, and α-d-galacto-pyranoside as well as of d-mannose have been obtained with acetone and titanium(IV) chloride supported on modified silica gel.[4] $H_3PMo_{12}O_{40}$ was the most active of 18 heteropolyacids examined as catalysts in the acetonation of l-sorbose.[5]

Isopropylidenation of methyl α-d-glucoseptanoside with limited quantities of 2,2-dimethoxypropane and TsOH in DMF gave 5 as the major product.[6] The two octonolactones obtained in near equal proportions by Kiliani ascent from

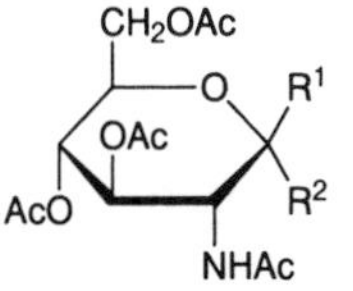

1 R^1 = OMe, R^2 = H
2 R^1 = H, R^2 = OMe
3 R^1 = H, R^2 = OAc

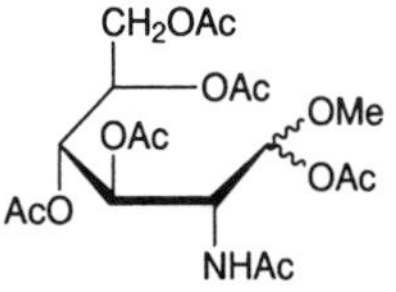

4

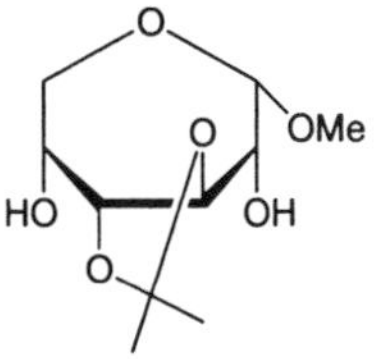

5

Carbohydrate Chemistry, Volume 30
© The Royal Society of Chemistry, 1998

D-*glycero*-D-*gulo*-heptose were easily separated by chromatography after acetonation, since the D-*erythro*-L-*galacto* product formed the diacetonide **6**, whereas the D-*erythro*-L-*talo* isomer gave the tri-acetonide **7**.[7] Treatment of α,α'-trehalose, its mono-*galacto* and its di-*galacto* equivalent with 2,2-dimethoxypropane and catalytic TsOH in DMF gave diacetonides in all three cases, the substitution occurring at the 4- and 6-positions of the glucopyranose and at the 3- and 4-positions of the galactopyranose residues.[8]

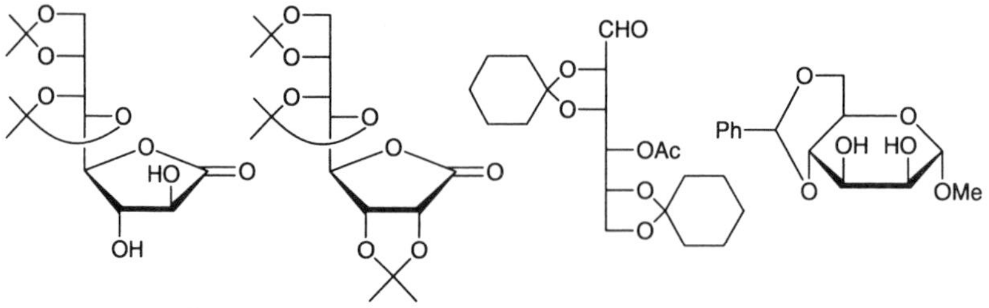

6   7   8   9

The cyclohexylidenation of D-mannose diethyl dithioacetal aimed at the preparation of the fully *O*-protected *aldehydo*-D-mannose derivative **8** has been described.[9]

A detailed comparative study on the benzylidenation of D-arabinose diethyldithioacetal with benzaldehyde dimethylacetal and catalytic TsOH or with benzaldehyde in the presence of HCl or ZnCl₂ has been published,[10] and the benzylidenation of methyl α-D-mannopyranoside with benzaldehyde and TsOH has been improved, with yields of acetal **9** >72%.[11] Transacetalation of fully 6-*O*-pivaloylated α-, β-, and γ-cyclodextrins with benzaldehyde dimethylacetal and catalytic camphorsulfonic acid gave mono-benzylidene products **10**. Reductive ring-opening (LAH/AlCl₃) of compounds **10** afforded predominantly the 2'-*O*-unprotected derivatives **11**.[12]

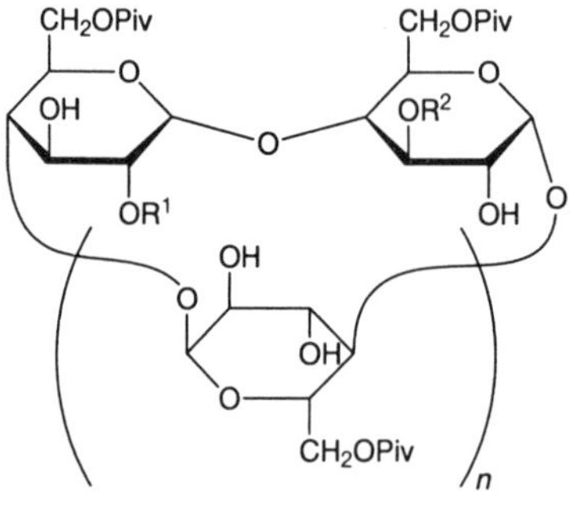

n = 4,5,6
10 R¹, R² = PhHC
11 R¹ = Bn, R² = H

3 Other Acetals

The *p*-nitrobenzylmethoxy group[13] and the 1-[(trimethylsilyl)ethoxy]ethyl group[14] have been introduced for hydroxyl protection, especially 2'-OH in solid-phase oligoribonucleoside synthesis. The low yields achieved in the preparation of the diastereomeric pyruvic acid acetals **14** from the corresponding 3,4-di-*O*-trimethyl-silyl ether **12** by use of the established procedure prompted the authors to devise an alternative route to **14** *via* the acetoxyacetonide **13**, as shown in Scheme 1.[15] The pyruvic acid acetal imidate **15** was obtained in 4 steps from *S*-phenyl 4',6'-*O*-benzylidene-1-thiolactoside for use in the synthesis of pyruvated trisaccharides.[16] New examples of the concomitant acetal formation and introduction of a carbamoyl function on treatment of *cis-trans*-configured 1,2,3-triols with chloral/ DCC (*e.g.*, **16** → **17**, see Vol. 28, p. 97, Refs. 16–18) have been reported.[17–19]

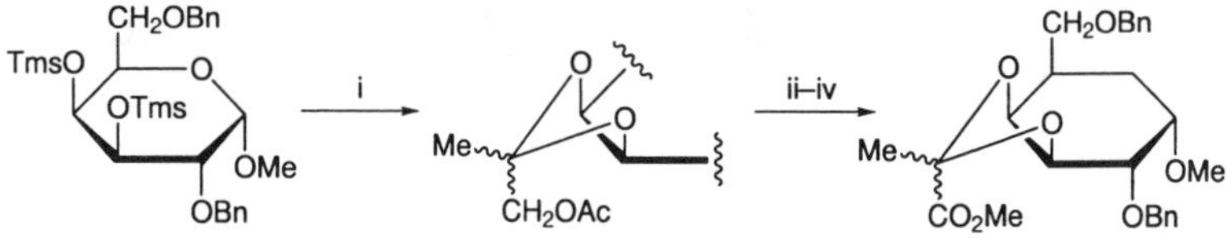

Reagents: i, MeC(O)CH₂OAc, TmsOTf, TfOH; ii, Et₃N, aq. MeOH; Swern oxidation; iv, Br₂, MeOH

Scheme 1

Condensation of methyl β-D-galactopyranoside with 1,1-dimethoxyhexadecane under TsOH-catalysis in DMF gave a mixture of acetals **18–21**. NMR spectro-scopic data of the four isomers were used in the structural assignments of plasmalopsychosines, members of a new class of human brain glycosphingolipids containing hexadecylidene groups.[20] Cyclohexane-1,2-diacetals derived from vicinal diols and 1,1,2,2-tetramethoxycyclohexane have been used as disarming protecting groups for selenoglycosides in the synthesis of high-mannose oligosaccharides.[21] Reaction of D-mannitol with pinacolone dimethylacetal in the presence of SnCl₂ gave products **22** as a mixture of separable diastereoisomers in 96% combined yield.[22]

Four oleanolic acid glycosides (*e.g.*, glucoside **23**) with unusual acidic acetal moieties have been isolated from *Beta vulgaris* (sugar beet).[23]

4 Reactions of Acetals

Various aspects of the ring-opening reactions of monosaccharide acetonides with Grignard reagents and with hydride nucleophiles have been discussed in a review article (12 refs.).[24]

The reduction of trichloroethylidene acetals to give ethylidene acetals is referred to above (Ref. 2). Isopropylidene groups have been removed from sugar

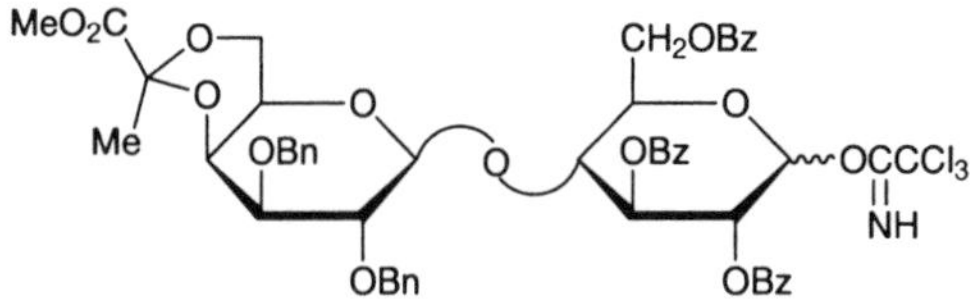

15

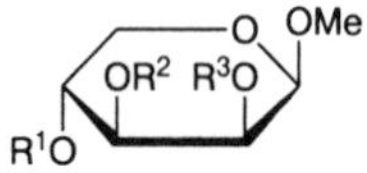

16 R^1 = R^2 = R^3 = H
17 R^1 = CONHC$_6$H$_{11}$

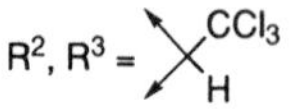

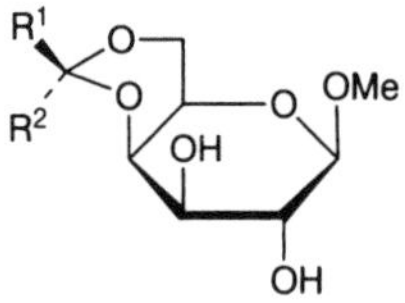

R^2, R^3 =

18 R^1 = C$_{16}$H$_{31}$, R^2 = H
19 R^1 = H, R^2 = C$_{16}$H$_{31}$

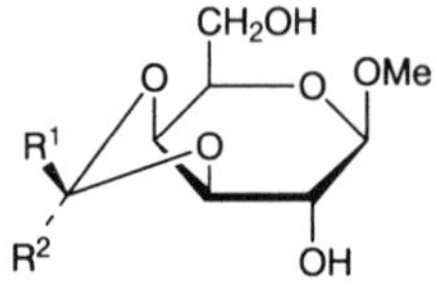

20 R^1 = C$_{16}$H$_{31}$, R^2 = H
21 R^1 = H, R^2 = C$_{16}$H$_{31}$

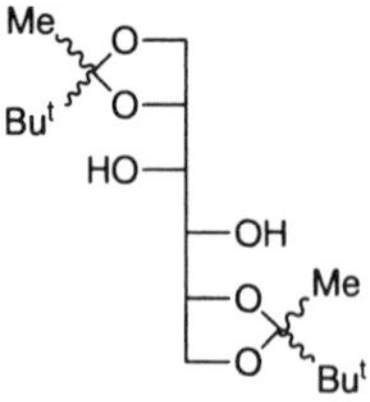

22

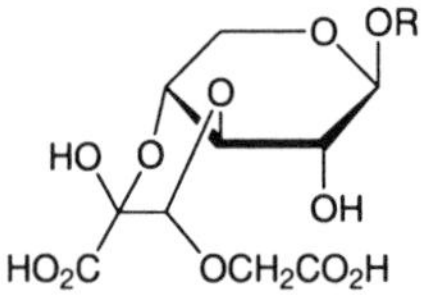

23 R = oleanoyl

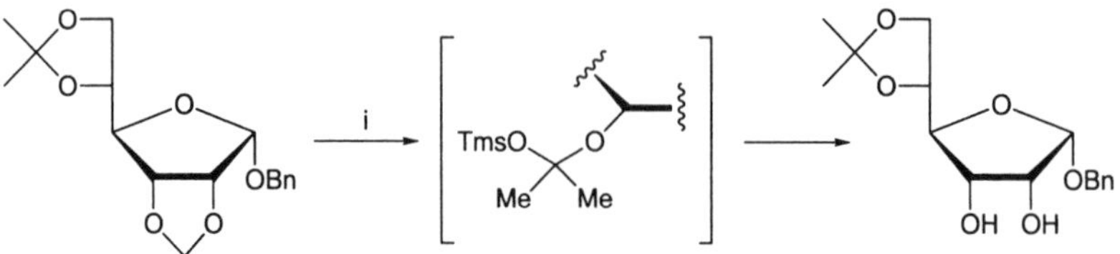

24

Reagents: i, TmsCH$_2$MgCl, PhH

Scheme 2

acetonides, including nucleosides, with K 10 clay in aqueous methanol,[25] and efficient debenzylidenation has been achieved by use of SnCl$_2$ in moist dichloromethane.[26] (Trimethylsilylmethyl)magnesium chloride in benzene liberates diols from bisacetonides selectively. A typical example is shown in Scheme 2. The reaction was assumed to proceed by way of a silylated acetal intermediate, such as **24**, and chelation-control was used to rationalize the selectivity.[27]

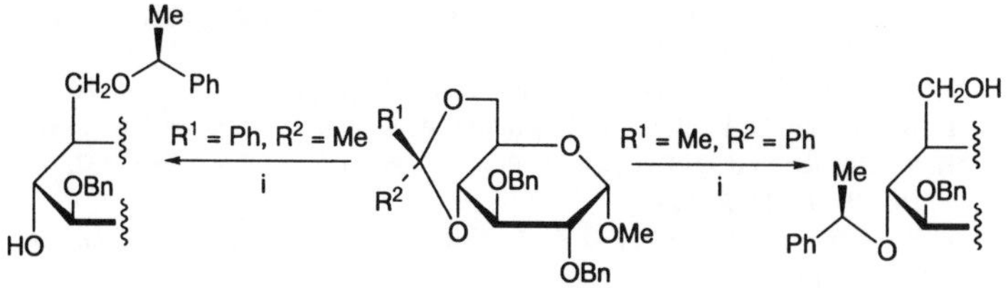

25

Reagents: i, LAH/AlCl₃

Scheme 3

A new, free-radical method for cleaving *O*-benzylidene-sugars to give bromo-benzoates employs NBS in the presence of catalytic AIBN in dry, refluxing chloroform.[28] Treatment with DDQ in toluene in the presence of a few equivalents of water converted the 4,6-*O*-*p*-methoxybenzylidene derivatives of mono- and di-saccharides very efficiently to *ca.* 4:1 mixtures of 6-*O*- and 4-*O*-*p*-methoxybenzoates.[29] Reductive cleavage of the diastereomeric 4,6-*O*-acetophenone acetals **25** with LAH/AlCl₃ gave single products, as indicated in Scheme 3.[30] Borane dimethylamine complex in the presence of BF₃ etherate in acetonitrile cleaves 4,6-*O*-benzylidene acetals of glucose and 2-amino-2-deoxyglucose to give predominantly the 6-benzyl ethers.[31] Reductive ring-opening of the benzylidene acetals of cyclodextrins is mentioned above (Ref. 12).

References

1 L.-X. Wang and Y.C. Lee, *J. Chem. Soc., Perkin Trans. I,* 1996, 581.

2 D. Rentsch and R. Miethchen, *Carbohydr. Res.,* 1996, **293**, 139.

3 J. Asakura, Y. Matsubara and M.Yoshihara, *J. Carbohydr. Chem.,* 1996, **15**, 231.

4 Z. Gan, F. Kong and Q. Wang, *Huanjing Huaxue,* 1996, **15**, 179 (*Chem. Abstr.,* 1996, **125**, 11 244).

5 G.M. Maksimov and M.N. Timofeeva, *React. Kinet. Catal. Lett.,* 1995, **56**, 191 (*Chem. Abstr.,* 1996, **124**, 176 715).

6 C.J. Ng and J.D. Stevens, *Carbohydr. Res.,* 1996, **284**, 241.

7 A.A. Bell, R.J. Nash and G.W.J. Fleet, *Tetrahedron: Asymm.,* 1996, **7**, 595.

8 R.W. Bassily, R.H. Youssef, R.I. El-Sokkary and M.A. Nashed, *J. Carbohydr. Chem.,* 1996, **15**, 653.

9 I.I. Bicherova, S.V. Turik, V.I. Kornilov and Yu.A. Zhdanov, *Zh. Obshch. Khim.,* 1996, **66**, 697 (*Chem. Abstr.,* 1996, **125**, 276 310).

10 J. Kuszmann and E. Gacs-Baitz, *Aust. J. Chem.,* 1996, **49**, 273.

11 T. Bhattacharyya and S. Basu, *Indian J. Chem., Sect. B:* 1996, **35B**, 397 (*Chem. Abstr.,* 1996, **124**, 317 638).

12 N. Sakairi, N. Nishi and S. Tokura, *Carbohydr. Res.,* 1996, **291**, 53.

13 G.R. Gough, T.J. Miller and L.A. Mantick, *TetrahedronLett.,* 1996, **37**, 981.

14 J. Wu, B.K. Shull and M. Koreeda, *Teterahedron Lett.,* 1996, **37**, 3647.

15 K. Hiruma, J. Tamura and H. Hashimoto, *Carbohydr. Res.,* 1996, **282**, 299.

16   T. Ziegler and G. Schuele, *J. Prakt. Chem./Chem.-Ztg.*, 1996, **338**, 238 (*Chem. Abstr.*, 1996, **125**, 33 969).

17   R. Miethchen, D. Rentsch and M. Frank, *J. Carbohydr. Chem.*, 1996, **15**, 15.

18   R. Miethchen, D. Rentsch, M. Frank and A. Liptak, *Carbohydr. Res.*, 1996, **281**, 61.

19   R. Miethchen and D. Rentsch, *J. Prakt. Chem./Chem.-Ztg.*, 1995, **337**, 422 (*Chem. Abstr.*, 1996, **124**, 30 140).

20   K.K. Sadozai, S.B. Levery, J.K. Anand and S. Hakomori, *J. Carbohydr. Chem.*, 1996, **15**, 715.

21   P. Grice, S.V. Ley, J. Pietruszka and H.W.M. Priepke, *Angew. Chem., Int. Ed. Engl.*, 1996, **35**, 197.

22   P. Renaud and S. Abazi, *Helv. Chim. Acta*, 1996, **79**, 1696.

23   M. Yoshikawa, T. Murakami, M. Kadoya, H. Matsuda, O. Muraoka, J. Yamahara and N. Murakami, *Chem. Pharm. Bull.*, 1996, **44**, 1212.

24   T.-Y. Luh, *Pure Appl. Chem.*, 1996, **68**, 635 (*Chem. Abstr.*, 1996, **125**, 86 987).

25   J.-i. Asakura, M.J. Robins, Y. Asaka and T.H. Kim, *J. Org. Chem.*, 1996, **61**, 9026.

26   J. Xia and Y. Hui, *Synth. Commun.*, 1996, **26**, 881.

27   Y.-H. Chen, Y.-T. Tseng and T.-Y. Luh, *J. Chem. Soc., Chem. Commun.*, 1996, 327.

28   S. Raina, K.K. Srivastava, E. Sampath Kumar and V.K. Singh, *Synth. Commun.*, 1996, **26**, 121.

29   Z. Zhang and G. Magnusson, *J. Org. Chem.*, 1996, **61**, 2394.

30   J. Hajko, G. Szabovik, J. Kerekgyarto, M. Kajtar and A. Liptak, *Aust. J. Chem.*, 1996, **49**, 357.

31   M. Oikawa, W.-C. Liu, Y. Nakai, S. Koshida, K. Fukase and S. Kusomoto, *Synlett*, 1996, 1179.

7
Esters

1 Carboxylic Esters

1.1 Synthesis – A short review (5 refs.) on the recently introduced 2-(chloro-acetoxymethyl) benzoyl- (CAMB) and 2-(2-chloroacetoxyethyl)benzoyl- (CAEB) protecting groups has been published.[1]

The stereochemistry of the *C*-methylation of 1,2:5,6-di-*O*-isopropylidene-α-D-gulofuranose 3-esters (**1** → **2**) using various lithium-containing bases has been examined. Results were inconsistent with the Li-chelate formation proposed by J. Mohr (Vol. 22, Chapter 7, Ref. 1) but indicative of a Li-bridged post-enolization complex as proposed by Seebach (*Helv. Chim. Acta*, 1985, **68**, 1373).[2]

Whereas acetolysis (H_2SO_4, Ac_2O) of peracetylated D-galactose diethyl dithioacetal **3** furnishes product **4**, as expected, the unprotected diethyl dithioacetal **5** gave penta-*O*-acetyl-D-galactofuranose **6** under these conditions.[3]

A convenient new procedure for the preparation of 1-*O*-trichloroacetimidates that requires neither an inert atmosphere nor anhydrous conditions and furnishes products pure enough for use without chromatographic purification, involves treatment of the *O*-benzyl-protected free sugar in dichloromethane at ambient temperature with trichloroacetonitrile and 50% aqueous KOH containing catalytic amounts of Bu_4NHSO_4.[4] The use of tetra-*O*-benzyl-α-D-mannopyranosyl trichloroacetate as glycosyl donor is referred to in Chapter 3, and a highly sensitive method for determining the absolute configuration of sugars by HPLC which relies on the formation of glycosyl 2-*tert*-butyl-2-methyl-1,3-benzodioxole-4-carboxylic acid esters, such as compound **7**, is covered in Chapter 23.

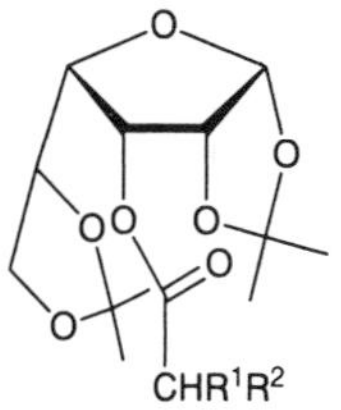

R^1 = Et, Pr, Pr^i, Bu

1 R^2 = H
2 R^2 = Me

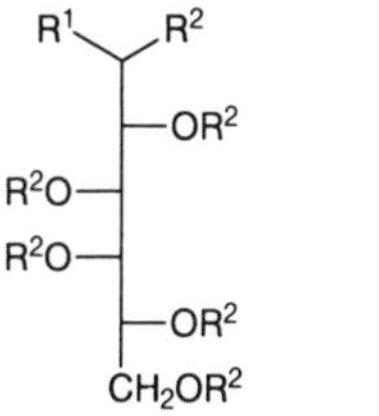

3 R^1 = SEt, R^2 = Ac
4 R^1 = OAc, R^2 = Ac
5 R^1 = SEt, R^2 = H

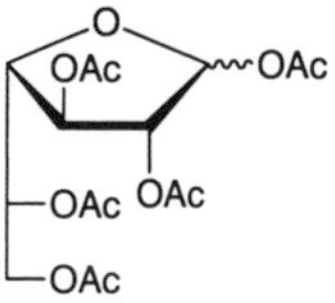

6

In connection with a study of its metabolism, abscisic acid (**8**) has been converted to esters **9** by reaction of its caesium salt with the appropriate peracetylated glycosyl halides.[5]

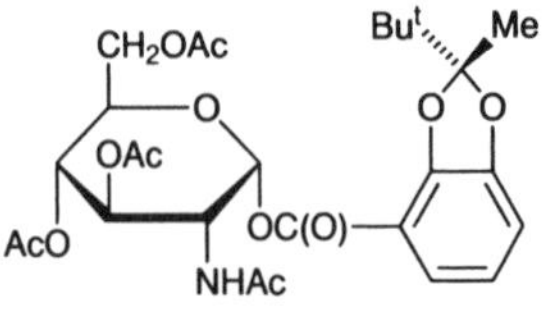

7

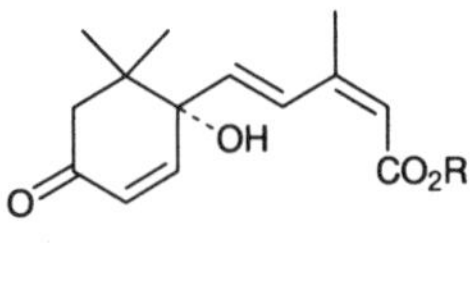

8 R = H
9 R = β-D-Glc*p*(OAc)₄ or
     β-D-Lactose(OAc)₇ *etc*

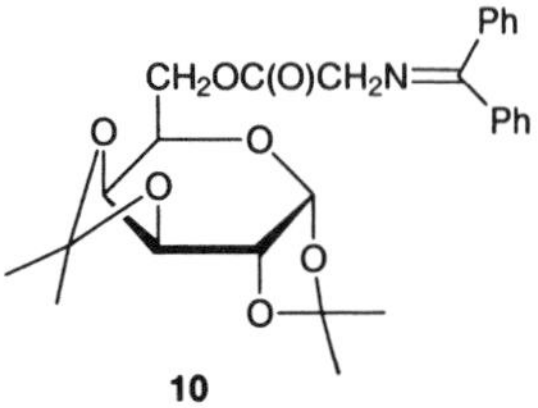

10

Dibutyltin oxide-mediated acetylation, benzoylation, *m*-bromobenzoylation and *p*-nitrobenzoylation of benzyl α-L-lyxopyranoside gave the 3-esters regioselectively in good yields.[6] (*N*-Diphenylmethylene)glycinates, *e.g.*, compound **10**, have been prepared from primary sugar alcohols by transesterification with ethyl *N*-diphenylmethyleneglycinate. The sterically hindered secondary hydroxyl group of diacetone glucose (**11**) was inert to these conditions but (*N*-diphenylmethylene)glycinate **12** was obtained by an esterification/transimidation sequence, as shown in Scheme 1. Products **11** and **12** were used in asymmetric aldol condensation, as indicated in the Scheme.[7]

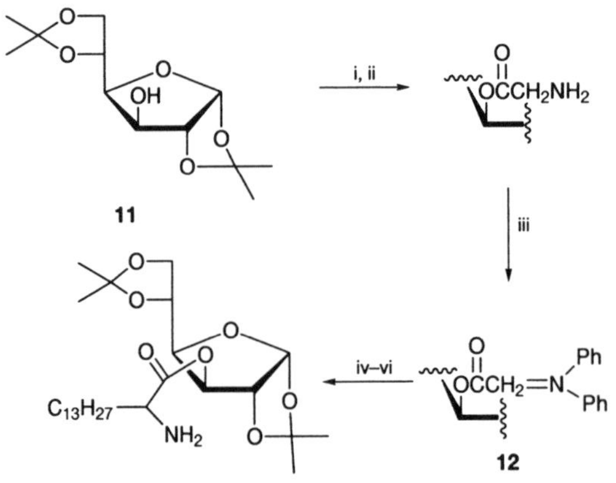

Reagents: i, CbzHNCH₂CO₂H, DCC, DMAP; ii, H₂ Pd/C; iii, Ph₂C= NHHCl;
iv, C₁₃H₂₇CHO, phase transfer; v, HCl; vi, NH₃

Scheme 1

The regioselective primary malonylation of phenyl β-D-glucopyranoside, serving as model substrate for natural glycoconjugates, has been accomplished by treatment with malonic acid in an aprotic solvent in the presence of *tert*-butyl isocyanide (see also Ref. 34 below).[8] Transesterifications of D-glucose with methyl 2,4-dihydroxybenzoate, methyl salicylate and methyl cinnamate to give the 6-monoesters have been carried out in DMF in the presence of potassium carbonate,[9] and the selectivity of the pivaloylation of methyl α-D-mannopyranoside has been studied under a variety of conditions.[10]

2,3,3′,4,4′-Penta-*O*-benzylsucrose (**13**) reacted selectively with 1 molar equivalent of pivaloyl chloride to give the 6′-monoester **14**; under Mitsunobo conditions, 1′,6-diesters, such as **15**, were favoured.[11] Sucrose polyesters, *e.g.*, **17**, which are naturally present on the surface of tobacco leaves, have been prepared from the known precursor **16** by DCC-mediated coupling with the appropriate, enantiomerically pure acids and subsequent removal of the ether protecting groups.[12] The synthesis of niruriside (**18**), a plant-derived sucrose tetra-*O*-acetate-di-*O*-cinnamate with HIV inhibitory properties, in five steps from 2,1′:4,6-di-*O*-isopropylidene sucrose has been reported.[13] A template technique for the selective acetylation of sucrose at the secondary-positions has been developed based on a cross-linked polystyrene matrix containing triads of trityl chloride groups strategically spaced to react with and thus block the primary hydroxyl groups of sucrose, leaving the secondary hydroxyls free for acetylation.[14] Acetylation of sucrose with Ac_2O in DMF using the polymer-supported butyl tin reagent **19** as catalyst afforded the 6-*O*-acetate regioselectively in 59% yield.[15]

	R¹	R²	R³	R⁴	R⁵	R⁶	R⁷	R⁸
13	H	H	H	Bn	Bn	Bn	Bn	Bn
14	H	H	Piv	Bn	Bn	Bn	Bn	Bn
15	p-NO₂Bz	p-NO₂Bz	H	Bn	Bn	Bn	Bn	Bn
16	Ac	Bn	Bn	p-MBn	All	H	Bn	Bn
17	Ac	H	H	X	X	Y	H	H
18	Ac	Ac	Z	Ac	H	Ac	Z	H

X = C(O)CH₂CH(Me)Et   Y = C(O)CH(Me)Et
Z = cinnamoyl

Benzoylation of the 3,4:3′,4′-diacetonide of α,α′-di-*galacto*-trehalose (see Chapter 6, Ref. 8), followed by mild acid hydrolysis, gave the 2,2′,6,6′-tetra-*O*-benzoate; acetylation of this compound by use of the orthoester method furnished the 4,4′-di-*O*-acetate-2,2′,6,6′-tetra-*O*-benzoate selectively.[16] Four isomeric trehalose 6,6′-dicorynomycolates were prepared by DCC/DMAP.HCl-promoted condensations of 2,2′,3,3′,4,4′-hexa-*O*-benzyl-α,α′-trehalose with the four diastereo- and enantio-merically pure corynomycolic acids, followed by hydogenolysis.[17] A number of cord factor analogues have been synthesized by tributylstannylation of unprotected α,α-trehalose, followed by regioselective reactions with long-chain fatty acid chlorides.[18] The possible disaccharide fragments of a polysaccharide with novel bioactivity, 3,3′- and 4,4′-di-*O*-caproylgentibiose, were synthesized conventionally from appropriately functionalized monosaccharide components.[19]

19

20

Dialkylstannylene acetals derived from a number of carbohydrate diols reacted with diacyl chlorides in the presence of a tertiary amine to give symmetrical, non-glycosidically linked disaccharides, such as compound **20**.[20] The tungsten complex-catalysed metathesis reaction of glycosides carrying ω-unsaturated substituents (see Vol. 28, Chapter 3, Ref. 217) has been used for intramolecular macrocyclization (*e.g.*, **21** → **22**).[21]

Syntheses of glycolipid GLA-60 and of a carboxylic acid analogue are covered in Chapters 10 and 16, respectively, and new fluorine-containing GLA-60 analogues, prepared as potential lipopolysaccharide antagonists, are referred to in Chapters 9 and 10.

All 12 regioisomeric tribenzoates of *myo*-inositol have been synthesized as precursors of *myo*-inositol trisphosphates by multi-step sequences involving standard protection/deprotection techniques and controlled benzoyl group migrations.[22,23] The quantitative acylation of 1-D-*myo*-inositol 1,2,6-*tris*-(dihydrogen phosphate) is covered in Chapter 18.

21   **22**   **23** R = C(O)CH$_2$OMe
                                                                                                24 R = H

Ytterbium(III) triflate in methanol is an efficient catalyst for cleaving methoxy-acetates (*e.g.*, **23** → **24**). The methoxyacetate protecting group was introduced 30 years ago (C.B. Reese and J.C.M. Stewart, *Tetrahedron Lett.*, 1968, 4723) but has found little application so far, due to difficulties associated with its removal.[24] Efficient, selective anomeric deacetylation of peracetylated reducing disaccharides has been achieved with hydrazine hydrate (1.2 equivalent) in acetonitrile.[25] It has been observed that the BF$_3$.Et$_2$O-catalysed glycosidation of several pento- and hexo-pyranose peracetates with simple alcohols is accompanied by selective deacetylation at the 2-position.[26,27]

Many new, selective acylation and deacylation reactions by use of enzymes have been reported. A comparative study of the enzyme-catalysed acetylation of monosaccharides in organic solvents found a *Pseudomonas* lipase the most efficient enzyme and D-mannitol the best substrate.[28] Benzyl 6-*O*-acetyl-β-D-glucopyranoside and its L-enantiomer were differently recognized by subtilisin in the enzyme-catalysed butanoylation in anhydrous acetone, the 3-ester of the D- and the 2-ester of the L-sugar being formed preferentially.[29] The monoacetate **26**, required as building block in a trisaccharide synthesis, was prepared from diol **25** by regioselective acetylation using a *P. fluorescens* lipase.[30]

The alcalase-mediated transesterification of methyl α-D-glucopyranoside and octyl α- and β-D-glucopyranoside with vinyl octanoate or vinyl laurate in anhydrous pyridine furnished predominantly the 6-monoesters.[31] Esterification of D-glucose and methyl α-D-glucopyranoside with 3 equivalents of dodecanoic acid catalysed by Novozym 345 (an immobilized *Candida antarctica* lipase) and enhanced by microwave irradiation, gave the 6-*O*-decanoates in 70 and 90% yield, respectively; maltose furnished as the major product (68%) the 6,6'-diester.[32] D-Glucose-6-acrylate, -6-methylacrylate and -6-crotonate were synthesized by exposure of D-glucose to *O*-alkenoyloximes in the presence of a protease.[33] A simple and high-yielding protocol for the primary malonylation of flavonoid glycosides involved the regioselective, *C. antarctica* lipase-mediated introduction of a benzylmalonyl group, followed by hydrogenolysis of the benzyl moiety (see also Ref. 10 above).[34] Efficient selective primary acetylation of neuraminic acid, both as the free sugar and as the non-reducing end-group of a melanoma-associated ganglioside, has been achieved by acyl transfer from vinyl acetate under catalysis by subtilisin in DMF.[35]

27 R = O-pentenyl
28 R = CH₂Cl or

(CH₂)CMe
       ‖
       O

25 R = H
26 R = Ac

29

The *C. antarctica* lipase-catalysed acylation of several disaccharides by ethyl butanoate or ethyl dodecanoate in *tert*-butanol at temperatures between 40 and 82 °C was examined; primary esterification was preferred, the ratio of 6'-mono- to 6,6'-di-ester depending on the structure of the disaccharide.[36] In another comparative investigation involving the benzyl and dodecyl β-glycosides of some common disaccharides, a variety of acyl donors and 15 different lipases in amyl alcohol, very high but somewhat different results were obtained with carbonate **27** and esters **28** and the enzymes from *C. antarctica* and *P. cepacia*. By careful choice of reaction conditions, mixed diesters such as the benzyl β-lactose derivative **29** were prepared in good yield.[37] 1'-*O*-Capryloyl-6-*O*-myristoyl-sucrose, accompanied by the 1'-*O*-capryloyl-6'-*O*-myristoyl isomer, was obtained by similar successive exposure of sucrose to trichloroethyl caprylate in the presence of subtilisin on celite and to trichloroethyl myristate in the presence of Lipase SP-435-L.[38] 1-*O*-(β-D-Galactopyranosyl)-*sn*-glycerol was regioselectively acylated at the primary hydroxyl groups of both the glycerol moiety and the galactosyl moiety, by sequential use of lipases from *Achromobacter* sp. and *Mucor javanicus* with vinyl esters as donors. Structure **30** shows a typical product.[39] The (2S)-1-monoesters **31** were the major products (50–70%) in the transesterification of 2-O-(β-D-glucopyranosyl)-*sn*-glycerol with the 2,2,2-tri-fluoroethylesters of long-chain fatty acids under catalysis by *P. cepacia* lipase in pyridine which was both, regio- and diastereo-selective.[40]

CH₂O-myristoyl structure **30**; b-D-GlcpO structure **31** n = 8, 10, 12, 14 or 16; b-D-Glcp-OR structures **32**, **33**, **34**

CH₂O-myristoyl

HO—O　O—CH₂
  OH　ǁ∼OH
      CH₂O-oleoyl
  OH

30

CH₂OC(O)(CH₂)ₙMe

b-D-GlcpO—|

CH₂OH

31 n = 8, 10, 12, 14 or 16

b-D-Glcp-OR

32 R = （2′,3′ epoxide）—OAc

33 R = —OAc

34 R = —OAc

The enzyme-catalysed regio- and enantio-selective esterification of *rac*-1,2-O-hexylidene-*myo*-inositol at the 5-position is covered in Chapter 18.

Various degrees of regioselectivity were observed in the rabbit serum esterase-catalysed hydrolysis of selectively pivaloylated methyl α-D-mannopyranosides.[10] Regioselective ethanolysis of peracylated methyl α- and β-D-glucopyranosides and methyl α-D-mannopyranoside in anhydrous solvents (*e.g.*, *n*-hexane-ethanol, 99:1) under the influence of a lipase from *C. rugosa* gave the 6-monohydroxy derivatives exclusively; no 4→6 migration was observed under these anhydrous conditions.[41] Exposure of a 1:1 mixture of (2′R,3′S) and (2′S,3′R) epoxides **32** to *P. cepacia* lipase caused selective hydrolysis of the acetate in the aglycon moiety of the (2′R,3′S) component to give a readily separable mixture of **33** and **34**. Similar results were obtained with the corresponding α-glycosides.[42]

1.2　Natural Products – Some natural sucrose and trehalose polyesters are referred to above. β-Sophorosyl octadec-11-enoate was the major organic component extracted from the membranes of the bacterium *Sarcina ventricula*.[43]

Efficient galloylation by use of tri-O-benzylgalloyl chloride in the presence of DMAP allowed the preparation of galloylglucoses with up to five galloyl groups, suitable as models for investigating certain properties of tannins.[44] The plant ellagitannin pedunculagin[45] and its trideca-O-methyl derivative[46] have been synthesized, the former by a biomimetic route involving oxidative coupling of the galloyl moieties after attachment to the glucose core, the latter relying on diastereoselective reaction of the glucose moiety with hexamethoxydiphenyl chloride. Syzyginin B (**35**), a new ellagitannin isolated from the leaves of clove (*Syzygium aromaticum*) contains a novel aromatic diacyl group with a dibenzo-1,4-dioxin structure.[47] The hydolysable tannin **36** has been shown to display algicidal properties.[48] A two-step mechanism involving complexation before

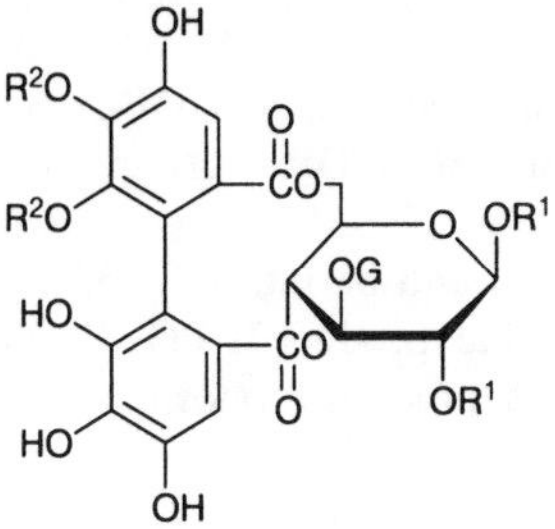

G = galloyl

35 R^1 = H, R^2 R^2 =

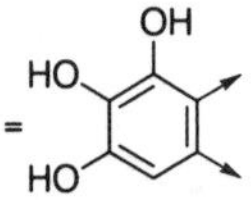

36 R^1 = G, R^2 = H

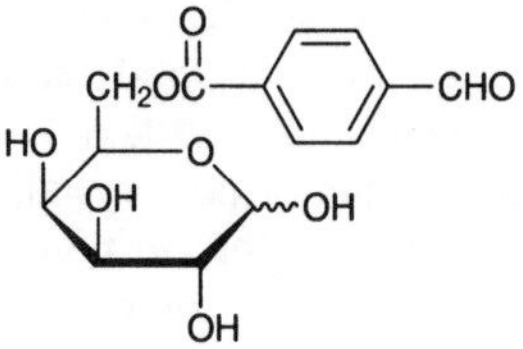

37

precipitation has been proposed for the co-precipitation of protein, *e.g.,* BSA, with hydrolysable tannins (see Vol. 29, Chapter 7, Ref. 43).[49]

Three novel triterpenoid saponins with hypoglucaemic activity, isolated from a Colombian climbing plant and named tuberosides A, B, and C, contain β-D-glucopyranose esterified at the anomeric centre with oleanic acid derivatives.[50] Porphyrine-carboxylate esters of D-galactopyranose have been synthesized either by transesterification of a porphyrine methyl ester with 1,2:3,4-di-*O*-isopropylidene-α-D-galactopyranose, followed by acetal hydrolysis or by condensation of 6-*O*-(4-formylbenzoyl)-D-galactose (**37**) with pyrrole.[51]

2   Phosphates and Related Esters

Several phosphorylations of nucleosides are referred to in Chapter 20, and the use of protected 1-dibenzylphosphates as glycosyl donors under neutral conditions is covered in Chapter 3.

A new synthetic strategy in which vinyl groups serve as leaving groups has been applied to the preparation of perbenzylated D-gluco- and -galacto-pyranoside 1-phosphate.[52] Anomerically pure β-L-fucopyranose 1-dibenzylphosphate has been synthesized by reaction of benzoyl-, rather than acetyl-, protected fuzosyl donors with dibenzylphosphate; no (2→1)-acyl migration was observed.[53]

2,3,4,6-Tetra-*O*-acetyl-β-D-gluco- and -galactopyranosyl H-phosphonates were converted into citronellyl and dolichyl glycosyl phosphates by reaction with the respective alcohols, followed by oxidation and deacetylation,[54] and dolichyl β-D-mannopyranosyl phosphate was prepared by reaction of a protected α-D-mannopyranosyl bromide with dolichyl phosphate.[55] Ceramide-1-phosphate sugars, *e.g.,* β-D-Glc*p*-OPO$_3$-Cer, a new type of glycophospholipid, have been obtained by direct glycosylphosphite/phosphate- and *O*-glycosyl trichloroacetimidate/ phosphate-exchange reactions.[56]

Experiments with an ^{18}O-labelled sugar indicated that the UDP-*N*-acetylglucosamine 2-epimerase-catalysed equilibration of UDP-GlcNAc with UDP-ManNAc proceeds *via* 2- acetamido-D-glucal.[57]

X-Ray irradiation of polycrystalline sodium and barium salts of D-glucose 1- and 6-phosphate and of D-ribose 5-phosphate produced ·PO$_3^{2-}$ radicals, as evidenced by ESR spectroscopic studies; no precursor phosphoranyl radicals were detected.[58]

The five benzyl L-*glycero*-α-D-*manno*-heptopyranosides having one free OH group (for synthesis see Chapter 2, Ref. 35) were phosphitylated with reagent 38 and subsequently converted to the monophosphates by oxidation and debenzylation.[59] Several ascorbic acid derivatives, such as 39, with a phosphodiester-linkage at C-2 to another biologically active compound, have been prepared by use of the phosphoramidate approach.[60] Compound 41, a D-glucose-based analogue of 1D-*myo*-inositol 1,4,5-trisphospate, has been synthesized in nine steps from D-glucose *via* the hydroxyethyl glucoside intermediate 41, using standard reactions and protection strategies.[61] New alkyl and aryl cyclophosphates were obtained on treatment of 3,5-cyclophosphorochloridate 42 with a range of alcohols and phenols (see Vol. 29, Chapter 7, Ref. 55).[62]

p-Nitrophenyl β-D-ribofuranoside was phosphorylated to give in 33% yield the 6-phosphate 43 required for probing the mechanism of *N*-ribohydrolases and purine nucleoside phosphorylase.[63] Selective primary phosphorylation of unprotected D-glucose and D-mannose to obtain phosphate diesters, such as cholesterol derivative 44, has been accomplished by condensing the sugars with the appropriate alkyl H-phosphonate monoesters using pivaloyl chloride as condensing agent.[64] Methyl 6-*O*-diphenylphosphinoyl-2,3-*O*-isopropylidene-α-D-mannofuranoside (46), formed on treatment of precursor 45 with Ph$_2$PCl and subsequent exposure to air, has been subjected to X-ray analysis, together with its 6-*C*-diphenylphosphinoyl analogue (see Chapter 17).[65] A new synthesis of Dahp, as its methylester/methyl β-glycopyranoside 49, involved the phosphorylation 47 → 48 → 49 as the final steps.[66] Mannose 6-phosphorothioate (51), required for a study on the biochemistry of phosphorylated mannose, was prepared from the fully trimethylsilylated precursor 50, as shown in Scheme 2.[67]

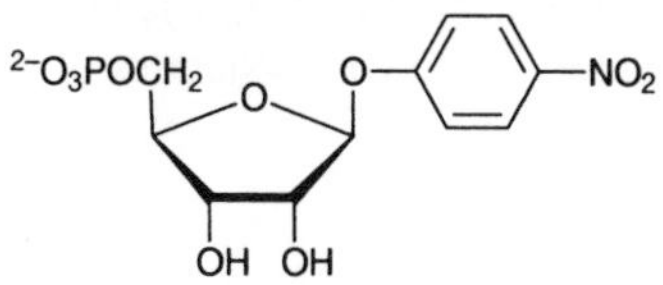

43

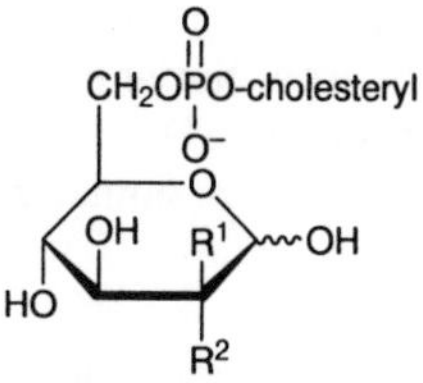

R^1 = H, R^2 = OH or
R^1 = OH, R^2 = H

44

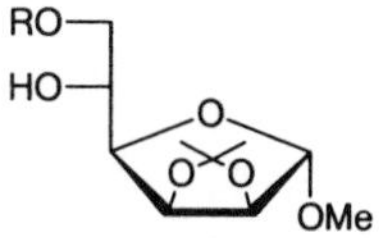

45 R = H
46 R = $\overset{\text{O}}{\underset{\parallel}{P}}(Ph)_2$

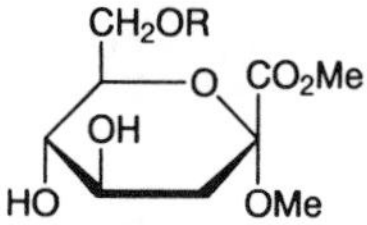

47 R = H
48 R = PO$_3$Bn$_2$
49 R = PO$_3$H$_2$

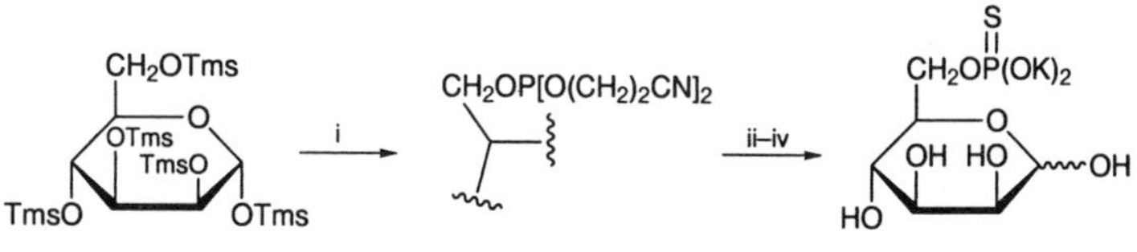

50

51

Reagents: i, Pr^{i}P[O(CH$_2$)$_2$CN]$_2$, tetrazole; ii, S, Py; iii, MeOH; iv, KOH, MeOH

Scheme 2

Phosphorylated lipophilic derivatives of ara C as potential pro-drugs, spin-labelled ATP derivatives and phosphorylation products of 5'-AMP are referred to in Chapter 20, inositol phosphate derivatives in Chapter 18, and ^{3}H-labelled Lipid A analogues in Chapter 3. Analogues of 2-deoxy-D-ribose with phosphorous in the ring are covered in Chapter 17.

Nine new zwitterionic digalactosyl ceramides isolated from the earthworm *Pheretima asiatica* all had a choline phosphate ester group at the 6-position of the terminal galactosyl moiety.[68] A sialyl Lewis X analogue in which the sialic acid residue is replaced by a galactosyl residue bearing a phosphate group is referred to in Part 3 below (Ref.78).

3 Sulfates and Related Esters

Sodium benzyl β-D-glucoside 2-sulfate, recently isolated from the Egyptian plant *Salvadora persica,* and named salvanoside, has been synthesized in four steps from D-glucose.[69] The D-glucose-based, polarized, anionic surfactants **54** and **55** were obtained from the unprotected glycoside **52** and thioglycoside **53**, respectively by exposure to SO_3/pyridine complex in DMF. 6-Sulfates of 2-acylamido-2-deoxy-D-glucose were similarly prepared.[70]

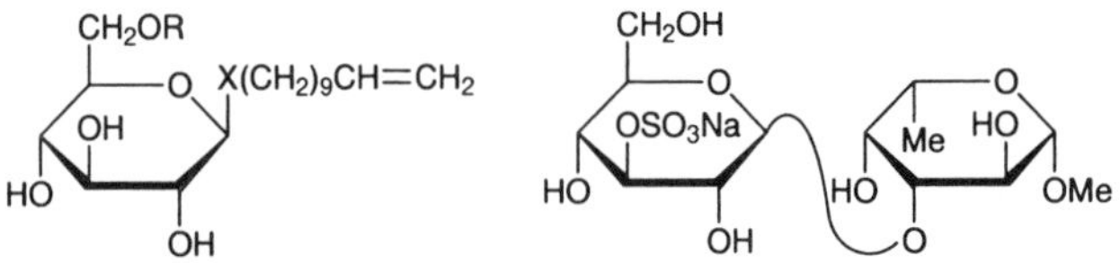

52 X = O, R = H
53 X = S, R = H
54 X = O, R = SO$_3$Na
55 X = S, R = SO$_3$Na

56

The sulfation of sucrose in a three phase system with SO_3 in DMF and tetrachloroethylene has been described.[71] Synthesis by standard procedures, with introduction of the sulfate group as the last step before deprotection, was employed in the synthesis of the sulfated disaccharide **56**.[72]

A mechanism involving chelation of magnesium to the negatively charged sulfite-oxygen atom has been proposed for the stereoselective opening of the D-mannitol-derived cyclic sulfite **57** with *tert*-butylmagnesium chloride to furnish sulfinate **58**.[73] Opening of the cyclic sulfate **59** with azide ion furnished **60** in 80% yield, although a reaction temperature of >100 °C was required.[74] The regioselective opening of cyclic sulfates as key-steps in the synthesis of 3,6-dideoxy-sugars is referred to in Chapter 12.

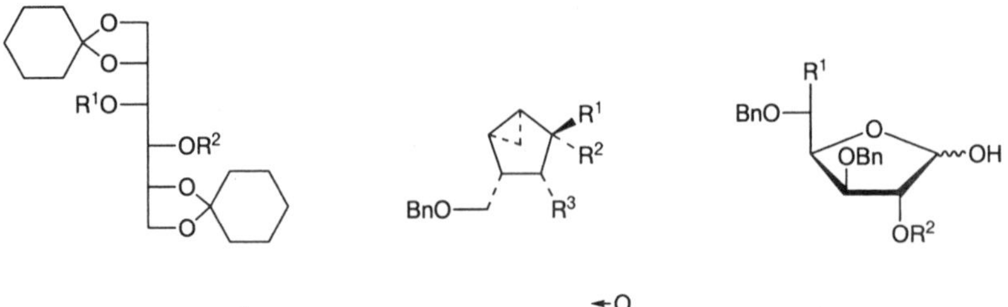

57 R^{1}R^2 = $^-$O$-$S̈
58 R^1 = $^-$O$-$S$^+$, R^2 = H
        Bu^t

59 R^1 = H, R^{2}R^3 = $-$O\\S=O
60 R^1 = N^3, R^2 = H, R^3 = OH

R^1 = H or CH$_2$OBn
61 R^2 = H
62 R^2 = Ts

Model experiments with methyl α-D-galactopyranoside 6-sulfate showed that sulfate is not lost under the Carlson condition for releasing *O*-linked oligoaccharides from glycoprotein; the more vigorous conditions used for releasing *N*-linked oligoaccharides, on the other hand, cause significant 3,6-anhydride formation.[75]

Lewis X 3'-sulfate and 3',6'-disulfate,[76] three sialyl Lewis X analogues in which the sialic acid residue is replaced by a galactosyl residue bearing an anionic substituent (SO_3Na, PO_3Na_2, CH_2CO_2Na)[77] and sulfated laminari-oligosaccharides with high anti-HIV activity[78] have been synthesized. Sulfated glucuronyl paraglobosides and glycoconjugates of sulfated and methylated oligosaccharides are referred to in Chapter 4.

Four new steroid glycosides containing a 6-deoxy-β-D-glucose 4-sulfate moiety have been isolated from starfish.[79]

4 Sulfonates

Phase transfer conditions gave the best results in the preparation of tosylate 62 from diol 61; no chloride formation was observed.[80] The imidazole-1-sulfonate group allowed S_N2 displacements at crowded carbon atoms by good nucleophiles under relatively mild conditions (e.g., 63 → 64).[81] The use of primary and secondary monosaccharide triflates for modifying erythromycin and other macrolides is referred to in Chapter 19.

63 R^1 = OSO$_2$—N⌒N , R^2 = R^3 = H   69   70 X = H or Cl

64 R^1 = H, R^2 = N$_3$, R^3 = H
65 R^1 = OTf, R^2 = R^3 = H
66 R^1 = H, R^2 = OTf, R^3 = H
67 R^1 = Bu, R^2 = OH, R^3 = H
68 R^1 = OTf, R^2 = H, R^3 = D

Triflate 65 reacted with butyl lithium in ether to give the elimination product 69; under the same conditions the epimeric triflate 66 underwent C-alkylation to furnish 67. Experiments with the deuterated triflate 68 showed that the elimination proceeds by abstraction of the α-hydrogen, followed by a hydrogen shift from C-4 to C-3.[82] The mechanism of the photolysis of 1,2:3,4-di-O-isopropylidene-6-O-tosyl-α-D-galactopyranose in methanol in the presence of a base (NaOH, DABCO), which produces the corresponding 6-alcohol and toluene, has been investigated.[83]

5 Other Esters

Orthoacetates 70 were formed quantitatively by reaction of the corresponding diol with $(MeO)_3CCH_2Cl$ or $(MeO)_3CCHCl_2$ in the presence of catalytic TsOH.

Opening with aqueous TFA was non-selective to give 4-O-/6-O mixtures of (monochloro)acetates and (dichloro)acetates, respectively.[84]

A new procedure for cleaving allyloxycarbonates using Pd(0), formed *in situ* from Pd(OAc)$_2$ and trisodium tris(*m*-sulfophenyl)phosphine, with sodium azide as allyl scavenger, has been applied to a number of carbohydrate derivatives.[85]

71 R = H

72 R = CONH—⟨C$_6$H$_4$⟩—Cl

73 R = H

74 R = [structure with CH$_2$OBn, OR^1]

R^1 = Me or Bn

Glycopyranosyl carbamates were obtained with almost 100% β-selectivity by treatment of anomerically unprotected precursors with aryl isocyanates (*e.g.*, 71 → 72).[86]

The terphenyl boronic acid 73 reacted selectively with 5-*O*-protected β-glyco-sides of 2-deoxy-D-ribose to form boronate esters 74.[87]

References

1   T. Ziegler, G. Pantkowski and G. Schuele, *GIT Fachz. Lab,* 1996, **40**, 46 (*Chem. Abstr.*, 1996, **124**, 290 003).

2   J. Mulzer, M. Hiersemann, J. Buschmann and P. Luger, *Liebigs Ann. Chem.,* 1996, 649.

3   L.M. Lerner, *Carbohydr. Res.*, 1996, **282**, 189.

4   V. J. Patil, *Tetrahedron Lett.*, 1996, **37**, 1481.

5   J. Balsevich, G. Bishop, S.L. Jacques. L.R. Hogge, D.J.H. Olson and N. Laganiere, *Can. J. Chem.* 1996, **74**, 238.

6   A.K.M.S. Kabir, M.A. Sattar, A.Z.M.S Chowdhury, M.A. Sattar and M.M. Rahman, *J. Bangladesh Chem. Soc.*, 1995, **8**, 95 (*Chem. Abstr.*, 1996, **125**, 301 356).

7   G. Solladié, J.-F. Saint Clair, M. Philippe, D. Semeria and J. Maignan, *Tetrahedron: Asymm.,* 1996, **7**, 2359.

8   R. Roscher, J.-P. Steffen, M. Herderich, W. Schwab and P. Schreier, *J. Agric. Food Chem.,* 1996, **44**, 1626 (*Chem. Abstr.*, 1996, **125**, 114 968).

9   A.F. Artamonov, L.F. Burkovskaya, and G.K. Nikonov, *Khim. Prir. Soedin.*, 1994, 561 (*Chem. Abstr.*, 1996, **124**, 87 500).

10   D. Ljevakovic, D. Parat, J. Tomasic and S. Tomic, *Croat. Chem. Acta*, 1995, **68**, 477 (*Chem. Abstr.*, 1996, **124**, 56 446).

11   S. Jarosz, *J. Carbohydr. Chem.*, 1996, **15**, 73.

12   S. Oscarson and H. Ritzen, *Carbohydr. Res.*, 1996, **284**, 271.

13   H.I. Duynstee, H. Ovaa, G.A. van der Marel and J.H. van Boom, *Recl. Trav. Chim. Pays-Bas*, 1996, **115**, 339.

14   W.M. Macindoe, M. Jenner and A. Williams, *Carbohydr. Res.*, 1996, **289**, 151.

15   W.M. Macindoe, A. Williams and R. Khan, *Carbohydr. Res.*, 1996, **283**, 17.

16   R.W. Bassily, R.H. Youssef, R.I. El-Sokkary and M.A. Nashed, *J. Carbohydr. Chem.*, 1996, **15**, 653.

17   M. Nishizawa, D.M. García, R. Minagawa, Y. Noguchi, H. Imagawa, H. Yamada, R. Watanabe, Y.-C. Yoo and I. Azuma, *Synlett*, 1996, 452.

18   Y. Gama, *Yukagaku*, 1995, **44**, 671 (*Chem. Abstr.*, 1996, **124**, 56 459).

19   Y. Qiu, Y. Nakahara and T. Ogawa, *Biosci. Biotech. Biochem.*, 1996, **60**, 986.

20   T.B. Grindley and H. Namazi, *Tetrahedron Lett.*, 1996, **37**, 991.

21   G. Descotes, J. Ramza, J.-M. Basset, S. Pagano, E. Gentil and J. Banoub, *Tetrahedron*, 1996, **52**, 10903.

22   S.-K. Chung, Y.-T. Chang and K.-H. Sohn, *J. Chem. Soc., Che., Commun.*, 1996, 163.

23   S.-K. Chung, Y.-T. Chang and Y. Ryu, *Pure Appl. Chem.*, 1996, **68**, 931 (*Chem. Abstr.*, 1996, **125**, 143 168).

24   T. Hanamoto, Y. Sugimoto, Y. Yokoyama and J. Inanaga, *J. Org. Chem.*, 1996, **61**, 4491.

25   R. Khan, P.A, Konowicz, L. Gardossi, M. Matulova and S. de Gennero, *Aust. J. Chem.*, 1996, **49**, 293.

26   M.-Z. Liu, H.-N. Fan, Z.-W. Guo and Y.-Z. Hui, *Carbohydr. Res.*, 1996, **290**, 233.

27   M.-Z. Liu, H.-N. Fan, Z.-W. Guo and Y.-Z. Hui, *Chin. Chem. Lett.*, 1996, **14**, 190 (*Chem. Abstr.*, 1996, **125**, 33 959).

28   N.-X. Zhang, S.-G. Cao, H. Dong, Y.-B. Lin, Y.-Q. Ren, S.-P. Han and H. Yang, *Gaodeng Xuexiao Huaxue Xuebao*, 1996, **17**, 1404 (*Chem. Abstr.*, 1996, **125**, 301 341).

29   B. Danieli, F. Peri, G. Carrea, D. Monti and S. Riva, *Carbohydr. Lett.*, 1995, **1**, 363 (*Chem. Abstr.*, 1996, **124**, 317 623).

30   I. Matsuo, M. Isomura, R. Walton and K. Ajisaka, *Tetrahedron Lett.*, 1996, **37**, 8795.

31   M.S. Soedjak and J.E. Spradlin, *Biocatalysis*, 1994, **11**, 241 (*Chem. Abstr.*, 1996, **124**, 343 872).

32   M. Gelo-Pujic, E. Guibé-Jampel, A. Loupy, S.A. Galema and D. Mathé, *J. Chem. Soc., Perkin Trans. 1*, 1996, 2777.

33   N.P. Ivanova, E.F. Panarin, G.A. Kazanina, E.E. Kever I.I. Malakhova and V.M. Denisov, *Zh. Obshch. Khim.*, 1995, **65**, 1885 (*Chem. Abstr.*, 1996, **125**, 11 234).

34   S. Riva, B. Danieli and M. Luisetti, *J. Nat. Prod.*, 1996, **59**, 618 (*Chem. Abstr.*, 1996, **125**, 58 896).

35   S. Takayama, P.O. Livingston and C.-H. Wong, *Tetrahedron Lett.*, 1996, **37**, 9271.

36   M. Woudenberg-van Oosterom, P. van Rantwijk and R.A. Sheldon, *Biotech. Bioeng.*, 1996, **49**, 328 (*Chem. Abstr.*, 1996, **124**, 232 921).

37   L. Lay, L. Panza, S. Riva, M. Khitri and S. Tirendi, *Carbohydr. Res.*, 1996, **291**, 197.

38   F.J. Plou, M.A. Cruces, M. Bernabe. M. Martin-Lomas, J.L. Parra and A. Ballestros, *Ann. N. Y. Acad. Sci.*, 1995, **750**, 332 (*Chem. Abstr.*, 1996, **124**, 87 558).

39   H. Shirahashi, T. Morimoto, A. Nagatsu, N. Murakami, K. Tatta, J. Sakakibara, H. Tokuda and H. Nishino, *Chem. Pharm. Bull.*, 1996, **44**, 1404.

40   D. Colombo, F. Roncchetti, A. Scala, I.M. Taino and L. Toma, *Tetrahedron: Asymm.*, 1996, **7**, 771.

41   K.-F. Hsiao, F.-L. Yang, S.-H. Wu and K.-T. Wang, *Biotechnol. Lett.*, 1995, **17**, 963 (*Chem. Abstr.*, 1996, **125**, 87 515).

42   A. Trincone and E. Pagnotta, *Tetrahedron: Asymm.*, 1996, **7**, 2773.
43   J. Lee and R.I. Hollingsworth, *Tetrahedron*, 1996, **52**, 3873.
44   H. Kawamoto, S. Iwatsuru, F. Nakatsubo and K. Fumiaki, *Mokuzai Gakkaishi*, 1996, **42**, 868 (*Chem. Abstr.*, 1996, **125**, 329 155).
45   K.S. Feldman and R.S. Smith, *J. Org. Chem.*, 1996, **61**, 2606.
46   T. Itoh, J.-i. Chika, S. Shirakami, H. Ito, T. Yoshida, Y. Kubo and J.-i. Uenishi, *J. Org. Chem.*, 1996, **61**, 3700.
47   T. Tanaka, Y. Orii, G. Nonaka, I. Nishioka and I. Kouno, *Phytochemistry.*, 1996, **43**, 1345.
48   E.M. Gross, H. Meyer and G. Schilling, *Phytochemistry.*, 1996, **41**, 133.
49   H. Kawamoto, F. Nakatsubo and K. Murakami, *Phytochemistry.*, 1996, **41**, 1427.
50   A. Espada, C. Jiménez, J. Dopeso and R. Riguera, *Liebigs Ann. Chem.*, 1996, 781.
51   H.K. Hombrecher, S. Ohm and D. Koll, *Tetrahedron*, 1996, **52**, 5441.
52   G.-J. Boons, A. Burton and P. Wyatt, *Synlett*, 1996, 310.
53   B.M. Heskamp, H.J.G. Broxterman, G.A. van der Marel and J.H. van Boom, *J. Carbohydr. Chem.*, 1996, **15**, 611.
54   N.S. Utkina, S.D. Mal'tsev, L.L. Danilov and V.N. Shibaev, *Bioorg. Khim.*, 1995, **21**, 376 (*Chem. Abstr.*, 1996, **124**, 30 128).
55   N.S. Utkina, S.D. Mal'tsev, L.L. Danilov and V.N. Shibaev, *Bioorg. Khim.*, 1996, **22**, 314 (*Chem. Abstr.*, 1996, **125**, 301 333).
56   T.S. Martin, G. Dufner, B. Kratzer and R.R. Schmidt, *Glycoconjugate J.*, 1996, **13**, 547 (*Chem. Abstr.*, 1996, **125**, 301 410).
57   R.F. Sala, P.M. Morgan and M.E. Tanner, *J. Am. Chem. Soc.*, 1996, **118**, 3033.
58   A. Sanderud and E. Sagstuen, *J. Chem. Soc., Faraday Trans.*, 1996, **92**, 995 (*Chem. Abstr.*, 1996, **125**, 11 270).
59   B. Grzescyk, O. Holst and A. Zamojski, *Carbohydr. Res.*, 1996, **290**, 1.
60   K. Morisaki and S. Ozaki, *Carbohydr. Res.*, 1996, **286**, 123.
61   D.J. Jenkins and B.V.L Potter, *Carbohydr. Res.*, 1996, **287**, 169.
62   S.B. Khrebtova, M.P. Korteev, N.M. Pugashova, A.R. Bekker, V.K. Bel'skii, A.I. Stash, A.A. Nazarov, A.V. Ignatenko and E.E. Nifant'ev, *Zh. Obshch. Khim.*, 1996, **66**, 752 (*Chem. Abstr.*, 1996, **125**, 276 312).
63   L.J. Mazzella, D.W. Parkin, P.C. Tyler, R.H. Furneaux and V. Schramm, *J. Am. Chem. Soc.*, 1996, **118**, 2111.
64   L. Knerr, X. Paneecoucke, G. Schmitt and B. Zuu, *Tetrahedron Lett.*, 1996, **37**, 5123.
65   M.A. Brown, P.J. Cox, R.A. Howie, O.A. Melvin and J.L.Wardell, *J. Chem. Soc., Perkin Trans. 1*, 1996, 809.
66   J. Mlynarski and A. Banaszek, *Carbohydr. Res.*, 1996, **295**, 69.
67   A.H. Haines and D.J.R. Massy, *Synthesis*, 1996, 1422.
68   R. Tanaka, K. Miyahara and N. Noda, *Chem. Pharm. Bull.*, 1996, **44**, 1152.
69   M.S. Kamel, A.-N. El-Shorbagi, E. Karvinen and A. Koskinen, *Bull. Pharm. Sci.,Assiut Univ.*, 1995, **18**, 87 (*Chem. Abstr.*, 1996, **125**, 34 006).
70   A. Leydet, C. Jeantet-Segonds, P. Barthélémy, B. Boyer and J.P. Roque, *Recl. Trav. Chim. Pays-Bas*, 1996, **115**, 421.
71   W. Szeia and C. Niedzielski, *Wiad. Chem.*, 1995, **49**, 733 (*Chem. Abstr.*, 1996, **125**, 58 893).
72   Z.-W. Guo, S.-J. Deng and Y.-Z. Hui, *J. Carbohydr. Chem.*, 1996, **15**, 965.
73   N. Pelloux-León, I. Gauthier-Luneau, S. Wendt and Y. Vallée, *Tetrahedron: Asymm.*, 1996, **7**, 1007.
74   L.S. Jeong and V.E. Marquez, *Tetrahedron Lett.*, 1996, **37**, 2353.

75   K.R. King, J.M. Williams, J.R. Clamp and A.P. Corfield, *J. Carbohydr. Chem.*, 1996, **15**, 41.
76   Y.-M. Zhang, A. Brodzki, P. Sinay, F. Uzabiaga and C. Picard, *Carbohydr. Lett.*, 1996, **2**, 67 (*Chem. Abstr.*, 1996, **125**, 222 301).
77   M. Yoshida, Y. Kawakami, H. Ishida, M. Kiso and A. Hasegawa, *J. Carbohydr. Chem.*, 1996, **15**, 399.
78   K. Katsuraya, K. Inazawa, H. Nakashima and T. Uryu, *Front. Biomed. Biotechnol.*, 1996, 3 195 (*Chem. Abstr.*, 1996, **125**, 222 308).
79   R. Higuchi, M. Fujita, S. Matsumoto, K. Yamada, T. Miyamoto and T. Sasaki, *Liebigs Ann. Chem.*, 1996, 837.
80   Y. Du and F. Kong, *J. Carbohydr. Chem.*, 1996, **15**, 797.
81   J.-M. Vatèle and S. Hanessian, *Tetrahedron*, 1996, **52**, 10557
82   A. El Nemr, T. Tsuchiya and Y. Kobayashi, *Carbohydr. Res.*, 1996, **293**, 31.
83   R.J. Berki, E.R. Binkley, R.W. Binkley, D.G. Hehemann, D.J. Koholic and J. Masnovi, *J. Carbohydr. Chem.*, 1996, **15**, 33.
84   S. Oscarson and U. Tedebark, *J. Carbohydr. Chem.*, 1996, **15**, 507.
85   S. Sigismondi and P. Sinou, *J. Chem. Res. (S)*, 1996, 46.
86   R.G.G. Leenders, R. Ruytenbeek, E.W.P. Damen and H.W. Scheeren, *Synthesis*, 1996, 1309.
87   H. Yamashita, K. Amano, S. Shimada and K. Narasaka, *Chem. Lett.*, 1996, 537.

8
Halogeno-sugars

1 Fluoro-sugars

A review on the synthesis of *gem*-difluoromethylene compounds has described carbohydrate (including nucleoside) examples.[1] The synthesis of acetylated glycosyl fluorides directly from the unprotected methyl glycosides has been effected as the first step of a GLC analytical method. First the glycoside is treated with HF followed by the addition of acetic anhydride.[2] Full details (cf. Vol. 25, p.105) on the preparation of glycosyl fluorides by reaction of aryl 1-thioglycosides with 4-difluoroiodotoluene have been described.[3]

The synthesis of 3,4-di-*O*-acetyl-2,6-dideoxy-6,6,6-trifluoro-α-L-*lyxo*-hexopyranosyl bromide from D-lyxose features the use of timethylsilyltrifluoromethane in the presence of fluoride ion to effect the addition of the trifluoromethyl group to an aldehyde.[4] Similar methodology has been utilized in the synthesis of 5-deoxy-5,5,5-trifluoro-D-lyxose, -L-ribose, -L-arabinose and -D-xylose.[5] A symposium report has looked at synthetic routes to 6-deoxy-6,6,6-trifluorosugars[6] and the synthesis of racemic 3-deoxy-2-*C*-trifluoromethyl-*erythro* and -*threo* -pentofuranoses is covered in Chapter 14.

L-Xylose has been converted into a 2-deoxy-2-fluoro-β-L-arabinofuranose derivative, by way of a fluoride ion displacement of an imidazolylsulfonate, and then into a range of pyrimidine nucleosides.[7] Displacement of another imidazolylsulfonate with Et$_3$N.3HF initially formed the corresponding sulfonyl fluoride which was subsequently displaced by fluoride ion.[8] Methyl 2,3-dideoxy-3-fluoro-5-*O*-(4-methylbenzoyl)-α-D-ribofuranoside, an intermediate for the syntheses of some anti-HIV nucleosides, has been synthesized from 2-deoxy-D-ribose in five steps and in 24% yield.[9]

Treatment of the D-arabino- and D-xylo-furanosides **1** and **2** with DAST afforded the deoxyfluoro derivatives **3** and **4** in the former case, whereas the latter gave the anhydro-sugar **5** and the deoxyfluoro compound **6** (Scheme 1).[10] Osmium tetroxide-catalysed dihydroxylation of the chiral unsaturated derivative **7** led to the synthesis of 2-deoxy-2-fluoro-D-xylose and -L-lyxose.[11] 1,2,5-Trideoxy-2-fluoro-1,5-imino-D-glucitol and 1,2,5-trideoxy-1-fluoro-2,5-imino-D-mannitol have been synthesized as potential glucosidase inhibitors.[12] Other glucosidase inhibitors, 5-fluoro glucosides, have been prepared by photobromination (NBS, hv) of glycosyl fluoride **8** to give bromide **9** which was transformed into difluoride **10** (AgF, CH$_3$CN).[13]

Carbohydrate Chemistry, Volume 30
© The Royal Society of Chemistry, 1998

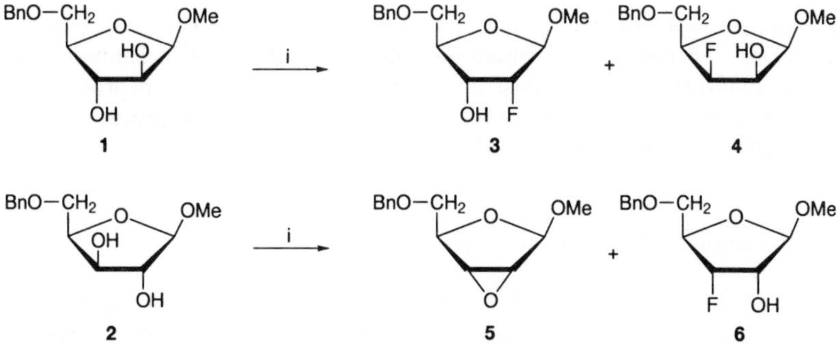

Reagents: i, DAST, CH₂Cl₂/Py

Scheme 1

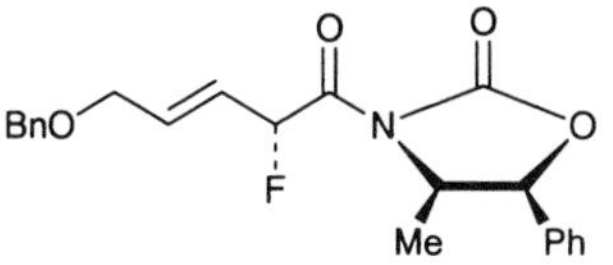

7 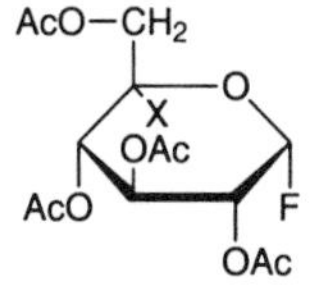   8 X = H   11
         9 X = Br
         10 X = F

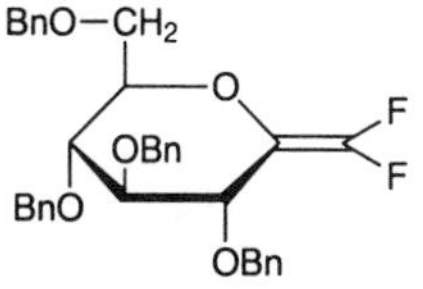

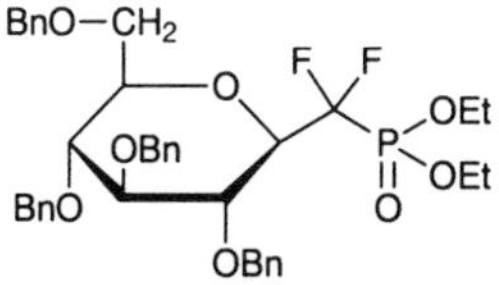

12

Some difluoromethylene phosphonates have been prepared by phosphonyl radical addition to difluoroenol ethers, *e.g.* **11** gave **12**.[14] Treatment of unsaturated 1,6-anhydro sugar **13** with DAST has afforded **14**, and the epoxide **15** was opened (KHF₂, ethylene glycol) to give deoxyfluoro sugar **16**.[15]

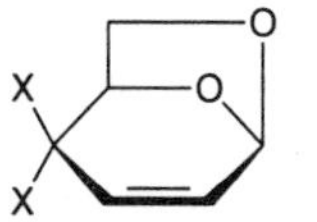

13 X = OH, Y = H   15   16
14 X = H, Y = F

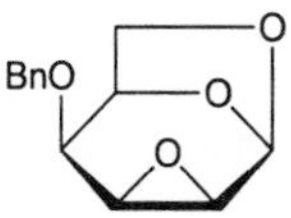

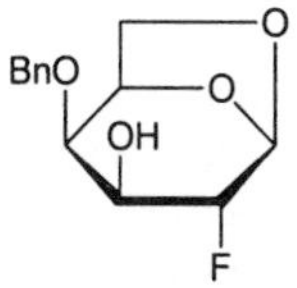

The basic hydrolysis of 1,3,4,6-tetra-*O*-acetyl-2-deoxy-2-[¹⁸F]fluoro-D-glucose for the preparation of 2-deoxy-2-[¹⁸F]fluoro-D-glucose has been described as an advantageous alternative to the commonly used acid hydrolysis approach.[16] However, the product epimerizes to 2-deoxy-2-[^{18}F] fluoro-D-mannose under basic conditions[17] so care must be taken with base-dependent deprotection of *O*-substituted derivatives. The synthesis of an antifreeze glycoprotein analogue using glycosyl fluorides is covered in Chapter 3.

2   Chloro-, Bromo- and Iodo-sugars

Glycosylation reactions using 2,3,4-tri-*O*-chlorosulfonyl-β-L-fuco- and -β-D-xylo-pyranosyl chloride have been studied. Previous reports suggested that both are good glycosyl donors affording β-glycosides selectively, but those findings are not general and in most cases the α-glycoside is the major product.[18] Chloride ion displacement applied to the sucrose 1′,6′-ditriflate **17** gave the dichloro-dideoxy derivative **18** or, under milder conditions, the monochloro compound **19** which was exposed to other nucleophiles to give **20**.[19]

17 X = Y = OTf
18 X = Y = Cl
19 X = OTf, Y = Cl
20 X = SAc, N₃, F, OMe, Y = Cl

21

Nucleophilic epoxide ring opening of benzyl 2,3-anhydro-α-D-ribopyranoside occurs exclusively at C-3 to give, for example, the 3-bromo-3-deoxy-D-xylose derivative,[20] whereas for benzyl 3,4-anhydro-β-D-ribopyranoside nucleophilic attack can occur at C-3 or C-4 depending on the choice of metal counterion.[21]

Selective chlorination (4-Me₂NC₆H₄PPh₂, CCl₄) of methyl 3,4-anhydro-β-D-tagatofuranoside successfully distinguished between the two primary hydroxyl groups to give the chloride **21**.[22] Unprotected D-aldoses have been selectively halogenated (Ph₃P, CX₄ or *N*-halosuccinimide in DMF) at the primary position.[23] When d-galactochloralose **22** was heated under reflux in DMF with the Vilsmeier reagent a mixture of **23-25** was produced.[24]

A number of 6-substituted D-fructose derivatives (including the chlorodeoxy- and deoxyiodo-compounds) have been prepared from the corresponding 3-substituted D-glyceraldehydes using an aldolase and dihydroxyacetone mono-phosphate.[25]

Photobromination (NBS, hv, CCl₄) of the acetylated 5-thio-xylopyranosyl β-bromide **26** affords the dibromide **27** whereas the α-bromide **28** gives a mixture of **27** and the 5-bromo-compound **29**. The 5-thio-glucosyl bromide **30**, under the

22   **23**   **24** R = CHO
                               25 R = H

same conditions, generates the tribromide **31**.[26] Similar treatment of the 2,6-anhydro-heptonamide derivative **32** gave rise to the glycosyl bromide **33**,[27] while the 2,5-anhydro-heptonate ester **34** gave **35**.[28] Photobromination, at C-2, of a 2,5-anhydro-heptonic acid derivative is involved in the synthesis of a hydantocidin analogue covered in Chapter 10 and the radical bromination of benzylidene derivatives in the presence of methanesulfonate esters is mentioned in Chapter 6.

26 X = Br, Y = H   **29**   **30**   **31**
27 X = Y = Br
28 X = H, Y = Br

32 X = H   **34** X = H
33 X = Br   **35** X = Br

2,5-Anhydro-L-idose has been converted into 2,5-anhydro-4-*O*-benzoyl-1-iodo-1,3,6-trideoxy-L-*arabino*-hexitol, a key synthon in the synthesis of (-)-*allo*-muscarine.[29]

References

1   M.J. Tozer and T.F. Herpin, *Tetrahedron*, 1996, **52**, 8619.
2   B.A. Bergamaschi and J.I. Hedges, *Carbohydr. Res.*, 1996, **280**, 345.
3   S. Caddick, L. Gazzard, W.B. Motherwell and J.A. Wilkinson, *Tetrahedron*, 1996, **52**, 149.
4   Y. Takagi, K. Nakai, T. Tsuchiya and T. Takeuchi, *J. Med. Chem.*, 1996, **39**, 1582.
5   P. Munier, M.-B. Giudicelli, D. Picq and D. Anker, *J. Carbohydr. Chem.*, 1996, **15**, 739.

6   T. Yamazaki, K. Mizutani and T. Kitazume, *A.C.S. Symp. Ser.*, 1996, **639**, 105
       (*Chem. Abstr.*, 1996, **125**, 301 338).
7   T. Ma, S.B. Pai, Y.L. Zhu, J.S. Lin, K. Shanmuganathan, J. Du, C. Wang, H. Kim,
       M.G. Newton, Y.C. Cheng and C.K. Chu, *J. Med. Chem.*, 1996, **39**, 2835.
8   T.S. Chou, L.M. Becke, J.C. O'Toole, M.A. Carr and B.E. Parker, *Tetrahedron
       Lett.*, 1996, **37**, 17.
9   Z.-W. Guo, B.-G. Huang, W.-J. Xiao and Y.-Z. Hui, *Chin. J. Chem.*, 1995, **13**, 363
       (*Chem. Abstr.*, 1996, **124**, 30 228).
10   I.A. Mikhailopulo and G.G. Sivets, *Synlett*, 1996, 173.
11   F.A. Davis and H. Qi, *Tetrahedron Lett.*, 1996, **37**, 4345.
12   S. Andersen, A. de Raadt, M. Ebner, C. Ekhart, G. Gradnig, G. Legler, I. Lundt,
       M. Schichl, A.E. Stutz and S. Withers, *Electron. Conf. Trends Org. Chem.*, 1995
       (Pub. 1996), Paper 19 (*Chem. Abstr.*, 1996, **125**, 33 993).
13   J.D. McCarter and S.G. Withers, *J. Am. Chem. Soc.*, 1996, **118**, 241.
14   T.F. Herpin, J.S. Houlton, W.B. Motherwell, B.P. Roberts and J.-M. Weibel,
       J. Chem. Soc., Chem. Commun., 1996, 613.
15   R. Haeckel, G. Lauer and F. Oberdorfer, *Synlett*, 1996, 21.
16   F. Fuechtner, J. Steinbach, P. Paeding and B. Johannsen, *Appl. Radiat. Isot.*, 1996,
       47, 61 (*Chem. Abstr.*, 1996, **124**, 202 753).
17   P. Varelis and R.K. Barnes, *Appl. Radiat. Isot*, 1996, **47**, 731 (*Chem. Abstr.*, 1996,
       125, 301 343).
18   R.E. Hubbard, J.G. Montana, P.V. Murphy and R.J.K. Taylor, *Carbohydr. Res.*,
       1996, **287**, 247.
19   H. Kakinuma, H. Yuasa and H. Hashimoto, *Carbohydr. Res.*, 1996, **284**, 61.
20   P.K. Vasudeva and M. Nagarajan, *Tetrahedron*, 1996, **52**, 5607.
21   P.K. Vasudeva and M. Nagarajan, *Tetrahedron*, 1996, **52**, 1747.
22   A.K. Norton and M. von Itzstein, *Aust. J. Chem.*, 1996, **49**, 281.
23   M.-L. Larabi, C. Frechu and G. Demailly, *Carbohydr. Lett.*, 1996, **2**, 61 (*Chem.
       Abstr.*, 1996, **125**, 222 268).
24   O. Makinabakan, Y.G. Salman and L. Yuceer, *Carbohydr. Res.*, 1996, **280**, 339.
25   P. Page, C. Blonski and J. Périé, *Tetrahedron*, 1996, **52**, 1557.
26   M. Baudry, M.-N. Bouchu, G. Descotes, J.-P. Praly and F. Bellamy, *Carbohydr.
       Res.*, 1996, **282**, 237.
27   L. Kiss and L. Somsak, *Carbohydr. Res.*, 1996, **291**, 43.
28   J.C. Estevez, J. Saunders, G.S. Besra, P.J. Brennan, R.J. Nash and G.W.J. Fleet,
       Tetrahedron: Asymm., 1996, **7**, 383.
29   V. Popsarin, O. Beric, J. Casanadi, M. Popsarin and D. Miljkovic, *J. Serb. Chem.
       Soc.*, 1995, **60**, 625 (*Chem. Abstr.*, 1996, **124**, 9 199).

9
Amino-sugars

1 Natural Products

A triglycosylceramide isolated from the marine sponge *Agelas dispar* contained a terminal 2-acetamido-2-deoxy-β-D-galactopyranosyl moiety.[1] The endiyne antibiotic C-1027 contained 4-deoxy-4-dimethylamino-5-*C*-methyl-D-allose.[2] A cytotoxic macrolide isolated from the marine sponge *Callipelta* sp. incorporated the carbamate-containing 4-amino-sugar moiety **1**, of undetermined absolute stereochemistry.[3] The polyhydroxylated *nor*-tropane glycoside **2** was isolated from the fruits of the *Nicandra physalodes* plant.[4]

2 Syntheses

Syntheses covered in this section are grouped according to the method used for introducing the amino-functionality.

2.1 By Chain Extension – 6-Amino-6-deoxy-D-fructose hydrochloride was synthesized by enzyme catalysed aldol condensation of 1,3-dihydroxyacetone with 3-amino-3-deoxy-D-glyceraldehyde, the latter being prepared by conversion of D-fructose into the known 2,3-anhydro-D-glyceraldehyde diethyl acetal and then subjection of this to ring-opening with azide ion followed by hydrogenolysis over a palladium catalyst.[5] 2-Acetamido-2-deoxy-D-mannono-1,5-lactone **4** was synthesized by a four-component Ugi reaction involving the D-arabinose-derived aldehyde **3** and 1-isocyanocyclohexene (Scheme 1).[6] The 6-acetamido-6,8-dideoxy-octos-7-ulose **7**, related to lincosamine, was obtained by chain extension of the dialdose derivative **5** at C-6 (Scheme 2). The intermediate **6** was obtained as a 1:1 mixture with its C-6 epimer.[7] In an alternative approach, the lincosamine

Carbohydrate Chemistry, Volume 30
© The Royal Society of Chemistry, 1998

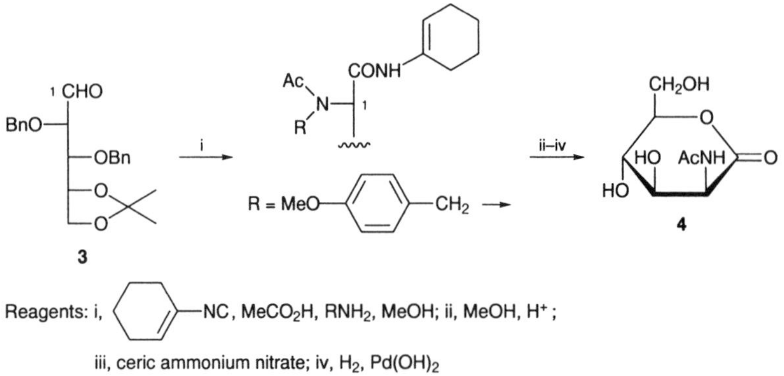

Reagents: i, —NC, MeCO₂H, RNH₂, MeOH; ii, MeOH, H⁺;

iii, ceric ammonium nitrate; iv, H₂, Pd(OH)₂

Scheme 1

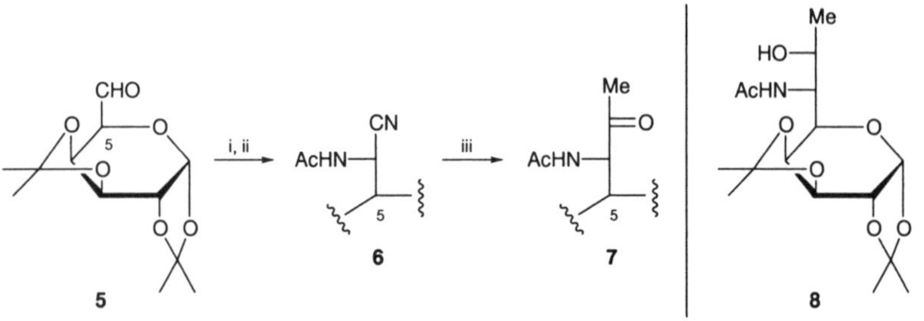

Reagents: i, KCN, NH₃, H₂O, NH₄Cl; ii, Ac₂O, Py; iii, MeMgI

Scheme 2

derivative **8** was obtained following hydroxyethylation of the *N*-benzylimino derivative of aldehyde **5**. Thus, addition of 1-(dimethylphenylsilyl)ethylmagnesium chloride to this imine in the presence of CuI and BF₃.OEt₂ gave largely the desired isomer, the silyl group of which was oxidatively replaced by a hydroxy group with retention of configuration.[8] The same approach was used for the highly diastereoselective synthesis of the precursor **9** of the 4-ethylamino-sugar of calicheamycin (Scheme 3).[9] Addition of phenylmagnesium bromide to 3-*O*-benzoyl-1,2-*O*-isopropylidene-α-D-xylofuranurononitrile gave 5-benzamido-5-deoxy-1,2-*O*-isopropylidene-5,5-di-*C*-phenyl-α-D-xylofuranose rather than the imine expected from addition of only one equivalent of this Grignard reagent.[10]

2.2   By Epoxide Ring Opening – The isonucleoside **10** was synthesized in nine steps from D-glucose, the key step being the regioselective addition of adenine to a 3,4-epoxide precursor.[11] Azide ring opening of a 2,3-anhydro-D-glyceraldehyde derivative was noted above.

2.3   By Nucleophilic Displacement – Reaction of 2,3:4,5-di-*O*-isopropylidene-β-D-fructopyranose 1-triflate with tyrosine benzyl ester gave an Amadori-type

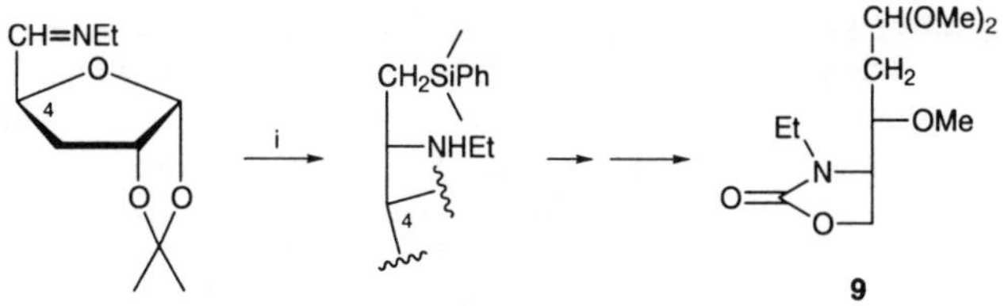

Reagents: i, PhSiCH$_2$Li, CeCl$_3$

Scheme 3

compound which was elaborated into *N*-(1-deoxy-D-fructos-1-yl)-peptides of the enkephalin series.[12] Displacements of 4-triflate groups by azide ions with inversion were utilized in the construction of a) the glycosyl bromides used in the preparation of the 3′-chloro-3′-deamino-4′-deoxy-4′-trifluoroacetamido-daunorubicin **11**;[13] b) *N*-(3-deoxy-L-*glycero*-tetronoyl)-D-perosamine-containing mono- and di-saccharides (e.g. **12**) which were identified as antigenic determinants of *Vibrio cholerae*;[14] and c) the 4-amino-4-deoxy-L-xylono-1,4-lactam derivative **13** from 2,3,5-tri-*O*-benzyl-D-arabinose.[15] *N*-Alkyl and *N*,*N*-dialkyl derivatives of 5-amino-5-deoxy-1,2-*O*-isopropylidene-D-xylofuranose were prepared by displacement of the corresponding 5-tosylate with primary and secondary amines, and evaluated as chiral catalysts for the addition of diethylzinc to aldehydes.[16]

All four isomeric 5-amino-5-deoxy-D-pentono-1,5-lactams have been prepared. D-Xylose and D-arabinose were converted into 5-azido-5-deoxy-1,2-*O*-isopropylidene derivatives via 5-tosylate intermediates. Oxidation and reduction of the free

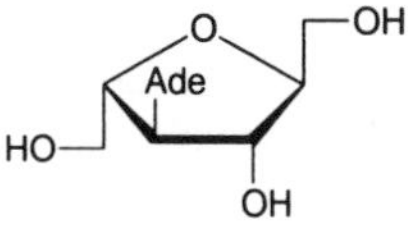

10

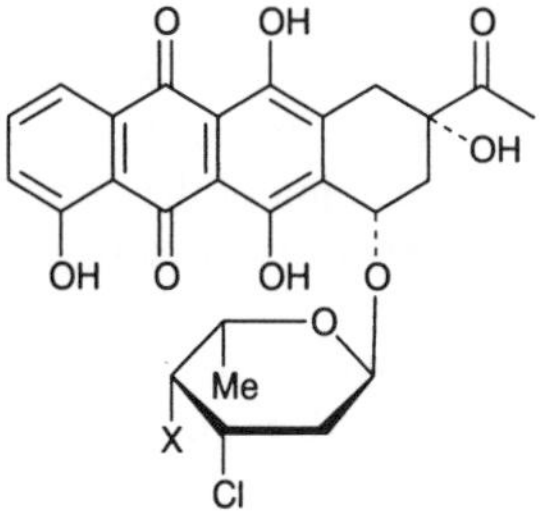

11 X = N$_3$ or NHCOCF$_3$

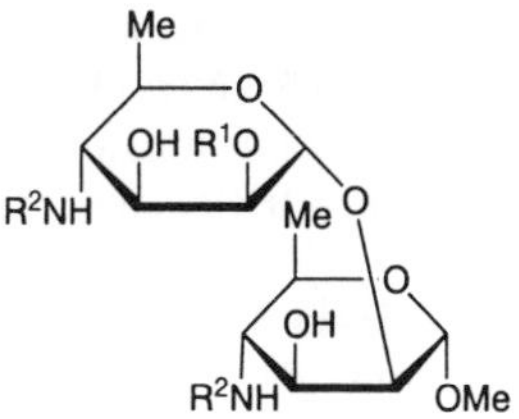

12 R^1 = H or Me

R^2 =

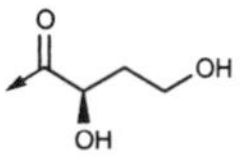

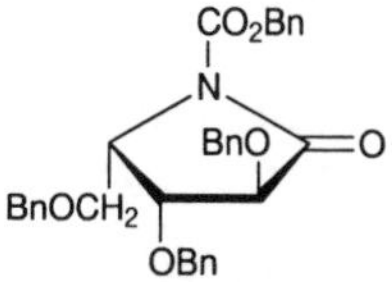

13

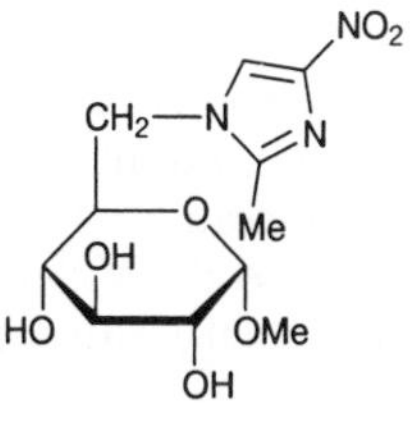

14

3-hydroxy group gave the corresponding D-ribose and D-lyxose isomers. Hydrolysis, bromine oxidation and catalytic hydrogenation gave the lactams.[17]

Several 'reversed azole nucleosides' such as the *N*-(glucos-6-yl)-imidazole **14** have been synthesized by displacement of a primary hydroxy group under Mitsunobu reaction conditions.[18] Methyl penta-*N,O*-acetyl-α-D-lincosaminide (see structure **8** for the stereochemistry of this compound) has been made from *myo*-inositol in a monumental 27 step synthesis, which involved optical resolution via a mandelate ester, ring cleavage following Baeyer-Villiger reaction, Wittig chain extension, osmium tetroxide induced *cis*-dihydroxylation, and introduction of the amino-group by mesylation and displacement with azide.[19] 6-Amino-6-deoxy-sucrose[20] and 6,6'-diamino-6,6'-dideoxy-cylcomaltoheptaose[21] have been synthesized following displacement of sulfonate groups with azide ion.

2.4   By Amadori Reaction – The enkastines, inhibitors of the bacterial endopeptidase enkephalinase, were confirmed as *N*-(1-deoxy-D-fructos-1-yl)-dipeptides by synthesis, through condensation of glucose with the dipeptides (e.g. Ile-Asp) and Amadori rearrangement of the products.[22] The *N*-(1-deoxy-D-fructos-1-yl) derivatives of various 6-substituted 3-amino-2-aryl-4(3H)-quinazolinones were obtained through condensation with glucose and Amadori rearrangement.[23] A kinetic study of the formation and degradation of 1-deoxy-1-*N*-morpholino-D-fructose by reaction of D-glucose with morpholine is covered in Chapter 10.

2.5   From Azido-sugars – *C*-Methyl, butyl and phenyl 2-azido-2-deoxy-β-D-gluco-and galacto-pyranosides, and thence the corresponding 2-acetamido analogues, were obtained from the corresponding 2-azido-2-deoxy-2,3,6-tri-*O*-benzyl-D-hexono-1,5-lactones by addition of the alkyl- or aryl-lithiums to the lactone moiety, and reduction of the resulting lactols with Et$_3$SiH and BF$_3$.OEt$_2$.[24] 4-Nitrophenyl 4-amino-4-deoxy-β-D-galactopyranoside was obtained by reaction of a 4-azido-2,3,6-tri-*O*-benzoyl-4-deoxy-α-D-galactopyranosyl chloride with 4-nitrophenoxide ion, and selective reduction of the azido group with 1,3-propane-dithiol and triethylamine; it was slowly hydrolysed by a β-galactosidase.[25]

2.6   From Unsaturated Sugars – Azido phenylselenation of the D-*arabino*-furanoid glycal **15** gave the inseparable anomeric 2-azido-2-deoxy-D-gluco-furanoses **16** as the major products (Scheme 4), whereas the corresponding pyranoid glycal is known to react more slowly and to give mixtures of *gluco*- and *manno*-isomers. The mixture **16** was converted to the corresponding separable 2-acetamido derivatives **17**, and also to 4,6-di-*O*-acetyl-2-azido-3-*O*-benzyl-2-deoxy-αβ-D-glucopyranose in three steps [i, Hg(OAc)$_2$; ii, HOAc, H$_2$O, iii, Ac$_2$O, Py, DMAP].[26] The aza-sugar analogue **20** of the chitinase inhibitor allosamidin has been obtained using a new modification of the sulfonamidoglycosylation procedure (cf. Vol.24, p.38) that employs the *N,N*-dibromide of 2-trimethylsilyl-ethylsulfonamide as a reagent to form the glycosyl donor **19** from glycal **18** (Scheme 5).[27]

2-Acetamido-2-deoxy-α-D-talopyranosides of serine and two hexose derivatives, e.g. **23**, were synthesized from tetra-*O*-benzoyl-2-hydroxy-D-galactal **21** via

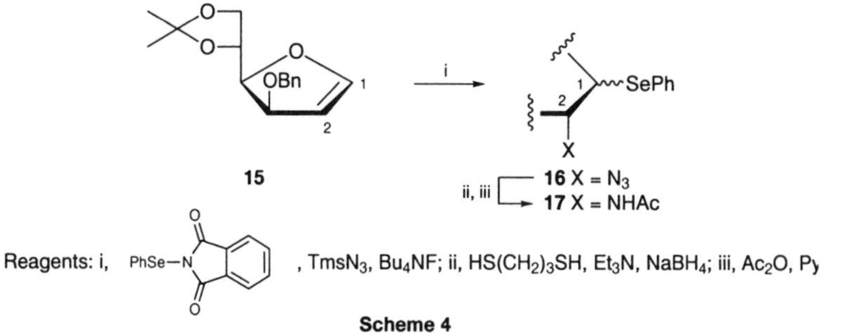
OBn
SePh
X
ii, iii
Reagents: i,
PhSe—N
, TmsN3, Bu4NF; ii, HS(CH2)3SH, Et3N, NaBH4; iii, Ac2O, Py

15 **16** X = N_3 **17** X = NHAc

Scheme 4

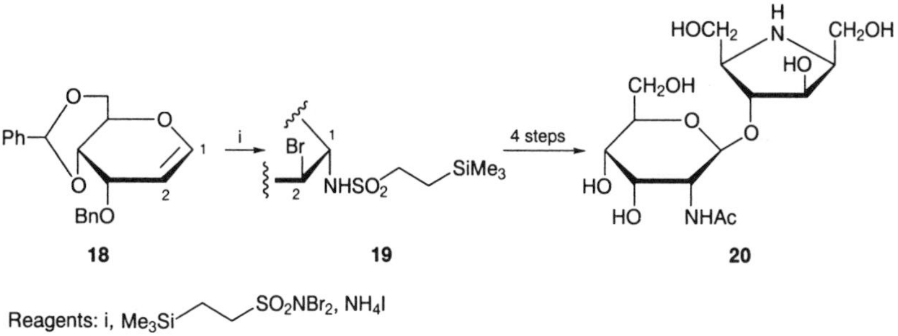
Ph
BnO
Br
NHSO2
SiMe3
4 steps
HOCH2
CH2OH
HO
NHAc
Reagents: i, Me3Si
SO2NBr2, NH4I

18 **19** **20**

Scheme 5

the 2-benzoyloximino-glycosyl bromide **22** which underwent α-selective glycosylation and *talo*-selective hydroboration (Scheme 6).[28] A mixture of the α-D-altropyranoside (68%) **26** and the corresponding α-D-taloside (14%) was obtained by allylic rearrangement of the 4-*O*-trichloroacetimidate **24** and *cis*-dihydroxylation of the resulting 2-deoxy-2-trichloroacetamido-hex-3-enopyranoside **25**. Alternatively, epoxidation of **25** and acetolysis yielded a mixture of the α-D-mannoside **27** (64%) and the corresponding α-D-idoside (13%) (Scheme 7).[29] Full details on the rearrangement of allylic cyanates to unsaturated amino-sugars (Vol.28, p.124) have been published.[30] The pyranosylphosphonic acid nucleotide analogues **30**, in which the distance between the base unit and the phosphorus atom approximates that in normal nucleotides, and its β-anomer, were obtained from tri-*O*-acetyl-D-glucal **28** (Scheme 8). Ferrier reaction gave **29** and its β-anomer in similar amounts, and these were separately subjected to a Mitsunobu reaction to replace the allylic 4-hydroxy group with nucleo-bases with inversion. The α-anomers of the products readily isomerized to the corresponding 3-deoxy-hex-1-enopyranose isomers on treatment with methanolic ammonia.[31] When the pentaacetate of 3-deoxy-D-*glycero*-D-*galacto*-nonulosonic acid (KDN) methyl ester was treated with trimethylsilyl triflate in acetonitrile, a Ritter-type reaction took place in which allyl carbocation reacted with solvent to form an acetonitrilium ion that reacted with water to give the epimeric acetamido-sugars **31** on work-up.[32] The synthesis of hydroxylamino-sugar derivatives by Michael additions to unsaturated lactones is covered in Chapter 10.

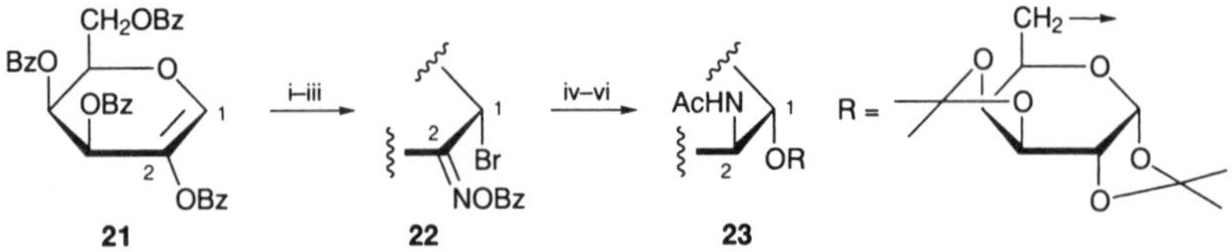

Reagents: i, NH$_2$OH, Py; ii, BzCl,Py; iii, NBS, *hv*;
iv, ROH, AgOTf, tetramethylurea; v, BH$_3$·THF; vi; Ac$_2$O,MeOH

Scheme 6

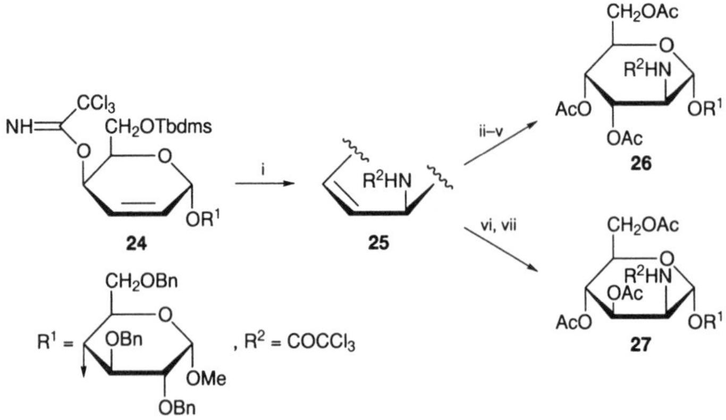

Reagents: i, DMF, 140 °C; ii, OsO$_4$, Py; iii, NaHSO$_4$; iv, Bu$_4$NF; v, Ac$_2$O, Py;
vi, MCPBA; vii, Ac$_2$O, AcOH, BF$_3$·OEt$_2$

Scheme 7

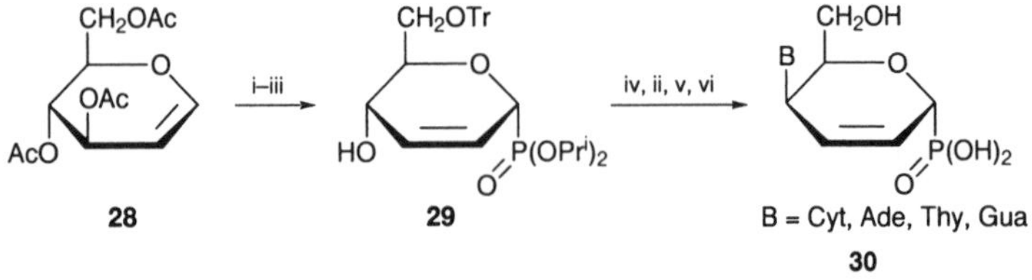

Reagents: i, P(OPr^i)$_3$, BF$_3$·OEt$_2$; ii, NH$_3$,MeOH; iii, TrCl, Py; iv, *N*-Bz-BH,
DEAD, Ph$_3$P; v, HOAc, H$_2$O; vi, Me$_3$SiBr

Scheme 8

2.7   From Aldosuloses – Reductive amination of the 1-aldehydo-derivative produced by oxidation of 2,3;4,5-di-*O*-isopropylidene-β-D-fructopyranose, with a variety of amino acids, followed by deprotection, gave Amadori rearrangement products such as **32**.[33] The 3,4-ditosylate **33** was obtained similarly by reductive amination with propylamine, selective hydrolysis of the 3,4-*O*-isopropylidene group, tosylation and acid hydrolysis. In aqueous buffer at pH 7.4, it gave the rearrangement products **34** and **35**.[34] The conversion of the D-galactopyranoside **36** into 2-amino-2-deoxy-D-talopyranoside **37** by an oxidation, oximation, reduction sequence has been optimized (Scheme 9). Compound **37** was obtained along with its D-*galacto*-isomer in the ratio 4:1, and the overall yield did not exceed 50%.[35]

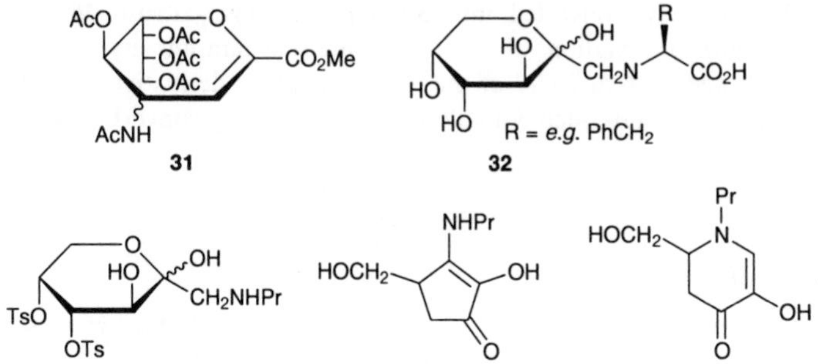

31

32

33

34

35

36

37

Reagents: i, Pr_4N^+ ruthenate, *N*-methylmorpholine *N*-oxide; ii, NH_2OH; iii, BzCl, Py; iv, LAH; v, Ac_2O, Py

Scheme 9

2.8 From Aminoacids – The synthesis of a constituent amino-sugar of calicheamicin by asymmetric allylboronation of an L-serinal derivative (Vol. 29, p. 131) has been reviewed.[36] (+)-Elsaminose **39**, a constituent of the antibiotic elsamicin A, was synthesized from the L-threonine-derived D-threose derivative **38** (Scheme 10).[37]

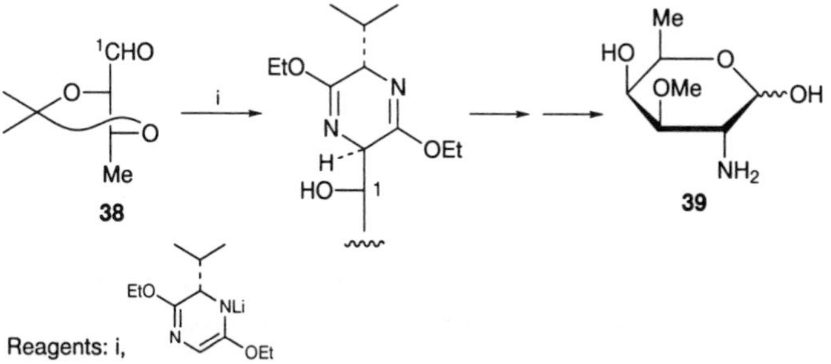

38

39

Reagents: i,

Scheme 10

2.9 From Chiral Non-carbohydrates – A section on amino-sugars has been included in a review of the synthesis of monosaccharides from non-carbohydrate sources.[38] L-Ristosamine **42** was synthesized by addition of the C_3-synthon **40** (which contains a masked aldehyde function that can be readily demasked by mild acid hydrolysis) to the L-lactaldehyde derivative **41** (Scheme 11).[39] The 'TBSOP' adduct **43** (Vol.27, p.114), derived from 2,3-O-isopropylidene-D-glyceraldehyde, has been converted into 3-amino-3-deoxy-D-altrose **46** by a route involving *cis*-hydroxylation of its unsaturated lactam moiety and periodate cleavage between C-6 and C-7 of the derived heptitol derivative **45** as key steps (Scheme 12). 3-Amino-3-deoxy-L-allose was obtained by converting **43** to its C-3

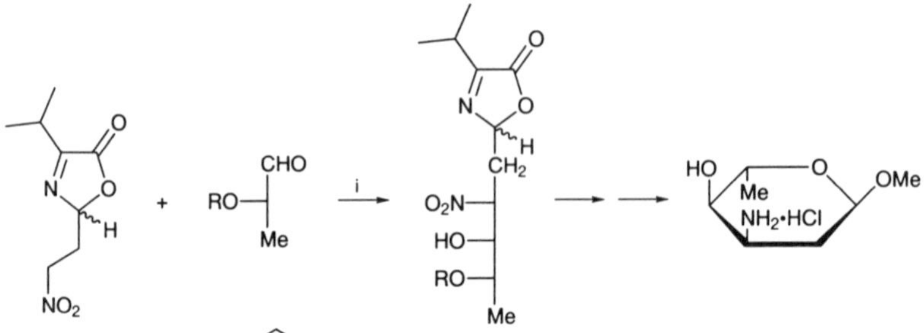

Reagents: i, KF / Al$_2$O$_3$

Scheme 11

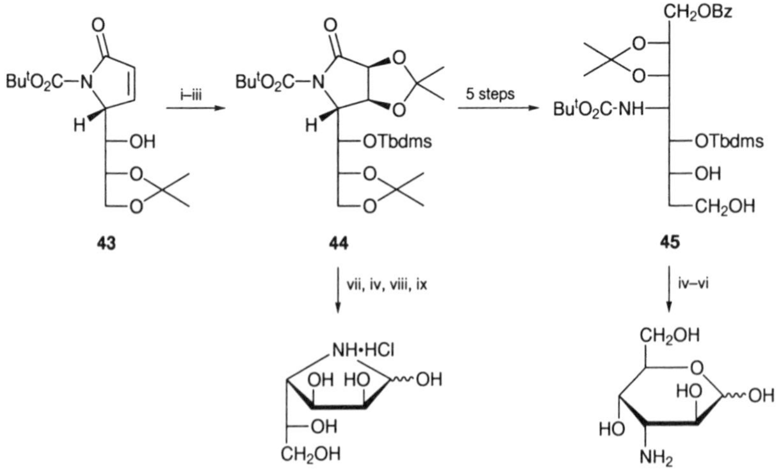

Reagents: i, TbdmsCl, imidazole, DMF; ii, KMnO$_4$, dicyclohexano-18-crown-6; iii, Me$_2$C(OMe)$_2$, H$^+$;
iv, NaIO$_4$; v, H$_3$O$^+$; vi, NaOMe; vii, HOAc, H$_2$O; viii, NaBH$_4$; ix, H$_3$O$^+$

Scheme 12

epimer by treatment with Et$_3$N and DMAP, and then using the same sequence of reactions. 4-Amino-4-deoxy-D-talose **47** was obtained from intermediate **44** by partial hydrolysis, periodate cleavage between C-6 and C-7 and concomitant reduction of the C-6 aldehyde and the lactam moiety with borohydride.[40] The 4-amino-4,5-dideoxy-D-xylose derivative **49** was obtained from the allylic alcohol **48** by Overman rearrangement followed by osmium tetroxide-catalysed *cis*-dihydroxylation and cleavage of the chiral auxiliary group from C-1 (Scheme 13).[41]

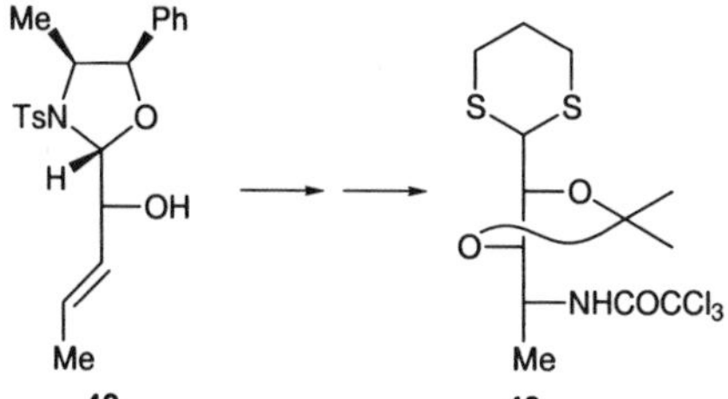

Scheme 13

2.10 From Achiral Non-carbohydrates – Methyl α-L-daunosaminide **53** has been synthesized from methyl sorbate **50** by addition of a chiral amine to give **51** in 95% d.e., followed by osmylation with concomitant lactonisation to give the lactone **52** and its isomer epimeric at C-4 and C-5, in the ratio 2:3 (Scheme 14). Osmylation in the presence of a chiral ligand altered this ratio to 1:3 in favour of the desired lactone.[42] 3-Amino-3-deoxy-L-taloside **56** was obtained from ethyl sorbate *via* the nitrone **54**, the preparation of which involved a Sharpless asymmetric dihydroxylation. Cycloaddition with vinylene carbonate gave the isoxazolidine **55** with high diastereoselectivity (Scheme 15). Compound **56** was converted to a protected glycosyl fluoride for use in the construction of the antifungal macrolide SCH 38516.[43] 5-Amino-5-deoxy-DL-arabinono-1,5-lactam and 5-amino-5,6-dideoxy-DL-altrono-1,5-lactam have been synthesized from dihydropyridines by routes reported earlier (see e.g. Vol. 28, p.227).[44]

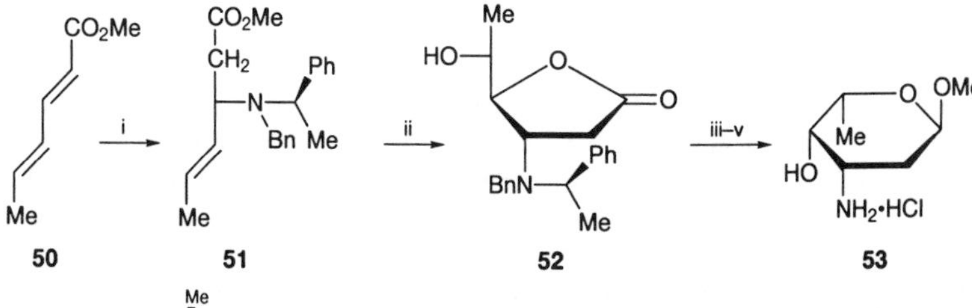

Reagents: i, Ph‑CH(Me)‑NBn‑Li , ii, OsO$_4$, K$_3$Fe(CN)$_6$, K$_2$CO$_3$; iii, H$_2$, Pd / C; iv, Bu$^i_2$AlH; v, MeOH, HCl

Scheme 14

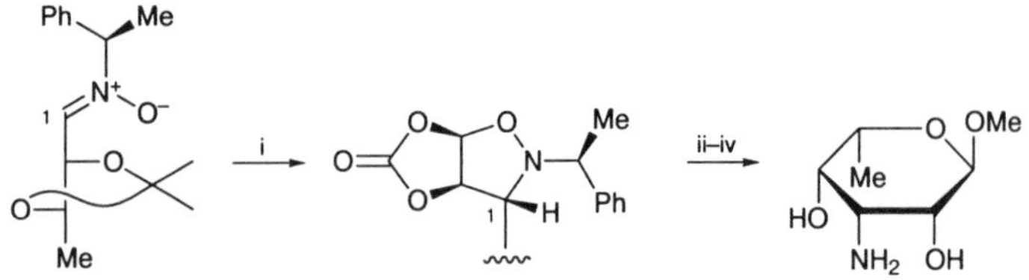

Reagents: i, [epoxide]=O ; ii, HCl; iii, H$_2$, Pd(OH)$_2$ /C; iv, MeOH, HCl

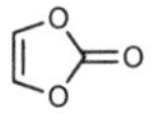

Scheme 15

3   Properties and Reactions

3.1   Enzyme Inhibition – 2-Acetamido-2-deoxy-D-glucopyranose, chitobiose and chitotriose were β-glycosidically linked to various exo-glycosidase inhibitors, and evaluated as chitinase inhibitors. Good inhibition was found for an *N*-formyl-pyrrolidine analogue (transition state mimic) and epoxybutyl chitobioside (an irreversible inhibitor).[45]

3.2   *N*-Acyl Derivatives – Chito-oligosaccharides, especially the pentamer, hexamer and heptamer, have been obtained by enzymic hydrolysis of chitosan and *N*-acetylation.[46] The *N*-trichloroethoxycarbonyl group, alternatively abbreviated as Troc or Teoc, has become a popular protecting group for 2-amino-sugar glycosyl donors, since these give high yields of β-glycosides, and the protecting group can be removed under non-basic conditions (e.g. Zn under acidic conditions).[47,48] Ethyl 2-acetamido-3,4,6-tri-*O*-acetyl-1-thio-2-*N*-trichloro-ethoxycarbonyl-β-D-glucopyranoside was also an effective glycosyl donor.[47] *N*-Tetrachlorophthaloyl or *N*-(4,5-dichlorophthaloyl) groups were useful alternatives to the *N*-phthaloyl for *N*-protection in 2-amino-sugar glycosyl donors because the donors gave products with high β-selectivity and the protecting groups could be removed under milder conditions (e.g. ethylenediamine).[49-51] *N*-Phthaloyl and *N*-tetrachlorophthaloyl groups could be cleaved by use of Merrifield resins bearing *N*-(*p*-linked-benzyl)-1,ω-ethylene-, butane- or hexane-diamine ligands.[52]

Three cyclic pentapeptide and two tetrapeptide mimics in which a central amino acid is replaced by a sugar amino acid, such as methyl 2-amino-2-deoxy-α-or β-D-glucopyranuronic acid, have been synthesized.[53] 'Glycotides', oligomeric analogues of nucleotides linked by amide bonds, e.g. the triglycotide **57**, have been constructed by standard, solution peptide construction methods. Twelve azidodeoxy-sugar acid ester building blocks were synthesized and coupled by reduction of the azido-group of one component to an amine which was then condensed with the acid released by saponification of another component. Reaction of such amines with 1,3,5-benzenetri(carbonyl chloride) was the basis for the preparation of combinatorial libraries; with a single amine, the trifunctional core **58** was obtained, while with 4 equivalents of three different amines at one time, a small library of nine products was produced.[54]

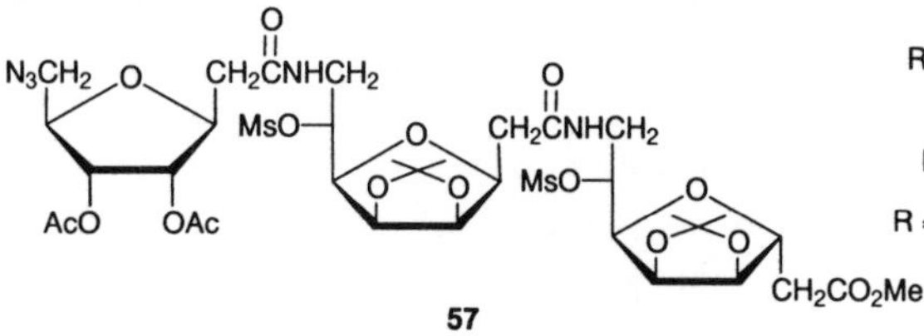

57 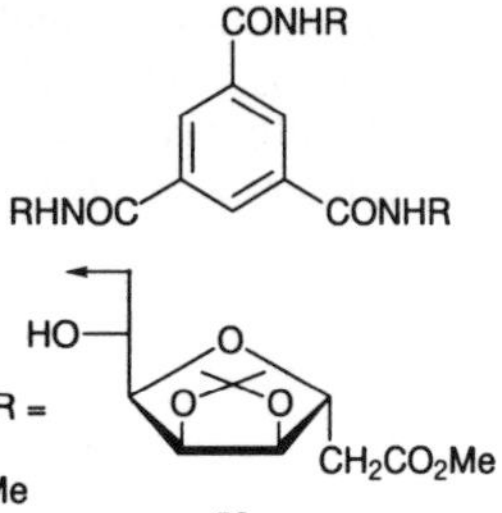   **58**

The uracil derivative **59** has been synthesized and found to be herbicidal,[55] while 5-fluorouracil linked to 2-amino-2-deoxy-D-glucose via various aminoacid linkers gives products, e.g. **60**, that have better anti-tumour activity than 5-fluorouracil itself.[56] *S*-Nitrosothiols such as **61** have been prepared from the corresponding unprotected amino-sugars by reaction with 3-acetamido-4,4-dimethylthietan-2-one, and their aqueous solubility and stability determined.[57] A further methyl 4,6-dideoxy-4-(3-deoxy-L-*glycero*-tetronamido)-α-D-manno-pyranoside derivative (cf. Vol. 29, p.138) has been synthesized, using new *N*-acylating agents derived from homoserine and L-malic acid,[58] and a number of (1→2)-linked di- and hexa-saccharides comprised of this sugar as the repeating unit and with functionalized spacer aglycons suitable for linking to proteins, have been prepared.[59,60]

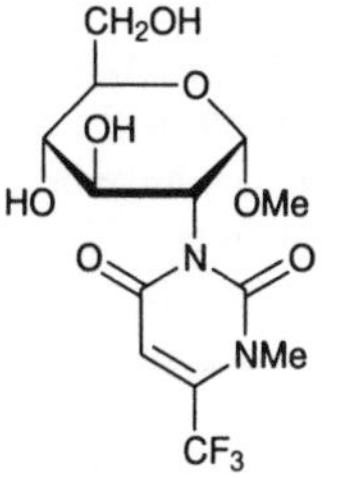

59 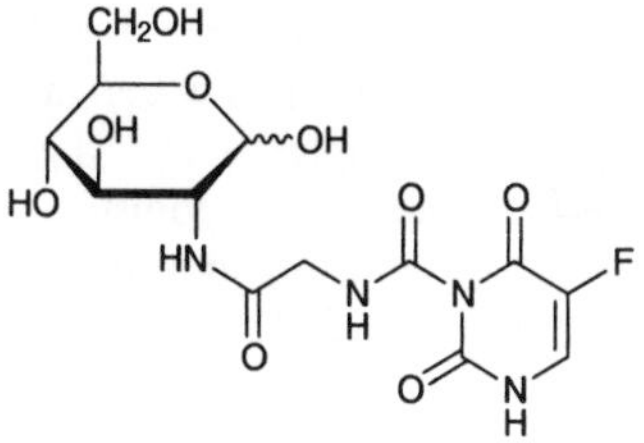   **60** 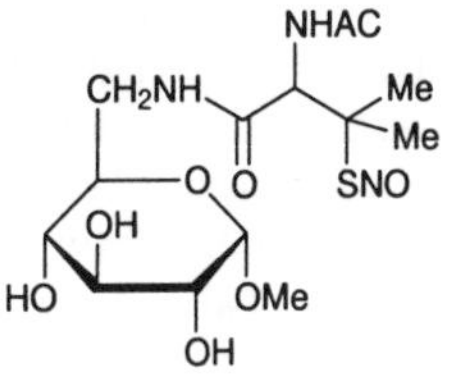   **61**

3.3 Phosphamides, *N*-Sulfates and Related Esters – 5-*N*-,3-*O*-Cyclic phosphamides have been prepared by reaction of 5-amino-5-deoxy-1,2-*O*-isopropylidene-α-D-xylofuranose with either phenylphosphonic dichloride or phosphoryl chloride, and their conformations determined.[61] Chemo-selective *N*-sulfation of 2-amino-2-deoxy-β-D-glucopyranose derivatives with a single free hydroxy-group was achieved with phenyl chlorosulfate followed by hydrolysis in aqueous NaHCO$_3$.[62]

3.4   Isothiocyanates and Related Compounds – Thiourea derivatives have been made by reaction of 6-deoxy-6-isothiocyanato-β-cyclodextrin with either methyl 6-amino-6-deoxy-β-D-glucopyranoside or 6-amino-6-deoxy-1-*N*-(*N-tert*-butoxyglycinyl)-β-D-glucopyranosylamine.[63] 3-Deoxy-3-thioureidoaldoses, formed on acid hydrolysis of N^1-(1,2:5,6-di-*O*-isopropylidene-α-D-allo- and -gluco-furanos-3-yl)-thioureas and their N^3-phenyl analogues, gave cyclic thiourea derivatives **62** on treatment with a base resin.[64]

62

3.5   *N*-Alkyl Derivatives – 2-Carboxymethylamino-2-deoxy-D-glucose, -mannose and -galactose were obtained by direct carboxmethylation of the free aminosugars.[65] Tetra-*O*-acetyl-2-deoxy-2-(4-nitro- and 2-methyl-4-nitro-imidazolyl)-D-glucopyranose have been obtained by reaction of 2-amino-2-deoxy-glucose with 1,4-dinitroimidazole and its 2-methyl analogue.[66] The radical-induced cyclization of a phenyl 2-allylamino-2-deoxy-1-seleno-α-D-glycopyranosides to give bicyclic *C*-glycosides is covered in Chapter 3. Syntheses of 6-(4-carboxamido-1,2,3-triazolyl)-6-deoxy-D-ribose and -D-xylose from the corresponding azido-sugar derivatives is covered in Chapter 10.

3.6   Lipid A Analogues – Carboxymethyl and 3-carboxypropyl 2-amino-2-deoxy-α- and β-D-glucopyranoside 4-phosphate derivatives bearing a 2,2-difluoro-fatty acyl group on N-2 and a complex fatty acyl group on O-3, have been synthesized as potential lipopolysaccharide antagonists.[67] Six analogues of Lipid A, comprising 2-amino-2-deoxy-β-D-glucopyranosides of *N*-myristoyl-L-serine with complex fatty acyl groups on N-2 or O-3 or both, were synthesized conventionally. Three were mitogenic, and one of these was able to stimulate cellular nitric oxide production.[68] Chitobiose derivatives with lauroyl and/or myristoyl groups on N-2 and O-3 in both sugar residues were weakly mitogenic but stimulated significant cellular nitric oxide production.[69]

3.7   Amidine and Guanidine Derivatives – The *N*-alkyl-D-glucosamidinium salts **63** have been synthesized by condensation of nojirimycin with the appropriate amine followed by oxidation with iodine.[70] Amidine-linked pseudo-disaccharide **66** has been synthesized by reaction of the thionolactam **64** with the 6-amino-sugar **65** (Scheme 16), and a (1→4)-linked analogue has been obtained similarly.[71]

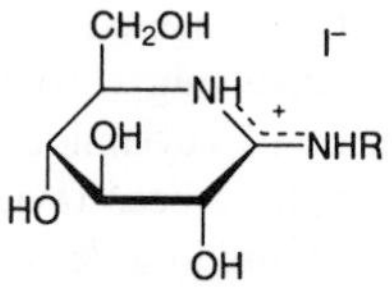

R = C₄H₉ or C₁₂H₂₅

63

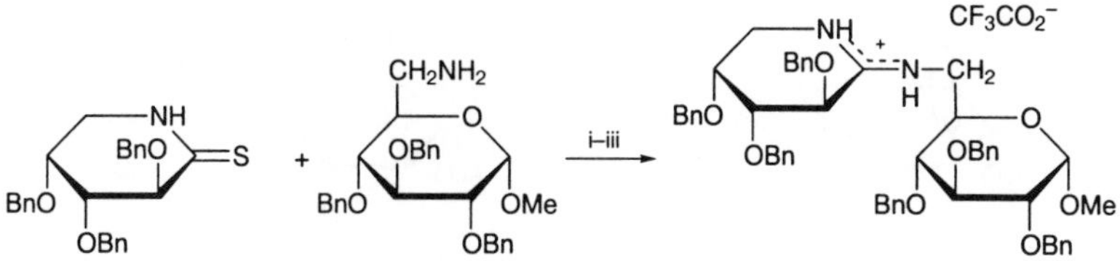

Reagents: i, HgCl₂, Et₃N; ii, (BuᵗO₂C)₂O, Py; iii, CF₃CO₂H

Scheme 16

3.8 Assorted Derivatives – Stability constants for copper(II) complexes of amino-sugars have been reviewed.[72] Copper (II) complexes of two 1,6-anhydro-amino-sugars are covered in Chapter 17. The thermal degradation (130 °C, 20 min) of the Amadori product 1-deoxy-1-*N*-prolinyl-D-fructose has been studied. Identification of various aldoses, including glucose, and (deoxy)aldonic acids by GLC–MS analysis of pertrimethylsilylated derivatives, was taken as further evidence of the reversibility of the Amadori reaction.[73] Selective 3,6-di-*O*-benzylation of 4-methoxyphenyl 2-deoxy-2-phthalimido-β-D-glucopyranoside was achieved after activation with dibutyltin oxide.[74] The mechanism of UDP-*N*-acetylglucosamine 2-epimerase has been deduced to involve formation and recapture of 2-acetamido-2-deoxy-D-glucal from studies employing substrate ¹⁸O-labelled at the glycosidic position.[75] A new synthesis of the potent β-glucuroni-dase inhibitor **67**, which also has antimetastatic activity, used the fermentation-derived *N*-acetyl-analogue siastatin B as the starting point.[76]

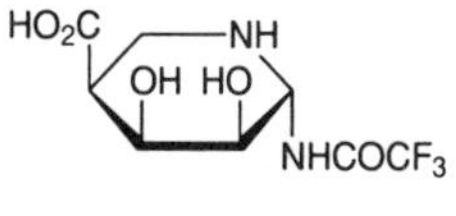

67

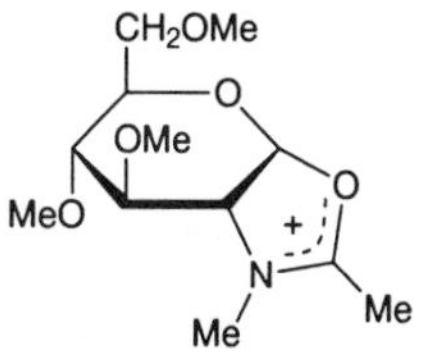

68

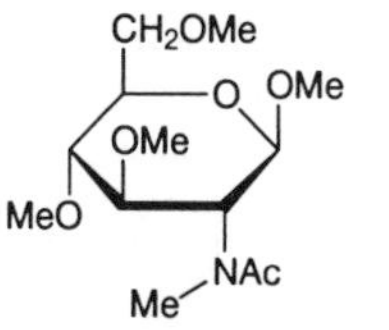

69

Conditions for use in the reductive-cleavage analysis of permethylated poly-saccharides (BF₃.OEt₂ or TmsOTf, 1,2-dichloroethane, 70 °C) wherein all 2-deoxy-2-acetamido-*N*-methyl-hexopyranoside residues are converted into oxazolinium ions, e.g. **68**, which give methyl glycosides, e.g. **69**, on quenching with methanol, have been found through model studies.[77] The preparation of sugar-containing macromolecules, involving co-polymerization of the acrylate macromonomer **70** with styrene, or of acylamide with poly-(*O*-glycosyl)serine macromonomers made from e.g. **71**, have been reviewed.[78] The preparation of a 2-*N*,3-*O*-cyclic sulfate of benzyl 2-acetamido-2-deoxy-4,6-*O*-benzylidene-β-D-allo-pyranoside and displacement with inversion at C-3 by a 1-thio-L-fucose nucleo-phile[79] is detailed in Chapter 11.

70   **71**

Syntheses of amino-sugar *C*-glycosides and dihydroxy-amino acids from 2-amino-2-deoxy-D-glucose are covered in Chapters 3 and 24, respectively, forma-tion of 1,4-pyridazines by dimerization of 2-amino-sugars is covered in Chapter 10, and [1]H-n.m.r. studies of 2-deoxy-2-ureido-β-glucosides in Chapter 21.

4   Diamino-sugars

The 2,5-anhydro-3,4-diamino-pentose diethyl acetal **72** was synthesized from L-xylose by sequential reaction of epoxide and triflate intermediates with azide ion. After reduction to a 3,4-diamino-2,5-anhydroalditol, oxoruthenium(V) com-plexes of it and related 2,3-diamino-nucleosides (see Chapter 17) were prepared and evaluated as ribonuclease inhibitors.[80] A number of routes were investigated for the preparation of the 2-amino-3-azido-2,3-dideoxy-D-glucoside **73**, a pre-cursor for a mimetic of the cyclopdepsipeptide didemnin B, from 2-acetamido-2-deoxy-D-glucose. Because *N*-deacetylation was much easier in the presence of a free 3-hydroxy-group, the best route involved preparation of a 2-deoxy-2-trifluoroacetamido-D-allopyranoside, and displacement of a mesylate group from C-3 by azide ion with inversion.[81]

72   **73**

2,4-Diamino-2,4-dideoxy-L-arabinose and -L-ribose, **75** and **76** respectively, were obtained in similar quantities by chain extension of the L-serine-derived aldehyde **74** as shown in Scheme 17.[82] The imino-sugar glycosylamide **78** was obtained along with a smaller amount of its C-2 epimer as shown in Scheme 18 from the α,β-unsaturated acid **77**, derived from the fermentation product siastatin B. These compounds and the guanadino-analogue **79** were prepared for evaluation as antimetastatic agents.[83] The racemic 3-guanidino-hex-4-enuronic acid derivatives **80**, **81** and 1,2-diepi-**80** were obtained by way of the hetero-Diels-Alder condensation shown in Scheme 19; they showed only weak sialidase inhibition.[84]

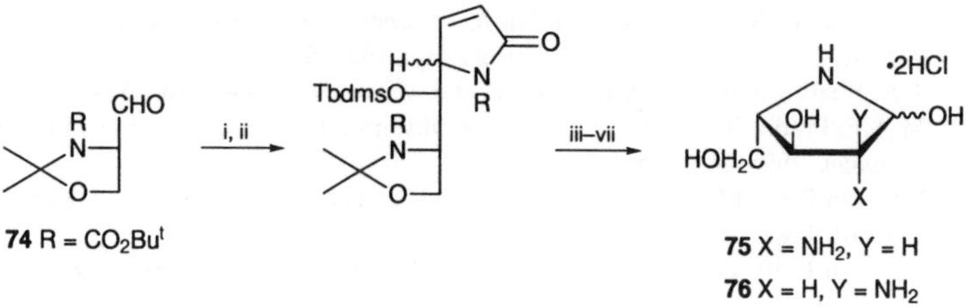

74 R = CO₂Buᵗ

75 X = NH₂, Y = H
76 X = H, Y = NH₂

Reagents: i, TbdmsO—⟨pyrrole⟩N-R , BF₃·OEt₂; ii, TbdmsCl; iii, KMnO₄; iv, LiOH; v, NaIO₄; vi, H⁺, MeOH; vii, H₃O⁺

Scheme 17

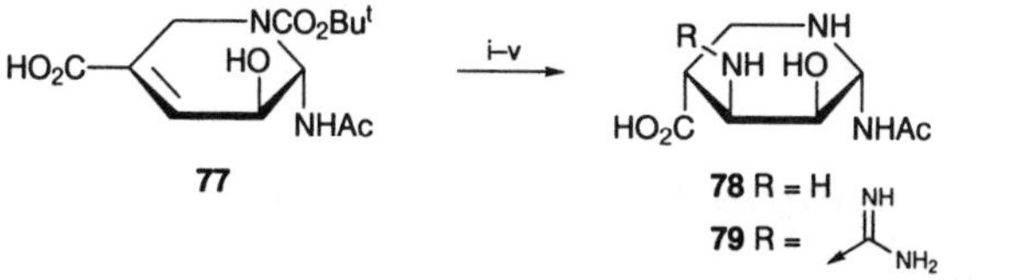

77

78 R = H
79 R = guanidino

Reagents: i, Ph₂C=N₂; ii, Cl₃CCN, DBU; iii, TsOH, H₂O; iv, NaBH₄; v, H₃O

Scheme 18

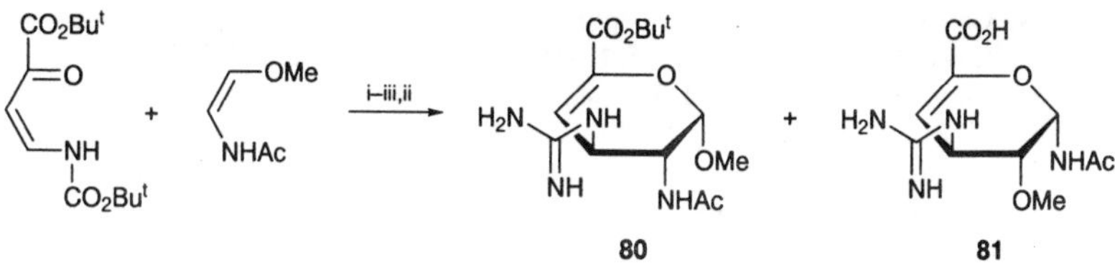

80

81

Reagents: i, SnCl₄; ii, CF₃CO₂H; iii, pyrazole-C(NCO₂Buᵗ)(NHCO₂Buᵗ)

Scheme 19

References

1   V. Constantino, E. Fattorusso, A. Mangoni, M. Di Rosa, A. Inaro and P. Maffia, *Tetrahedron*, 1996, **52**, 1573.
2   K. Iida, S. Fukuda, T. Tanaka, M. Hirama, S. Imajo, M. Ishiguro, K. Yoshida and T. Otani, *Tetrahedron Lett.*, 1996, **37**, 4997.
3   A. Zampella, M.V. D'Auria, L. Minale, C. Debitus and C. Roussakis, *J. Am. Chem. Soc.*, 1996, **118**, 11085.
4   R.C. Griffiths, A.A. Watson, H. Kizu, N. Asano, H.J. Sharp, M.G. Jones, M.R. Wormald, G.W.J. Fleet and R.J. Nash, *Tetrahedron Lett.*, 1996, **37**, 3207.
5   P. Page, C. Blonski and J. Périé, *Tetrahedron*, 1996, **52**, 1557.
6   T.A. Keating and R.W. Armstrong, *J. Am. Chem. Soc.*, 1996, **118**, 2574.
7   M.A.F. Prado, R.J. Alves, A. Braga de Oliveira and J. Dias de Souza Filho, *Synth. Commun.*, 1996, **26**, 1015.
8   F.L. van Delft, M. de Kort, G.A. van der Marel and J.H. van Boom, *J. Org. Chem.*, 1996, **61**, 1883.
9   F.L. van Delft, G.A. van der Marel and J.H. van Boom, *Carbohydr. Lett.*, 1996, **2**, 53 (*Chem. Abstr.*, 1996, **125**, 222 296).
10   M. Koos, B. Steiner and J. Alfoldi, *Chem. Pap.*, 1994, **48**, 426 (*Chem. Abstr.*, 1996, **124**, 9 210).
11   H.W. Yu, L.T. Ma and L.H. Zhang, *Chin. Chem. Lett.*, 1995, **6**, 655 (*Chem. Abstr.*, 1996, **124**, 56 526).
12   A. Jakas and Š. Horvat, *J. Chem. Soc., Perkin Trans. 2*, 1996, 789.
13   N. Aligiannis, N. Pouli, P. Marakos, A.-L. Skaltsounis, S. Leonce, A. Pierre and G. Atassi, *Bioorg. Med. Chem. Lett.*, 1996, **6**, 2473.
14   A. Arencibia-Mohar, Q. Madrazo-Alonso, A. Ariosa-Alvarez, J.S. Perez, M. Alfonso, J.L. Perez, M. Ramirez, R. Montes and V. Verez-Bencomo, *Carbohydr. Lett.*, 1995, **1**, 173 (*Chem. Abstr.*, 1996, **125**, 114 999).
15   Y. Konda, T. Machida, M. Akaiwa, K. Takeda and Y. Harigaya, *Heterocycles*, 1996, **43**, 555.
16   B.T. Cho and N. Kim, *J. Chem. Soc., Perkin Trans 1*, 1996, 2901.
17   K. Kefurt, K. Kefurtova, V. Markova and K. Slivova, *Collect. Czech. Chem. Commun.*, 1996, **61**, 1027 (*Chem. Abstr.*, 1996, **125**, 301 399).
18   K. Walczak and J. Suwinski, *Pol. J. Chem.*, 1996, **70** , 867 (*Chem. Abstr.*, 1996, **125**, 196 213).
19   N. Chida, K. Nakazawa, S. Ninomiya, S. Amano, K. Koizumi, J. Inaba and S. Ogawa, *Carbohydr. Lett.*, 1995, **1**, 335 (*Chem. Abstr.*, 1996, **124**, 317 686).
20   H. Kakinuma, H. Yuasa and H. Hashimoto, *Carbohydr. Res.*, 1996, **284**, 61.
21   B. Di Blasio, S. Galdiero, M. Saviano, C. Pedone, E. Benedetti, E. Rizzarelli, S. Pedotti, G. Vecchio and W.A. Gibbons, *Carbohydr. Res.*, 1996, **282**, 41.
22   L. Vértsey, H.-W. Fehlhaber, H. Kogler and P.W. Schindler, *Liebigs Ann. Chem.*, 1996, 121.
23   M.A. Saleh, M.A. Abdo, M.F. Abdel-Megeed and G.A. El-Hiti, *Indian J. Chem., Sect. B: Org. Chem. Incl. Med. Chem.*, 1996, **35B**, 147 (*Chem. Abstr.*, 1996, **124**, 232 946).
24   E. Ayadi, S. Czernecki and J. Xie, *J. Chem. Soc., Chem. Commun.*, 1996, 347.
25   S. Yoon, H.G. Kim, K.H. Chun and J.E.N. Shan, *Bull. Korean Chem. Soc.*, 1996, **17**, 599 (*Chem. Abstr.*, 1996, **125**, 329 142).
26   E. Chelain, S. Czernecki, M. Chmielewski and Z. Kaluza, *J. Carbohydr. Chem.*, 1996, **15**, 571.

27   D.A. Griffith and S.J. Danishefsky, *J. Am. Chem. Soc.*, 1996, **118**, 9526.
28   E. Kaji, Y. Osa, K. Takahashi and S. Zen, *Chem. Pharm. Bull.*, 1996, **44**, 15.
29   K. Takeda, E. Kaji, H. Nakamura, A. Kiyama, Y. Konda, Y. Mizuno, H. Takayanagi and Y. Harigaya, *Synthesis*, 1996, 341.
30   Y. Ichikawa, C. Kobayashi and M. Isobe, *J. Chem. Soc., Perkin Trans 1*, 1996, 377.
31   P. Alexander, V.V. Krishnamurthy and E.J. Prisbe, *J. Med. Chem.*, 1996, **39**, 1321.
32   G.B. Kok, B.L. Mackey and M. von Itzstein, *Carbohydr. Res.*, 1996, **289**, 67.
33   T. Iwamoto, T. Kan, S. Katsumura and Y. Ohfune, *Synlett*, 1996, 169.
34   X. Zhang and P. Ulrich, *Tetrahedron Lett.*, 1996, **37**, 4667.
35   P.L. Barili, G. Berti, G. Catelani, F. D'Andrea and V. Di Bussolo, *Carbohydr. Res.*, 1996, **290**, 17.
36   T.M. Mitzel and L.A. Paquette, *Chemtracts: Org. Chem.*, 1995, **8**, 279 (*Chem. Abstr.*, 1996, **124**, 202 735).
37   M. Ruiz, V. Ojea and J.M. Quintela, *Tetrahedron Lett.*, 1996, **37**, 5743.
38   T. Hudlicky, D.A. Entwistle, K.K. Pitzer and A.J. Thorpe, *Chem. Rev.*, 1996, **96**, 1195 (*Chem. Abstr.*, 1996, **124**, 290 002).
39   A. Barco, S. Benetti, C. De Risi, G.P. Pollini, G. Spalluto and V. Zanirato, *Tetrahedron*, 1996, **52**, 4719.
40   P. Spanu, G. Rassu, F. Ulgheri, F. Zanardi, L. Battistina and G. Casiraghi, *Tetrahedron*, 1996, **52**, 4829.
41   G. Poli, T. Pagni, S.I. Maffiolo, C. Scolastico and M. Zanda, *Gazz. Chim. Ital.*, 1995, **125**, 505 (*Chem. Abstr.*, 1996, **124**, 202 843).
42   S.G. Davies and G.D. Smyth, *Tetrahedron: Asymm.*, 1996, **7**, 1273.
43   Z. Xu, C.W. Johannes, S.S. Salman and A.H. Hoveyda, *J. Am. Chem. Soc.*, 1996, **118**, 10926.
44   T. Tschamber, E.-M. Rodriguez-Perez, P. Wolf and J. Streith, *Heterocycles*, 1996, **42**, 669.
45   M.G. Peter, J.P. Ley, S. Petersen, M.H.M.G. Schumacher-Wandersleb, K.-D. Spindler, M. Spindler-Barth, A. Turberg and M. Londershausen, *Chitin World*, [*Proc. Int. Conf. Chitin Chitosan*], *6th*, 1994, 359 (*Chem. Abstr.*, 1996, **125**, 58 928).
46   S. Aiba and E. Muraki, *Adv. Chitin Sci.*, 1996, **1**, 192 (*Chem. Abstr.*, 1996, **125**, 33 973).
47   W. Dullenkopf, J.C. Castro-Palomino, L. Manzoni and R.R. Schmidt, *Carbohydr. Res.*, 1996, **296**, 135.
48   U. Ellervik and G. Magnusson, *Carbohydr. Res.*, 1996, **280**, 251.
49   J.C. Castro-Palomino and R.R. Schmidt, *Liebigs Ann. Chem.*, 1996, 1623.
50   J.S. Debenham and B. Fraser-Reid, *J. Org. Chem.*, 1996, **61**, 432.
51   H. Shimizu, Y. Ito, Y. Matsuzaki, H. Iijima and T. Ogawa, *Biosci., Biotechnol., Biochem.*, 1996, **60**, 73 (*Chem. Abstr.*, 1996, **124**, 261 532).
52   P. Stangier and O. Hindsgaul, *Synlett*, 1996, 179.
53   E. Graf von Roedern, E. Lohof, G. Hessler, M. Hoffmann and H. Kessler, *J. Am. Chem. Soc.*, 1996, **118**, 10156.
54   J.P. McDevitt and P.T. Lansbury, Jr., *J. Am. Chem. Soc.*, 1996, **118**, 3818.
55   P. Kamireddy and W.M. Murray, *Pure Appl. Chem.*, 1996, **68**, 743 (*Chem. Abstr.*, 1996, **125**, 87 079).
56   X.-G. Luo, R.-X. Zhuo and M.-Q. Li, *Gaodeng Xuexiao Huaxue Xuebao*, 1996, **17**, 1416 (*Chem. Abstr.*, 1996, **125**, 301 460).
57   J. Ramirez, L. Lu, J. Li, P.G. Braunschweiger and P.G. Wang, *Bioorg. Med. Chem. Lett.*, 1996, **6**, 2575.
58   P. Lei, Y. Ogawa and P. Kovác, *J. Carbohydr. Chem.*, 1996, **15**, 485.

59   Y. Ogawa, P. Lei and P. Kovác, *Carbohydr. Res.*, 1996, **288**, 85.
60   Y. Ogawa, P. Lei and P. Kovác, *Carbohydr. Res.*, 1996, **293**, 173.
61   T. Oshikawa, M. Yamashita, K. Kaneoka, T. Usui, N. Osakabe, C. Takahashi and K. Seo, *Heterocycl. Commun.*, 1996, **2**, 261 (*Chem. Abstr.*, 1996, **125**, 301 406).
62   R.J. Kerns and R.J. Linhardt, *Synth. Commun.*, 1996, **26**, 2671.
63   J.M. García Fernández, C.O. Mellet, S. Maciejewski and J. Defaye, *J. Chem. Soc., Chem. Commun.*, 1996, 2741.
64   J.L. Jiménez Blanco, C.O. Mellet, J. Fuentes and J.M. García Fernández, *J. Chem. Soc., Chem. Commun.*, 1996, 2077.
65   V.V. Mossine, C.L. Barnes, G.V. Glinsky and M.S. Feather, *Carbohydr. Res.*, 1996, **284**, 11.
66   K. Walczak and J. Suwinski, *Pol. J. Appl. Chem.*, 1995, **39**, 87 (*Chem. Abstr.*, 1996, **125**, 11 317).
67   M. Shiozaki, N. Deguchi, W.M. Macindoe, M. Arai, H. Miyazaki, T. Mochizuki, T. Tatsuta, J. Ogawa, H. Maeda and S. Kurakata, *Carbohydr. Res.*, 1996, **283**, 27.
68   K. Miyajima, K. Ikeda and K. Achiwa, *Chem. Pharm. Bull.*, 1996, **44**, 2268.
69   K. Ikeda, K. Miyajima and K. Achiwa, *Chem. Pharm. Bull.*, 1996, **44**, 1985
70   G. Legler, M.-T. Finken and S. Felsch, *Carbohydr. Res.*, 1996, **292**, 91.
71   K. Suzuki, T. Fujii, K. Sato and H. Hashimoto, *Tetrahedron Lett.*, 1996, **37**, 5921.
72   G. Micera and H. Kozlowski, *Handb. Met.-Ligand Interact. Biol. Fluids: Bioinorg. Chem.*, 1995, **1**, 707 (*Chem. Abstr.*, 1996, **124**, 343 809).
73   V.A. Yaylayan and A. Huyghues-Despointes, *Carbohydr. Res.*, 1996, **286**, 179.
74   I. Robina, E. López-Barba and J. Fuentes, *Synth. Commun.*, 1996, 26.
75   R.F. Sala, P.M. Morgan and M.E. Tanner, *J. Am. Chem. Soc.*, 1996, **118**, 3033.
76   T. Satoh and Y. Nishimura, *Carbohydr. Res.*, 1996, **286**, 173.
77   Y.M. Ahn and G.R. Gray, *Carbohydr. Res.*, 1996, **296**, 215.
78   M. Okada and K. Aoi, *Macromol. Rep.*, 1995, **A32**, 907 (*Chem. Abstr.*, 1996, **124**, 30 122).
79   B. Aguilera and A. Fernández-Mayoralas, *J. Chem. Soc., Chem. Commun.*, 1996, 127.
80   P. Wentworth, Jr., T. Wiemann and K.D. Janda, *J. Am. Chem. Soc.*, 1996, **118**, 12521.
81   J.M. Ramanjulu and M.M. Joullié, *J. Carbohydr. Chem.*, 1996, **15**, 371.
82   P. Soro, G. Rassu, P. Spanu, L. Pinna, F. Zanardi and G. Casiraghi, *J. Org. Chem.*, 1996, **61**, 5172.
83   Y. Nishimura, T. Satoh, H. Adachi, S. Kondo, T. Takeuchi, M. Azetaka, H. Fukuyasu and Y. Iizuka, *J. Am. Chem. Soc.*, 1996, **118**, 3051.
84   P.D. Howes and P.W. Smith, *Tetrahedron Lett.*, 1996, **37**, 6595.

10
Miscellaneous Nitrogen-containing Derivatives

1 Glycosylamines and Related Glycosyl-N-bonded Compounds

1.1 Glycosylamines – In a new method, dissociation constants of carboxylic acids and phenols in methanol have been determined from the catalytic constants for the weak acid-catalysed mutarotation of N-(4-chlorophenyl)-β-D-glucopyranosylamine.[1] Bicyclic oxazinones **2** were prepared by cyclization of N-(β-D-glucopyranosyl)-glycine ethyl ester **1** (Scheme 1). C-Alkylation gave β-amino acid derivatives, e.g. **3**, with high diastereoselectivity, but these were accompanied in some cases by unwanted dialkylated products, e.g. **4**.[2] Reaction of β-D-glucosylamine with 3-acetamido-4,4-dimethylthietane-2-one followed by nitrosation (NaNO₂, HCl) gave the nitrosothiol derivative **5**. A similar process gave related analogues from 2- and 6-amino-sugars. The aqueous solubility and stability of these novel NO donors were studied.[3]

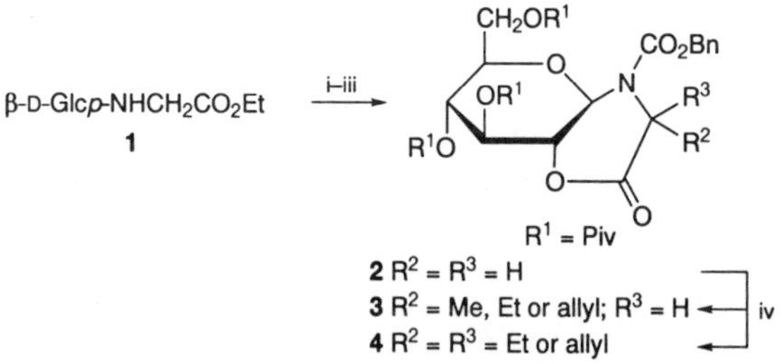

β-D-Glc*p*-NHCH₂CO₂Et   **1** $\xrightarrow{\text{i-iii}}$

R¹ = Piv

2 R² = R³ = H
3 R² = Me, Et or allyl; R³ = H
4 R² = R³ = Et or allyl

Reagents: i, BnO₂CCl, Pri₂NEt; ii, NaOH, H₂O; iii, PivCl, Py;
iv, Li HMDS, HMPA, R²X

Scheme 1

The configuration and solution conformation of various glucosyl amidines **6–8** has been studied by NMR techniques and semi-empirical calculations.[4] Regio-isomeric 'pseudonucleosides' **9** with sulfahydantoins as the base moiety were synthesized by conventional nucleoside synthesis reactions.[5] Glycosylaminopyrimidines **10** were converted into glycosylaminopyridines **11** by a Diels-Alder – retro-Diels-Alder reaction sequence involving addition of dimethyl

Carbohydrate Chemistry, Volume 30
© The Royal Society of Chemistry, 1998

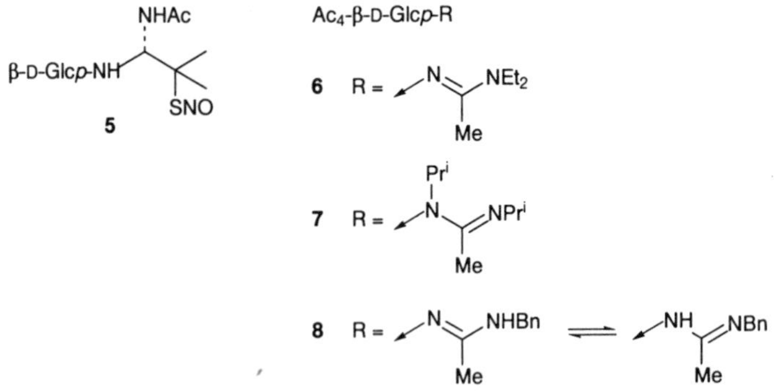

5

6 R = (structure with N, NEt₂, Me)

7 R = (structure with Prⁱ, N, NPrⁱ, Me)

8 R = (structure with N, NHBn, Me ⇌ NH, NBn, Me)

Ac₄-β-D-Glcp-R

acetylenedicarboxylate to the diene of the heterocycle and elimination of MeNCO (Scheme 2).[6] 1-*N*-Glycosyl-triazoles such as **12** were synthesized by dipolar cycloaddition of dimethyl acetylenedicarboxylate to various disaccharide α- or β-glycosyl azides, and proved to be effective glycosyl donors in the presence of TmsOTf.[7]

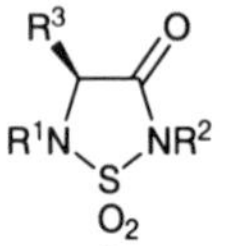

9

R¹, R² = β-D-Ribf, H; R³ = Bn or Buⁱ

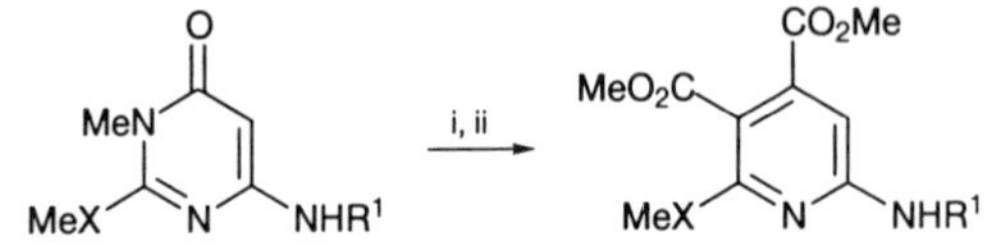

10

R¹ = Ac₄-β-D-Glcp or
Ac₃-β-D-Xylp; X = S or O

11

R¹ = β-D-Glcp or
β-D-Xylp; X = S or O

Reagents: i, MeO₂CC≡CCO₂Me; ii, NaOMe, MeOH

Scheme 2

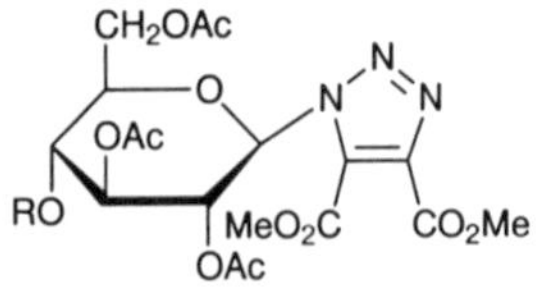

12

R = Ac₄-α-or β-D-Glcp

A kinetic analysis of the Amadori reaction of glucose with morpholine has been reported,[8] and a study that provides further evidence for the reversibility of the Amadori reaction is covered in Chapter 9 (Section 3.8).

1.2   Glycosylamides Including *N*-Glycopeptides – The *N*-glycosyl-maleimide **13** was isolated from the leaves of *Garania mangostana*.[9] The positional isomer **14**

of the lipid A component GLA-60 (a 2-*N*,3-*O*-di-fatty acid acylated 2-amino-2-deoxy-D-glucose-4-phosphate) was synthesized by conventional reactions starting from a glycosyl azide. It had much stronger lipopolysaccharide antagonist activity than GLA-60.[10] *N*-Acetyl-*N*-allyl-β-glycosylamines were synthesized directly from unprotected mono- and oligo-saccharides by reaction with allylamine then Ac$_2$O.[11] The allyl group is suitable to allow for specific linking reactions. Thus condensation with a thiol was used in the preparation of the sialyl-Lewis^x mimic **15**, which was polymerized with other diacetylenic components to give polydiacetylenes that formed liposomes displaying polyvalent oligosaccharide assemblies and were nanomolar inhibitors of P-selectin.[12]

13 **14**

15

The *N*-β-chitobiosyl and -chitotriosyl derivatives of histidine and β-) spinacine (e.g. **16**), wherein the sugar is linked to the α-amino-group of the amino acid, have been synthesized and shown to be weak inhibitors of brine shrimp chitinase (see Vol. 28, p.39).[13] The enzyme glycosylasparaginase, which normally cleaves the glycosylamide bond in glycoproteins, has been shown to be capable of catalysing the synthesis of glycosylasparagines, using L-aspartic acid methyl ester as the donor and 2-acetamido-2-deoxy-β-D-glucopyranosylamine as the acceptor.[14] The *N*-(2-acetoxy-4-methoxybenzyl) group has been introduced for protection of the aspartyl amide bond during the solid-phase synthesis of *N*-linked glycoproteins. It prevents unwanted aspartimide formation, and can be removed by treatment with trifluoroacetic acid.[15] In an extension of work on asparagine-linked sialyl-Lewis^x glycopeptides (cf. Vol. 29, p.148), a cyclic heptapeptide containing three *N*-(sialyl-Lewis^x)-asparagine residues has been synthesized and shown to be a sub-micromolar inhibitor of cell adhesion, but the increase in potency due to the multi-valent nature of this glycopeptide was not dramatic.[16] The synthesis of reversed amide bond analogues of 2-acetamido-2-deoxy-glucose linked to asparagine, which contain a *C*-glycosidic bond, are covered in Chapter 3.

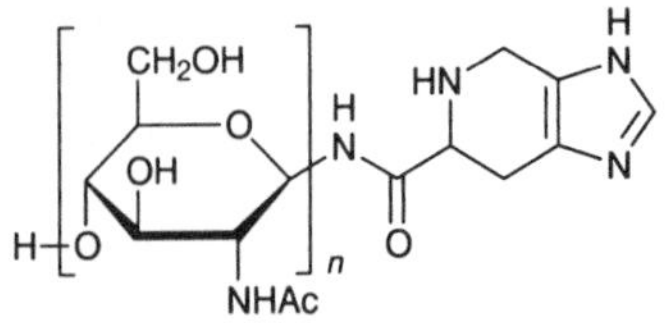

n = 2 or 3

16

1.3   *N*-Glycosyl-carbamates, -isothiocyanates, -thioureas and Related Compounds

The use of D-xylose-derived 1,2-fused oxazolidin-2-ones as chiral auxiliaries is covered in Chapter 24.

Hydantocidin and its analogues continue to be popular targets for synthesis. Full details (cf. Vol. 27, p.237 and 339) of the synthesis of (+)-hydantocidin and 5-*epi*-hydantocidin from D-fructose have been published.[17] The biologically inactive D-galactopyranose analogue **18** was obtained from the 1-*C*-methoxycarbonyl-D-galactose derivative **17** (Scheme 3), itself derived from D-galactose and

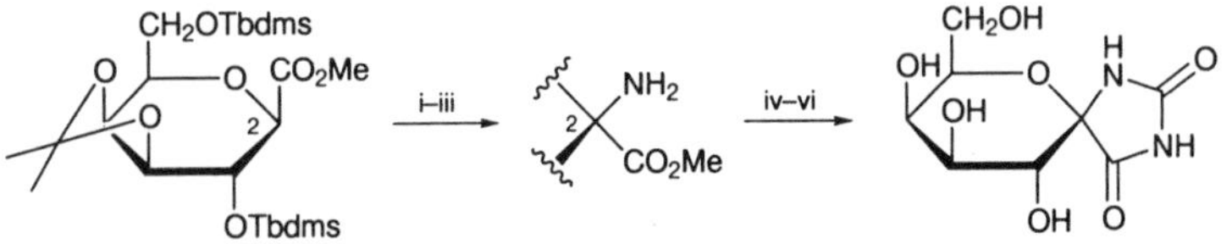

17   **18**

Reagents: i, NBS, CCl₄, (PhCO)₂O₂; ii, NaN₃, DMF; iii, H₂, Pd / C; iv, KOCN; v, KOBuᵗ; vi, H₃O

Scheme 3

nitromethane. The key steps involved radical bromination at the anomeric centre and displacement by azide ion with inversion.[18] The L-rhamno-furanose and -pyranose analogues **23** and **27** have been synthesized as shown in Scheme 4 from the triflate **19**, which was made from L-rhamnose by Kiliani ascent. The furanose isomer **23** was obtained by ring contraction of **19** to give the 1-*C*-methoxycarbonyl-L-rhamnose derivative **20**, followed by application of the same radical bromination route used in the synthesis of **18**, this time via the azido-sugar intermediate **21**. The pyranose isomer **27** was obtained by azide displacement reaction on the triflate **19**, reduction, and ionic bromination of the resulting amine **25** to give the bicyclic intermediate **26**. The diketopiperazines analogue **24** was obtained by coupling the amine **22** with an *N*-protected glycine derivative followed by cyclization. The C-2 epimer of **24** was obtained when the coupling with the glycine derivative (step vii) was effected by use of DCC rather than ClCO₂Et. The corresponding pyranose diketopiperazines analogue was similarly obtained from the amine **26**.[19-21]

Condensation of tetra-*O*-acetyl-β-D-glucopyranosyl isothiocyanate with 2-amino-2-deoxy-D-glucose gave a 9:1 mixture of the adduct **28** and its C-5 epimer,

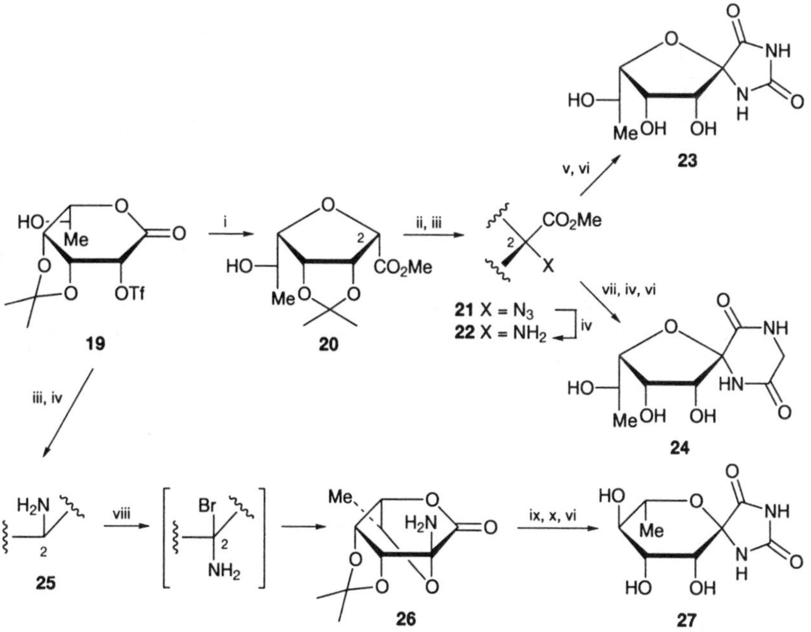

Reagents: i, K_2CO_3, MeOH; ii, NBS, $(PhCO)_2O_2$, CCl_4; iii, NaN_3, DMF; iv, H_2, Pd; v, KOCN; vi, H_3O^+; vii, $BnO_2CNHCH_2CO_2H$, $ClCO_2Et$, Et_3N; viii, NBS, NaOAc, MeCN; ix, PhNCO; x, Δ, MeOH

Scheme 4

which cyclized on silica gel with loss of water to give **29**. Similar chemistry was reported for β-D-ribopyranosyl and gentio-biosyl and -triosyl isothiocyanate derivatives, as well as 2-deoxy-2-isothiocyanato-β-D-glucopyranose.[22] The reactions of a variety of sugar thiourea derivatives from the corresponding isothiocyanates has been reported, as exemplified by the conversions shown in Scheme 5.[23] The formation of cyclic derivatives from 3-deoxy-3-thioureido-D-glucose is covered in Chapter 9, Section 3.4. *N*-Glycosyl-thiourea analogues of the antibiotic rifamycin have been made by coupling peracetylated β-D-gluco- or α-D-

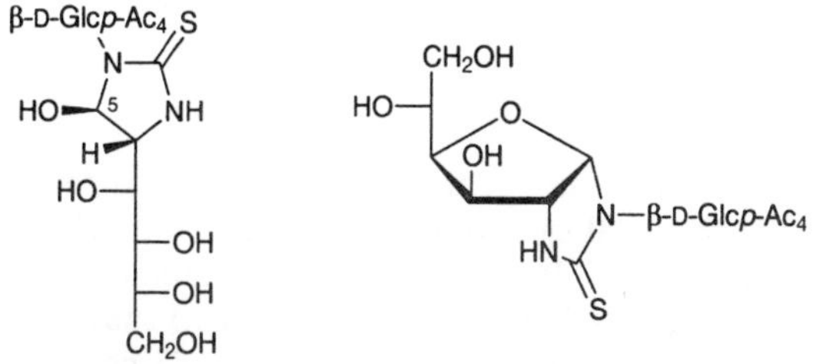

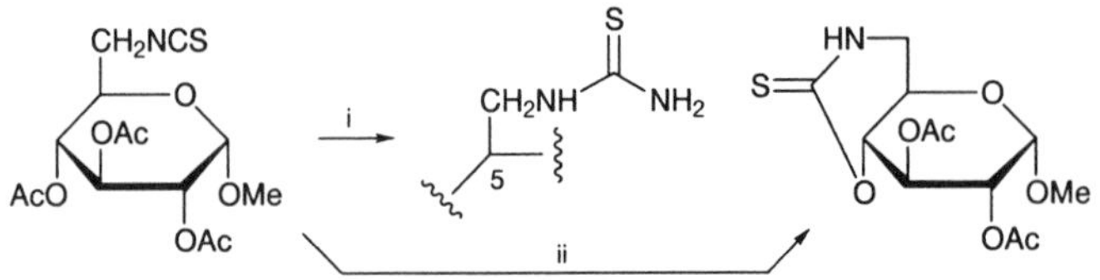

Reagents: i, NH₃, Py; ii, NH₃, Et₂O

Scheme 5

arabino-pyranosyl isothiocyanate with a pendant 2-aminoethylthio-substituent on a rifamycin derivative, and found to have some activity against Gram-positive bacteria.[24] Dendrimers terminated with 4, 6 and 8 α-D-mannopyranosyl residues were prepared by reacting tetra-*O*-acetyl-α-D-mannopyranosyl isothiocyanate with dendrimeric polyamines.[25] Condensation of 6-deoxy-6-isothiocyanato-α-cyclodextrin with either methyl 6-amino-6-deoxy-β-D-glucopyranoside or 6-amino-6-deoxy-*N*-(*N*-*tert*-butoxyglycinyl)-β-D-glucopyranosylamine gave thiourea-linked derivatives.[26]

The tetrahydropyrano[2,3-*d*]oxazole **30**, a trehazoline analogue that contains a *trans*-fused five membered ring, was obtained by coupling tetra-*O*-benzyl-β-D-glucopyranosyl isothiocyanate with tetra-*O*-acetyl-β-D-glucopyranosylamine, treating the resulting di-*N*-glucosyl-thiourea with (a) base to remove the acetyl groups and (b) 2-chloro-3-ethylbenzoxazolinium tetrafluoroborate and Et₃N to induce cyclization, and then removing the benzyl ether protecting groups by hydrogenolysis. The *cis*-fused oxazole **31** was similarly obtained using tetra-*O*-acetyl-α-D-glucopyranosyl isothiocyanate in place of its β-anomer, but during debenzylation in this case, ring contraction to the furano-oxazole **32** occurred. The α-D-glucosylated analogue of **31** was obtained using tetra-*O*-acetyl-α-D-glucopyranosylamine in place of its β-anomer, but this intermediate could not be deprotected.[27]

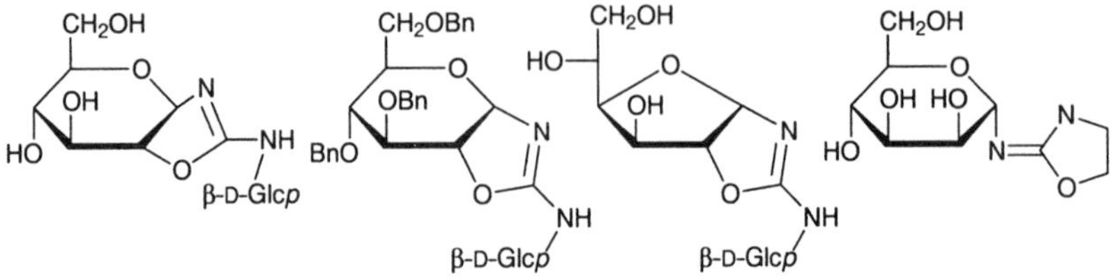

Glycosylimino-oxazolines such as the α-D-mannosylimino-derivative **33** have been prepared as further analogues of trehazoline, and their glycosidase inhibitory properties determined.[28] Reaction of the glucofurano-isoxazole **34** with amines is known to cause displacement of the thiobenzyl substituent, but it has now been found that unsaturated derivatives such as **35** can be the products formed under forcing conditions (Scheme 6).[29]

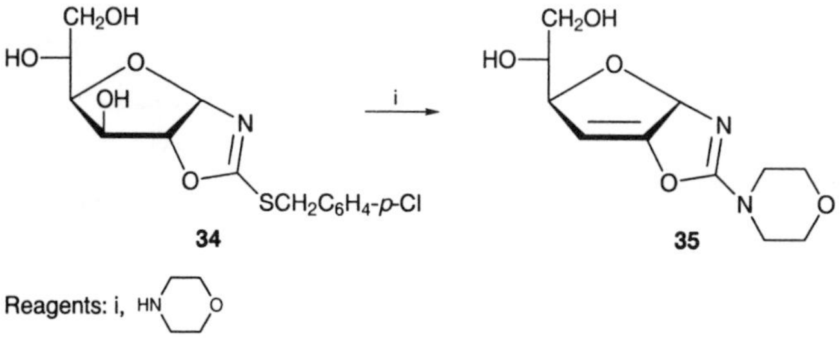

Scheme 6

2 Azido-sugars

The synthesis and characteristic transformations of 1,2-*cis*- and 1,2-*trans*-glyco-pyranosyl azides have been reviewed.[30] Phase transfer catalysed reactions of peracetylated α-L-fucosyl, α-D-mannosyl and α-L-rhamnosyl bromides with sodium azide provided the corresponding β-glycosyl azides with inversion of configuration, suggesting that the displacements proceeded by S_N2 mechanisms.[31]

In simple effective procedures, the β-neuraminosyl azide **36** was prepared by $SnCl_4$-catalysed reaction of the corresponding α,β-glycosyl acetate with trimethyl-

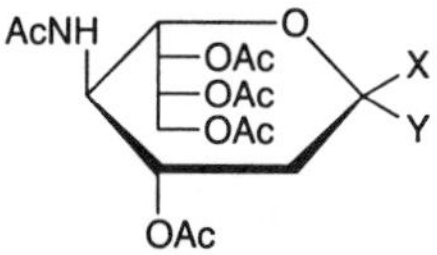

36 X = N_3, Y = CO_2Me
37 X = CO_2Me, Y = N_3

silyl azide, while the β-anomer **37** was obtained by reaction of the corresponding β-glycosyl chloride with lithium azide in HMPA.[32] The synthesis of the 6-deoxy-L-*manno*-hept-2-ulosonyl azide derivative **21**, intended for use in the generation of rhamnofuranose combinatorial libraries, is detailed in Scheme 4 above. The synthesis and photolysis of D-hex-2-ulopyranosyl azides have been investigated, as shown in Scheme 7 for the 3-deoxy and 3-*O*-methyl-D-fructose derivatives. Photolysis of **38** led to nitrene insertions with cleavage of either the C-2–C-3 bond or the C-1–C-2 bond to give the unstable compounds **39** (major) and **40** (minor), respectively.[33]

The tetra-*O*-acetyl-glycopyranosylidene 1,1-diazides of D-glucose, D-galactose and D-mannose (e.g. **41**) were synthesized by reaction of the corresponding 1-α-bromo-1-β-chlorides with sodium azide, the yields being better under conditions of phase transfer catalysis than in DMSO. Tetra-*O*-benzyl-glucopyranosylidene 1,1-diazide was obtained alternatively by reaction of tetra-*O*-benzyl-D-glucono-1,5-lactone with $TmsN_3$ and $BF_3.OEt_2$.[34] Syntheses and reactions of 1-cyano-glycopyranosyl azides have been studied, as exemplified for the D-galactose

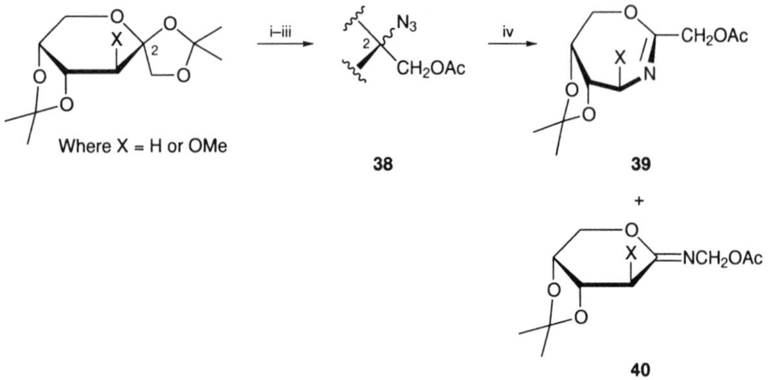

Where X = H or OMe

38   **39**

+

40

Reagents: i, TmsN₃, TmsOTf, MeCN, ii, Et₃N, aq. KF; iii, Ac₂O, Py; iv, *hv*

Scheme 7

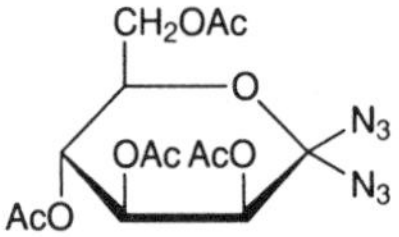

41

derivative **43** in Scheme 8, which was obtained from the bromide **42** by displacement with inversion by azide ion, and in a sequential fashion was condensed with azide ion to give tetrazole **44**. In the preparation of the α-azide anomer of **43** (not shown), by reaction of **42** with LiCl then NaN₃ with double inversion, there was concomitant formation of the α-anomer of **44**. These products were converted into heterocycles such as **45** and **46**.[35] The reaction of the 2-*O*-triflate esters of 2,3,4-all-*cis*-substituted aldono-1,4-lactone derivatives

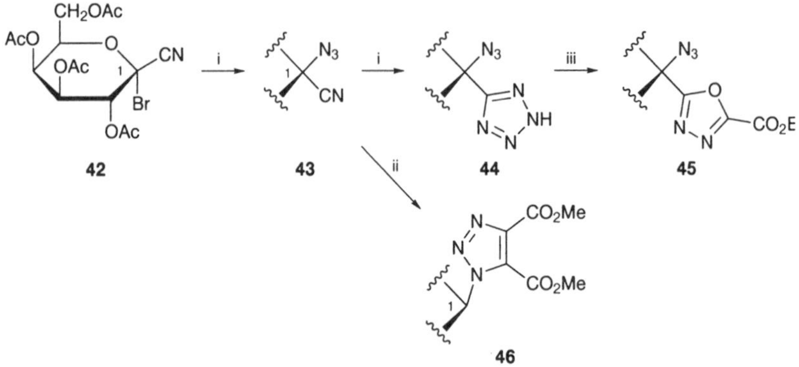

42   **43**   **44**   **45**

46

Reagents: i, NaN₃, DMSO; ii, MeO₂CC≡CCO₂Me; iii, EtO₂CCOCl

Scheme 8

with azide ion (five examples reported) gave the 2-azides with inversion as the kinetic products, but these epimerized *in situ* to give the thermodynamic product with overall retention, as exemplified in Scheme 9 for the conversion of triflate **47** into **48** then **49**.[36]

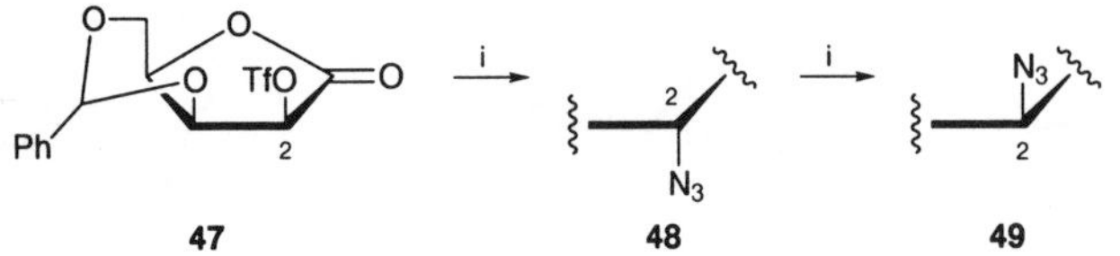

47   **48**   **49**

Reagents: i, NaN₃

Scheme 9

Improved conditions, and **safety considerations**, for the conversion of amines to azides by diazo transfer from triflyl azide have been reported. Thus reaction of 2-amino-2-deoxy-D-glucose with TfN₃, DMAP and NaOMe, followed by acetylation, gave tetra-*O*-acetyl-2-azido-2-deoxy-D-glucopyranose in 78% yield, and a pseudo-trisaccharide triamine was converted into a triazide in 49% yield. Metal ion catalysis was found to be helpful in some cases.[37] The azido phenylselenation of a D-*arabino*-furanoid glycal to give inseparable anomeric phenyl 2-azido-2-deoxy-1-seleno-D-glucofuranoses as the major products is covered in Chapter 9, Section 2.6. Benzyl 3-azido-3-deoxy-α-D-xylopyranoside resulted from exclusive opening of the epoxide ring of benzyl 2,3-anhydro-α-D-ribopyranoside at C-3 on reaction with azide ion.[38] 6-Azido-6-deoxy-D-fructose was synthesized by enzyme-catalysed aldol condensation of 1,3-dihydroxyacetone with 3-azido-3-deoxy-D-glyceraldehyde, the latter being prepared by conversion of D-fructose into the known 2,3-anhydro-D-glyceraldehyde diethyl acetal and then ring-opening of this with azide ion.[39] An 6-azido-2,3,6-trideoxy-hexonic acid derivative features in the synthesis of a dipeptide isostere reported in Chapter 24.

3   Nitro-sugars

Mixtures of C-2 epimeric 1-deoxy-1-nitroalditols have been prepared in 48-91% yields by addition of nitromethane to pentoses or hexoses in aqueous methanol in the presence of a strongly basic anion-exchange resin in the OH⁻-form, the reactions being quenched with dry ice. In some cases, 2,5- and 2,6-anhydro-1-deoxy-1-nitroalditols were also present in the products. The practical use of this procedure was exemplified by the preparation of the 1-deoxy-1-nitro-L-mannitol and -L-glucitol, and thence L-mannose and L-glucose, starting from L-arabinose.[40] The isomersization of the 3-deoxy-3-nitro-D-xylofuranose derivative **50** in acidic, basic and neutral conditions has been studied, e.g. when heated under reflux in methanol a mixture of **50** and its C-3 epimer **51** is formed.[41] Reduction of methyl 2,3-dideoxy-3-nitro-α- or β-D-*erythryo*- or D-*threo*-hex-2-enopyranosides (e.g. **52**, Scheme 10) with sodium borodeuteride resulted in allylic displacement of the 4-acetoxy group with introduction of a deuterium atom at C-2, *trans* to the glycosidic methoxy group as in product **53**.[42] The synthesis of C-2 branched-chain sugars by Michael addition of formaldehyde dimethylhydrazone to nitro-alkenes is covered in Chapter 14.

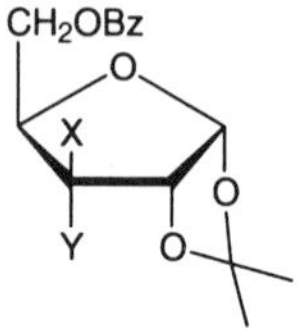

50 X = NO$_2$, Y = H
51 X = H, Y = NO$_2$

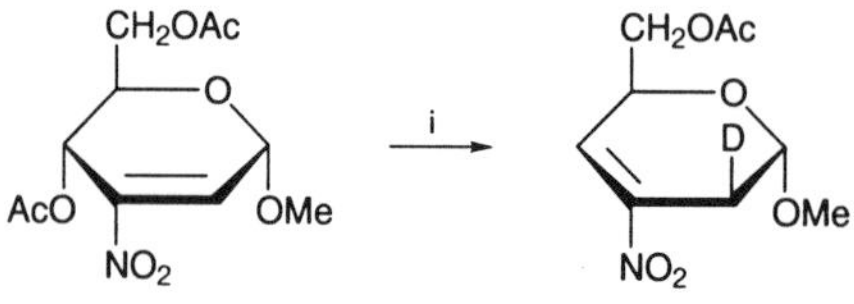

52 → **53**

Reagents: i, NaBD$_4$

Scheme 10

4   Diazirino-sugars

The carbene-generating biotinylated 2-acetamido-2-deoxy-D-glucopyranosyl-amide diazirine derivative **54** and the lactoside **55** have been synthesized for photoaffinity labelling of β-(1→4)-galactosyl transferase[43] and GM$_3$ synthase.[44]

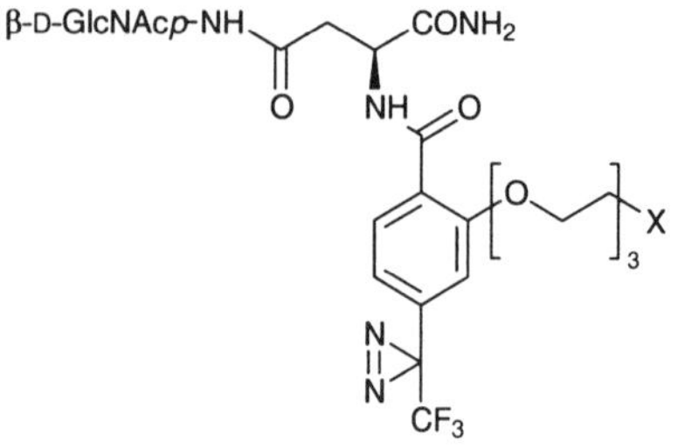

54

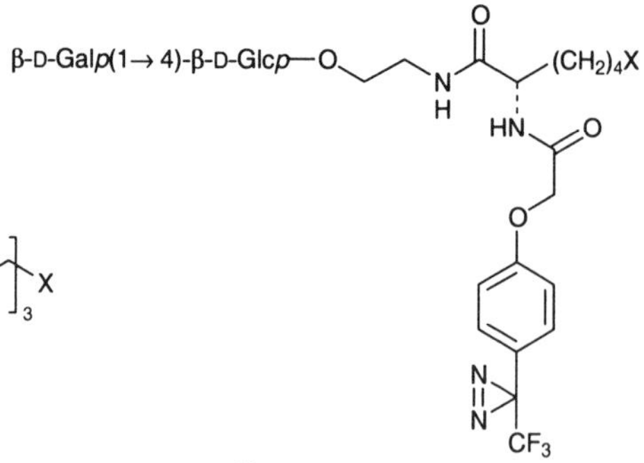

55

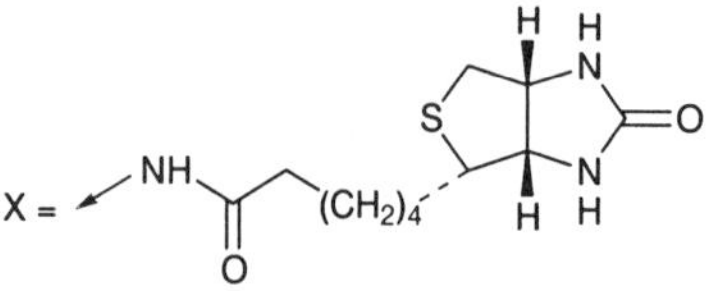

5   Oximes, Hydroxylamines, Nitriles, Imines and Amidines

The preparation of sugar amidines, amidrazones and amidoximes as inhibitors of carbohydrate-processing enzymes has been reviewed.[45] Treatment of 2-deoxy-D-*arabino*-hexose oxime with chloramine-T gave 2-deoxy-D-*arabino*-hexono-1,4-oximinolactone, but the intermediate nitrones formed from this and 2-deoxy-D-

erythro-pentose oxime underwent dipolar cycloaddition in the presence of alkenes, e.g. with styrene, to give the epimeric oxazolines such as **56**.[46] The *N*-phenylcarba-moyloximolactone **57**, synthesized from sodium D-glucuronate in eight steps, is a potent inhibitor of β-glucuronidases from bovine liver (K_i 0.2 μM) and *E. coli* (K_i 8 μM).[47] Derivatives of erythromycin A 9-oxime in which the oxime oxygen atom is linked to a secondary carbon atom of a sugar are covered in Chapter 19.

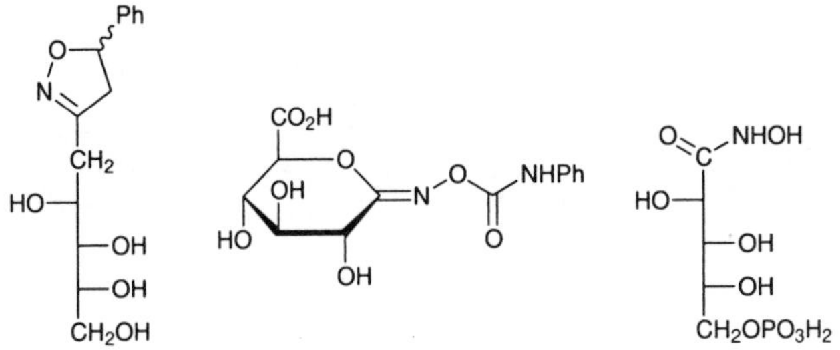

56   **57**   **58**

D-Arabinonohydroxamic acid 5-phosphate **58**, a potent transition state inhi-bitor of D-glucose 6-phosphate isomerase, was made from 2,3,4-tri-*O*-benzyl-D-arabinonic acid 5-(dibenzyl phosphate) by condensation with *O*-benzylhydroxyla-mine (using carbonyldiimidazole) followed by hydrogenolysis.[48] D-Arabinono-hydroxamic acid itself and D-threonohydroxamic acid, potent inhibitors of D-xylose isomerase, were obtained similarly.[49] As exemplified in Scheme 11, the addition of *N*-benzylhydroxylamine to unsaturated lactones such as **59** provided easy access to precursors of 3-amino-2,3-dideoxy-sugars (e.g. **60**). These could be inverted at C-5 by Mitsunobu or sulfonate displacement reactions, or converted to isomeric isoxazolidines, e.g. **62**, via epoxide **61**.[50]

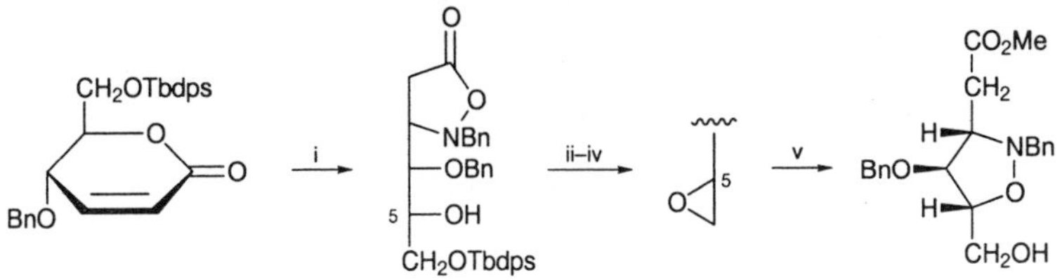

59   **60**   **61**   **62**

Reagents: i, BnNHOH; ii, MsCl, Py; iii, HF, Py; iv, OH⁻; v, K₂CO₃, MeOH

Scheme 11

The inseparable isomeric nitrones formed by reaction of deoxynojirimycin (DNJ) derivative **63** with dimethyldioxirane, gave the 1,3-dipolar cycloadducts **64** and **65** on reaction with trichloroacetonitrile (Scheme 12). The β-1-*C*-substituted-DNJ derivative **66** was obtained from the reaction of a related nitrone with methyl 3-butenoate under high pressure (15 kbar), but overall the access to this compound from DNJ was poor.[51] Syntheses of 3-*C*-branched-aldose and -nucleo-side derivatives, by addition of the anion of ethyl acetate to nitrones prepared by reaction of *N*-methylhydroxylamine to 3-keto-precursors,[52] are detailed in Chapters 14 and 20, respectively. The intramolecular cyclization of 3-*O*-(cyclohex-2-enyl)-1,2-*O*-isopropylidene-pentofuranose 5-nitrones, resulting in pentacyclic derivatives, is covered in Chapter 24.

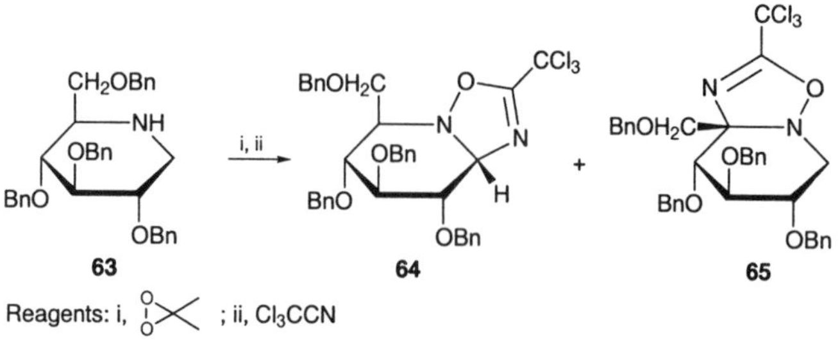

Reagents: i, ; ii, Cl₃CCN

Scheme 12

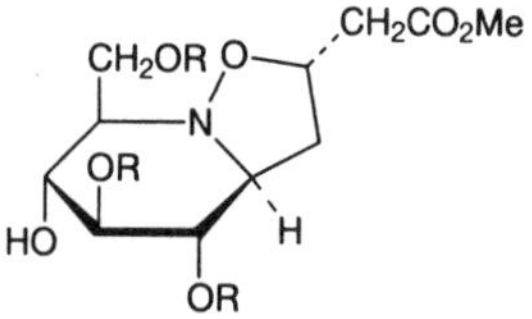

66   R = Tbdms

Tetra-*O*-acetyl-β-D-galactopyranosyl cyanide and the unsaturated nitrile **67** could be deacetylated under mild conditions (dilute NaOMe in MeOH at 0°C). Under less gentle conditions, **67** was converted into the crystalline imidate **68**, and thence into the amidine **69**, amidrazone **70**, imidazole **71** and the corresponding benzimidazole (Scheme 13).[53]

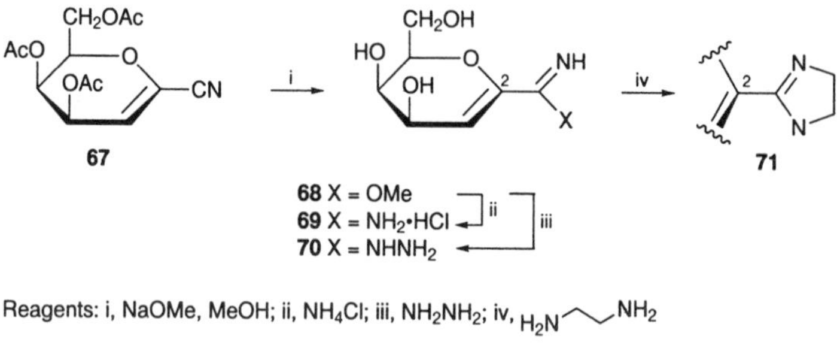

68 X = OMe
69 X = NH₂•HCl
70 X = NHNH₂

Reagents: i, NaOMe, MeOH; ii, NH₄Cl; iii, NH₂NH₂; iv, H₂N‿NH₂

Scheme 13

The amidines **72** were shown to be only marginally better than *N*-substituted β-D-glucosylamines as inhibitors of four β-glucosidases.[54]

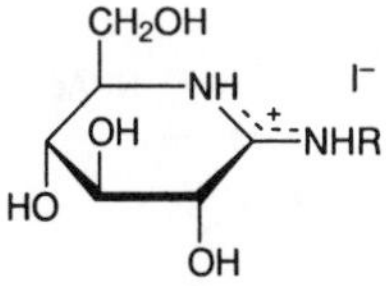

72 R = Bu^n or $C_{12}H_{25}$

6   Hydrazones and Related Compounds

The synthesis of oct-2-ulosono-1,4-lactone phenylhydrazones from D-gluconic acid is covered in Chapter 16, while the formation of a 1-*C*-substituted 1,5-dideoxy-1,5-iminopentitol by reduction of 6-azido-6-deoxy-D-galactose diphenylformazan with triphenylphosphine is detailed in Chapter 18.

7   Other Heterocycles

Dimerization of 2-amino-3,4,6-tri-*O*-benzyl-2-deoxy-D-glucose with 1,1'-thionyl- or sulfonyl-di-imidazole gave the 'fructosazine' and bis-tetrahydropyrano-piperazine derivatives **73** and **74** in a ratio of 1:5, respectively;[55] a related compound appeared in Vol. 25, p.132. The β-carboline derivative **75**, isolated as a natural product from a hybrid plant cell culture product, was synthesized in six steps from tryptamine and (*E*)-4,5,6-tri-*O*-acetyl-2,3-dideoxy-D-*erythro*-hex-2-enose.[56] Thiazolo-triazoles such as **76**, termed acyclo-*C*-nucleosides, were obtained on deacetylation of the product from condensation of peracetylated D-gluconic or galactaric acids with 4-amino-3-aryl-1,2,4-triazole-5-thiols in the presence of POCl$_3$.[57] The related 1,2,4-triazole **77** and dihydroimidazole **78** were obtained by condensation of D-glucono-1,5-lactone with aminoguanidine and ethylenediamine, respectively, followed by acetylation then *O*-deacetylation.[58]

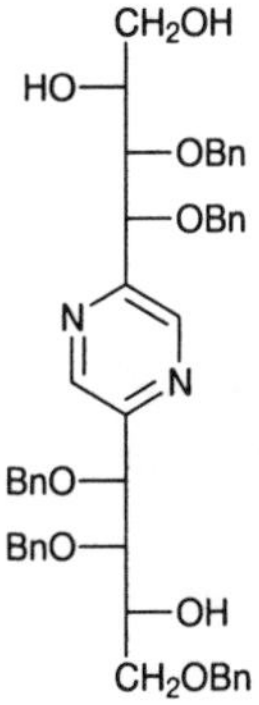

73

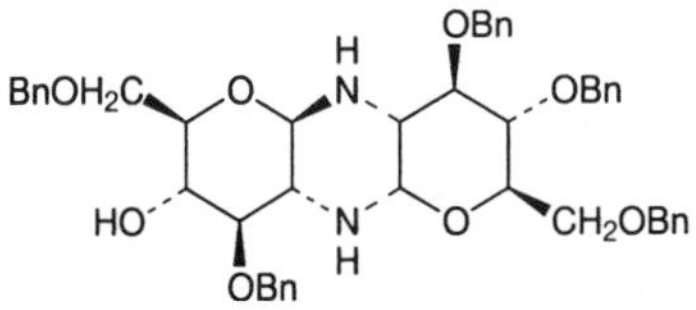

74

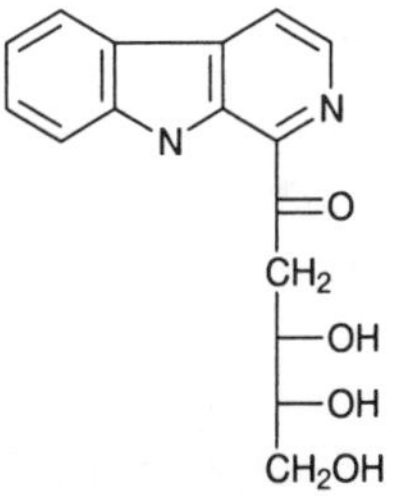

75

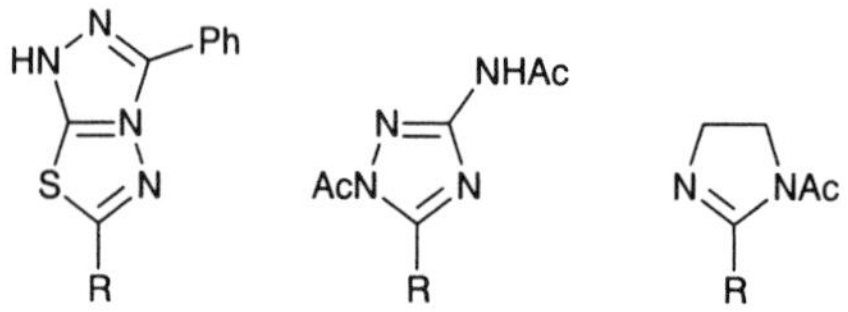

76   77   78

R = D-*gluco*-pentitol-l-yl

Tatsuta has reviewed his approach to the synthesis of nagstatin and its analogues and their glycosidase inhibitory properties[59] and has extended this approach (cf. Vol. 29, p.163) to the synthesis of the 1,2,4-triazolo-fused D-*galacto*-deoxynojirimycin 79 and its D-*manno*-deoxynojirimycin analogue from 2,3,4-tri-*O*-benzyl-L-ribose and -L-xylose, respectively.[60] The isomeric 1,2,3-triazolo-fused D-*gluco*- and D-*manno*-deoxynojirimycins 81 were synthesized from the epimeric acetylides 80 (Scheme 14), prepared by addition of lithium trimethylsilyl-acetylide to 2,3,5-tri-*O*-benzyl-L-xylono-1,4-lactone followed by borohydride reduction; they did not inhibit β-glycosidases.[61] The tetrazolo-fused 5-deoxy-D-*gulo*-deoxynojirimycin analogue 83 (referred to as a 5-epi-L-*rhamno*-derivative) was synthesized from the 5-azido-lactone 82 (Scheme 15).[62]

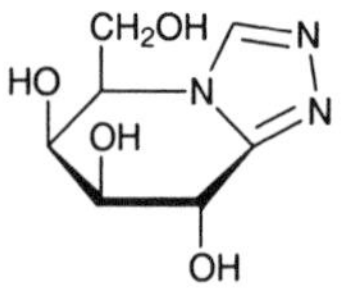

79 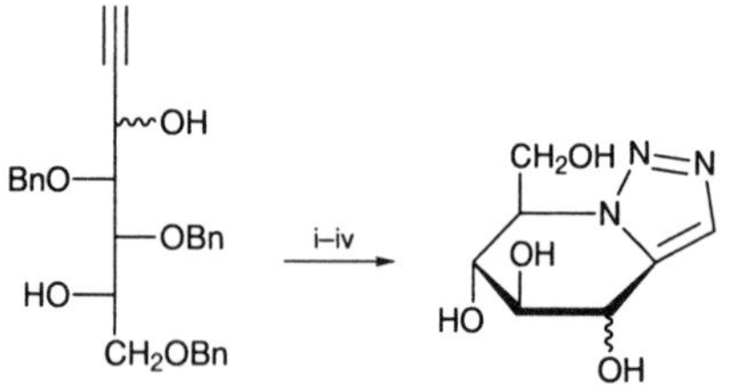   80   81

Reagents: i, BnBr, NaH; ii, TsCl, Py, DMAP;
            iii, NaN₃, DMSO, Δ; iv, H₂, Pd/C, HOAc

Scheme 14

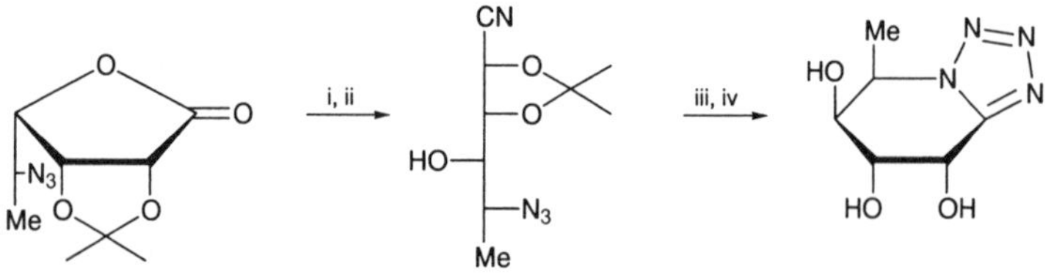

82   83

Reagents: i, NH₃, MeOH; ii, (CF₃CO)₂O, Py; iii, Δ; iv, H₃O⁺

Scheme 15

The fused 1,2,3-triazolino-pyrrolidine **84** was obtained by oxidation of 2,3,4-tri-*O*-acetyl-5-azido-5-deoxy-D-xylose dibenzyl dithioacetal with MCPBA, the reaction involving intramolecular dipolar cycloaddition of the azido group onto a 1,2-double bond formed by loss of acetic acid. A similar product was obtained from the D-*ribo*-analogue.[63] Nucleotide analogues based on hexenopyranosyl phosphonic acid derivatives with a nucleoside base moiety linked at C-4 are covered in Chapters 9 (Section 2.6) and 17.

The methylumbelliferyl glucuronoside analogue **85** in which the C-6 carboxylic acid group is replaced by tetrazole group, was synthesized from D-glucurono-3,6-lactone, but was neither a substrate nor an inhibitor of bovine liver β-glucuronidase.[47] Dihydropyrazoles such as **86** have been obtained by dipolar cycloaddition of diazomethane to sugar-derived α,β-unsaturated esters and lactones.[64] 4-Carboxamido-1-(5-deoxy-D-pentos-5-yl)-1,2,3-triazoles such as **87** have been synthesized by cycloaddition of ethyl propiolate to 5-azido-5-deoxy-pentose derivatives, followed by reaction with ammonia and deprotection.[65] The tetracyclic derivative **88** was obtained by [2+2]cycloaddition of tosyl isocyanate to a 3-*O*-(*cis*-1-butenyl) ether derivative of D-xylose, then introduction and intramolecular displacement of a 5-*O*-tosyl group. The epimer at the position marked with an asterisk was obtained from the corresponding *trans*-1-butenyl ether.[66]

84   85   86

87   88

The cyclic guanidino-sugar analogues **91**–**93** have been synthesized as potential transition state analogues of glycosidase substrates. The epoxide **89**, obtained by enantioselective lipase-catalysed hydrolysis of a racemic mixture, was converted to the thioureido derivative **90** then to variously *N*-substituted-guanidino derivatives with the aid of mercury(II) chloride, and allowed to cyclize on hydrolysis (Scheme 16). While the *N*-benzyloxy-analogue **91** was present only in the form shown at several different pH values, the other two analogues were in equilibrium with furanose forms.[67] The preparation of such compounds has also been covered in a review.[68]

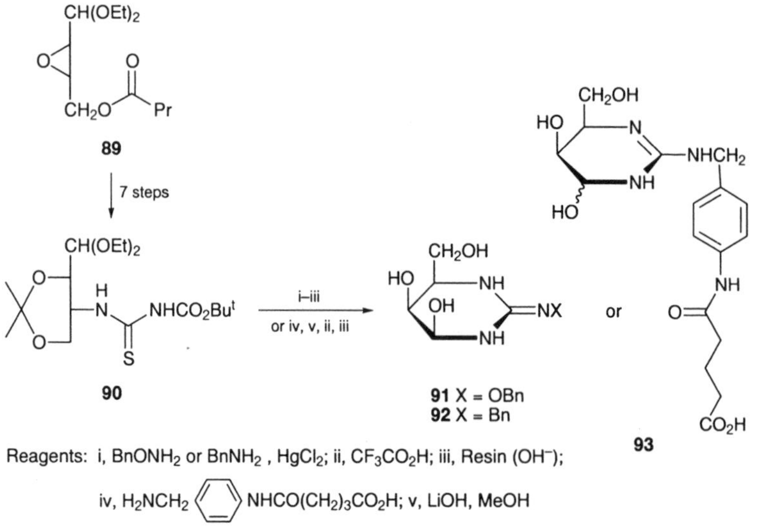
CH(OEt)2
O
CH2O
Pr
89
7 steps
CH(OEt)2
H
N
NHCO2Bu^t
S
90
i–iii
or iv, v, ii, iii
CH2OH
HO
OH
NH
=NX
NH
91 X = OBn
92 X = Bn
or
CH2OH
HO
N
NHCH2
NH
HO
NH
O
CO2H
93
Reagents: i, BnONH2 or BnNH2 , HgCl2; ii, CF3CO2H; iii, Resin (OH⁻);
iv, H2NCH2 NHCO(CH2)3CO2H; v, LiOH, MeOH

Scheme 16

3-Deoxy-3-imidazolyl-substituted cyclodextrins are covered in Chapter 4, and bis-(glucos-3-yl)-bipyridine in Chapter 14.

References

1 K. Swiataczowa, A. Wawrzynow and R. Korewa, *Pol. J.Chem.*, 1995, **69**, 1306 (*Chem. Abstr.*, 1996, **124**, 87 539).

2 M.N. Keynes, M.A. Earle, M. Sudharshan and P.G. Hultin, *Tetrahedron*, 1996, **52**, 8685.

3 J. Ramirez, L. Yu, J. Li, P.G. Braunschweiger and P.G. Wang, *Bloorg. Med. Chem. Lett.*, 1996, **6**, 2575.

4 M. Avalos, R. Babiano, P. Cintas, C.J. Duran, J.L. Jiminez and J.C. Palacios, *Tetrahedron*, 1996, **52**, 9263.

5 G. Dewynter, N. Aouf, Z. Regainia and J.-L. Montero, *Tetrahedron*, 1996, **52**, 993.

6 J. Cobo, M. Melguizo, A. Sanchez, M. Nogueras and E. De Clercq, *Tetrahedron*, 1996, **52**, 5845.

7 C. Peto, G. Batta, Z. Györgydeák and F. Sztaricskai, *J. Carbohydr. Chem.*, 1996, **15**, 465.

8 A. Huyghes-Despointes and V.A. Yaylayan, *J. Agric. Food Chem.*, 1996, **44**, 1464 (*Chem. Abstr.*, 1996, **125**, 11 277).

9 D. Krajewski, G. Toth and P. Schreier, *Phytochemistry*, 1996, **43**, 141.

10 M. Shiozaki, M. Arai, W.M. Macindoe, T. Mochizuki, S. Kurakata, H. Maeda and M. Nishijima, *Chem. Lett*, 1996, 735.

11 W. Spevak, F. Dasgupta, C.J. Hobbs and J.O. Nagy, *J. Org. Chem.*, 1996, **61**, 3417.

12 W. Spevak, C. Foxall, P.H. Charych, F. Dasgupta and J.O. Nagy, *J. Med. Chem.*, 1996, **39**, 1018.

13   J.P. Ley and M.G. Peter, *J. Carbohydr. Chem.*, 1996, **15**, 51.
14   I. Mononen, G.I. Ivanov, I.B. Stoineva, T. Noronkoski and D.D. Petkov, *Biochem. Biophys. Res. Commun.*, 1996, **218**, 510 (*Chem. Abstr.*, 1996, **124**, 202 850).
15   J. Offer, M. Quibell and T. Johnson, *J. Chem. Soc., Perkin Trans. 1*, 1996, 175.
16   U. Sprengard, M. Schudok, W. Schmidt, G. Kretzschmar, *Angew. Chem., Int. Ed. Engl.*, 1996, **35**, 321.
17   N. Nakajima, M.Matsumoto, M. Kirihara, M. Hashimoto, T. Katoh and S. Terashima, *Tetrahedron*, 1996, **52**, 1177.
18   T.W. Brandstetter, M.R. Wormald, R.A. Dwek, T.D. Butters, F.M. Platt, K.T. Tsitsanou, S.E. Zographos, N.G. Oikonomakos and G.W.J. Fleet, *Tetrahedron: Asymm.*, 1996, **7**, 157.
19   J.C. Estevez, J. Saunders, G.S. Besra, P.J. Brennan, R.J. Nash and G.W.J. Fleet, *Tetrahedron: Asymm.*, 1996, **7**, 383.
20   J.C. Estevez, M.D. Smith, A.L. Lane, S. Crook, D.J. Watkin, G.S. Besra, P.J. Brennan, R.J. Nash and G.W.J. Fleet, *Tetrahedron: Asymm.*, 1996, **7**, 387.
21   J.C. Estevez, M.D. Smith, M.R. Wormald, G.S. Besra, P.J. Brennan, R.J. Nash and G.W.J. Fleet, *Tetrahedron: Asymm.*, 1996, **7**, 391.
22   J. Fuentas, J.L. Molina, D. Olano and M. Angeles Pradero, *Tetrahedron: Asymm.*, 1996, **7**, 203.
23   J.M. García Fernández, C. Ortiz Mellet, V.M. Díaz Pérez, J.L. Jiménez Blanco and J. Fuentes, *Tetrahedron*, 1996, **52**, 12947.
24   C. Bartolucci, L. Cellai, C. Martuccio, A. Rossi, A.L. Segre, S.R. Savu and L. Silvesto, *Helv. Chim. Acta*, 1996, **79**, 1611.
25   T.K. Lindhorst and C. Kieburg, *Angew. Chem., Int. Ed. Engl.*, 1996, **35**, 1953.
26   J.M. García Fernández, C. Ortiz Mellet, S. Maciejewski and J. Defaye, *J. Chem. Soc., Chem. Commun.*, 1996, 2741.
27   M. Shiozaki, T. Mochizuki, H. Hanzawa and H. Haruyama, *Carbohydr. Res.*, 1996, **288**, 99.
28   C. Uchida and S. Ogawa, *Bioorg. Med. Chem.*, 1996, **4**, 275 (*Chem. Abstr.*, 1996, **124**, 343 903).
29   P. Meszaros, I. Pinter and G. Toth, *Aust. J. Chem.*, 1996, **49**, 409.
30   Z. Györgydeák, *Acros Org. Acta*, 1995, **1**, 74 (*Chem. Abstr.*, 1996, **124**, 87 594).
31   S. Cao and R. Roy, *Carbohydr. Lett.*, 1996, **2**, 27 (*Chem. Abstr.*, 1996, **125**, 196 173).
32   Z. Györgydeák, L. Szylágyi, Z. Dinya and J. Jekö, *Carbohydr. Res.*, 1996, **291**, 183.
33   J.-P. Praly, C. Bonnevie, P. Haug and G. Descotes, *Tetrahedron*, 1996, **52**, 9057.
34   J.-P. Praly, F. Péquery, C. Di Stèfano and G. Descotes, *Synthesis*, 1996, 577.
35   L. Somsak, E. Sos, Z. Györgydeák, J.-P. Praly and G. Descotes, *Tetrahedron*, 1996, **52**, 9121.
36   T.M. Krüle, B. Davis, H. Ardon, D.D. Long, N.A. Hindle, C. Smith, D. Brown, A.L. Lane, D.J. Watkin, D.G. Marquess and G.W.J. Fleet, *J. Chem.Soc., Chem. Commun.*, 1996, 1271.
37   P.B. Alper, S.-C. Hung and C.-H. Wong, *Tetrahedron Lett.*, 1996, **37**, 6029.
38   P.K. Vasudeva and N. Nagarajan, *Tetrahedron*, 1996, **52**, 5607.
39   P. Page, C. Blonski and J. Périé, *Tetrahedron*, 1996, **52**, 1557.
40   E. Lattova, M. Petrusova, A. Gaplovsky and L. Petrus, *Chem. Pap.*, 1996, **50**, 97 (*Chem. Abstr.*, 1996, **125**, 87 044).
41   G. Dai, D. Shi and L. Zhou, *Huaxue Tongbao*, 1995, 38 (*Chem. Abstr.*, 1996, **124**, 317 625).
42   A. Seta, K. Tokuda, M. Kaiwa and T. Sakakibara, *Carbohydr. Res.*, 1996, **281**, 129.

43   Y. Hatanaka, M. Hashimoto, S. Nishihara, H. Narimatsu and Y. Kanoaka, *Carbohydr. Res.*, 1996, **294**, 95.
44   Y. Hatanaka, M. Hashimoto, K.I.-P. Hidari, Y. Sanai, Y. Tezuka, Y. Nagai and Y. Kanoaka, *Chem. Pharm. Bull.*, 1996, **44**, 1111.
45   B. Ganem, *Acc. Chem. Res.*, 1996, **29**, 340 (*Chem. Abstr.*, 1996, **125**, 58 858).
46   M.L. Fascio and N.B. D'Accorso, *J. Heterocycl. Chem.*, 1996, **33**, 1573.
47   R. Hoos, J. Huixin, A. Vasella and P. Weiss, *Helv. Chim. Acta*, 1996, **79**, 1757.
48   C. Bonnette, L. Salmon and A. Gaudemer, *Tetrahedron Lett.*, 1996, **37**, 1221.
49   A. Gaudemer, C. Fanet, F. Gaudemer and L. Salmon, *Tetrahedron Lett.*, 1996, **37**, 2237.
50   M. Jurczak, D. Socha and M. Chmielewski, *Tetrahedron*, 1996, **52**, 1411.
51   L.A.G.M. van den Broek, *Tetrahedron*, 1996, **52**, 4467.
52   J.M.J. Tonchet, I. Kovacs, F. Barbalat-Rey and N. Dolatshahi, *Nucleosides Nucleotides*, 1996, **15**, 337.
53   L. Somsak, *Carbohydr. Res.*, 1996, **286**, 167.
54   G. Legler and M.-T. Finken, *Carbohydr. Res.*, 1996, **292**, 103.
55   R.J. Kerns, T. Toida and R.J. Linhardt, *J. Carbohydr. Chem.*, 1996, **15**, 581.
56   M. Kitajima, S. Shirakawa, S.G.A. Abdel-Moty, H. Takayama, S. Sakai, N.Ami and J. Stöckigt, *Chem. Pharm. Bull.*, 1996, **44**, 2195.
57   M.A.E. Shaban, A.Z. Nasr and M.A.M. Taha, *Pharmazie*, 1995, **50**, 534 (*Chem. Abstr.*, 1996, **124**, 56 544).
58   A.Z. Nasr, *Alexandrai J. Pharm. Sci.*, 1996, **10**, 139 (*Chem. Abstr.*, 1996, **125**, 301 457).
59   K. Tatsuta, *Pure Appl. Chem.*, 1996, **68**, 1341 (*Chem. Abstr.*, 1996, **125**, 222 299).
60   K. Tatsuta, Y. Ikeda and S. Miura, *J. Antibiotics*, 1996, **49**, 836.
61   T.D. Heightman, M. Locatelli and A. Vasella, *Helv. Chim. Acta*, 1996, **79**, 2191.
62   J.P. Shilvock, J.P. Wheatley, B. Davis, R.J. Nash, R.C. Griffith, M.G. Jones, M.Müller, S. Crook, D.J. Watkin, C. Smith, G.S. Besra, P.J. Brennan and G.W.J. Fleet, *Tetrahedron Lett.*, 1996, **37**, 8569.
63   P. Norris, D. Horton and D.E. Girdhar, *Tetrahedron Lett.*, 1996, **37**, 3925.
64   S. Baskaran, J. Vasu, R. Prasad, G.K. Trivedi and J. Chandrasekhar, *Tetrahedron*, 1996, **52**, 4515.
65   P. Norris, D. Horton and B. R. Levine, *Heterocycles*, 1996, **43**, 2643.
66   B. Furman, Z. Kaluza and M. Chmielewski, *Tetrahedron*, 1996, **52**, 6019.
67   J.-H. Jeong, B.W. Murray, S. Takayama and C.-H. Wong, *J. Am. Chem. Soc.*, 1996, **118**, 4227.
68   H.J.M. Gijsen, L. Qiao, W. Fitz and C.-H. Wong, *Chem. Rev.*, 1996, **96**, 443 (*Chem. Abstr.*, 1996, **124**, 87 497).

11
Thio- and Seleno-sugars

A review on the enzymic and chemo-enzymic synthesis of carbohydrates includes the preparation of thio-sugars by use of aldolase-catalysed condensations.[1]

Some reactions of the dithioacetal mono-S-oxide **1** with reducing agents have been explored.[2] The 2-thio-diethyl dithioacetal **3**, formed from **2** by exposure to ethanethiol under acidic conditions, reacted with 1.2 and 2.0 molar equivalents of $HgCl_2$ in the presence of $BaCO_3$ to give the furanoside **4** and pyranoside **5**, respectively. The latter readily equilibrates with its 2-epimer **6** under the reaction conditions.[3]

1	**2** $R^1 = H, R^2 = OH$	**4**	**5** $R^1 = SEt, R^2 = H$
	3 $R^1 = EtS, R^2 = H$		**6** $R^1 = H, R^2 = SEt$

The isopropylidenation of aldose trimethylene dithioacetals is covered in Chapter 6 and the acetolysis of aldose diethyl dithioacetals in Chapter 7. The C-C-1(C-2 bond cleavage of the α-oxoketene dithioacetal **7** on exposure to mineral acid is referred to in Chapters 13 and 14. The phenylthio precursor **8** of 2-keto-3-deoxy-D-*arabino*-heptonic acid has been prepared by chain-extension of 2,3:4,5-di-O-cyclohexylidene-D-arabinose with $PhSCH_2CO_2Me$, followed by deprotection.[4]

Nucleophilic substitutions of both 1,2-*cis*- and 1,2-*trans*-disposed peracetylated glycopyranosyl bromides with thiophenol under phase transfer conditions proceeded by the S_N2-type mechanism (*e.g.*, **9** →**10**).[5] A practical large scale synthesis of the glucosinolate sinigrin (**12**) from 1-thio-α-D-glucopyranose per-O-acetate (**11**) has been reported.[6] A new route to 2-deoxysugars from 1-xanthates is covered in Chapter 12. Chemoselective deprotection of thioacetates in the presence of O- and N-acetates has been achieved by use of hydrazinium acetate.[7]

Carbohydrate Chemistry, Volume 30
© The Royal Society of Chemistry, 1998

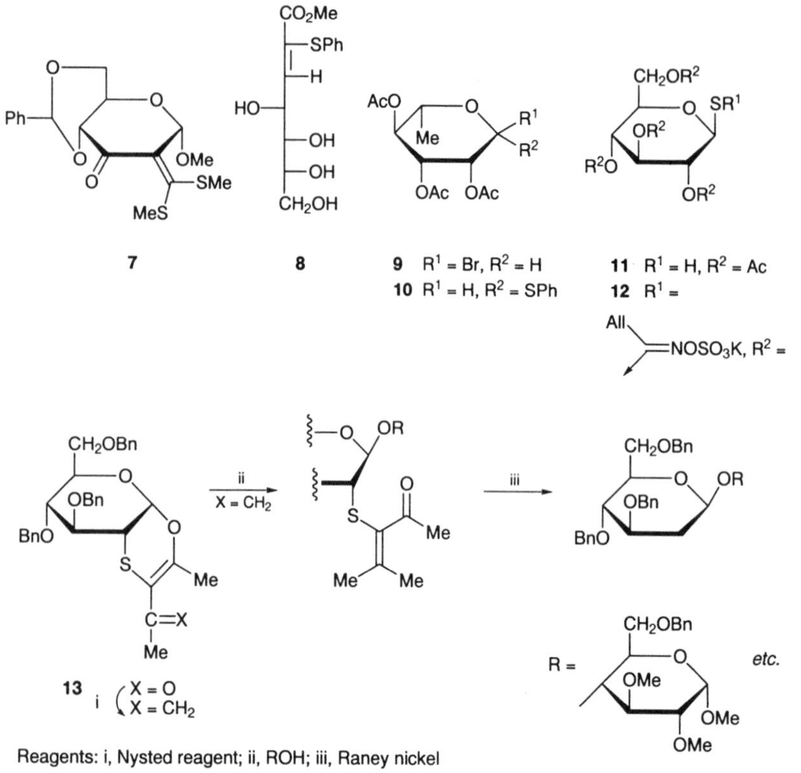

Scheme 1

Compound **13**, the product of cycloaddition between α,α-dioxothiones and tri-*O*-benzyl-D-glucal (see Vol. 29, Chapter 11, Ref. 11) has been used to prepare 2-deoxy-glycosides, as shown in Scheme 1.[8] The cycloaddition reaction has now been applied to several other glycal- and α,α-dioxothione-derivatives.[9]

Addition of 2,4-dimethylbenzenesulfenyl chloride to the glycal derived from tetra-*O*-acetyl-neuraminic acid methyl ester, followed by reaction with sodium methylthiolate, furnished the 2-methylthio-3-thiosialic acid derivative **14**.[10] Its use as a glycosyl donor is referred to in Chapter 4, and the azido phenylselenation of a furanoid glycal is covered in Chapter 9.

The 3-thioacetate **17** has been obtained by opening of the cyclic sulfamidate **16**, which was formed on exposure of the *cis*-vicinal hydroxy acetamide **15** to 1,1'-sulfonyldiimidazole.[11] Regioselective opening of benzyl 2,3-anhydro-α-D-ribo-pyranoside gave access to benzyl 3-(phenylthio)-α-D-xyloside (**18**).[12] The 3-thio- and 6-thio-analogues **20** and **21**, respectively, of amiprilose (**19**), have been synthesized by standard methods.[13] 3,6-Thio-anhydro-D-glucose derivatives **24**

14

15 $R^1 = R^2 = H$
16 $R^1R^2 =$

17

18

and **25** were made by treatment of the 3-*O*-sulfonate-6-thioacetate **22** and 6-bromo-6-deoxy-3-thioacetate **23**, respectively, with methanolic sodium methoxide.[14,15] Compound **25** was later converted to 8-*O*-methylthioswainsonine (**26**), C-4 and C-5 of the thiosugar precursor becoming C-1 and C-2, respectively, in the product.[15] Similar methoxide treatment of a *gluco*-configured 2,3-di-*O*-sulfonate-6-thioacetate and a *galacto*-configured 3-*O*-sulfonate-6-thioacetate gave 4,6-thioanhydroguloside **27** and 4,6-thioanhydroglucoside **28**, respectively, presumably *via* epoxide intermediates.[16,17]

The key-step in the synthesis of the 4-thio-α,β-D-*erythro*-pentofuranose derivative **31** from L-arabinose was the well-established ring-closure **29** →**30** (see Vol. 29, Chapter 11, Ref. 14).[18] A range of other 2-deoxy-D-*erythro*-pentose dibenzyl-dithioacetals have been prepared and similarly cyclized to 2-deoxy-4-thio-D-*erythro*-pentofuranose derivatives.[19] A synthesis of 4-thio-1,4-anhydro-L-ribitol (**32**) from D-arabinitol was based on the episulfonium ion rearrangement shown in Scheme 2.[20] The conformation of 4-thio-L-lyxono-1,4-lactone is referred to in Chapter 21.

A new method for the conversion of aldopyranosides to 5-thioaldopyranosides is exemplified in Scheme 3; the isomeric acyclic intermediates, *e.g.* compound **33**, could be separated, inverted at C-5 and cyclized stereoselectively, thus offering access to several different products.[21,22] 5-Thio-D-arabinose (**34**) has been synthesized from D-arabinose in six standard reaction steps. Synthesis of 5-thio-L-fucopyranose, as the tetraacetate **35**, from D-arabinose required chain-extension at the non reducing end which was achieved by diastereoselective reaction of aldehyde **36** with methyl lithium; the D-*altro*-product **37** was then treated sequentially with tosyl chloride and potassium thioacetate.[23] A number of aryl α-5-thiofucosides, as well as *p*-nitrophenyl α-1,5-dithiofucoside, were prepared from 1-acetate **35** directly or *via* a trichloroacetimidate.[24] These compounds were used to demonstrate the importance of sulfur in the ring and oxygen at the

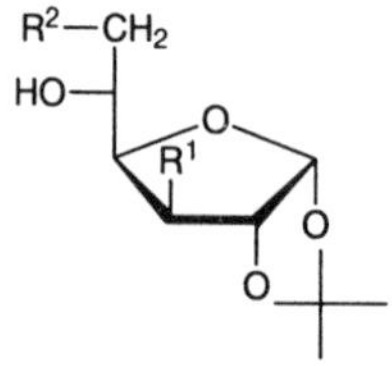

19 $R^1 = O(CH_2)_3NMe_2$, $R^2 = OH$
20 $R^1 = S(CH_2)_3NMe_2$, $R^2 = OH$
21 $R^1 = OH$, $R^2 = S(CH_2)_3NMe_2$

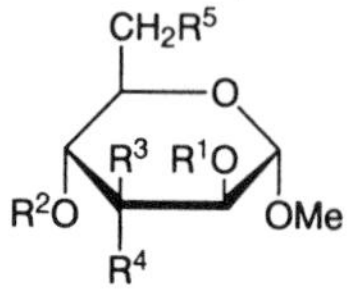

	R^1	R^2	R^3	R^4	R^5
22	Me	Bn	H	OMs	SAc
23	Bn	Bz	SAc	H	Br

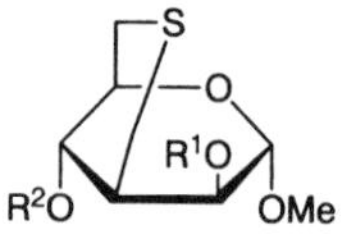

24 $R^1 = Bn$, $R^2 = Bz$
25 $R^1 = Me$, $R^2 = Bn$

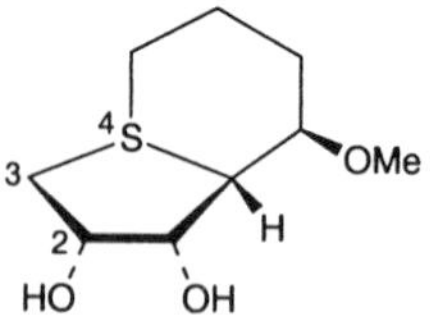

26

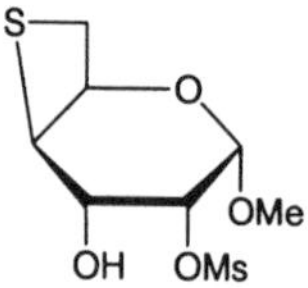

27

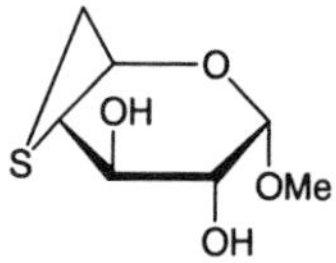

28

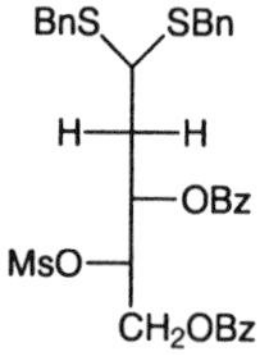

29

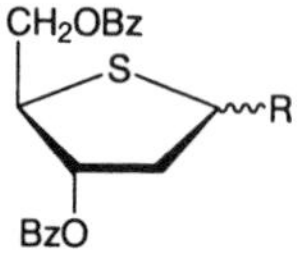

30 R = SBn
31 R = OAc

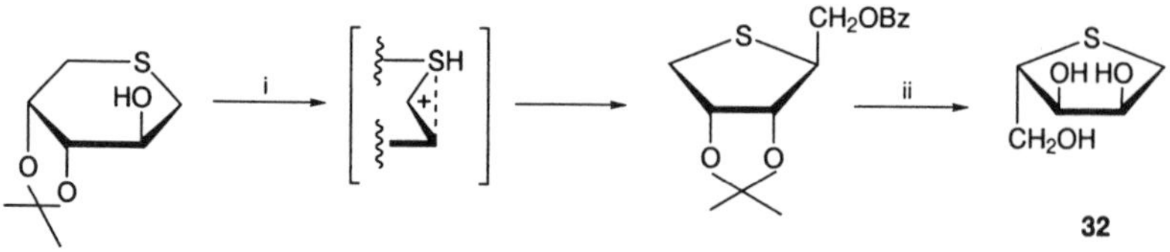

32

Reagents: i, TPP, DEAD, BzOH; ii. deprotection

Scheme 2

anomeric position for α-fucosidase inhibition.[25] Several approaches to 6-deoxy-5-thio-D-glucose from 5,6-anhydro-1,2-*O*-isopropylidene-α-D-glucose were hampered by side reactions such as polymerization of thiolate ions and rearrangements *via* episulfonium ions.[26]

Full details on the synthesis of 5-thio-D-glucose based on the radical promoted rearrangement of cyclic thionocarbonate **38** to 5,6-*S*,*O*-thiolcarbonate **39** have

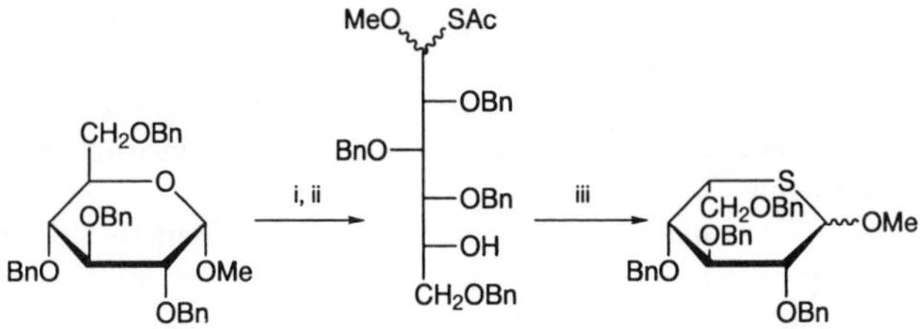

33

Reagents: i, Me₂BBr; ii, AcSH, Pr^i₂NEt; iii, TPP, DEAD

Scheme 3

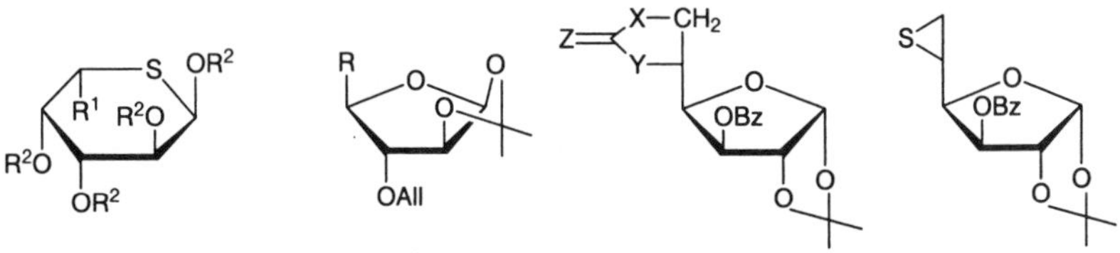

34 R¹ = R² = H
35 R¹ = Me, R² = Ac

36 R = CHO
37 R = Me

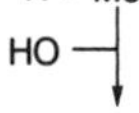

38 X = Y = O, Z = S
39 X = O, Y = S, Z = O
40 X = S, Y = Z = O

41

been published (see Vol. 21, Chapter 11, refs. 16,17).[27] Methanolysis (MeO⁻/ MeOH) of **39** caused rapid formation of episulfide **41**, whereas the 5,6-*O,S*-isomer **40** underwent simple hydrolysis under the same conditions.[28] Radical bromination of per-*O*-acetylated 5-thio-pentopyranosyl bromides gave mixtures of dibromides and tribromides, such as **42**, as the main products.[29]

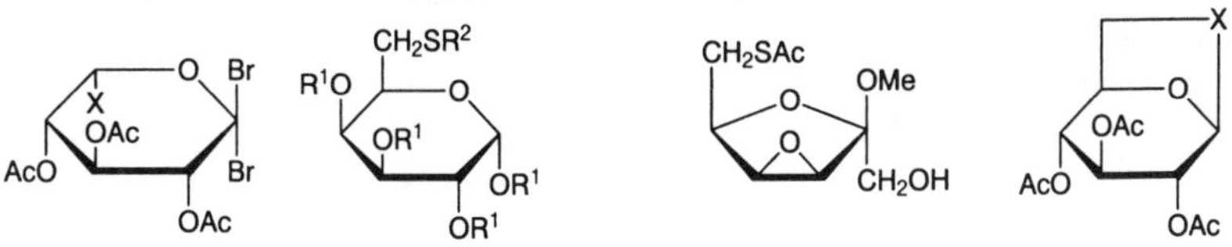

42 X = H or Br

R¹ = protecting groups
43 R² = Ac
44 R² = [furanyl] , [triazolyl] *etc.*
45 R² = [benzoxazolyl]

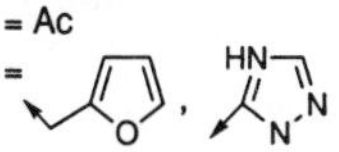

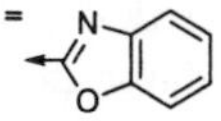

46

47 X = S or Se

The primary thiosugars **43**,[30] **44**,[31] **45**,[32] and **46**[33] as well as 6'-thiosucrose,[34] have all been prepared from precursors carrying a good leaving group at the primary position. Treatment of hexopyranosyl bromides with a good leaving group at C-6 with Na_2S or NaHSe gave 1,6-epithio- and episeleno-sugars, respectively, such as compounds **47**; especially high yields were effected by use of benzylethylammonium tetrathio- or seleno-molybdate instead of Na_2S and NaHSe, respectively.[35]

S-Isosters of 4-amino-4-deoxy- and 4-deoxy-4-guanidino-Neu5Ac2en are covered in Chapter 16 and sulfur-containing uridine analogues in Chapter 20.

Exposure of the acyclic dimesylate **48** to sodium selenide in aqueous acetone gave the protected *carba*-2-deoxy-D-ribose analogue **49** with a selenium atom in place of the 3-hydroxymethylene group in almost quantitative yield.[36] A uridine analogue containing sulfur in place of the 3-hydroxymethylene group is referred to in Chapter 20.

The following sulfur-linked di-, tri- and tetra-saccharides have been prepared by nucleophilic substituion of a good leaving group on an appropriately functionalized monosaccharide by a protected 1-thio-sugar with inversion: 3-thionigerose, 3-thiolaminaribiose, 6-thiogentiobiose, 3,3′-dithiolaminaritriose, 6,6′-dithiolaminaritriose and 6-*S*-(β-D-glucopyranosyl)-3,6-dithiolaminaribiose,[37] 3′-*S*-(β-D-glucopyranosyl)-3,3′,6′-trithiolaminaritriose,[38] 4-thiogalactobiose,[39] *N*,*N*′-diacetyl-4-thiochitobiose,[40] *N*,*N*′,*N*″-triacetyl-4,4′-dithiochitotriose and *N*,*N*′,*N*″,*N*‴-tetraacetyl-4,4′,4″-trithiochitotetraose.[41] The 6-seleno-isomaltose derivative **50** was prepared by displacement of a leaving group at C-6 by a glycosyl selenate as well as by displacement of an anomeric leaving group by a 6-selenide.[42]

Scheme 4 shows a novel route to 2-thiosophorose (**51**) which equilibrates readily with 2-thioepisophorose (**52**).[43] Preparation of diheterodisaccharide **53** and related glycosidase inhibitors involved opening of bis-aziridine **54** with 1,2:5,6-di-*O*-isopropylidene-3-thio-α-D-glucofuranose.[44]

Amphiphilic cyclodextrins **56** have been synthesized in high yields in one step from the corresponding bromides **55**.[45] In a similar preparation of peracetylated heptakis [6-*S*-(2,3-dihydroxypropyl)-6-thio]-β-cyclodextrin (**58**), the heptakis iodide **57** was used as precursor. Oxidation with MCPBA furnished heptakis sulfone **59**.[46] Hemithiocyclodextrins (cyclodextrins in which half of the interglycosidic oxygen atoms are replaced by sulfur atoms) and hemithiocellodextrins have been obtained in yields of *ca.* 10 % by exposure of 4-thio-α-maltosyl fluoride to cyclodextringlycosyltransferases[47] and 4-thio-β-cellobiosyl fluoride to cellulases,[48] respectively.

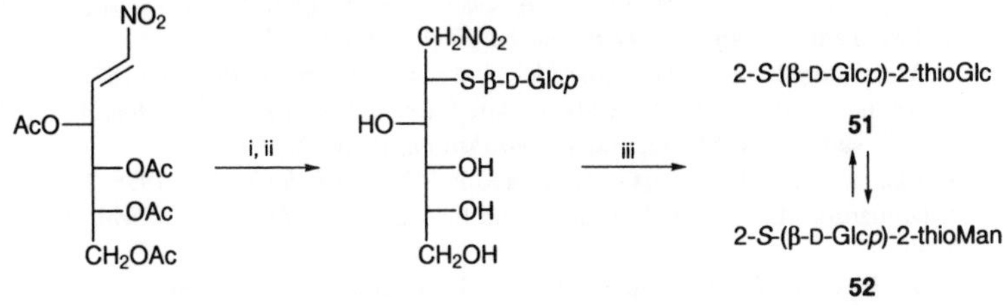

Reagents: i, β-D-Glcp-SNa, MeOH; ii, HCl, MeOH; iii, Nef reaction

Scheme 4

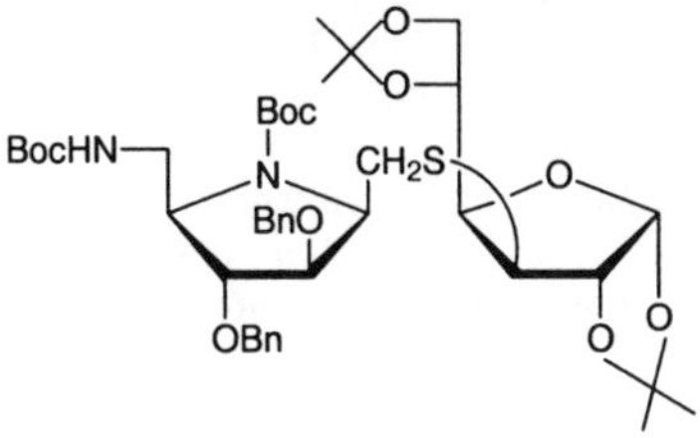

	n	R	X
55	6–8	H	Br
56	6–8	H	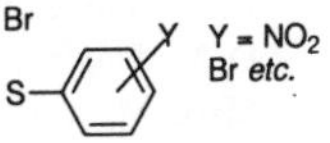 S— Y = NO₂ Br *etc.*
57	7	Ac	I
58	7	Ac	SCH₂CH(OAc)CH₂OAc
59	7	Ac	SO₂CH(OAc)CH₂OAc

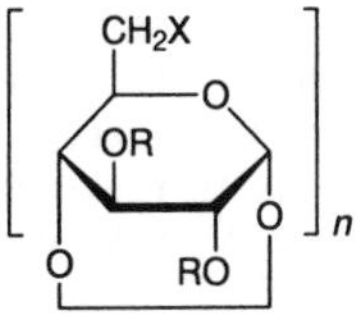

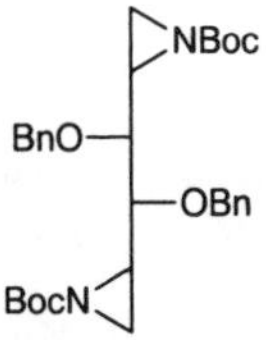

References

1   C.-H. Wong, *Pure Appl. Chem.*, 1995, **67**, 1609 (*Chem. Abstr.,* 1996, **124**, 87 494).

2   Y. Arroyo-Gómez, J.A. López-Sastre, J.F. Rodriguez-Amo and M.A. Sanz-Tejedor, *J. Chem. Soc., Perkin Trans. 1*, 1996, 2933.

3   D. Horton, P. Norris and B. Berrang, *Carbohydr. Res.*, 1996, **283**, 53.

4   B.B. Paidak, Yu.M. Mikshiev and V.I. Kornilov, *Zh. Obshch. Khim.*, 1995, **65**, 349 (*Chem. Abstr.*, 1996, **124**, 9105).

5   S. Cao and R. Roy, *Carbohydr. Lett.*, 1996, **2**, 27 (*Chem. Abstr.*, 1996, **125**, 196 173).

6   W. Abramski and M. Chmielewski, *J. Carbohydr. Chem.*, 1996, **15**, 109.

7   W.K.C. Park, S.J. Meunier, D. Zanini and R. Roy, *Carbohydr. Lett.*, 1995, **1**, 27 (*Chem. Abstr.*, 1996, **125**, 87 066).

8   C.H. Marzabadi and R.W. Franck, *J. Chem. Soc., Chem. Commun.*, 1996, 2651.

9   C. Capozzi, A. Dios, R.W. Franck, A. Geer, C. Marzabadi, S. Menichetti, C. Nativi
     and M. Tamarez, *Angew. Chem., Int. Ed. Engl.*, 1996, **35**, 777.
10   V. Martichonok and G.M. Whitesides, *J. Am. Chem. Soc.*, 1996, **118**, 8187.
11   B. Aguilera and A. Fernández-Mayoralas, *J. Chem. Soc., Chem. Commun.*, 1996, 127.
12   P.K. Vasudeva and M. Nagarajan, *Tetrahedron*, 1996, **52**, 5607.
13   P. Vanlemmens, D. Postel, G. Ronco and P. Villa, *Carbohydr. Res.*, 1996, **289**, 171.
14   I. Izquierdo, M.T. Plaza, A. Ramrez and F. Aragón, *J. Heterocycl. Chem.*, 1996, **33**,
     1239.
15   I. Izquierdo, M.T. Plaza and F. Aragón, *Tetrahedron: Asymm.*, 1996, **7**, 2567.
16   I. Izquierdo Cubero and M.T.Plaza Lopez-Espinosa, *Carbohydr. Lett.*, 1995, **1**, 191
     (*Chem. Abstr.*, 1996, **125**, 87 011).
17   I. Izquierdo Cubero, M.T. Plaza Lopez-Espinosa and A. Saenz de Burnaga Molina,
     Carbohydr. Res., 1996, **280**, 145.
18   H. Ait-Sir, N.-E. Fahme, G. Goethals, G. Ronco, B. Tber, P. Villa, D.F. Ewing and
     G. Mackenzie, *J. Chem. Soc., Perkin Trans 1*, 1996, 1665.
19   S. Shaw-Ponter, P. Rider and R.J. Young, *Tetrahedron Lett.*, 1996, **37**, 1871.
20   H.-J. Altenbach and G.F. Merhof, *Tetrahedron: Asymm.*, 1996, **7**, 3087.
21   H. Hashimoto, M. Kawanishi and H. Yuasa, *Carbohydr. Res.*, 1996, **282**, 207.
22   H. Hashimoto, M. Kawanishi and H. Yuasa, *Chem. Eur. J.*, 1996, **35**, 556.
23   M. Izumi, O. Tsuruta and H. Hashimoto, *Carbohydr. Res.*, 1996, **280**, 237.
24   O. Tsuruta, H. Yuasa and H. Hashimoto, *Bioorg. Med. Chem. Lett.*, 1996, **6**, 1989.
25   H. Yuasa, Y. Nakano and H. Hashimoto, *Carbohydr. Lett.*, 1996, **2**, 23 (*Chem.
     Abstr.*, 1996, **125**, 196 125).
26   E. Bozó, S. Boros, J. Kuszmann and E. Gács-Baitz, *Carbohydr. Res.*, 1996, **290**, 159.
27   Y. Tsuda, Y. Sato, K. Kanemitsu, S. Hosoi, K. Shibayama, K. Nakao and Y.
     Ishikawa, *Chem. Pharm. Bull.*, 1996, **44**, 1465.
28   Y. Tsuda and K. Shibayama, *Chem. Pharm. Bull.*, 1996, **44**, 1476.
29   M. Bandry, M.-N. Bouchu, G. Descotes, J.-P. Praly and F. Bellamy, *Carbohydr.
     Res.*, 1996, **282**, 237.
30   D.A. Yeagley, A.J. Benesi and M. Miljkovic, *Carbohydr. Res.*, 1996, **289**, 189.
31   M.A. Martins Alho, N.B. D'Accorso and I.M.E. Thiel, *J. Heterocycl. Chem.*, 1996,
     33, 1339.
32   M.T. Lakin, N. Mouhous-Riou, C. Lorin, P. Rollin, J. Kroon and S. Pérez,
     Carbohydr. Res., 1996, **290**, 125.
33   A.K. Norton and M. von Itzstein, *Aust. J. Chem.*, 1996, **49**, 281.
34   H. Kakinuma, H. Yuasa and H. Hashimoto, *Carbohydr. Res.*, 1996, **284**, 61.
35   H. Driguez, J.C. McAuliffe, R.V. Stick, D.M.G. Tilbrook and S.J. Williams, *Aust. J.
     Chem,*. 1996, **49**, 343.
36   K. Schürrle and W. Piepersberg, *J. Carbohydr. Chem.*, 1996, **15**, 435.
37   M.-O. Cantour-Galcera, J.-M. Guillot, C. Ortiz-Mellet, F. Pfleiger-Carrara,
     J. Defaye and J. Gelas, *Carbohydr. Res.*, 1996, **281**, 99.
38   M.-O. Cantour-Galcera, C. Ortiz-Mellet and J. Defaye, *Carbohydr. Res.*, 1996, **281**,
     119.
39   U. Nilsson, R. Johansson and G. Magnusson, *Chem. Eur. J.*, 1996, **2**, 295
40   L.-X. Wang and Y.C. Lee, *Carbohydr. Lett.*, 1995, **1**, 185 (*Chem. Abstr.*, 1996, **125**,
     87 049).
41   L.-X. Wang and Y.C. Lee, *J. Chem. Soc., Perkin Trans 1*, 1996, 581.
42   S. Czernecki and D. Randriamandimby, *J. Carbohydr. Chem.*, 1996, **15**, 183.
43   M. Petrusova, E. Lattova, M. Matulova, L.Petrus and J.N. BeMiller, *Carbohydr.
     Res.*, 1996, **283**, 73.

44   L. Campanini, A. Duréault and J.-C. Depezay, *Tetrahedron Lett.*, 1996, **37**, 5095.
45   K. Chmurski, A.W. Coleman and J. Jurcak, *J. Carbohydr. Chem.*, 1996, **15**, 787.
46   H.H. Baer and F. Santoyo-Gonzales, *Carbohydr. Res.*, 1996, **280**, 315.
47   L. Bornaghi, J.-P. White, D. Penninga, A.K. Schmidt, L. Dijkhuizen, G.E. Schulz
      and H. Driguez, *J. Chem. Soc., Chem. Commun.*, 1996, 2541.
48   V. Moreau and H. Driguez, *J. Chem. Soc., Perkin Trans. 1*, 1996, 525.

12
Deoxy-sugars

Osmylation of the chirally masked 2-alkoxypent-3-enal **1** and subsequent functional group manipulations furnished enantiopure 5-deoxy-L-lyxose and -xylose derivatives.[1] A new preparation of 5-deoxyribonolactone derivative **2** from ribonolactone by tributyltin hydride reduction of a 5-chloro-5-deoxy intermediate has been published.[2] Conversion of compound **2** to a 5′-deoxy-nucleoside analogue is covered in Chapter 20. A multistep synthesis of methyl 5-*O*-benzoyl-2,3-dideoxy-β-D-*glycero*-pentofuranose (**4**) from D-xylose involving Raney nickel desulfurization of compound **3** (see Chapter 11) has been described.[3] Several deoxy-ketoses have been prepared by Raney nickel desulfurization of 1-thioketose derivatives obtained by enzyme-catalysed aldol condensations (see Vol. 29, Chapter 2, Scheme 1). An example is given in Scheme 1.[4]

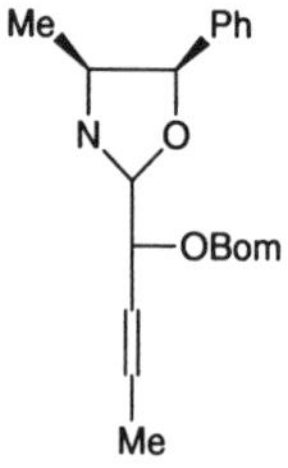

1

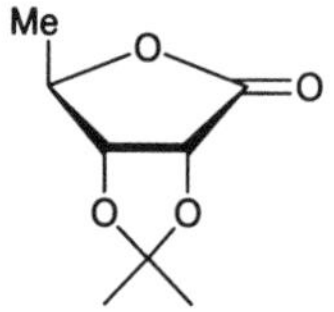

2

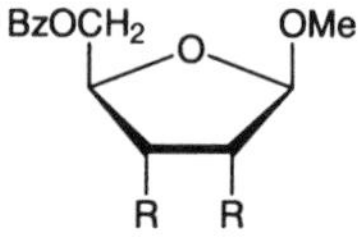

3 R = SEt
4 R = H

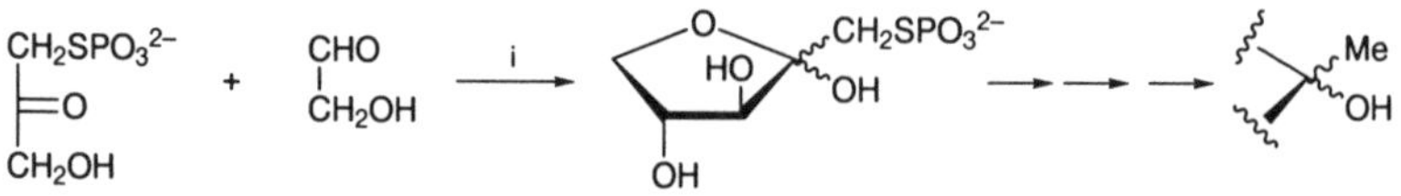

Regents: i, Fructosediphosphate aldolase

Scheme 1

In a *de novo* synthesis of the 2-deoxy-L-*arabino*-hexose derivative **7** the dihydropyran ring was formed by asymmetric [2+4]-heterocycloaddition of enone **5** to chiral vinyl ether **6**, as shown in Scheme 2.[5] Chain-extension at C-5 of methyl 2,3-*O*-isopropylidene-β-D-*ribo*-pentodialdo-1,4-furanoside with methoxymethylenetriphenylphosphorane and subsequent hydrolysis, reduction and

Carbohydrate Chemistry, Volume 30

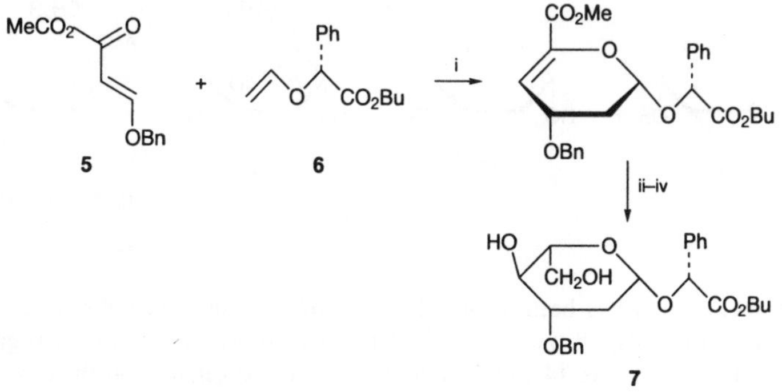

Regents: i, Eu(fod)$_3$; ii, LAH; iii, BH$_3$, Me$_2$S; iv, H$_2$O$_2$, NaOH

Scheme 2

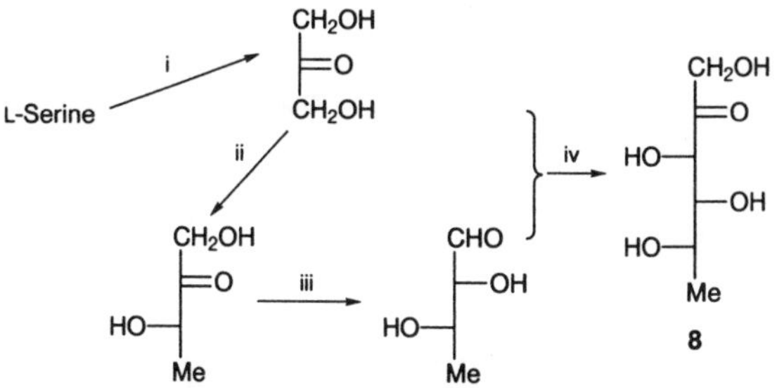

Regents: i, Aminotransferase; ii, MeCHO, transketolase; iii, microbial isomerase; iv, transketolase

Scheme 3

deprotection gave 5-deoxy-D-ribohexose in nearly 30% overall yield.[6] The preparation of 6-deoxy-L-sorbose (**8**) by a 'natural' route from L-serine and acetaldehyde is outlined in Scheme 3.[7] A facile 4-deoxygenation of α,β-unsaturated glyconic acid derivatives by reductive elimination, and the preparation of methyl α-L-daunosaminide from non-carbohydrate starting materials are referred to in Chapters 13 and 19, respectively.

Opening of epoxide **9** by lithium hydride in THF took place at C-3 exclusively to give benzyl 3-deoxy-α-D-*erythro*-pentopyranoside **10**. The selectivity is attributed to a repulsive effect of the pyranose ring oxygen on the approaching nucleophile.[8]

4-Deoxy-, 6-deoxy- and 4,6-dideoxy-D-mannose derivatives have been prepared by sodium borohydride reduction of the relevant tosyl esters and/or Barton deoxygenation at C-4 of compound **11**, followed by deprotection.[9] One-pot dideoxygenation of the fully protected *n*-pentenyl 3,6-di-*O*-thionocarbonyl β-D-galactopyranoside **12** to give the 3,6-dideoxy-sugar **13** (a derivative of *n*-pentyl

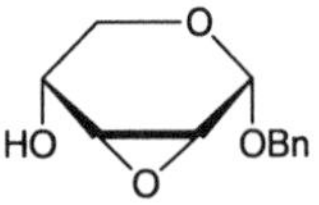

9 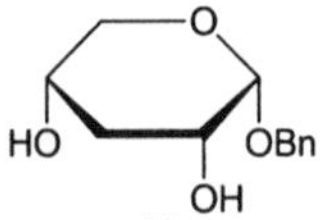   **10** 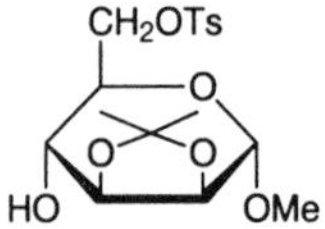   **11**

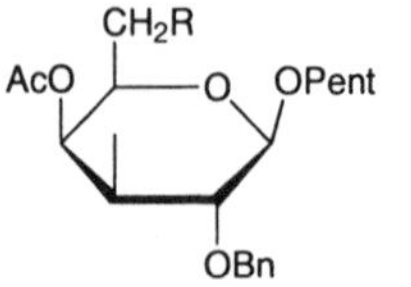

12 R = OC—⟨aryl⟩—F
              ‖
              S
13 R = H

β-D-abequoside) has been achieved by use of tris(trimethylsilyl)silane.[10] A new synthesis of methyl β-D-abequoside derivative **16** proceeded *via* the D-galactose-derived intermediates **14** and **15**, and involved deoxygenation at the 6-position by reduction of the tosylate with LAH and at the 3-positions by regioselective opening of the 3,4-cyclic sulfate (Scheme 4). The protected methyl 3,6-dideoxy-α-D-*arabino*-hexopyranoside **17** (methyl α-D-tyvelopyranoside) was similarly obtained by way of an α-D-mannose-derived 2,3-cyclic sulfate 6-tosylate.[11]

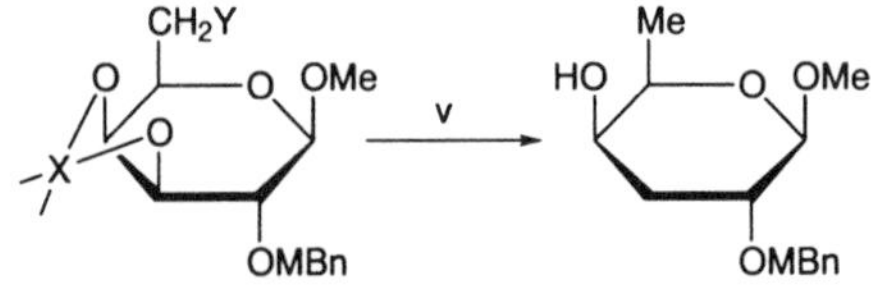

14 X = CMe₂, Y = OTs
15 X = SO₂, Y = H   **16**   **17**

Regents: i, LAH; ii, aq. HOAc; iii, SO₂Cl₂, Py; iv, RuCl₃, NaIO₄; v, Bu₄NBH₄

Scheme 4

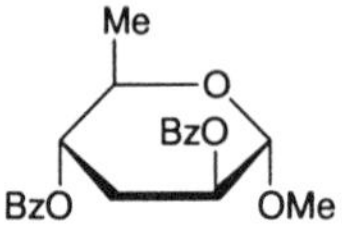

The preparation of 3-deoxy-D-*arabino*-hept-2-ulosonic acid (Dahp) from β-D-glucopyranosyl cyanide peracetate is referred to in Chapters 7 and 16, and the syntheses of 2-deoxy-α- and -β-D-*arabino*-hexopyranosyl phosphonic acids and related dicarboxylic acids from protected glycals are covered in Chapters 7 and 16, respectively.

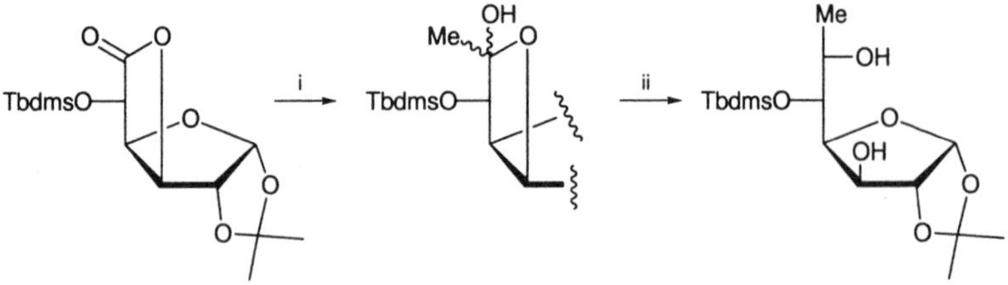

18   **19**

Reagents: i, MeLi; ii, NaBH₄, H₃O⁺

Scheme 5

The synthesis of 7-deoxy-L-*glycero*-D-*gluco*-heptose (**19**), an inhibitor of gluco-kinase and glucose 6-phosphatase, from 1,2-*O*-isopropylidene-D-glucuronolac-tone derivative **18** is shown in Scheme 5.[12] The 2'-*O*-cyclohexylcarbamoyl disaccharide **20** (see Chapter 3 for formation) was readily converted to the corresponding xanthate which was deoxygenated with tributyltin hydride to the 2'-deoxy disaccharide **21**.[13] Radical deoxygenation with tributyltin hydride has also been use to prepare 1',6'-dideoxysucrose from the corresponding 1',6'-ditriflate.[14]

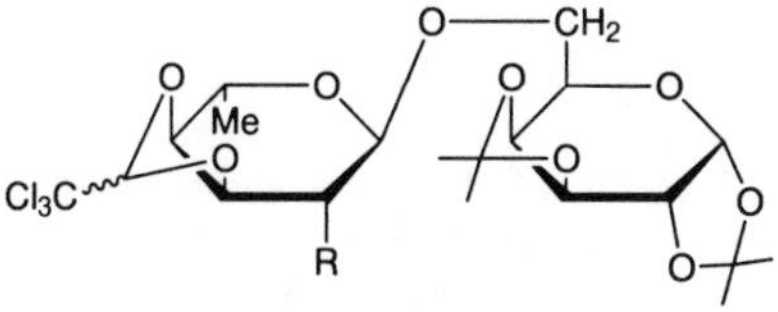

20 R = OCNHC$_6$H$_{11}$
          ‖
          O

21 R = H

α- or β-D-Tyv*p*-(1→3)-β-D-Gal*p*NAc-OR
R = (CH$_2$)$_2$Tms or (CH$_2$)$_8$CO$_2$Me

22

A number of deoxy analogues of natural di- and oligo-saccharides have been prepared from suitable deoxy-sugar components by conventional methods as models for glycans or for use in enzyme inhibition studies. They included the tyvelose-containing disaccharides **22**;[15] the seven mono-deoxy analogues of methyl β-isomaltoside;[16] the methyl α-glycosides of isomalto-oligosaccharides deoxygenated at C-3 of one of the units;[17] five different monodeoxy-analogues of the acceptor disaccharide **23** of the Lewis α-(1→3/4)-fucosyltransferase;[18] and three sialyl Lewis X-analogues containing 4-deoxy-, 6-deoxy- or 4,6-dideoxy-D-galactopyranose in place of d-galactopyranose.[19] Six analogues of β-maltosyl-(1→4)-trehalose, selectively deoxygenated either at one of the primary positions[20] or at C-4 or C-4''',[21] were obtained by [2+2] block syntheses, with catalytic reduction of a 6-iodide or radical deoxygenation at a 4-position either at the di- or the tetra-saccharide stage.

β-D-Gal*p*-(1→3)-β-D-Glc*p*NAc-O(CH$_2$)$_8$CO$_2$Me

23

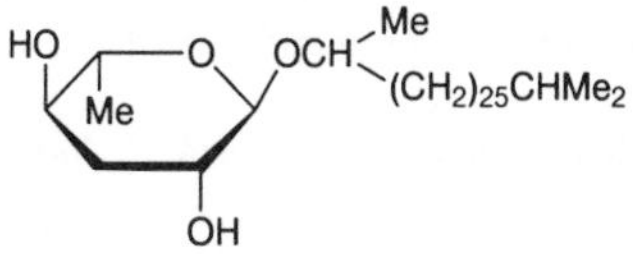

24

6-Deoxy-3-*O*-methyl-β-D-allopyranose was identified as constituent sugar of oxypregnane-oligoglycoside saponins ('stephanosides') from *Stephanotis lutchuensis*, together with the more common deoxy-sugars cymarose, oleandrose and thevetose.[22] Glycoside **24** and five closely related glycosides of 3,6-dideoxy-L-mannose (ascarylose) have been isolated from the nematode *Ascaris suum*.[23]
A mechanistic study of the fragmentation reactions of dideoxysugars in negative ion mass spectrometry is referred to in Chapter 22.

References

1   G. Poli, I. Pagni, S.I. Maffioli, C. Scolastico and M. Zanda, *Gaz. Chim. Ital.*, 1995, **125**, 505 (*Chem. Abstr.*, 1996, **124**, 202 843).
2   P.E. Joos, A. De Groot, E.L. Esmans, F.C. Alderweireldt, A. De Bruyn, J. Balzarini and E. De Clercq, *Heterocycles*, 1996, **42**, 173.
3   S. Lajsic, G. Cetkovic, V. Popsavin and D. Milijkovic, *J. Serb. Chem. Soc.*, 1995, **60**, 629 (*Chem. Abstr.*, 1996, **124**, 9124).
4   R. Duncan and D.G. Drueckhammer, *J. Org. Chem.*, 1996, **61**, 438.
5   G. Dujardin, S. Rossignol and E. Brown, *Tetrahedron Lett.*, 1996, **37**, 4007.
6   Z. Pakulski and A. Zamojski, *Pol. J. Chem.*, 1995, **69**, 912 (*Chem. Abstr.*, 1996, **124**, 30 132).
7   L. Hecquet, J. Botte and C. Demuynck, *Tetrahedron*, 1996, **52**, 8223.
8   P.K. Vasudeva and M. Nagarajan, *Tetrahedron*, 1996, **52**, 5607.
9   T. Nishio, Y. Miyake, K. Kubota, M. Yamai, S. Miki, T. Ito and T. Oku, *Carbohydr. Res.*, 1996, **280**, 357.
10   B. Wilson and B. Fraser-Reid, *Carbohydr. Lett.*, 1996, **2**, 1 (*Chem. Abstr.*, 1996, **125**, 196 124).
11   K. Zeegelaar-Jaarsveld, S.C. van der Plas, G.A. van der Marel and J.H. van Boom, *J. Carbohydr. Chem.*, 1996, **15**, 665.
12   Y. Blériot, K.H. Smelt, J. Cadefau, M. Bollen, W. Stalmans, K. Biggadike, L.N. Johnson, N.G. Oikonomakos, A.L. Lane, S. Crook, D.J. Watkin and G.W.J. Fleet, *Tetrahedron Lett.*, 1996, **37**, 7155.
13   R. Miethchen and D. Rentsch, *Liebigs Ann. Chem.*, 1996, 539.
14   H. Kakinuma, H. Yuasa and H. Hashimoto, *Carbohydr. Res.*, 1996, **284**, 61.
15   M. A. Probert, J. Zhang and D.R. Bundle, *Carbohydr. Res.*, 1996, **296**, 149.
16   R.U. Lemieux, U. Spohr, M. Bach, D.R. Cameron, T.P. Frandsen, B.B. Stoffer, B. Svensson and M.M. Palcic, *Can. J. Chem.*, 1996, **74**, 319.
17   E. Petrakova and C.P.J. Glaudemans, *Carbohydr. Res.*, 1996, **284**, 191.
18   M. Du and O. Hindsgaul, *Carbohydr. Res.*, 1996, **286**, 87.
19   S. Komba, H. Ishida, M. Kiso and A. Hasegawa, *Glycoconjugate J.*, 1996, **13**, 241 (*Chem. Abstr.*, 1996, **125**, 11 312).
20   H.P. Wessel, M. Trumtel and R. Minder, *J. Carbohydr. Chem.*, 1996, **15**, 523.
21   H.P. Wessel, M.-C. Viaud and M. Trumtel, *J. Carbohydr. Chem.*, 1996, **15**, 769.
22   K. Yoshikawa, N. Okada, Y. Kann and S. Arihara, *Chem. Pharm. Bull.*, 1996, **44**, 1790.
23   J.P. Bartley, E.A. Bennett and P.A. Darben, *J. Nat. Prod.*, 1996, **59**, 921 (*Chem. Abstr.*, 1996, **125**, 248 244).

13
Unsaturated Derivatives

1 General

The general area of research that has evolved from a study of 2,3-unsaturated sugars has been reviewed. Attention was given, in part, to the preparation of sucrose involving a non-acidic Ferrier rearrangement.[1] The use of glycals in the syntheses of complex oligosaccharides and glycoconjugates has also been reviewed.[2,3]

2 Pyranoid Derivatives

2.1 Syntheses of 1,2-Unsaturated Cyclic Compounds and Related Derivatives – An easy and efficient 'one pot' synthesis of peracetylated glycals from unprotected sugars has been described. In this method the sugar was sequentially treated with acetic anhydride and catalytic hydrobromic acid-acetic acid (to effect per-acetylation), more hydrobromic acid-acetic acid (to form the glycosyl bromide), sodium acetate (to neutralize excess hydrobromic acid) and finally a buffered mixture of sodium acetate, acetic acid, zinc-copper sulfate and water (to effect a reductive-elimination). The method was applied to the syntheses of glycals derived from D-glucose, D-galactose, L-rhamnose, L-arabinose, maltose, lactose and maltotriose.[4]

The previously reported (Vol. 29, p. 182, ref. 2) preparation of glycals by reaction of per-acetylated glycosyl bromides with bis(titanocenechloride) $(Cp_2TiCl)_2$ has been extended and shown to tolerate a variety of protecting groups such as benzoyl, benzyl, benzylidene or Tbdps in the O-4 and O-6 positions.[5] Tetra-*O*-acetyl-D-glucopyranosyl bromide has also been converted in high yield into tri-*O*-acetyl-D-glucal upon reaction with the chromium-ethylenediamine tetraacetate (EDTA) complex, $[Cr(II)(EDTA)]^{2-}$ under acidic conditions. The reaction was thought to proceed by way of a glycosyl-chromium(III)(EDTA) intermediate.[6]

Simply heating the 2-deoxy-sulfoxide **1** in benzene caused elimination of benzenesulfenic acid, producing the corresponding glycal in 98 % yield.[7]

Good yields of pyranoid glycals were obtained from acetal or ether protected D-*gluco*-, D-*galacto*- or D-*manno*- configured glycosyl sulfoxides by reaction with aryl- or alkyl-lithiums. It was determined that the nature of the lithium reagents

Carbohydrate Chemistry, Volume 30
© The Royal Society of Chemistry, 1998

had little effect on the reaction and that α-sulfoxides gave highest yields regardless of the orientation of the substituent at C-2.[8] In contrast, the corresponding D-*gluco*- or D-*manno*-configured glycopyranosyl chlorides reacted with aryl- or alkyl-lithiums giving moderate to good yields of *C*-aryl or *C*-alkyl glycals. For example 2,3,4,6-tetra-*O*-benzyl-α-D-glucopyranosyl chloride afforded compounds **2**.[9]

1   **2** R = Me, Bu^n, Bu^t, Ph   **3** R = NHAc
                                                                          4 R = OAc

5   **6** X = O, Y = NH$_2$ **7** OAc
                         8 X = NH, Y = NHBn
                         9 X = O, Y = NHBn

The glycal derivatives **3-5** were readily obtained from the per-acetylated methyl esters of NeuNAc, Kdn and Kdo, respectively, by treatment with trimethylsilyl triflate as a result of a reinvestigation of this reaction. It was found that the solvent and temperature used were important factors.[10] Glycal analogue **6** has been prepared, as a potential inhibitor of a β-D-galactosidase from *E.coli*, by sequential treatment of **7**, first with NBS (to give a glycosyl bromide) then with zinc (to form the glycal) and finally de-acetylated. The effect of basicity and hydrophobicity induced by the carboxamide group on enzyme inhibition was further explored by the preparation of derivatives **8** and **9**.[11]

The C-2 branched-glycal **10** has been synthesized by treatment of **11** (see Chapter 14, ref. 49 for its preparation) with mesyl anhydride-triethylamine.[12]

The mechanism of UDP-*N*-acetylglucosamine 2-epimerase, an important enzyme used by bacteria for preparing activated UDP-*N*-acetylmannosamine residues for use in cell wall formation, has been investigated using an ^{18}O-labelled UDP-GlcNAc derivative and shown to involve the intermediacy of 2-acetamido-2-deoxy-D-glucal.[13]

The unsaturated branched-sugar **12** was the product formed when α-oxoketene dithioacetal **13** (see Vol. 26, p. 162, ref. 47) was treated with dilute mineral acid.[14]

It has previously been reported that 3,4-bis-*O*-(Tbdms)-6-deoxy-L-glucal was cleanly deprotonated by *tert*-butyllithium at C-1 (see Vol. 26, p. 148, ref. 7; Vol. 25, p. 229, ref. 50; Vol. 25, p. 54, ref. 237 and Vol. 23, p. 136, ref. 11) but that 3,4,6-tri-*O*-(Tbdms)-D-glucal underwent competing lithiations α to silicon atoms

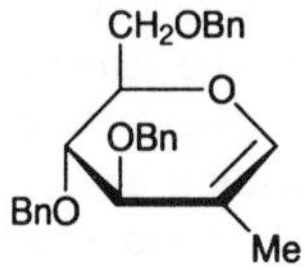

10 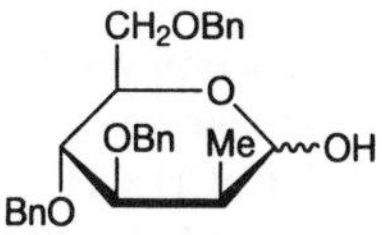   11 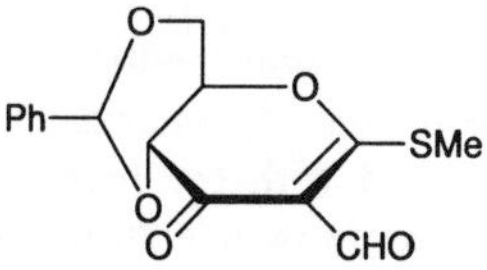   12

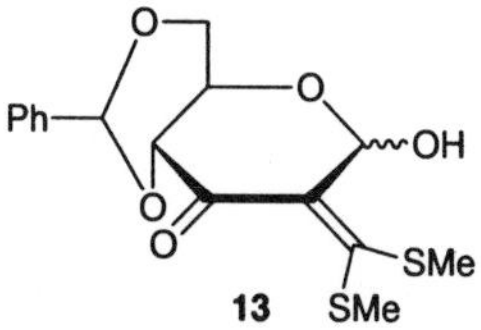

13

(see Vol. 25, p.151, ref. 2). Now it has been discovered that both compounds reacted with varying amounts (2.2 - 8.0 eq.) of *tert*-butyllithium, regioselectively lithiating C-1 and the 6-*O*-Tbdms group (when the latter is present), followed by the 4-*O*-Tbdms group.[15]

The synthesis and use of glycal derivatives in the preparation of staurosporine and *ent*-staurosporine is mentioned in Chapter 24 and an investigation into the isomerization of 2-oxo-pyranosides in pyridine affording a hydroxyenone product is covered in Chapter 15. The synthesis of a 1,2-unsaturated C-2 branched, 3,4-annulated sugar from a 2,3-unsaturated pyranose bearing a 4-*O*-allyl substituent is covered in Chapter 14.

2.2   Syntheses of 2,3-Unsaturated Cyclic Compounds

2.2.1 Syntheses Involving Allylic Rearrangement of Glycals – *C*-Allyl glycosides such as **14** and **15** were formed in high yield and with good α-selectivity when the corresponding per-acetylated glycal was treated with allyltrimethylsilane in the presence of the acidic clay, montmorrilonite K-10. (Several other examples were also described.[16]) A similar kind of reaction took place when alkylzinc reagents were reacted with variously configured per-acetylated or per-benzylated glycals in the presence of a Lewis acid. (Scheme 1 illustrates the process with tri-*O*-acetyl-D-glucal). The α,β-ratios of anomers ranged from exclusively α to 1.5:1.[17] Likewise, the D-glutamic acid derived zinc reagent **16** reacted with 3,4,6-tri-*O*-acetyl-D-glucal in the presence of a Lewis acid affording **17** as a 9:1 mixture of α:β-anomers.[18] In contrast to these reactions, the zinc-copper reagent NCCH₂CH₂Cu(CN)ZnI reacted with tri-*O*-acetyl- or benzyl-D-glucal affording

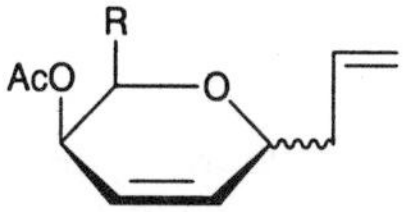

14 R = CH₂OAc,
       α:β = 18:1

15 R = Me,
       α:β = 67:1

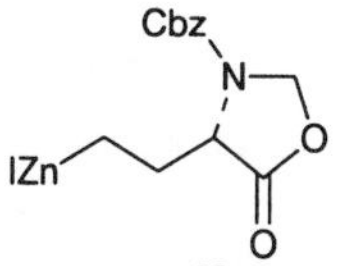

16

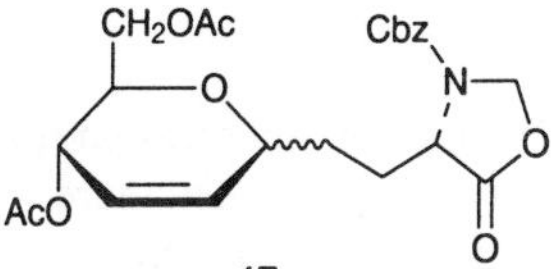

17

only small quantities of C-1 bonded derivatives, the major product being the unsaturated branched-sugar **18**. This reaction was found to be quite general for a range of zinc-copper reagents.[19] In a similar way, the 3-*O*-mesyl glycal compound **19** reacted with the reagent $IZn(NC)CuCH_2CH_2CO_2Me$, yielding unsaturated branched-derivative **20** in 49 % yield together with 16 % of rearranged product **21**.[18]

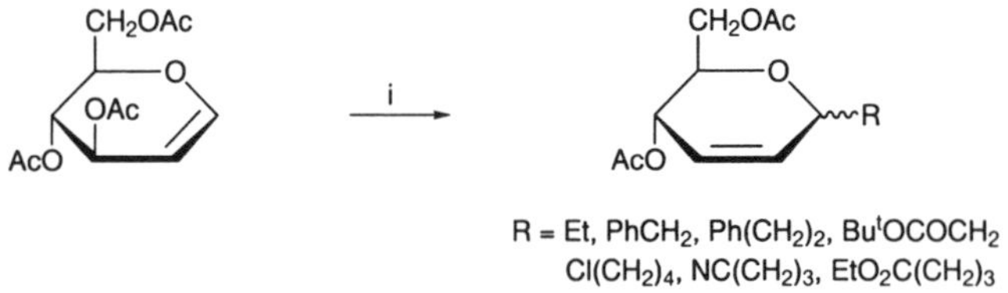

R = Et, PhCH$_2$, Ph(CH$_2$)$_2$, Bu^{t}OCOCH$_2$
Cl(CH$_2$)$_4$, NC(CH$_2$)$_3$, EtO$_2$C(CH$_2$)$_3$

Reagents: i, RZnI(Br), BF$_3$·OEt$_2$

Scheme 1

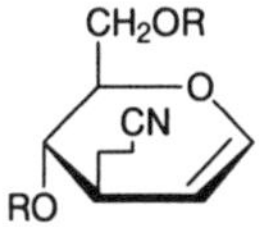

18 R = Ac or Bn 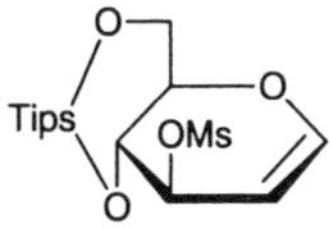   **19** 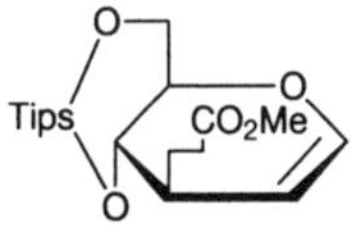   **20** 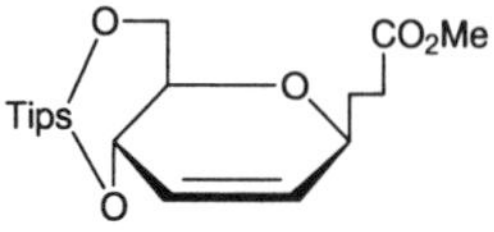   **21**

O-Trimethylsilyl ketene acetals, which are synthetically equivalent to Reformatsky reagents, reacted with glycals in the presence of a Lewis acid as depicted in Scheme 2. Enones, such as 1,5-anhydro-4,6-*O*-benzylidene-2-deoxy-D-*erythro*-hex-1-en-3-ulose, also reacted with the same reagents producing branched-chain products (see Chapter 14).[20]

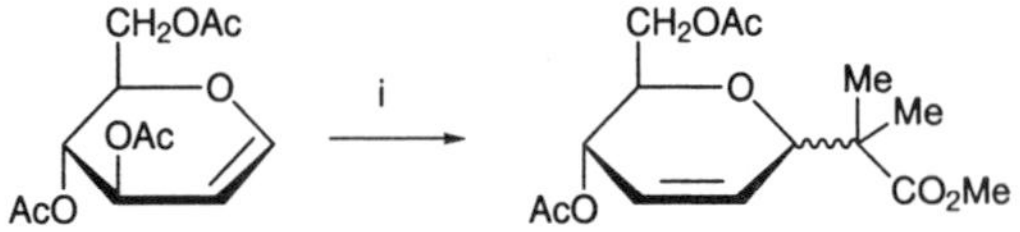

Reagents: i,

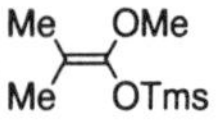

, Tms·OTf

Scheme 2

The known dimerization reaction that takes place when glycals are treated with Lewis acids proceeded more efficiently with acetyl perchlorate (a known polymerization initiator of dihydropyran) in dichloromethane at −78 °C. For example tri-*O*-acetyl-D-galactal gave a 77 % yield of dimer **22**.[21] The report that tri-*O*-benzyl-D-glucal gives a cyclopentane derivative under these conditions has been corrected.[22]

22   **23**

The formation of *O*-glycoside **23** was accomplished by reacting tri-*O*-acetyl-D-galactal with methyl 2,3,6-tri-*O*-benzyl-α-D-glucopyranoside in the presence of Pd(0) or Lewis acid catalysts.[23]

The unsaturated azides **24** and **25** resulted when 1,5-anhydro-2,3-dideoxy-4,6-di-*O*-Tbdms-D-*eyrthro*-hex-1-enitol was treated with the reagent combination (PhIO)$_n$-trimethylsilyl azide.[24]

Further details have appeared on the products formed when tri-*O*-acetyl-D-glucal was heated in water: initially, 4,6-di-*O*-acetyl-2,3-dideoxy-α- and β-D-*erythro*-hex-2-enopyranose were produced, which reacted further giving 4,6-di-*O*-acetyl-D-allal, (*E*)-4,6-di-*O*-acetyl-2,3-dideoxy-D-hex-2-enose and 2,3-di-*O*-acetyl-3-hydroxy-2-(hydroxymethyl)-2*H*-pyran (see Vol. 29, p. 184, ref. 16). Transition state kinetic barriers were determined and AM1 calculations performed to characterize the thermodynamics of the reaction.[25]

24   **25**   **26**

In a radical-based method for preparing C-2 branched-chain sugars by way of addition of dimethyl malonate to tri-*O*-acetyl-D-glucal (see Chapter 14), 1,4,6-tri-*O*-acetyl-2,3-dideoxy-α,β-D-*erythro*-hex-2-enopyranose was formed as a by-product in a small amount.[26]

2.2.2 Other Methods – The 2,3-unsaturated compound **26**, an intermediate in the preparation of an analogue of the antibiotic PA-48153 C (an immunosupressant) was prepared by deoxygenation of the corresponding 2,3-di-xanthate using diphenylsilane-AIBN. (See also Vol. 25, p. 158, ref. 24 for the preparation of a 5,6-unsaturated furanoid derivative using the same technique).[27]

Deoximation (using CAN or acid-catalysed transfer to levulinic acid) of compound **27**, itself prepared by reduction of 3,4,5,6-tetra-*O*-acetyl-1,2-dideoxy-1-nitro-D-*arabino*-hex-1-enitol, gave 6-*O*-acetyl-2,3-dideoxy-D-*erythro*-hex-2-eno-pyranose as product.[28]

The branched-chain sugars **28** (R^1 and R^2, various ester, keto or amido groups) were the major products from the Pd(0)-catalysed alkylation of 1,6-anhydro-2-chloro-2,3,4-trideoxy-β-D-*erythro*-hex-3-enopyranose with various active methylene reagents.[29]

Full details (see Vol. 28, p. 178, ref. 32 for a preliminary communication) have appeared on the synthesis of *N*-containing unsaturated sugars, *e.g.* isocyanate **29**, by thermal rearrangement of the corresponding 2-cyanato-3,4-unsaturated derivative.[30]

An investigation into the isomerization of a 3-oxo-pyranosides in pyridine where a hydroxyenone intermediate is observed is mentioned in Chapter 15.

2.3   Syntheses of 3,4-Unsaturated Cyclic Compounds – Compound **30**, available in three steps from 5-hydroxymethyl-1,6-anhydro-α-L-*altro*-hexopyranose was converted into *ent*-levoglucosenone by reaction with copper-quinoline (decarboxylation at C-5), zirconium oxide induced olefination at C-3 (reductive decarboxylation) then deacylation and oxidation. The 5-hydroxymethyl analogue of *ent*-levoglucosenone was also made from the same starting materials.[31]

A non-carbohydrate route to both enantiomers of levoglucosenone has also been reported in which the furan derivative **31** was desymmetrized using Sharpless asymmetric dihydroxylation methodology.[32]

Full details have appeared (see Vol. 28, p. 178, ref. 32 for a preliminary account) on the synthesis of isocyanates such as **32** (useful for preparing *N*-containing derivatives) by a thermal rearrangement of the corresponding 4-cyanato-2,3-unsaturated derivative.[30]

The hydride reductive rearrangement of some methyl 4-*O*-acetyl-3-deoxy-3-*C*-nitro-2-enopyranoside derivatives to methyl 3-*C*-nitro-3-enopyranosides is covered in Chapter 10.

2.4   Syntheses of 4,5-Unsaturated Cyclic Compounds – Swern oxidation of methyl 2,3,4-tri-*O*-acetyl-α-D-glucopyranoside followed by base-induced elimination of acetic acid afforded unsaturated aldehyde **33**. Further transformation of the aldehyde group into a methylene unit by way of a Wittig olefination and replacement of the acyl groups with benzyls or methyls yielded **34** which underwent a Diels-Alder reaction (the acyl protected variant of **34** was unreactive) with 2-methoxycarbonyl-*p*-benzoquinone, followed by acid-catalysed double

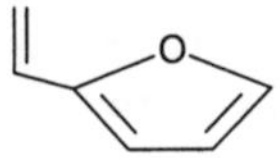

31

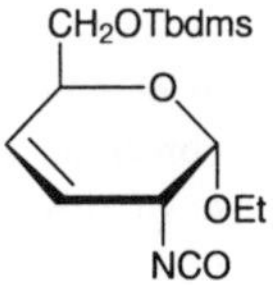

32

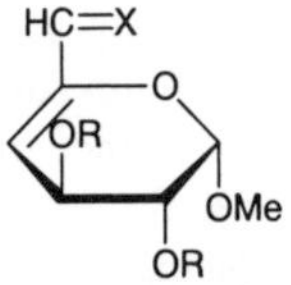

33 R = Ac, X = O
34 R = Bn or Me, X = CH$_2$

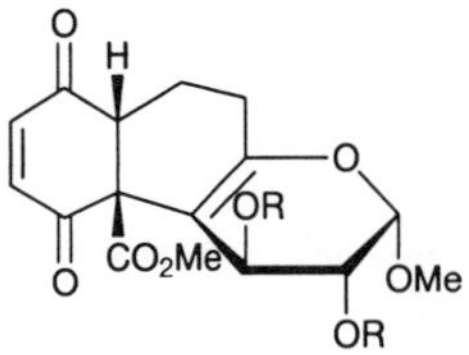

35 R = Bn or Me

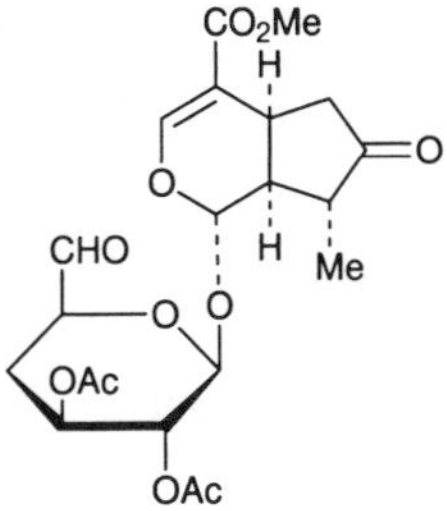

36

bond migration to give adducts **35**.[33] In a related way the 4,5-unsaturated iridoid derivative **36** was formed on PCC oxidation of the corresponding 2,3,4-tri-*O*-acetyl-β-D-glucopyranoside.[34]

2.5 Syntheses of 5,6- and Other Unsaturated Cyclic Compounds – A general approach to enol ethers, illustrated by conversion of nitro-alkene **37** into **38** by treatment with tetrabutylammonium hydrogen sulfate-potassium fluoride, has been described. The reaction was also found to be applicable to a wide range of acyclic sugars (see Section 4 below).[35]

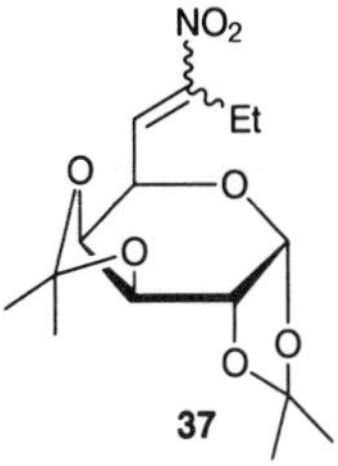

37

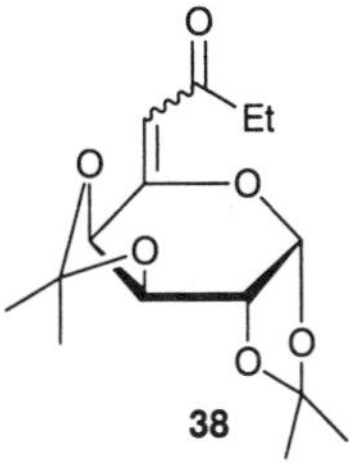

38

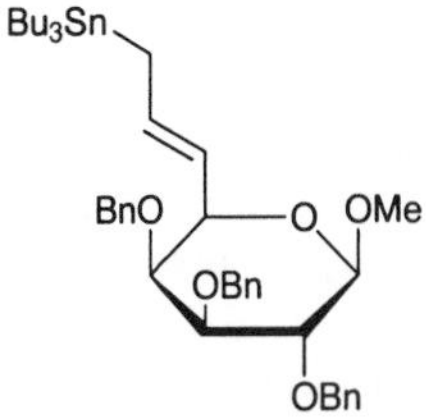

39

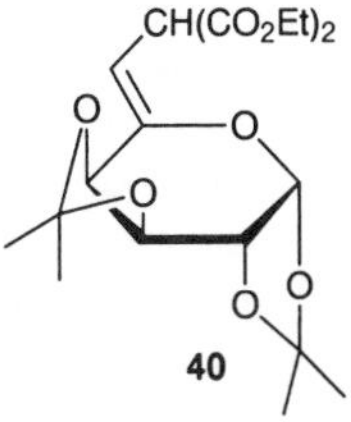

40

Tributyltin cuprate has been prepared and used as a milder alternative to tributyltin lithium for application in sugar chemistry. Thus tin derivative **39** was prepared from the corresponding allylic bromide without any allylic rearrangement.[36] (See Section 4 below for the use of this reagent in the preparation of acyclic alkenes).

6-Deoxy 1,2:3,4-di-*O*-isopropylidene-β-L-*arabino*-hex-5-enopyranose reacted with the radical generated from ethyl chloromalonate by tributyltin hydride, producing a small quantity of compound **40** together with products derived from reduction of the double bond in **40**.[37] (See Chapter 3 for an application of this method in preparing *C*-glycosides).

3   Furanoid Derivatives

3.1   Syntheses of 1,2-Unsaturated, Cyclic Compounds – A multigram synthesis of ribofuranoid glycals, which may prove useful in the synthesis of 2'-deoxy-nucleosides, has been achieved by utilizing as starting materials the readily available 2-deoxy-D-ribonolactone derivatives **41**. Reduction to the lactols with DIBAL, followed by reaction with mesyl chloride-triethylamine, gave the corresponding silyl protected glycals **42**. The bis-Tbdms protected compound was the easiest to work with, surprisingly being the most stable derivative, and the bis-Tbdps glycal could be selectively desilylated with TBAF at the O-5 position.[38]

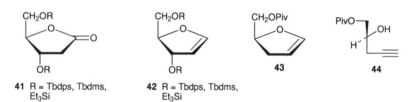

41 R = Tbdps, Tbdms,   **42** R = Tbdps, Tbdms,
     Et₃Si   Et₃Si

The asymmetric synthesis of compound **43**, a precursor for the preparation of 2',3'-didehydro-2',3'-dideoxy thymidine and cordecepin, has been accomplished by reacting the chiral alkynol **44** with the reagent combination molydenum pentacarbonyl-triethylamine. Several other glycals were prepared in a similar way, with the formation of furan by-products sometimes observed.[39]

3.2   Syntheses and Reactions of 2,3- and 3,4-Unsaturated Cyclic Compounds – Refluxing the 1,2-cyclic isourea **45** in pyridine caused elimination of water, affording the corresponding 2,3-unsaturated derivative.[40]

Further details (see Vol. 29, p. 189, ref. 30 for earlier report) on the mechanistically interesting reaction of 1,2:5,6-di-*O*-isopropylidene-3-*O*-triflyl-α-D-gluco-furanose with alkyllithiums to give 3-deoxy-1,2:5,6-di-*O*-isopropylidene-α-D-*erythro*-hex-3-enofuranose have been disclosed.[41] This latter compound underwent a Diels-Alder reaction with cyclopentadiene to give, surprisingly, **46** as the main product plus three isomers, and full details of this work have now been reported. (See Vol. 26, p. 160, ref. 37, but note that structure **32** should have a carbonyl group at C-4). As expected, the Diels-Alder reaction of isolevoglucosenone with cyclopentadiene also led to the same products, but in a different ratio.[42] The preparations of [3.3.0]bicyclic isoxazolidinyl nucleosides, as well as intermediates in their formation, by dipolar nitrone addition to 2,3-dideoxy-5-*O*-Tbdps-D-*erythro*-pent-2-eno-1,4-lactone are covered in Chapters 20 and 14, respectively.

3.3 Syntheses of 4,5- and Other Unsaturated Cyclic Compounds

3.3 **Syntheses of 4,5- and Other Unsaturated Cyclic Compounds** – Tosylhydrazones of α-oxycarbonyl compounds are readily converted into vinyl ethers with sodium in ethylene glycol (Bamford-Stevens elimination) as illustrated by the conversion of **47** to **48**.[43]

Sugar lactones can be made to undergo the Wittig olefination with methoxycarbonylmethylene(triphenylphosphorane) in toluene in a sealed tube at 140 °C to give the products in 27-90% yields, depending on the sugar configuration. Mainly 1,4-lactones were used and gave alkenes with *E:Z* selectivity ranging from 3:1 to 1:1.[44]

4 Acyclic Derivatives

A facile synthesis of a variety of 5-hydroxy-3-aldenoic acid derivatives derived initially from D-ribose has been reported involving a reductive elimination process (Scheme 3). A number of examples were also described with acetoxy as the leaving group at C-4.[45]

The alkyne **49** reacted with samarium diiodide in the presence of Pd(0) (conditions that usually produce allenes) producing ene-yne **50** as the major product (47%) with only 9% of the expected allene.[46]

As noted in Section 2.5 above, a general approach to enol ethers from nitroalkenes, as depicted by the conversion of **37** into **38**, has also been applied to the preparation of a range of acyclic derivatives. Here, the pyranose rings (of **37** and **38**) were replaced with various benzyl or cyclohexylidene protected L-*glycero*-

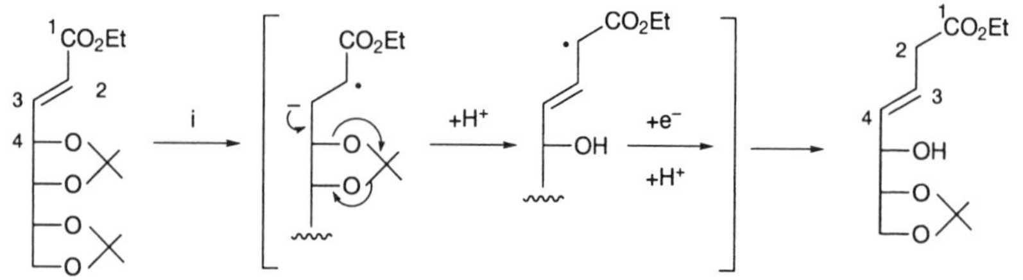

Reagents: i, activated Zn, EtOH

Scheme 3

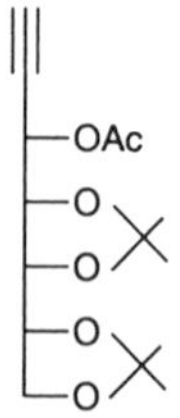

49 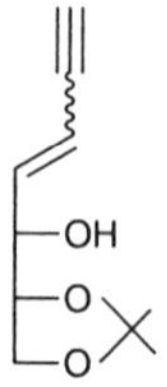   50 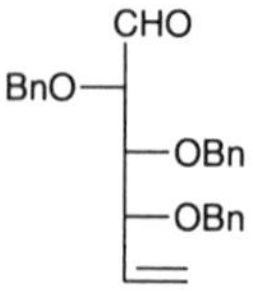   51

hydroxyethyl, D-*glycero*-dihydroxyethyl, D-*arabino*-tetrahydroxybutyl or D-*gluco*-pentahydroxypentyl appendages.[35] The use of tributyltin cuprate as a milder alternative to tributyltin hydride has been further illustrated (see also Section 2.5) by its reaction with methyl 2,3,4-tri-*O*-benzyl-6-*O*-tosyl-α-D-glucopyranoside, followed by treatment of the intermediate 6-*C*-tributyltin derivative with zinc to afford unsaturated aldehyde **51**.[36]

In a useful variant of the Wittig olefination process, unprotected aldoses reacted in dioxane at around reflux temperature with stabilized ylides of the type $Ph_3P=CHCO_2R$, where R is a bulky substituent (Bu^t, $CHPh_2$), giving high yields of acyclic products with *E*-selectivity. No cyclic products resulting from intramolecular Michael reaction were formed.[47]

Structural features of pyranoses unprotected at O-1 are important in determining product outcome on reaction with the anion derived from diethylmethyl-

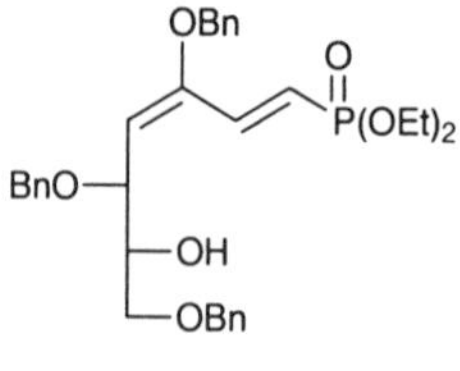

52 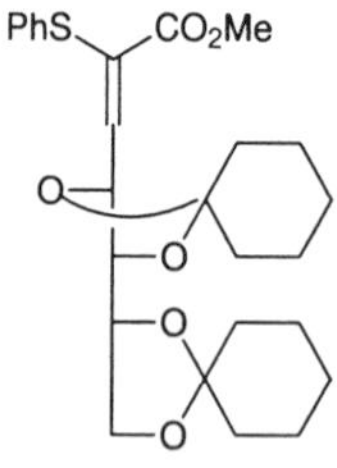   53

phosphonate. Thus, reaction of the latter with 2,3:4,6-di-O-isopropylidene-D-glucopyranose for example gave the corresponding glucopyranosyl methylphosphonate, while reaction with 2,3,4,6-tera-O-benzyl-D-galactopyranose gave the acyclic derivative **52**.[48]

The synthesis of the 2-phenylthio-hept-2-enoic acid **53**, a precursor of 2-keto-3-deoxy-D-*arabino*-heptonic acid has been achieved by addition of the anion derived from $PhSCH_2CO_2Me$ to 2,3:4,5-di-O-cyclohexylidene-D-arabinose.[49]

References

1   B. Fraser-Reid, *Acc. Chem. Res.*, 1996, **29**, 57 (*Chem. Abstr.*, 1996, **124**, 176 622).

2   S.J. Danishefsky and J.Y. Roberge, *Pure Appl. Chem.*, 1995, **67**, 1647 (*Chem Abstr.*, 1996, **124**, 87 495).

3   S.J. Danishefsky and M.T. Bilodeau, *Angew. Chem., Int. Ed. Engl.*, 1996, **35**, 1381.

4   B.K. Shull, Z. Wu and M. Koreeda, *J. Carbohydr. Chem.*, 1996, **15**, 955.

5   R.P. Spencer and J. Schwartz, *Tetrahedron Lett.*, 1996, **37**, 4357.

6   G. Kovács, J. Gyarmati, L. Somák and K. Micskei, *Tetrahedron Lett.*, 1996, **37**, 1293.

7   P. Garner, R. Leslie and J.T. Anderson, *J. Org. Chem.*, 1996, **61**, 6754.

8   M.Casillas, A.M. Gómez, J.C. López and S. Valverde, *Synlett*, 1996, 628.

9   A.M. Gómez, M. Casillas, S. Valverde and J.C. López, *Chem. Commun.*, 1996, 2357.

10   G.B. Kok, B.L. Mackey and M. von Itzstein, *Carbohydr. Res.*, 1996, **289**, 67.

11   L. Kiss and L. Somsák, *Carbohydr. Res.*, 1996, **291**, 43.

12   R.W. Scott and C.H. Heathcock, *Carbohydr. Res.*, 1996, **291**, 205.

13   R.F. Sala, P.M. Morgan and M.E. Tanner, *J. Am. Chem. Soc.*, 1996, **118**, 3033.

14   K. Peseke, S. Aldinger and H. Reinke, *Liebigs Ann. Chem.*, 1996, 953.

15   R.W. Friesen and L.A. Trimble, *J. Org. Chem.*, 1996, **61**, 1165.

16   K. Toshima, N. Miyamoto, G. Matsuo M. Nakata and S. Matsumura, *Chem. Commun.*, 1996, 1379.

17   S.N. Thorn and T. Gallagher, *Synlett*, 1996, 185.

18   B.J. Dorgan and R.F.W. Jackson, *Synlett*, 1996, 859.

19   S.N. Thorn and T. Gallagher, *Synlett*, 1996, 856.

20   R. Csuk, M. Schaade and C. Krieger, *Tetrahedron*, 1996, **52**, 6397.

21   A.L.J. Byerley, A.M. Kenwright and P.G. Steel, *Tetrahedron Lett.*, 1996, **37**, 9093.

22   A.L.J. Byerley, A.M. Kenwright and P.G. Steel, *Tetrahedron Lett.*, 1997, **38**, 2195.

23   K. Takeda, E. Kaji, H. Nakamura, A. Kiyama, Y. Konda, Y. Mizuno, H. Takayanagi and Y. Harigaya, *Synthesis*, 1996, 341.

24   P. Magnus and M.B. Roe, *Tetrahedron Lett.*, 1996, **37**, 303.

25   J. Madaj, J. Rak, J. Sokolowski and A. Wisniewski, *J. Org. Chem.*, 1996, **61**, 2988.

26   T. Linker, K. Hartmann, T. Sommermann, D. Scheutzow and E. Ruckdeschel, *Angew. Chem., Int. Ed. Engl.*, 1996, **35**, 1730.

27   K. Yasui, Y. Tamara, T. Nakatani, I. Horibe, K. Kawada, K. Keizumi, R. Suzuki and M. Ohtani, *J. Antibiotics*, 1996, **49**, 173.

28   M. Koóš, *Tetrahedron Lett.*, 1996, **37**, 415.

29   M. Matsumoto, H. Ishikawa and T. Ozawa, *Synlett*, 1996, 366.

30   Y. Ichikawa, C. Kobayashi and M. Isobe, *J. Chem Soc., Perkin Trans. 1*, 1996, 377.

31   Z. J. Witczak and R. Mielguj, *Synlett*, 1996, 108.

32   T. Taniguchi, K. Nakamura and K. Ogasawara, *Synlett*, 1996, 971.

33   A. Rani, A. Mukhopadhyay, S. Paul, R. Roy, S.N. Suryawanshi and D.S. Bhakuni, *Carbohydr. Lett.*, 1995, **1**, 207 (*Chem Abstr.*, 1996, **125**, 114 961).

34   K. Raj, V.T. Mathad and A.P. Bhaduri, *Nat. Prod. Lett.*, 1995, 7, 51 (*Chem. Abstr.*, 1996, **125**, 196 132).

35   J.-C. Bernal-Montes, P. Borrachero-Moya, F. Cabrera-Escribano, M. Gómez-Guillén, F. Madrid-Díaz and J.-M. Moreno-Martínez, *Carbohydr. Res.*, 1996, **285**, 49.

36   S. Jarosz, *Tetrahedron Lett.*, 1996, **37**, 3063.

37   L.Cipolla, L. Liguori, F. Nicotra, G. Torri and E. Vismara, *Chem. Commun.*, 1996, 1253.

38   J. A. Walker, J.J. Chen, D.S. Wise and L.R. Townsend, *J. Org. Chem.*, 1996, **61**, 2219.

39   F.E. McDonald and M.M. Gleason, *J. Am. Chem. Soc.*, 1996, **118**, 6648.

40   P. Meszaros, I. Pinter and G. Toth, *Aust. J. Chem.*, 1996, **49**, 409.

41   A. El Nemr, T. Tsuchiya and Y. Kobayashi, *Carbohydr. Res.*, 1996, **293**, 31.

42   D. Horton, J. P. Roski and P. Norris, *J. Org. Chem.*, 1996, **61**, 3783.

43   S. Chandrasekhar, S. Mohapatra and S. Lakshman, *Chem. Lett.*, 1996, 211.

44   M. Lakhrissi and Y. Chapleur, *Angew. Chem., Int. Ed. Engl.*, 1996, **35**, 750.

45   J.S. Yadav and D.K. Barma, *Tetrahedron*, 1996, 52, 4457.

46   J. Marco-Contelles, C. Destabel and J. Chiara, *Tetrahedron: Asymm.*, 1996, **7**, 105.

47   C.J. Railton and D.L.J. Clive, *Carbohydr. Res.*, 1996, **281**, 69.

48   Y.J. Lee, S.-H. Moon and S.-K. Chung, *Korean J. Med. Chem.*, 1995, **5**, 80 (*Chem. Abstr.*, 1996, **124**, 202 750).

49   B.B. Paidak, Yu. M. Mikshiev and V.I. Kornilov, *Zh. Obshch. Khim.*, 1995, **65**, 349 (*Chem. Abstr.*, 1996, **124**, 9 105).

14
Branched-chain Sugars

1   **Compounds with a C – C – C Branch-point**

$$\begin{array}{c} R \\ | \\ C-C-C \\ | \\ O \end{array}$$

1.1   **Branch at C-2 or C-3** – The β-oxo-Grignard reagent derived from 2-bromomethyl-1,3-dioxolane (normally considered difficult to prepare) has been added to a variety of protected 2- and 3-keto-sugars affording good yields of branched-derivatives such as the illustrated C-2 compounds **1**. Some similar C-3 pyranose and furanose derivatives were also described.[1]

The metal carbene-functionalized spirocyclic compound **2** was readily formed in good yield when 1,2:5,6-di-O-isopropylidene-3-C-(3-propynyl)-α-D-allofuranose was treated with the appropriate metal pentacarbonyl. Other similar examples were described which led to C-2 branched-derivatives.[2]

Further details have been reported on the mechanistically interesting formation of 3-C-butyl-1,2:5,6-di-O-isopropylidene-α-D-allofuranose by the addition of butyllithium to 1,2:5,6-di-O-isopropylidene-3-O-triflyl-α-D-allofuranose which proceeded through a 3-keto intermediate. (See Vol. 29, p. 192, ref. 6).[3]

Methyl 3,3-trifluoropyruvate was used as an achiral starting material for conversion in a straightforward way into 3-deoxy-2-C-trifluoromethyl-α-D/L-*erythro*- and β-D/L-*threo*-pentofuranosides.[4]

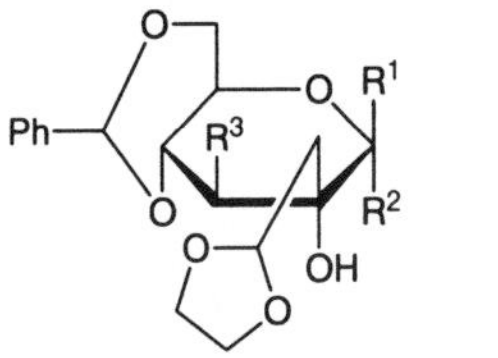

1 R^1 = OMe, R^2 = H, R^3 = OBz
or R^1 = H, R^2 = OMe, R^3 = OBz
or R^1 = H, R^2 = OMe, R^3 = H

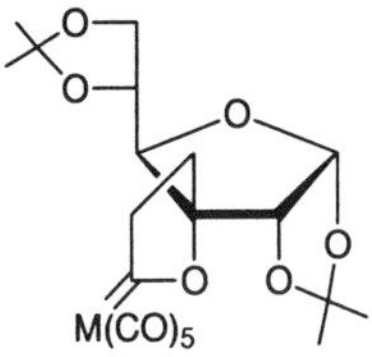

2 M = Cr or W

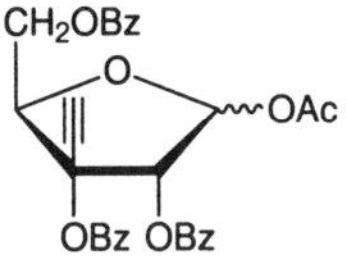

3

The previously described 3-*C*-ethynyl-1,2-*O*-isopropylidene-α-D-ribofuranose (see Vol. 28, p. 183, ref. 9) has been converted into tri-*O*-benzoyl glycosyl acetate **3**, using standard chemical transformations, and used as a precursor for the synthesis of the nucleoside 1-(3-*C*-ethynyl-β-D-*ribo*-pentofuranosyl)uracil, a potent anti-tumour agent.[5,6]

The preparation of the branched-chain sugar found in the naturally occurring nucleoside antibiotic amipurimycin is depicted in Scheme 1. The synthesis involved deoxygenation at C-4 of a 3-ulose derivative with triphenyphosphine-iodine-imidazole in the first step.[7]

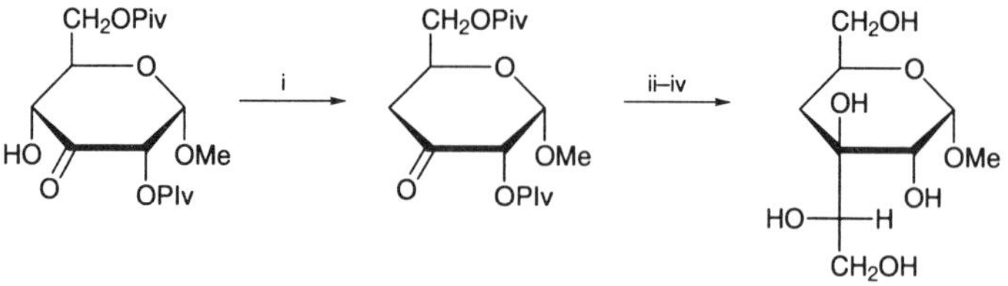

Reagents: i, Ph₃P, I₂, Imidazole; ii, Ph₃P=CHCO₂Et; iii, OsO₄; iv, LiAlH₄

Scheme 1

Ascorbic acid readily reacted in Michael fashion with various nitro-styrenes to afford adducts **4**.[8]

In continuation of the use of dichloromethyllithium in the synthesis of branched-chain sugars (see also Vol. 29, p. 192, ref. 8 and Vol. 25, p. 162, ref. 9) 3-*C*-(dichloromethyl)-1,2:5,6-di-*O*-isopropylidene-α-D-allofuranose reacted with tetrabutylammonium hydroxide in DMSO to produce **5**, but with caesium acetate-18-crown-6 in toluene to produce the 'inversion' product **6**. The former was thought to arise by conventional hydrolysis and the latter by formation of a *spiro*-chloroepoxide intermediate followed by nucleophilic ring opening at the quaternary centre. Other C-CHCl₂ secondary compounds behaved similarly.[9]

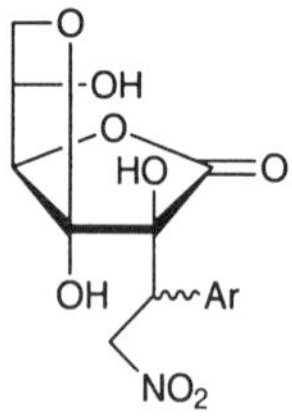

4

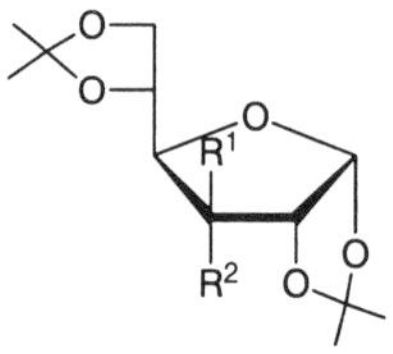

5 R¹ = CHO, R² = OH
6 R¹ = OH, R² = CHO

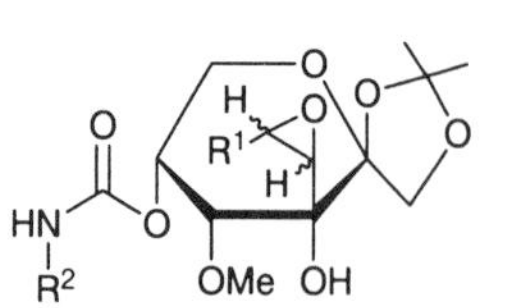

7 R¹ = (CH₂)₃Ph, CH₂CHMe₂
R² = ClCH₂CO, CO₂Et *etc.*

(−)-Ovilicin, a natural product with antiangiogenic activity, has inspired the synthesis of modified analogues such as **7**, which were prepared from the known 1,2:4,5-di-*O*-isopropylidene-β-D-*erythro*-hex-2,3-diulo-2,6-pyranose.[10]

The protected apiose derivatives **8** (some other R¹,R² combinations were described) have been made by reaction of the corresponding glycosyl bromide with the *cis*-alkenes $R^1CH=CHR^2$ in the presence of tris(trimethylsilyl)silane-AIBN.[11]

The branched-chain sugar **9** (prepared by intramolecular cycloaddition of a terminal deoxy-*N*-hydroxyaminopentofuranose bearing a 3-*C*-prop-2-enyl branch) on UV irradiation afforded a radical observed by ESR spectroscopy. The information obtained was used for configurational and conformational assignments.[12]

The bipyridyl derivative **10** was prepared by treating bromopyridine derivative **11** (made from diacetone-D-glucose) with nickel(II) chloride-zinc-triphenyl-phosphine in DMF. The products were regarded as potential chiral ligands.[13]

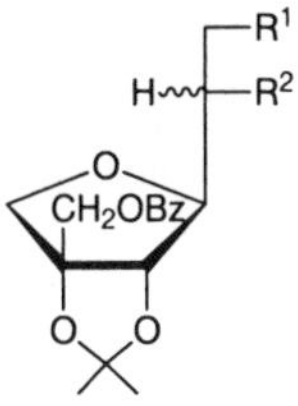

8 $R^1 = R^2 = CO_2Me$
$R^1 = CO_2Me, R^2 = H$
$R^1 = CN, R^2 = H$

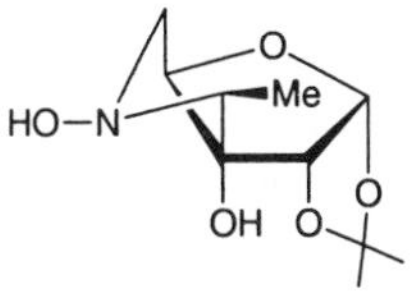

9

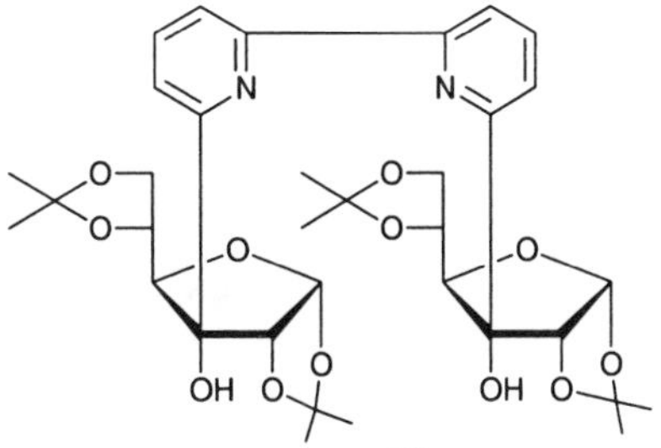

10 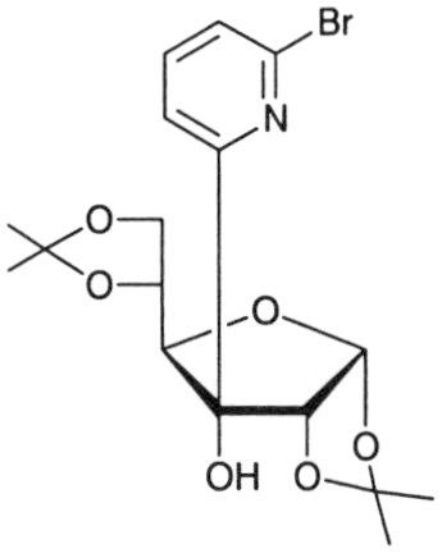   **11**

The synthesis of functionalized 9-membered ethers from 1,2:5,6-di-*O*-cyclohexylidene-3-*C*-vinyl-D-allofuranose is covered in Chapter 24.

1.2 Branch at C-4 – The syntheses of some C-4 branched derivatives of NeuNAc have been achieved from a protected 4-keto analogue, methyl (methyl 5-acetamido-3,5-dideoxy-8,9-*O*-isopropylidene-β-D-*manno*-2,4-nonudiulopyranosid)onate. For example conversion of the latter into *spiro*-epoxide **12** by reaction with zirconocene dichloride-Zn-diiodomethane, then peracid, provided a reactive centre for the addition of carbon centred nucleophiles giving products such as **13**.[14] The synthesis of furo[2,3-*c*]pyran-β-D-thymidine, a potential antiviral agent, is mentioned in Chapter 20.

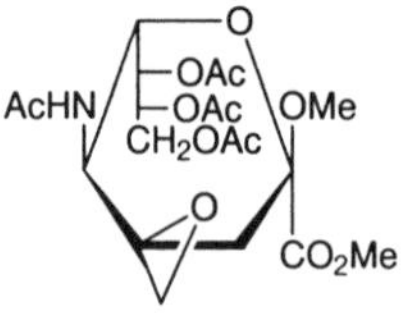

12 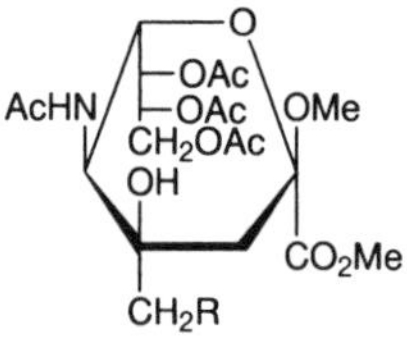   **13** R = OMe, CN, Cl, NHAc, N$_3$ 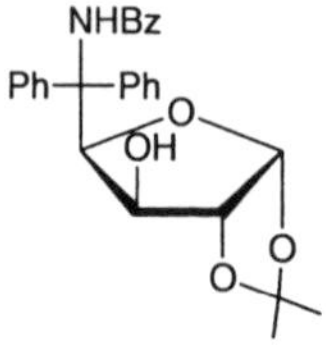   **14**

2   **Compounds with a C – $\overset{\displaystyle R}{\underset{\displaystyle N}{C}}$ – C Branch-point**

The addition of phenylmagnesium bromide to 3-*O*-benzoyl-1,2-*O*-isopropylidene-α-D-xylofuranurononitrile unexpectedly formed the branched compound **14**.[15]

Work directed towards the synthesis of a novel type of spironucleoside is illustrated in Scheme 2. The carbon atoms introduced in steps ii and iii coming from the ethyl acetate and acetonitrile, respectively.[16]

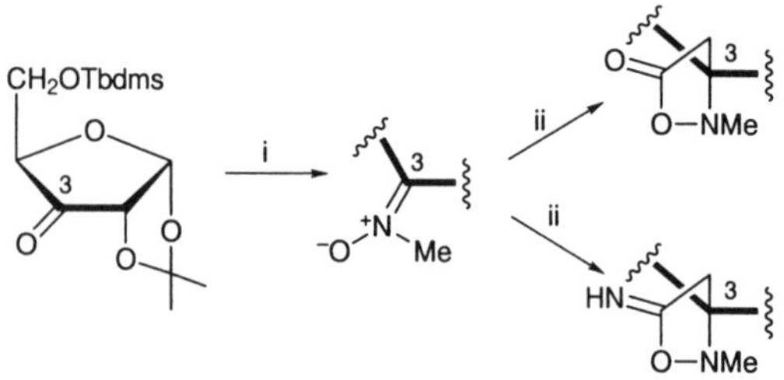

Reagents: i, MeNHOH; ii, Li(Tms)$_2$N, EtOAc; iii, LDA, MeCN

Scheme 2

[3+2] Cycloaddition of 1,3-dipoles to nitro-activated alkenes provided access to branched-chain sugars as depicted in Scheme 3.[17]

3   **Compounds with a C – $\overset{\displaystyle R}{C}$H – C Branch-point**

A review has been published on the oxidation of the C-Si to C-OH bond – a method that can be used to generate hydroxymethyl branches in sugars.[18]

3.1   Branch at C-2 – This section initially deals with compounds containing a single branch at C-2; compounds containing branch points at other positions in addition to C-2 appear at the end of the section.

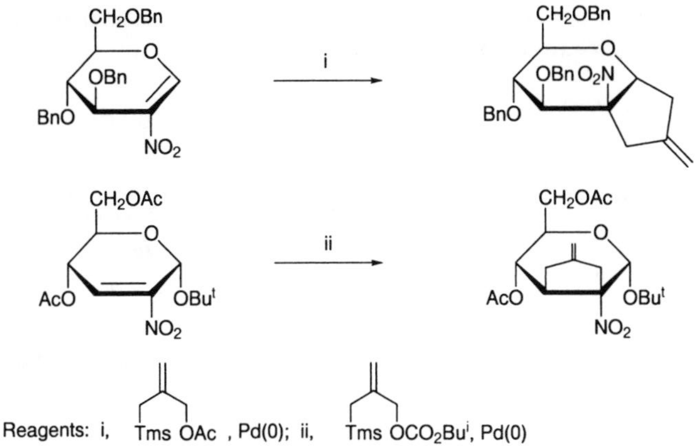

Reagents: i, Tms OAc , Pd(0); ii, Tms OCO₂Buⁱ, Pd(0)

Scheme 3

A radical induced addition of dimethyl malonate to tri-*O*-acetyl-D-glucal under two sets of conditions has been described. Thus, use of ceric ammonium nitrate in MeOH afforded the major product **15** and a smaller ammount of **16** along with the α-glycosyl nitrate analogue of **15**. Whereas, use of manganese triacetate in acetic acid gave **15** and its α-anomer as the major products together with the glycosyl acetate equivalent of **16**.[19]

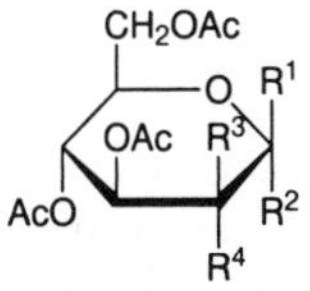

15 R¹ = OMe, R² = H, R³ = H, R⁴ = CH(CO₂Me)₂
16 R¹ = H, R² = OMe, R⁴ = H, R³ = CH(CO₂Me)₂

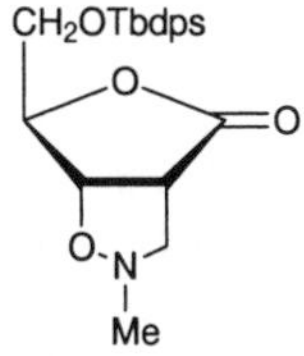

17

Cycloaddition of *N*-methylnitrone to 2,3-dideoxy-6-*O*-Tbdps-D-*erythro*-pent-2-enono-1,4-lactone afforded derivative **17** as a useful intermediate for making [3.3.0]bicyclic isoxazolidinyl nucleosides.[20]

Rhodium-catalysed addition of the carbene generated from ethyl diazoacetate to tri-*O*-acetyl-D-galactal gave an 8:1 mixture of α-*exo* **18** and β-*exo* **19** with no products formed with the ethoxycarbonyl group in the *endo*-orientation. Using tri-*O*-benzyl-, tri-*O*-acetyl- and tri-*O*-(triisopropylsilyl)-D-glucals led to all four possible isomers, with the α-*exo* forms predominating. Further it was noted that the '*gluco*' analogue of **18** reacted with hydrobromic acid to give compound **20**; no ring expansion leading to oxepanes, as seen earlier, took place (see Vol. 29, p. 197, ref. 28 and p. 187, ref. 24).[21] The same reaction has also been described by a different group but using tri-*O*-Tbdms-, tri-*O*-benzoyl- and tri-*O*-benzyl-D-glucals as well as per-benzylated derivatives of D-xylal, D-galactal and L-fucal. In these cases only α-*exo* compounds were reported as products.[22]

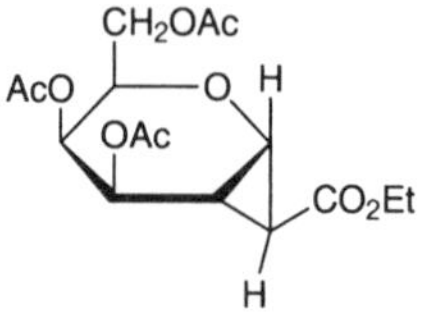

The photoreductive cyclization of 2,3-unsaturated-*C*-glycosyl carbonyl compounds is depicted in Scheme 4.[23]

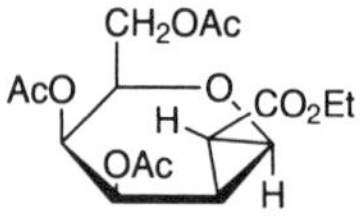

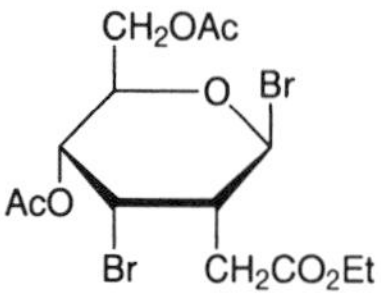

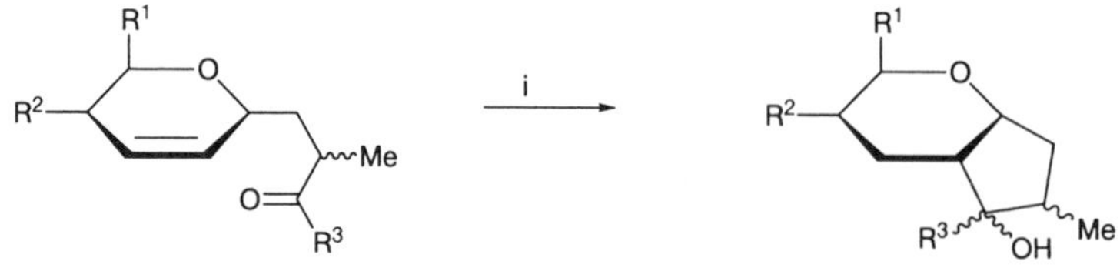

R¹ = Me, H, CH₂OAc
R² = OAc
R³ = H, Me

Reagents: i, Et₃N, *hv*

Scheme 4

Formaldehyde dimethylhydrazone readily added to nitro-alkene **21** under neutral conditions and gave branched-derivative **22** as the major diastereomer. The hydrazone group in **22** was further transformed to an aldehyde or nitrile functional group by treatment with ozone or magnesium monoperoxyphthalate, respectively. The same reaction with other sugar nitro-alkenes has been previously described in preliminary form. (See J.-M. Lassaletta and R. Fernández, *Tetrahedron Lett.*, 1992, **32**, 3691).[24]

Under catalysis by tris(dimethylaminosulfonium)difluorotrimethylsilicate, 1,5-anhydro-4,6-*O*-benzylidene-2-deoxy-D-*erythro*-hex-3-ulo-1-enitol reacted with the ketene acetal, 1-methoxy-2-methyl-1-(trimethylsilyloxy)-propene, by Michael reaction at C-1 and gave **23**. When the catalyst involved was trimethylsilyl triflate the branched derivatives **24** and **25** unexpectedly resulted. These latter products were thought to arise by a trimethylsilyl triflate induced non-selective cleavage of the benzylidene acetal of the initially formed intermediate **23** affording 4-*O*- and 6-*O*-benzylic carbenium ions which underwent a subsequent ring closure at C-2. (See Chapter 13 for reaction of the same ketene acetal with glycals.)[25]

Methyl 2-deoxy-2-*C*-formyl-3,4,6-tri-*O*-benzoyl-α-D-mannopyranoside reacted with a protected glycosyl 2-pyridyl sulfone in the presence of samarium diiodide and produced the disaccharide **26**. (See Vol. 29, p. 197, ref. 32 for an analogous

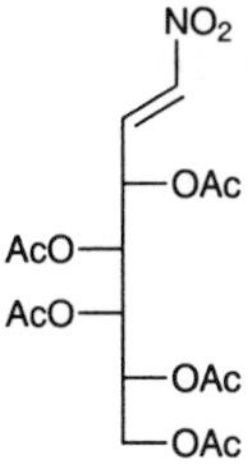

21

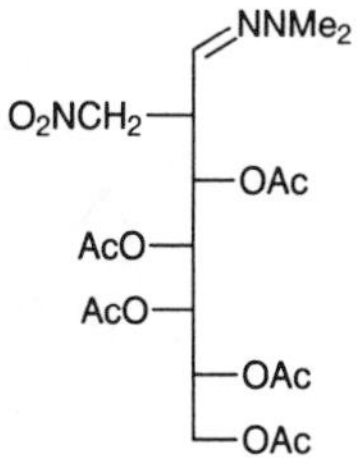

22

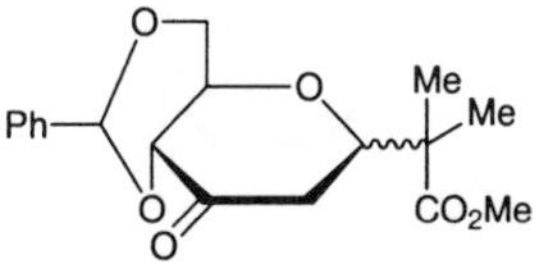

23

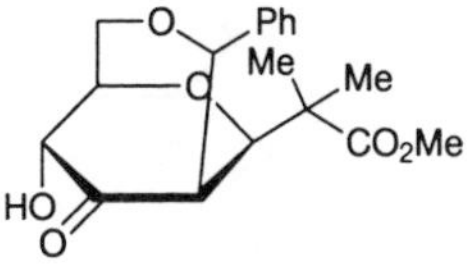

24

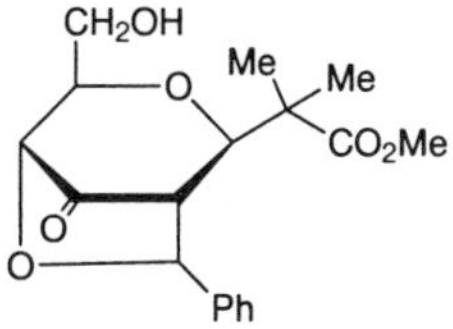

25

reaction). Deoxygenation of the free hydroxyl group of **26**, followed by deprotection gave a *C*-glycoside analogue of α-D-Man*p*-(1→2)-D-Man*p*-OH, a mycobacterial disaccharide, and a potential inhibitor of *M. tuberculosis*.[26]

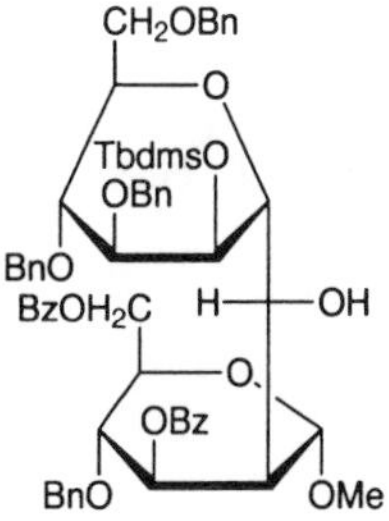

26

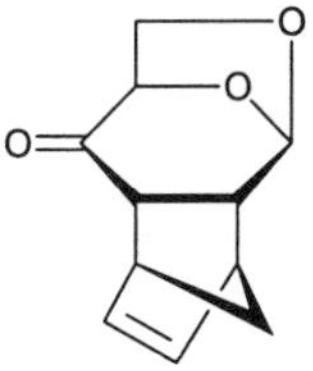

27

3-Deoxy-1,2:5,6-di-*O*-isopropylidene-α-D-*erythro*-hex-3-enofuranoside underwent a Diels-Alder reaction with cyclopentadiene to give **27** as the main product together with three minor isomers, and full details of this surprising reaction have now been reported. (See Vol. 26, p. 160, ref. 37, but note that structure **32** should have a carbonyl group at C-4).[27]

An efficient new route to bis-annulated pyranoid derivatives is shown in Scheme 5. Product yields were between 70 and 80%.[28]

In an alternative strategy for making bis-annulated pyranoid derivatives, addition of sodio dimethyl propargylmalonate to benzyl 2-*O*-tosyl-3,4-dideoxy-α-D-*glycero*-pent-3-enopyranoside gave unsaturated compound **28** as the major product. This, on treatment with the reagent $Co_2(CO)_8$ (to produce a

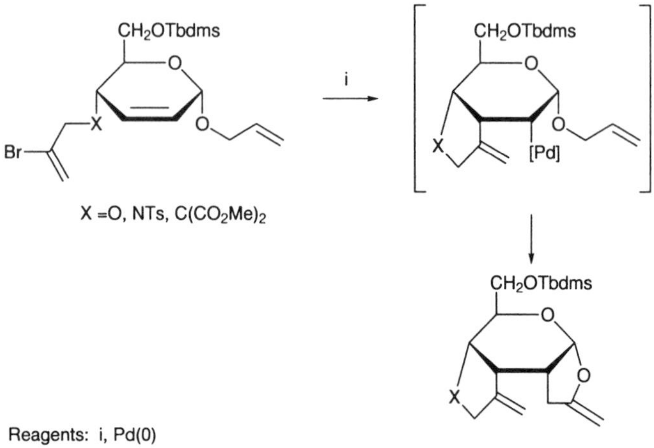

Reagents: i, Pd(0)

Scheme 5

hexacarbonyldicobalt complex with the alkyne group), afforded **29**.[29] In a similar way, and in addition to earlier work (see Vol. 28, p. 186, ref. 22), several examples of the reaction of $Co_2(CO)_8$ with various unsaturated pyranose sugars containing adjacent *O*-propargyl side chains, leading to bis-annulated derivatives, have also been described.[30]

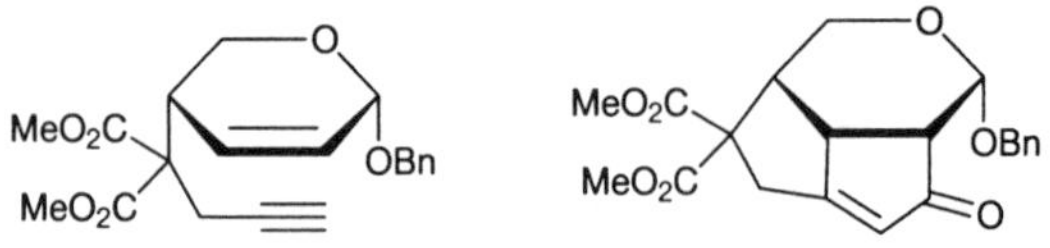

3.2   Branch at C-3 – The Lewis acid catalysed reaction of the zinc-copper reagents $R^2Cu(CN)ZnI$ with tri-*O*-acetyl- or tri-*O*-benzyl-D-glucal afforded the branched-chain derivatives **30** as the main products.[31] In a similar way, a protected pyranoid glycal containing an 3-*O*-mesyl group also reacted with the reagent $IZn(NC)Cu(CH_2)_2CO_2Me$ to afford **31** as the main product.[32] The use of copper-zinc reagents in this way, compliments the reaction of dialkylzincs with glycals which usually produces 2,3-unsaturated, C-1 substituted products by allylic rearrangement processes. (See Chapter 13).

The branched-3-*C*-methyl derivative **32** was formed stereoselectivity and in high yield when dimethylcopper lithium in the presence of TmsCl added 1,4-fashion to the corresponding 2,3-unsaturated lactone. Several similar examples were also described including addition of active methylene compounds (*e.g.* diethyl malonate under basic conditions).[33]

Addition of trifluoromethyltrimethylsilane to 1,2-*O*-isopropylidene-5-*O*-Tbdms-α-D-*erythro*-pentofuranos-3-ulose (prepared from D-xylose) afforded the corresponding 3-*C*-trifluoromethyl-ribose derivative **33**. After de-silylation at O-3 (NaOMe/MeOH) and radical-induced deoxygenation, the 3-deoxy-3-*C*-trifluoro-methyl-D-ribose derivative **34** was formed. D-Glucose was also used as starting material.[34]

In a new radical-based allylation procedure the allylsulfone **35** reacted with allyltosylsulfone-AIBN in cyclohexane or dibutylperoxide in chlorobenzene, affording a 2:3 mixture of *gluco:allo* products (**36**). See also Chapter 20 for an application in nucleoside synthesis.[35]

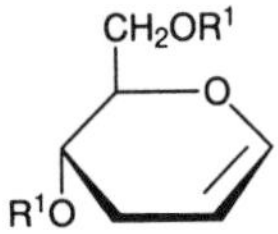

30 R¹ = Ac or Bn
R² = Bu, NC(CH₂)₃
NC(CH₂)₂

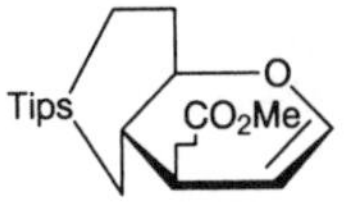

31

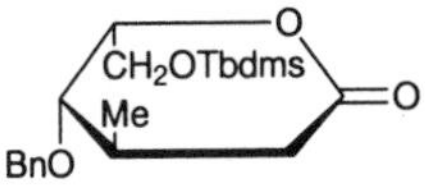

32

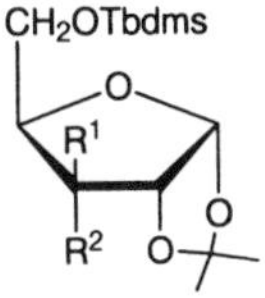

33 R¹ = CF₃, R² = OTms
34 R¹ = H, R² = CF₃

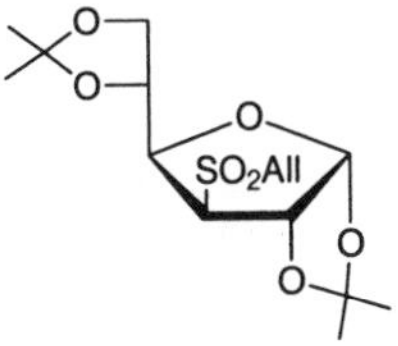

35

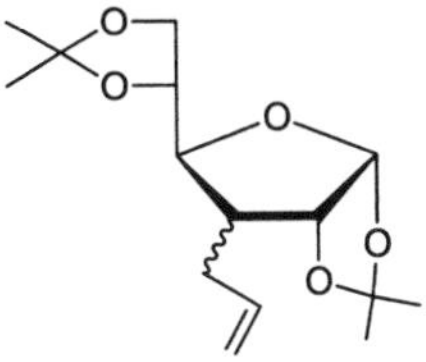

36

The 3-*C*-hydroxymethyl furanose derivative **37**, prepared by Peterson olefination of the corresponding 3-ulose compound followed by hydroboration, was readily converted into the anhydro derivative **38**. This was effected by successive treatment with TBAF, peracid, then acetic anhydride-pyridine and the product used to prepare 3'-deoxy-3-*C*'-hydroxymethyl-α-L-lyxopyranosyl thymine.[36]

3.3 Branch at C-4 –The 4-deoxy-4-*C*-hydroxymethyl compound **39** was pre-pared as an intermediate in the synthesis of 4-deoxy-4-*C*-hydroxymethyl-α-L-lyxopyranosylthymine. Its preparation started from the reaction of sodio dimethylmalonate with methyl 2,3-*O*-isopropylidene-4-*O*-triflyl-β-D-ribopyranose to give **40**. Subsequent standard chemistry transformed the C-4 branch into a hydroxymethyl group. The potentially shorter route to **39** involving hydro-boration of **41** (the 4-*C*-methylene compound itself prepared in two steps from methy 2,3-*O*-isopropylidene-β-D-ribopyranoside) gave instead the C-4 epimer of **39** as the major product.[37]

The reaction of anions derived from various β-dicarbonyl compounds with

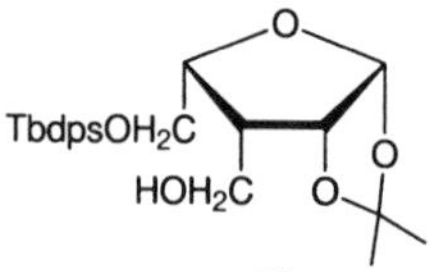

37

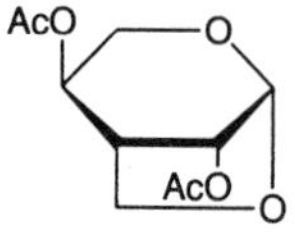

38

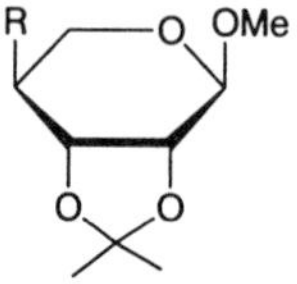

39 R = CH₂OH
40 R = CH(CO₂Et)₂

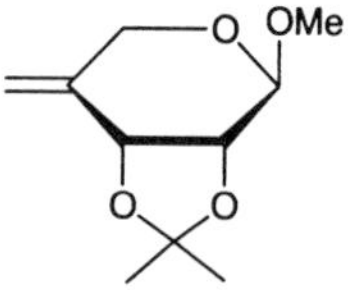

41

benzyl 2,3-anhydro-4-*O*-triflyl-β-L- and α-D-ribopyranoside afforded C-4 branched derivatives (essentially as described in Vol. 29, p. 200, ref. 44 and Vol. 28, p. 193, ref. 46).[38]

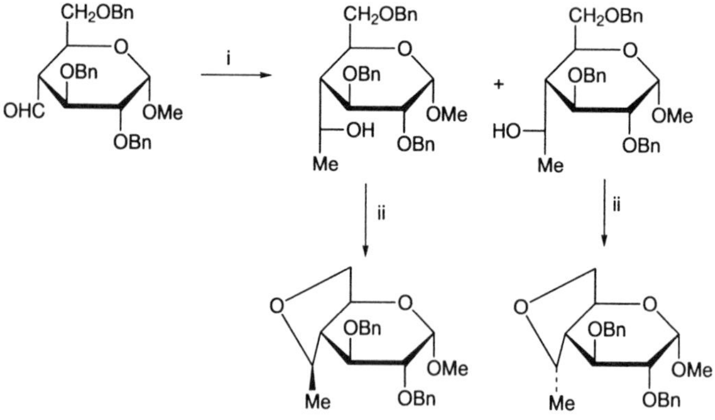

Reagents: i, MeMgBr, THF, –5 °C; ii, Tf₂O, Py, –10 °C

Scheme 6

The synthesis of some 2,4-dioxahydrindans is illustrated in Scheme 6.[39]

Oxidation of 4-*C*-hydroxymethyl-2,3,4-trideoxy-D-*erythro*-hex-2-enopyranose with silver carbonate on Celite produced 4-*C*-hydroxymethyl-2,3,4-trideoxy-D-*threo*-hex-2-eno-1,5-lactone together with the expected D-*erythro* lactone.[40]

Michael addition of various nitroalkanes to levoglucosenone under cathodic electrolysis conditions (circumstances that were found to be milder and higher yielding than the base-catalysed version) resulted in products **42**. Phthalimide, succinimide and saccharin were also nucleophiles in this process.[41]

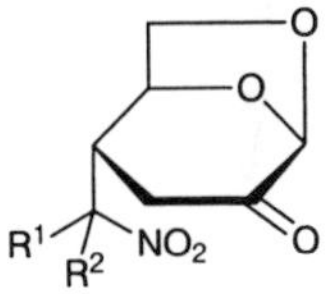

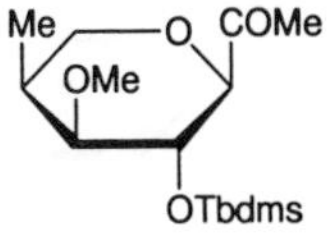

42   R^1 R^2
      H   CO$_2$Me
      Me   Me
      H   NO$_2$

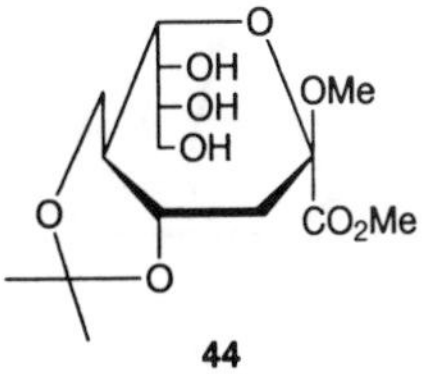

L-Quebrachitol has been converted in several steps into derivative **43**, a known intermediate for the synthesis of the polyene antibiotics restricticin and restrictinol. The key step involved a Baeyer-Villiger oxidation of a ketone-containing derivative.[42]

3.4   Branch at C-5 – The nonulosonic acid derivative **44** has been prepared by aldolase-catalysed reaction of pyruvate with 2-deoxy-2-*C*-hydroxymethyl-D-mannose as a possible synthon for the C-12 to C-20 sequence of amphotericin B. The latter compound was made in several steps from ethyl 2,3-dideoxy-4,5:6,7-di-*O*-isopropylidene-D-*arabino*-hept-2-enoate by conjugate addition of vinylmagnesium bromide followed by transformation of the vinyl group into a hydroxymethyl group in the key steps.[43]

4   Compounds with a C – C – C Branch-point

As part of an investigation into developing routes to natural sesquiterpinoid ring systems, the application of the Pausen-Khand reaction has led to some unexpected results. Thus the 'fused [3.3.0]bicyclooctenone-pyranose' system **45** was produced as expected when the 3-*C*-methylene derivative **46** was treated with Co(CO)$_8$ in an atmosphere of carbon monoxide. However, the 'fused [4.1.0]bicycloheptene-pyranose' system **47** was produced from the 4-*C*-methylene derivative **48** under similar conditions.[44]

The spirocyclopropane derivative **49** has been synthesized from methyl 3-*O*-benzyl-4,6-*O*-isopropylidene-β-D-glucopyranoside by reaction with TFAA-DMSO, lithium aluminium hydride, then diethylzinc-diiodomethane. Subsequent reaction with PDC followed by hydrogenolysis in the presence of triethylamine (OBn group survived) afforded **50**. Several further reactions, including a transformation of the formylmethyl group into a hydroxymethyl group, gave an appropriate subunit for the C-18 to C-23 fragment of the antibiotic lasonolide A.[45]

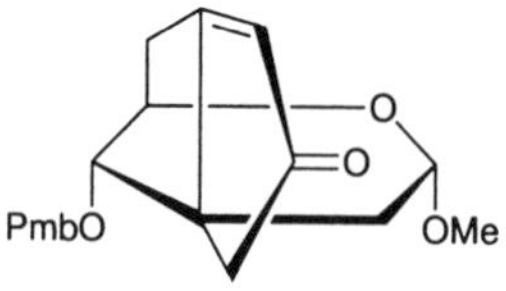

45

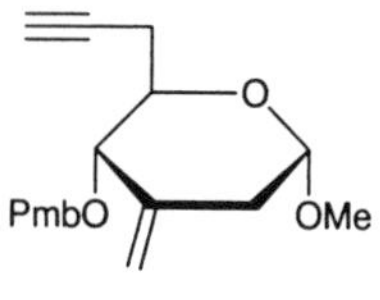

46

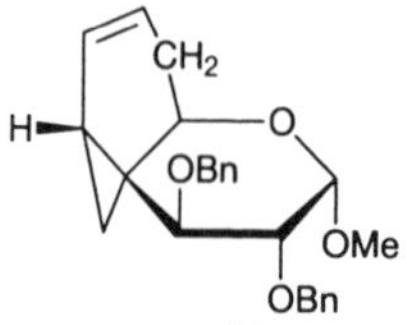

47

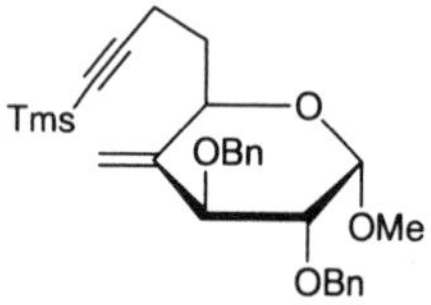

48

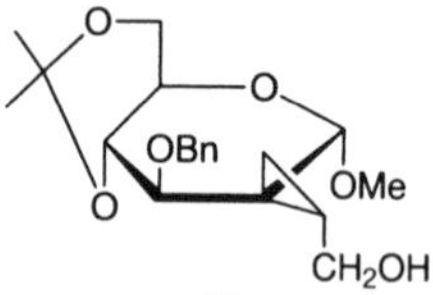

49

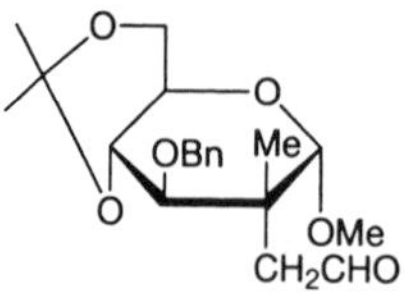

50

4   **Compounds with a C – C̈ – C or C – Ċ – C Branch-point**

Methyl 3-*C*-allyl-4,6-*O*-benzylidene-3-deoxy-α-D-glucopyranoside has been used
as a common starting material for preparing annulated sugars involving intramolecular Horner-Wadsworth-Emmons reactions. Thus successive reaction of
the allyl derivative with osmium tetroxide, lithium dimethylmethylphosphonate,
Dess-Martin periodane then potassium carbonate furnished **51**. Whereas successive reaction of the allyl derivative with borane-THF, PCC oxidation, lithium
dimethylmethylphosphonate, Dess-Martin periodane then potassium carbonate
afforded **52**. Analogous chemistry was also performed on methyl 2-*C*-allyl-4,6-*O*-
benzylidene-2-deoxy-α-D-allopyranoside.[46]

The unsaturated branched-sugar **53** was produced when α-oxoketene-dithioa-
cetal **54** (see Vol. 26, p. 162, ref. 47) was treated with dilute mineral acid.[47]

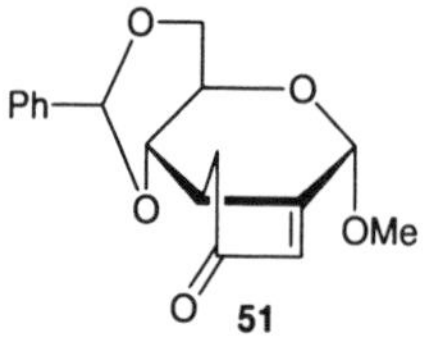

51

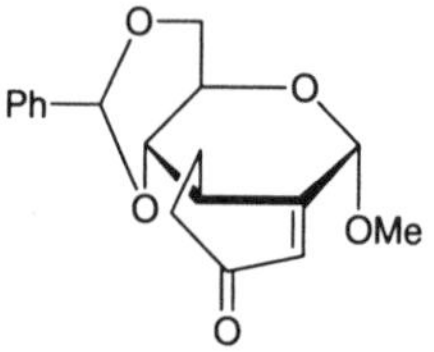

52

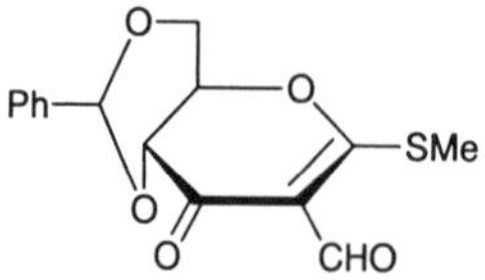

53

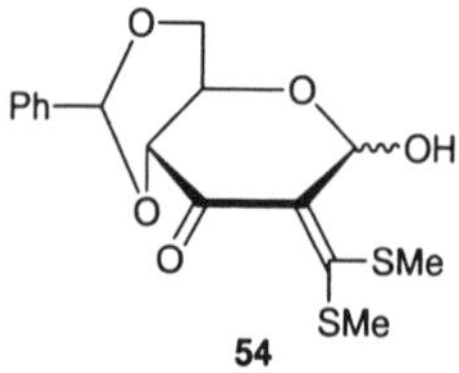

54

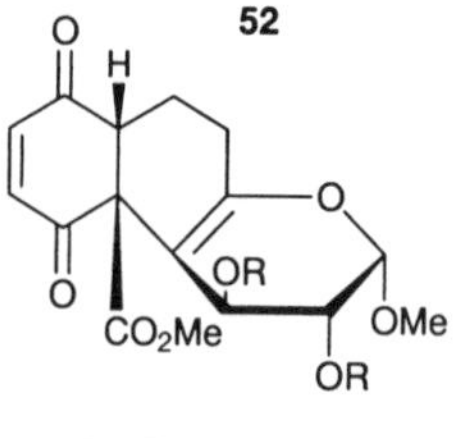

55 R = Me or Bn

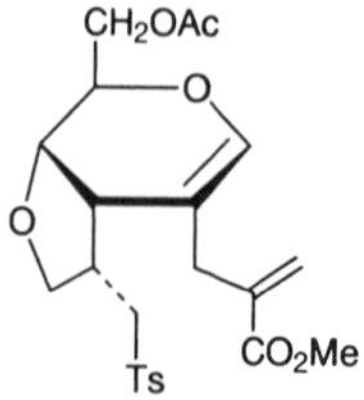

56

Tricyclic compounds **55** were formed by initial Diels-Alder reaction between methyl 2,3-di-*O*-benzyl (or methyl)-4,6,7-trideoxy-α-D-*threo*-hept-4,6-dienopyranoside and 2-methoxycarbonyl-*p*-quinone followed by a double bond migration.[48]

A staightforward synthesis of 3,4,6-tri-*O*-benzyl-2-*C*-methyl-D-glucal has been achieved from tri-*O*-benzyl-D-glucal by successive reaction of the latter with diethylzinc-diiodomethane (to effect a stereoselective cyclopropanation), mercury(II) acetate-water (hydroxymercuration induced opening of the cyclopropane ring to a *manno*-configured 2-*C*-methylmercury pyranose derivative), tributyltin hydride (formed a 2-*C*-methyl group) then mesylanhydride-triethylamine base-induced elimination.[49]

The stereocontrolled, sulfonyl radical-induced cyclopropanation of ethyl 6-*O*-acetyl-4-*O*-allyl-2,3-dideoxy-α-D-*erythro*-hex-2-enopyranoside in the presence of the reagent TsCH$_2$C(=CH$_2$)CO$_2$Me gave **56**.[50]

References

1   M. Schmeichel and H. Redlich, *Synthesis*, 1996, 1002.

2   K.H. Dötz, O. Neuss and M. Nieger, *Synlett*, 1996, 995.

3   A. El Nemr, T. Tsuchiya and Y. Kobayashi, *Carbohydr. Res.*, 1996, **293**, 31.

4   U. Wucherpfennig, T.A. Logothetis, U. Eilitz and K. Burger, *Tetrahedron*, 1996, **52**, 143.

5   A. Matsuda, H. Hattori, M. Tanaka and T. Sasaki, *Bioorg. Med. Chem. Lett.*, 1996, **6**, 1887.

6   H. Hattori, M. Tanaka, M. Fukushima, T. Sasaki and A. Matsuda, *J. Med. Chem.*, 1996, **39**, 5005.

7   A.P. Rauter, A.C. Fernandes, S. Czernecki and J.-M. Valery, *J. Org. Chem.*, 1996, **61**, 3594.

8   M. Schmidt and K. Eger, *Pharmazie*, 1996, **51**, 11 (*Chem. Abstr.*, 1996, **124**, 317 715).

9   K.-i. Sato, Y. Yamamoto and H. Hori, *Tetrahedron Lett.*, 1996, **37**, 2799.

10   G. Dorey, P. Léon, S. Sciberras, N. Guilbaud, A. Pierré, G. Atassi and D.C. Billington, *Bioorg. Med. Chem. Lett.*, 1996, **6**, 3045.

11   M. Drescher and F. Hammerschmidt, *Synthesis*, 1996, 1451.

12   J.M.J. Tronchet, M. Zsely, R.N. Yazji, F. Barbalat-Rey and M. Geoffroy, *Carbohydr. Lett.*, 1995, **1**, 343 (*Chem. Abstr.* 1996, **124**, 317 687).

13   M.A. Peterson and N.K. Dalley, *Synth. Commun.*, 1996, **26**, 2223.

14   D.R. Groves and M. von Itzstein, *J. Chem. Soc., Perkin Trans. 1*, 1996, 2817.

15   M. Koos, B. Steiner and J. Alfoldi, *Chem. Pap.*, 1994, **48**, 426 (*Chem. Abstr.*, 1996, **124**, 9 201).

16   J.M.J. Tronchet, I. Kovacs, F. Barbalat-Rey and N. Dolatshahi, *Nucleosides Nucleotides*, 1996, **15**, 337.

17   C.W. Holzapfel and T.L. van der Merwe, *Tetrahedron Lett.*, 1996, **37**, 2307.

18   G.R. Jones and Y. Landais, *Tetrahedron*, 1996, **52**, 7599.

19   T. Linker, K. Hartmann, T. Sommermann, D. Scheutzow and E. Ruckdeschel, *Angew. Chem., Int. Ed. Engl.*, 1996, **35**, 1730.

20   Y. Xiang, R.F. Schinazi and K. Zhao, *Bioorg. Med. Chem. Lett.* 1996, **6**, 1475.

21   J.D. Hoberg and D.J. Claffey, *Tetrahedron Lett.*, 1996, **37**, 2533.

22   C.M. Timmers, M.A. Leeuwenburgh, J.C. Verheijen, G.A. van der Marel and J.H. van Boom, *Tetrahedron: Asymm.*, 1996, **7**, 49.

23   J. Cossy and S. Ibhi, *Carbohydr. Res.*, 1996, **291**, 189.

24   J.-M. Lassaletta, R. Fernández, C. Gasch and J. Vázquez, *Tetrahedron*, 1996, **52**, 9143.

25   R. Csuk, M. Schaade and C. Krieger, *Tetrahedron*, 1996, **52**, 6397.

26   O. Jarreton, T. Skrydstrup and J.-M. Beau, *Chem. Commun.*, 1996, 1661.

27   D. Horton, J.P. Roski and P. Norris, *J. Org. Chem.*, 1996, **61**, 3783.

28   J.-F. Nguefack, V. Bolitt and D. Sinou, *Tetrahedron Lett.*, 1996, **37**, 59.

29   N. Naz, T.H. Al-Tel, Y. Al-Abed, W Voelter, R. Ficker and W. Hiller, *J. Org. Chem.*, 1996, **61**, 3250.

30   J. Marco-Contelles, *J. Org. Chem.*, 1996, **61**, 7666.

31   S.N. Thorn and T. Gallagher, *Synlett*, 1996, 856.

32   B.J. Dorgan and R.F.W. Jackson, *Synlett*, 1996, 859.

33   B. Herradón, E. Fenude, R. Bao and S. Valverde, *J. Org. Chem.*, 1996, **61**, 1143.

34   S. Lavaire, R. Plantier-Royon and C. Portella, *J. Carbohydr. Chem.*, 1996, **15**, 361.

35   B. Quiclet-Sire and S.Z. Zard, *J. Am. Chem. Soc.*, 1996, **118**, 1209.

36   B. Doboszewski and P. Herdewijn, *Tetrahedron*, 1996, **52**, 1651.

37   B. Doboszewski and P.A.M. Herdewijn, *Nucleosides Nucleotides*, 1996, **15**, 1495.

38   T.H. Al-Tel, R. Thuermer, R.A. Al-Qawasmeh and W. Voelter, *Prakt. Chem./ Chem.-Ztg.*, 1996, **338**, 320. (*Chem Abstr.*, 1996, **125**, 114 960).

39   A.T. Khan, H. Dietrich and R.R. Schmidt, *Synlett*, 1996, 131.

40   A.G.H. Wee and L. Zhang, *Org. Prep. Proceed. Int.* 1996, **28**, 339 (*Chem. Abstr.*, 1996, **125**, 87 058).

41   A.V. Samet, M.E. Niyazymbetov, V.V. Semenov, A.L. Laikhter and D.H. Evans, *J. Org. Chem.*, 1996, **61**, 8786.

42   N. Chida, T. Tobe, M. Yoshinaga, K. Osaka and S. Ogawa, *Carbohydr. Lett.*, 1996, **2**, 47 (*Chem. Abstr.*, 1996, **125**, 222 293)

43   A. Malleron and D. Serge, *New. J. Chem.*, 1996, **20**, 153 (*Chem. Abstr.*, 1996, **124**, 317 633).

44   V.S. Borodkin, N. Shpiro, V.A. Azov and N.K. Kochetkov, *Tetrahedron Lett.*, 1996, **37**, 1489.

45   M.K. Gurjar, P. Kumar and B.V. Rao, *Tetrahedron Lett.*, 1996, **37**, 8617.

46   M. Pipelier, M.S. Ermolenko, A. Zampella, A. Olesker and G. Lukacs, *Synlett*, 1996, 24.

47   K. Peseke, S. Aldinger and H. Reinke, *Liebigs Ann. Chem.*, 1996, 953.

48   A. Rani, A. Mukhopadhyay, S. Paul, R. Roy, S.N. Suryawanshi and D.S. Bhakuni, *Carbohydr. Lett.*, 1995, **1**, 207 (*Chem. Abstr.*, 1996, **125**, 114, 961).

49   R.W. Scott and C.H. Heathcock, *Carbohydr. Res.*, 1996, **291**, 205.

50   M.P. Bertrand, C. Lesueur and R. Nouguier, *Carbohydr. Lett.*, 1995, **1**, 393 (*Chem. Abstr.*, 1996, **124**, 317 624).

15
Aldosuloses and Other Dicarbonyl Compounds

1 Aldosuloses

The preparation and chemical properties of levoglucosenone as well as its use in the total synthesis of natural products have been reviewed,[1] and the stereoselective reduction of 3-bromolevoglucosenone to 1,6-anhydro-3-bromo-3,4-dideoxy-β-D-*threo*-hex-3-enopyranose 1 has been effected using zinc borohydride.[2] The known lactone 2 has been converted into the aldosulose derivative 3, a degradation product from the ansamycin antibiotic (+)-trienomycin A. This confirmed that the fragment of the natural product was the enantiomer of 3.[3]

1   2   3

An improved synthesis of 1,6-anhydro-2,3-di-*O*-benzyl-β-D-*xylo*-hexopyranos-4-ulose involves 1,6-anhydride formation from methyl 2,3-di-*O*-benzyl-α-D-glucopyranoside and subsequent Swern oxidation.[4]

Use of the enzyme dTDP-glucose-4,6-dehydratase on the substrate dTDP-glucose has generated the corresponding 6-deoxy-4-keto product – but it partly isomerized to the 3-keto compound during workup. 3-Azido-3-deoxy- and 3-deoxy-dTDP-glucose were also substrates affording the 3-modified 6-deoxy-4-keto products.[5]

Ozonolysis of the model compound methyl 4-*O*-ethyl-β-D-glucopyranoside has led to authentic samples of methyl, 3,6-di-*O*-acetyl-4-*O*-ethyl-β-D-*arabino*-hexopyranosid-2-ulose, methyl 2,6-di-*O*-acetyl-4-*O*-ethyl-β-D-*ribo*-hexopyranosid-3-ulose, and methyl 2,3-di-*O*-acetyl-4,*O*-ethyl-β-D-glucohexodialdo-1,5-pyranoside.[6]

The isomerization of 2-oxo- and 3-oxo-glycosides in pyridine has been elucidated. Methyl α-D-*arabino*-hexopyranosid-2-ulose 4 undergoes an intramolecular hydride shift leading to the α-D-*arabino*-hexopyranosid-3-ulose 5 which then converts to the thermodynamically more stable α-D-*ribo*-isomer 7 *via* an enediol intermediate 6 which also forms directly from 4 to a lesser extent.[7] The methyl α,β-D-*xylo*-hexopyranosid-4-uloses rearrange in pyridine exclusively to the methyl α,β-D-*ribo*-hexopyranosid-3-uloses *via* an enediol intermediate, whereas the corresponding D-*lyxo*-and D-*arabino*-4-uloses come into equilibrium, again *via* an enediol intermediate without leading to 3-keto compounds.[8]

Aldose reductase has been utilized for the selective reduction of aldos-2-uloses to 2-ketoses.[9] Singlet oxygen oxidation of alditol-1-*C*-ylfurans **8** has afforded quantitative yields of enosuloses **9**.[10] Conjugate addition to enone **10** readily afforded the 2-deoxy-α-hexos-3-ulosides **11** and thioglycosidic analogues.[11] 1-*C*-(2-Thiazolyl)-

α- or β-D-galactopyranosyl acetate derivatives **12** stereoselectively glycosylate primary or secondary sugar alcohols to give disaccharide derivatives *e.g.* **13** related to ketosyl compounds. The α-D-mannofuranosyl-based analogue similarly gives α-linked products, but the α-D-glucopyranosyl-related donor gives 1:1 α:β mixtures.[12]

2   Other Dicarbonyl Compounds

The 2,5-dimethoxytetrahydrofuran fulvanol **14**, an analogue of apiose, has been isolated from the plant *Hemerocallis fulva*.[13] Molecular modelling studies of the active site binding of the transition states of galactose oxidase, with D-galactose and D-glucose was carried out to investigate why the enzyme is so selective for galactose relative to glucose.[14]

Asymmetric epoxidation of *trans*-alkenes in the presence of the fructose-derived 2,3-diulose derivative **15** afforded products with e.e.'s of 70–95%,[15] and an improved method for the conversion of 1,2:4,5-di-*O*-cyclohexylidene-β-D-fructopyranose to the corresponding 3-ulose has been described.[16]

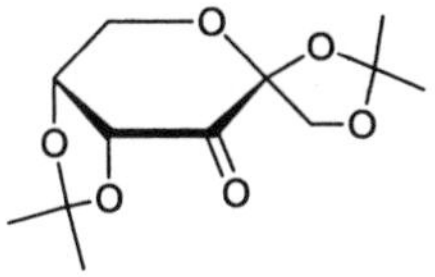

15

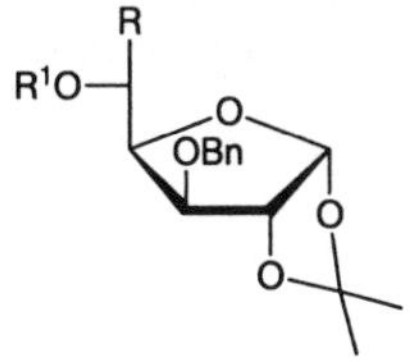

16 R = CH$_2$OH, R^1 = H
17 R = CHO, R^1 = Ac

The glucofuranose diol **16** has been converted into hexo-1,6-dialdose derivative **17** by standard methodology,[17] whereas the 1,5-dialdose **18** has been chain extended to the 1,6-dialdose derivative **19** (Scheme 1).[18] Oxidation of some 5,6-dideoxy-hex-5-enofuranosides has afforded the corresponding 5-deoxy-hexo-1,6-dialdose derivatives.[19]

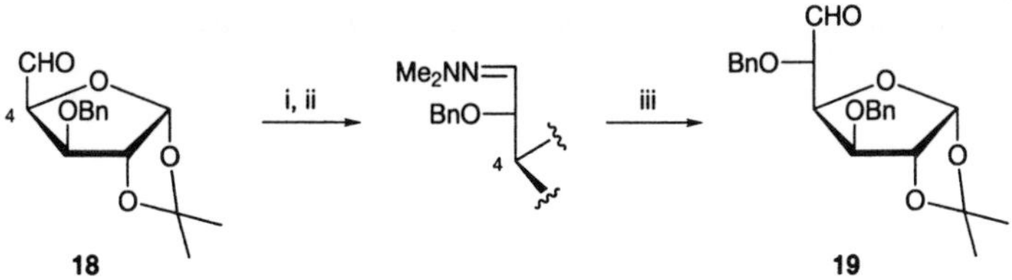

Reagents: i, CH$_2$=NNMe$_2$; ii, BnBr, NaH; iii, O$_3$, Me$_2$S

Scheme 1

Some strongly basic anion exchange resins in the periodate form have been used to selectively oxidize ribofuranosyl nucleosides, in the presence of glucopyranosides and galactopyranosides, to the corresponding ring-opened 2'3'-dialdehydes.[20]

References

1   M.S. Miftakhov, F.A. Valeev and N.I. Gaisina, *Usp. Khim.*, 1994, **63**, 922 (*Chem. Abstr.*, 1996, **124**, 87 489).
2   M.S. Miftakhov, I.N. Gaisina, F.A. Valeev and O.V. Shitikova, *Izv. Akad., Nauk, Ser. Khim.*, 1995, **12**, 2455 (*Chem. Abstr.*, 1996, **124**, 343 816).
3   A.B. Smith (III), J.L. Wood, W. Wong, A.E. Gould, C.J. Rizzo, J. Barbosa, K. Komiyama and S. Omura, *J. Am. Chem. Soc.*, 1996, **118**, 8308.
4   S. Caron, A.I. McDonald and C.H. Heathcock, *Carbohydr. Res.*, 1996, **281**, 179.
5   A. Naundorf and W. Klaffke, *Carbohydr. Res.*, 1996, **285**, 141.
6   T. Kishimoto, F. Nakatsuto, K. Murakami and T. Umezawa, *J. Wood Chem. Technol.*, 1995, **15**, 453 (*Chem. Abstr.*, 1996, **124**, 176 672).
7   H.-M. Liu and Y. Tsuda, *Chem. Pharm. Bull.*, 1996, **44**, 80.
8   H.-M. Liu and Y. Tsuda, *Chem. Pharm. Bull.*, 1996, **44**, 88.
9   J.A. Kotecha, M.S. Feather, T.J. Kubiseski and D.J. Walton, *Carbohydr. Res.*, 1996, **289**, 77.
10   J.M. Báòez Sanz, D. Galisteo González, J.A. Lopez Sastre, J.F. Rodríguez Amo, C. Romero-Avila García, M. Santos García and M.A. Sanz Tejedor, *Carbohydr. Res.*, 1996, **289**, 179.
11   K. Michael and H. Kessler, *Tetrahedron Lett.*, 1996, **37**, 3453.
12   A. Dondoni, A. Marra, I. Rojo and M.-C. Scherrmann, *Tetrahedron*, 1996, **52**, 3057.
13   T. Konishi, T. Inoue, S. Kiyosawa and Y. Fujiwara, *Phytochemistry*, 1996, **42**, 135.
14   R.M. Wachter and B.P. Branchaud, *J. Am. Chem. Soc.*, 1996, **118**, 2782.
15   Y. Tu, Z.-X. Wong and Y. Shi, *J. Am. Chem. Soc.*, 1996, **118**, 9806.
16   L. Tsao, C. Chzhou and Y. Lyu, *Izv. Akad. Nauk Ser. Khim.*, 1996, 1003 (*Chem. Abstr.*, 1996, **125**, 168 513).
17   S. Jarosz and E. Kozlowska, *Pol. J. Chem.*, 1996, **70**, 45 (*Chem. Abstr.*, 1996, **125**, 58 934).
18   J.M. Lassaletta, R. Fernández, E. Martín-Zamora and C. Pareja, *Tetrahedron Lett.*, 1996, **37**, 5787.
19   K. Krishnudu, P.R. Krishna and H.B. Mereyala, *Tetrahedron Lett.*, 1996, **37**, 6007.
20   N.A. Brusentsov and M.N. Preobrazhenskaya, *Bioorg. Khim.*, 1996, **22**, 215 (*Chem. Abstr.*, 1996, **125**, 248 299).

16
Sugar Acids and Lactones

1 Aldonic Acids and Lactones

A number of 2-azido-2-deoxy- and 2-acetamido-2-deoxy-3,4,6-tri-O-benzyl-D-hexono-1,5-lactones have been prepared by oxidation (DMSO, Ac$_2$O) of the corresponding free sugars.[1] The kinetics of the oxidation of L-rhamnose and D-mannose to the corresponding 1,4-hexonolactones by Cr(VI) in aqueous acetic acid has been studied,[2] while the rate constants and activation parameters have been determined for the oxidation of D-xylose D-lyxose, D-glucose, D-galactose, D-mannose and L-arabinose by palladium nitrate to their aldonic acids.[3,4] Glucose and gluconate aqueous solutions have been oxidized in air over various Pt/C catalysts. Some of the catalysts were activated with Bi or Au. The Bi-doped catalyst favoured the formation of 2-ketogluconate rather than glucarate.[5] The kinetics of glucose oxidation by persulfate ion, catalysed by iron salts (Fe(II) or Fe(III)), has been studied. A radical chain mechanism was proposed with the initial oxidation occurring at the anomeric centre to give the gluconic acid.[6]

In a Wittig chain extension of unprotected aldoses (using Ph$_3$P=CHCO$_2$R; and bulky R groups Bu^t, Ph$_2$CH), good yields of the acyclic α,β-unsaturated esters were obtained without recyclization to give pyranosyl-C-glycosides. The E-isomers were the predominant products.[7] A facile 4-deoxygenation of some of their derivatives is discussed in Chapter 13.

Aldol condensation of 2,2-diethyl-1,3-dioxolan-4-one lithium or zirconium enolates with aldehydo sugars has afforded higher carbon aldonic acid derivatives, e.g. 1.[8] The synthesis of L-ribono-1,4-lactone has been achieved from D-isoascorbic acid by way of the tetrose and pentitol derivatives 2 and 3 and the D-ribonolactone derivative 4 has been efficiently epimerized to the L-lyxonolactone 5 (Scheme 1).[9] A selective syn-epoxidation of racemic 2-O-benzyl-4-alkenamides followed by hydrolysis has afforded 3-deoxy-pentono-1,4-lactones.[10]

1   2   3

Carbohydrate Chemistry, Volume 30
© The Royal Society of Chemistry, 1998

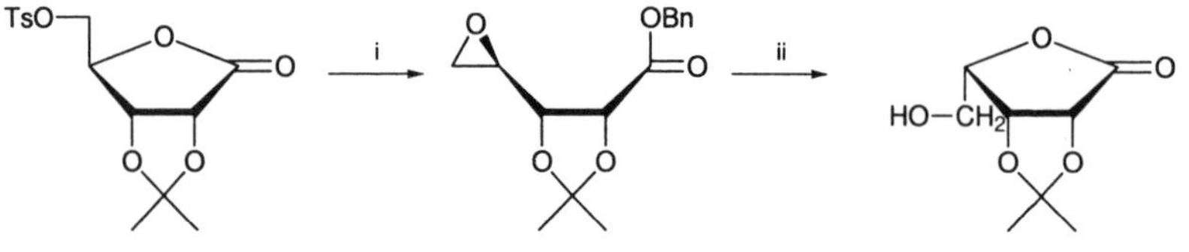

4   **5**

Reagents: i, KOBn, THF; ii, H$_2$, Pd/C

Scheme 1

Syntheses of some pentonohydroxamic acid derivatives are covered in Chapter 10. Alkaline degradation of 5-*O*-α-D-glucopyranosyl-D-fructopyranose (leucrose) has afforded two novel saccharinic acids, 2-*C*-(2-hydroxyethyl)erythronic acid and 2-*C*-(2-hydroxyethyl)threonic acid.[11] Treatment of 4,6-*O*-ethylidene-D-glucose with the sulfur ylid *N*,*N*-diethyl trimethylsulfonium-acetamide gave the epoxyaldonamide **6** as well as stereoisomers.[12]

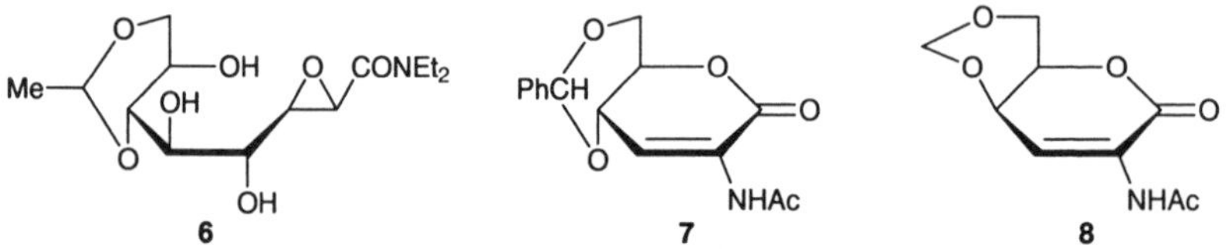

6   **7**   **8**

The acylation of aldonolactones for long periods leading to the formation of unsaturated lactones has been reviewed.[13] Treatment of unsaturated aldonolactone **7** with tin(IV) chloride has afforded a low yield of the most unusual rearrangement product **8**.[14] Methyl 2,3:5,6-di-*O*-isopropylidene-D-gluconate and -D- or -L-galactonate were deoxygenated at C-4 by standard conditions and then treated with samarium diiodide to give 2,4-dideoxyaldonates.[15] Amides of D-glucaric acid have been polymerized into hydroxylated nylons without OH group protection.[16] 2,3,4-Tri-*O*-methyl-pentonolactones when treated with DBU generate the 3-deoxy-2,3-unsaturated products, which on hydrogenation selectively afford 3-deoxy-2,4-di-*O*-methyl-D-or L-erythropentonolactone. The selectivity of the hydrogenation is governed by the stereochemistry at *C*-4.[17] A stereocontrolled oxidative addition of Mo(0) reagents to allylic acetate moieties of some unsaturated aldonolactones has afforded isolable organomolybdenum products (*e.g.* **9**).[18] Azide displacements applied to some 2-*O*-triflate derivatives of aldonolactones are discussed in Chapter 10, as is the addition of *N*-benzylhydroxylamine to 2,3-unsaturated aldonolactones. Aryllithium additions to aldonolactones in the synthesis of *C*-nucleosides is mentioned in Chapter 20.

Chemically defined glycolipids have been prepared by aminolysis of lactobiono-1,5-lactone and coupling of the products with long-chain dialkyl-L-glutamates.[19,20]

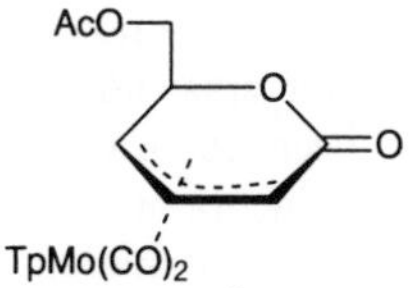

TpMo(CO)₂

9

Tp = hydridotris(1-pyrazolyl)borate

2 Anhydroaldonic Acids and Lactones

Some methyl 3,6- and 3,7-anhydro-8-azido-8-deoxy-octonic acid derivatives have been reduced to the corresponding 8-amino compounds and coupled *via* amide bond formation with the 8-azido precursors to give di- and tri-'glycotides'.[21] Some aldonolactone 2-*O*-triflates have been found to undergo ring contraction under acidic conditions in methanol to give methyl 2,5-anhydro-aldonates,[22] while others undergo the same ring contraction in methanol under basic conditions.[23]

The branched anhydroaldonic acid derivative **10** has been prepared from tri-*O*-benzyl-D-glucal (Scheme 2).[24] 5-Acetamido-4-*O*-acetyl-2,6-anhydro-2,3,5-tri-deoxy-nononic acid derivatives have been prepared as neuraminic acid analogues.[25] Treatment of acyclic aldose derivatives with sulfur ylids generates 2,3-epoxyaldonic acids (such as analogues of **6**) which may recyclize to give 3,6-anhydroaldonates.[26] A 3-amino-2,6-anhydro-3-deoxyheptonic acid derivative has been converted into a lipid A analogue by suitable acylation and phosphorylation procedures,[27] while the synthesis of isosteric phosphonate analogues of CMP-NeuAc have utilized anhydroaldonic acid derivatives.[28] The synthesis of some anhydroaldonamides linked to amino acids is covered in Chapter 10.

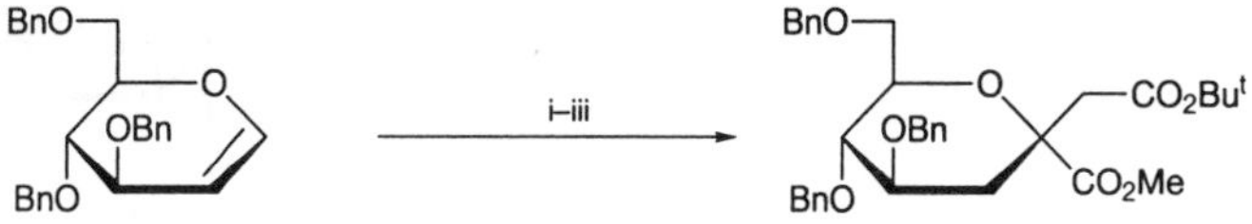

Reagents: i, Bu₃SnLi; ii, BuLi, ClCO₂Me; iii, NaHMDS, BrCH₂CO₂Buᵗ

Scheme 2

3 Ulosonic Acids

A review on the chemoenzymic synthesis of carbohydrates and carbohydrate mimetics features the use of aldolases in the synthesis of neuraminic acid and KDO analogues.[29]

A fully enzymic synthesis of *N*-acetylneuraminic acid from *N*-acetylglucosamine using *N*-acetylglucosamine 2-epimerase and *N*-acetylneuraminic acid aldolase on an industrial scale has been reported.[30] Some 9-amino- and 9-*N*-acylamino-9-deoxy-5-*N*-trifluoroacetylneuraminic acid analogues have been prepared by the aldolase catalysed reaction of sodium pyruvate with 6-azido-2-*N*-benzyloxycarbonylamino-2,6-dideoxy-D-mannose.[31] The enzyme glucarate dehydratase, which catalyses the conversion of D-gluconate to 3-deoxy-L-*threo*-hex-2-ulosarate, has been cloned, expressed in *E. coli*, and purified.[32] The pyruvate addition to aldehydes catalysed by 2-keto-3-deoxy-6-phosphogluconate aldolase has allowed the synthesis of 3-deoxy-D-*erythro*-hex-2-ulosonate, 3,6-dideoxy-D-*erythro*-hex-2-ulosonate and 3-deoxy-D-*ribo*-hept-2-ulosonate.[33,34] Dihydroxy acid dehydratase has been used to achieve the conversion of 4-deoxy-L-*threo*-pentonic acid into 3,4-dideoxy-pent-2-ulosonic acid on a preparative scale.[35]

Further work has been published on the oxidation (Pt, O_2) of L-sorbose to 2-keto-L-gulonic acid in the presence of added bases.[36] A 1-deoxy-1-nitroheptitol derivative has been converted to the 1-aldehyde and then chain-extended affording another synthesis of 3-deoxy-D-*glycero*-D-*galacto*-non-2-ulosonic acid.[37] A synthesis of 3-deoxy-L-*threo*-hex-2-ulosonic acid from L-(+)-diethyl tartrate and a dithioacetal of ethyl glyoxylate has been described.[38] A stereospecific synthesis of a Kdo derivative **11** featured an intramolecular addition step (Scheme 3) similar to that for an enzymic biosynthesis.[39]

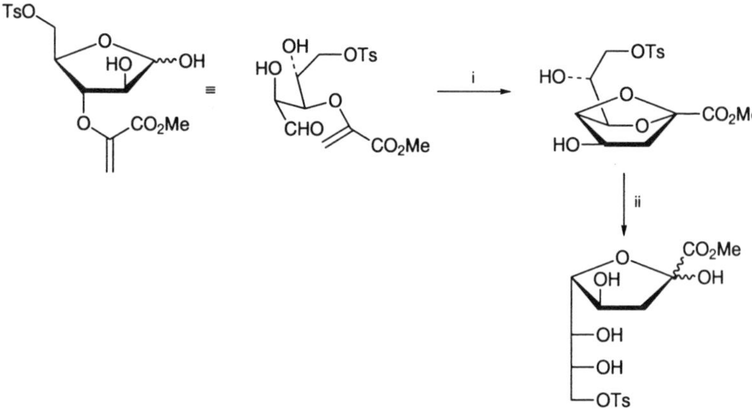

Reagents: i, SnCl₄; ii, H⁺/H₂O

Scheme 3

The 5-deoxy- and 5-epimer derivatives of 3-deoxy-D-*erythro*-hex-2-ulosonic acid have been shown to induce the expression of pectinase genes in the phytopathogenic bacterium *Erwinia chrysanthemi*, along with the parent compound. They were synthesized from a D-glucono-1,5-lactone derivative by way of a β-elimination and subsequent hydrolysis.[40] Treatment of tetra-*O*-acetyl-β-D-glucopyranosyl cyanide with DBU gave the corresponding 2-deoxyglycal derivative. The nitrile moiety was saponified to the acid, esterified to the methyl ester, and then methoxymercuration of the double bond and reduction of the organomercurial afforded 3-deoxy-D-*arabino*-hept-2-ulosonic acid as its methyl glycoside/methyl ester.[41]

The application of indium- and tin-mediated Barbier-Grignard reactions in aqueous solution to the chain extension of unprotected sugars has been reviewed.[42] Some C-5 variants of *N*-acetylneuraminic acid have been prepared by the indium-mediated addition of ethyl α-(bromomethyl)acrylate to variously *N*-substituted mannosamines,[43] and similar chemistry using (bromomethyl)vinyl phosphonate has led to the phosphonate analogue 12.[44] The reaction of *O*-protected aldono-1,4- and -1,5-lactones with (2-bromomethyl)-acrylate or -acrylonitrile or 2-bromobutyrolactone mediated by samarium diiodide led to chain-extended ulosonic acid derivatives.[45]

3-Deoxy-D-*threo*-hex-2-ulosonic acid has been synthesized by way of a Lewis acid catalysed ene reaction between methyl glyoxalate and 1-*tert*-butyldiphenylsilyloxy-but-3-ene.[46]

The disaccharide nucleoside 13 has been prepared as an analogue of CMP-NeuNAc,[47] and some thioglycoside disaccharides of neuraminic acid have been prepared as potential rotavirus inhibitors.[48] 2,3-Unsaturated neuraminic acid derivatives have been prepared from the corresponding methyl glycosides,[49] and the glucofuranose analogues of hydantocidin 14 and 15 have been synthesized. Treatment of 14 with Bu^{t}OK caused it to rearrange to 15.[50] The L-rhamnose hydantocidin mimic 16 has been prepared as well as the two closely related diketonepiperazines 17 and 18.[51] The use of bacterial α-(2→6)-sialyltransferase in the synthesis of oligosaccharides is covered in Chapter 4 and the utility of a glycosyl xanthate of neuraminic acid as a glycosyl donor is discussed in Chapter 3.

A new 8-step synthesis of the neuraminidase inhibitor 19 from neuraminic acid has been outlined,[52] and a number of analogues 18 with modified substituents on N-5,[53] as well as a 6-thio-analogue,[54] have been prepared and evaluated as neuraminidase inhibitors. The carboxamides 20 and 21 have also been synthesized as potential sialidase inhibitors,[55] and ^{11}C-labelled (in the guanidine moiety) 19 has been prepared for use in positron emission studies.[56]

Treatment of the peracetylated glycal methyl ester of neuraminic acid with *N*-iodosuccinimide in acetic acid followed by radical reduction afforded the peracylated methyl ester of neuraminic acid which allows the unwanted glycal byproducts of oligosaccharide synthesis to be reused.[57] Such glycosyl acetates with trimethylsilyl azide and stannic chloride afforded the β-glycosyl azide whereas the corresponding β-chloride with lithium azide gave the α-glycosyl azide.[58] Photobromination of methyl 2,5-anhydro-heptonate derivatives followed by displacement of the bromine with sodium azide and reduction has afforded the corresponding hept-2-ulsonic acid glycosylamines.[59]

Fischer methanolysis (HCl, MeOH) of Kdn gave mainly the methyl ester of the methyl β-pyranoside along with the α- and β-furanosides, the α-pyranoside and two furan elimination products.[60] The decarboxylation of peracylated neuraminic acid 22 with lead tetraacetate gave moderate yields of enone 23 and smaller quantities of 24 rather than the hoped for 25.[61] The thiomethylmercury derivative 26 has been prepared and found to be an inhibitor of the sialic acid 9-*O*-acetyl esterase from the influenza C virus[62] and a similar thiomethylmercury derivative has also been described.[63] The methylene Wittig product from the 4-keto-neuraminic acid derivative 27 has been epoxidized and the epoxide opened to give a number of branched derivatives 28.[64] Silylation of the primary hydroxy group of the β-methyl glycoside methyl ester of *N*-acetylneuraminic acid followed by per-*O*-acetylation, desilylation (80% HOAc), Swern oxidation and Wittig reactions has allowed the synthesis of C-9-chain-extended analogues of neuraminic acid.[65] The synthesis of 4-*N*-acetamido-4-deoxy derivatives of Kdo is discussed in Chapter 9 and the enzymic regioselective *O*-acetylation of *N*-acetylneuraminic acid is mentioned in Chapter 7.

22

23

24 X = H
25 X = OAc

26

27

28 X = OMe, CN, Cl, N_3, NHAc

4 Uronic Acids

The methods of synthesis of uronic acids by catalytic and nonspecific oxidation of sugars has been reviewed,[66] and a study of the oxidation of octyl α-D-glucopyranoside to octyl α-D-glucopyranosiduronic acid by sodium hypobromite catalysed by some ruthenium complexes has been reported.[67] The stability constants of metal ion-uronic acid complexes have been reviewed.[68]

The new sugar 3-*O*-[(R)-1-carboxyethyl]-D-glucuronic acid has been identified as a constituent of the exopolysaccharide produced by the bacterium *Altermonas* sp.1644 originally from deep sea hydrothermal vents.[69]

In a study of the relative reactivity of L-iduronic acid derivatives as glycosyl donors, trichloroacetimidates and *n*-pentenyl glycosides were the most effective, and better than the corresponding thioglycosides or glycosyl fluorides.[70] The reaction of 1,2:3,4-di-*O*-isopropylidene-α-D-galactopyranuronic acid with *p*-methoxybenzyl alcohol in the presence of *N,N*-dimethylformamide dineopentyl acetal gave the corresponding *p*-methoxybenzyl D-galactopyranuronate.[71] The synthesis of some *C*-glucuronosyl glycosides as cancer chemopreventives is mentioned in Chapter 3.

A route to hexuronic acid derivatives from either enantiomer of tartaric acid *via* **29** and **30** has been elaborated,[72] and the synthesis of peptides containing sugar amino acids such as **31-34** has been described.[73] The migration of C-1 of uronic acids in a remarkable tandem fragmentation-cyclization using diacetoxy-iodobenzene has, for example, led to **35** from **36** in 51% yield.[74] The two C-6 epimers of the 6-*N*-acetamido-4,6-dideoxyheptopyranosiduronic acid present in amipurimycin have been synthesized from a 4-deoxyglucose derivative.[75] Some 4-deoxy-hex-4-enopyranosiduronic acid derivatives have been prepared as potential sialidase inhibitors but they showed only slight inhibition.[76] Oxidation (PdCl$_2$, CuCl, O$_2$) of some 5,6-dideoxyhex-5-enofuranosides has led to 5-deoxyhexuronic acids,[77] while some glucuronic acid-substituted polymers have been prepared by the polymerization of unsaturated glycosides of glucuronic acid.[78]

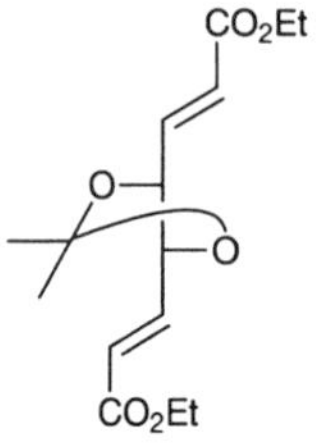

29

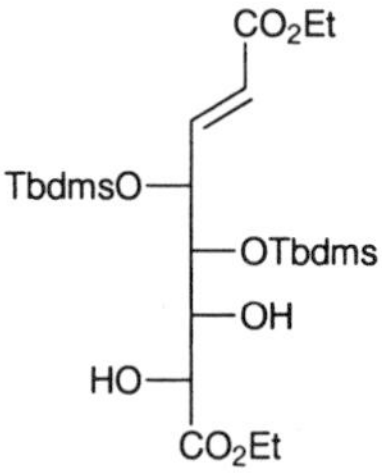

30

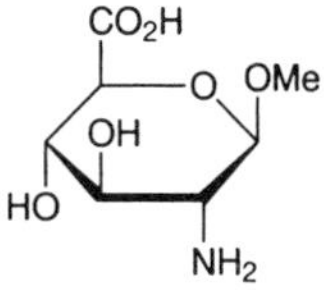

31

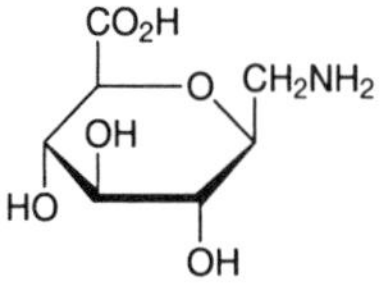

32

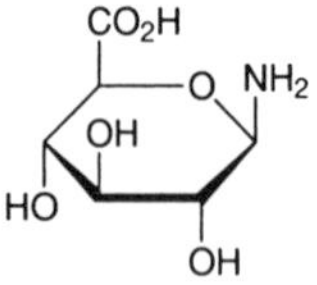

33

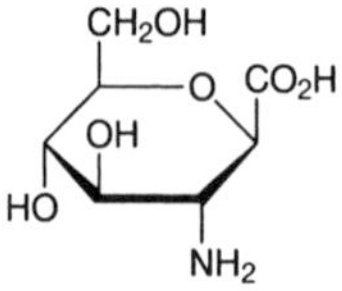

34

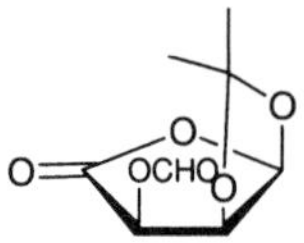

35

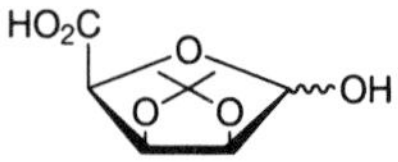

36

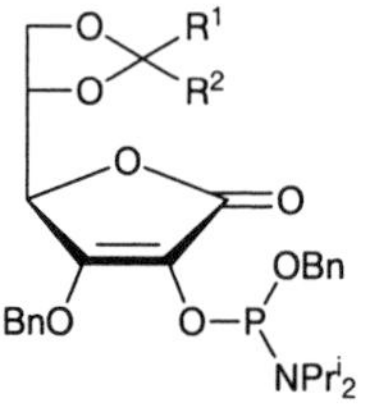

37 R^1 = R^2 = Me
38 R^1 = H, R^2 = Ph

5   Ascorbic Acids

The synthesis of 2-deoxy-L-ascorbic acid has been described,[79] and the dipole moment of L-ascorbic acid in dioxane has been calculated by using 'Guggenheim Theory'.[80] *O*-Benzyl ether and -benzylidene acetal derivatives of L-ascorbic acid and D-isoascorbic acid have been found to epimerize partly to the corresponding derivatives of L-isoascorbic acid and D-ascorbic acid, respectively, on treatment with triisobutylaluminium.[81] The synthesis of [1-^{13}C]- and [2-^{13}C]-L-ascorbic acid by application of the KCN chain-extension to suitable compounds has been outlined.[82]

The kinetics and mechanism of oxidation of L-ascorbic acid by some cobalt(III) complexes,[83,84] manganese(III) complexes[85] and a copper(II) complex[86] have been studied. Direct-current cyclic voltammetry has been used to investigate the suitability of some ferrocene derivatives as mediators for ascorbic acid oxidation in aqueous solution at low pH.[87]

5,6-*O*-Isopropylidene-L-ascorbic acid has been selectively alkylated at O-3

(K_2CO_3, acetone, alkyl halides). With excess alkyl halide the 2,3-di-*O*-alkylated derivatives could be obtained and by step-wise alkylation different alkyl moieties could be introduced at O-3 and O-2.[88] Alkylation of L-ascorbic acid derivatives at O-2 or O-3 with kojic acid and a tocopherol derivative afforded conjugates with enhanced thermal stability.[89,90]

The phosphoramidites **37** and **38** have been prepared and were converted into phosphate esters by condensing with alcohols (including some carbohydrate derivatives) and oxidation.[91] The synthesis of 6-*O*-palmitoyl-L-ascorbic acid was achieved in a non-aqueous medium with an immobilized lipase from *Candida antartica*,[92] while use of palmitoyl chloride on L-ascorbic acid gave the same product.[93]

Some 6-amino-6-deoxy-L-ascorbic acid derivatives have been synthesized and their cytotoxic properties evaluated,[94] and ($\pm$) 4-amino-4-deoxy-ascorbic acid has been prepared.[95] Michael-like reactions between L-ascorbic acid and some nitrostyrenes have led to adducts such as **39**.[96]

References

1   E. Ayadi, S. Czernecki and J. Xie, *J. Carbohydr. Chem.*, 1996, **15**, 191.
2   M. Rizzoto, M.I. Frascaroli, S. Signorella and L.F. Sala, *Polyhedron*, 1996, **15**, 1517 (*Chem. Abstr.*, 1996, **124**, 317 614).
3   V.I. Krupenskii, *Isv. Vyssh. Uchebn. Zaved., Khim. Khim. Tekhnol.*, 1995, **38**, 129 (*Chem. Abstr.*, 1996, **124**, 202 761).
4   V.I. Krupenskii, *Zh. Prikl. Khim. (S.-Petersburg)*, 1995, **68**, 869 (*Chem. Abstr.*, 1996, **124**, 176 652).
5   M. Besson, G. Flèche, P. Fuertes, P. Gallezot and F. Lahmer, *Recl. Trav. Chim. Pays-Bas*, 1996, **115**, 217.
6   V.N. Kislenko, A.A. Berlin and N.V. Litorchenko, *Zh. Obshch. Khim.*, 1995, **65**, 1197 (*Chem. Abstr.*, 1996, **124**, 146 605).
7   C.J. Railton and D.L.J. Clive, *Carbohydr. Res.*, 1996, **281**, 69.
8   E. Untersteller, A.J. Fairbanks and P. Sinay, *Carbohydr. Lett.*, 1995, **1**, 217 (*Chem. Abstr.*, 1996, **125**, 115 019).
9   B.V. Rao and S. Lahiri, *J. Carbohydr. Chem.*, 1996, **15**, 975.
10   P.E. Maligres, S.A. Weissman, V. Upadhyay, S.J. Cianciosi, R.A. Reamer, R.M. Purick, J. Sager, K. Rossen, K.K. Eng, D. Askin, R.P. Volante and P.J. Reider, *Tetrahedron*, 1996, **52**, 3327.
11   K. Niemela, *Carbohydr. Res.*, 1996, **285**, 41.
12   F.J. López-Herrera, A.M. Heras-López, M.S. Pino-González and F.S. García, *J. Org. Chem.*, 1996, **61**, 8839.
13   R.M. de Lederkremer and O. Varela, *Trends Org. Chem.*, 1993, **4**, 227 (*Chem. Abstr.*, 1996, **125**, 58 865).
14   A.P. Nin, R.M. de Lederkremer and O. Varela, *Tetrahedron*, 1996, **52**, 12911.
15   C. Taillefumier, S. Colle and Y. Chapleur, *Carbohydr. Lett.*, 1996, **2**, 39, (*Chem. Abstr.*, 1996, **125**, 196 188).
16   L. Chen and D.E. Kiely, *J. Org. Chem.*, 1996, **61**, 5847.
17   F. Zamora and J.A. Galbis, *Carbohydr. Res.*, 1996, **293**, 251.

18   Y.D. Ward, L.A. Villanueva, G.O. Allred and L.S. Liebeskind, *J. Am. Chem. Soc.*, 1996, **118**, 897.
19   Z. Zhang, F. Fukunaga, Y. Sugimura and K. Nakao, *Carbohydr. Res.*, 1996, **290**, 225.
20   Z. Zhang, F Fukunaga, Y. Sugimura, K. Nakao and T. Shimizu, *Carbohydr. Res.*, 1996, **292**, 47.
21   J.P. McDevitt and P.T. Lansbury, Jr., *J. Am. Chem. Soc.*, 1996, **118**, 3818.
22   C.J.F. Bichard, T.W. Brandstetter, J.C. Estevez, G.W.J. Fleet, D.J. Hughes and J.R. Wheatley, *J. Chem. Soc., Perkin Trans. 1*, 1996, 2151.
23   J.C. Estevez, J. Saunders, G.S. Besra, P.J. Brennan, R.J. Nash and G.W. J. Fleet, *Tetrahedron: Asymm.*, 1996, **7**, 383.
24   N.J.Barnes, M.A. Probert and R.H. Wightman, *J. Chem. Soc., Perkin Trans. 1*, 1996, 431.
25   B.P. Bandgar, *Indian J. Chem., Sect. B: Org. Chem. Incl. Med. Chem.*, 1996, **35B**, 492 (*Chem. Abstr.*, 1996, **125**, 34 004).
26   F.J. López-Herrera, M.S. Pino-Gonzáles, F. Sarabia-García, A. Heras-López, J.J. Ortega-Alcántara and M.G. Pedraza-Cebrian, *Tetrahedron: Asymm.*, 1996, **7**, 2065.
27   M. Shiozaki, T. Mochizuki, T. Wakabayashi, S. Kurakata, T. Tatsuta and M. Nishijima, *Tetrahedron Lett.*, 1996, **37**, 7271, and Corrigendum, 8627.
28   M. Imamura and H. Hashimoto, *Chem. Lett.*, 1996, 1087.
29   H.J.M. Gijsen, L. Qiao, W. Fitz and C.-H. Wong, *Chem. Rev.*, 1996, **96**, 443 (*Chem. Abstr.*, 1996, **124**, 87 497).
30   U. Kragl, M. Kittelmann, O. Ghisalba and C. Wandrey, *Ann. N. Y. Acad. Sci.*, 1995, **750**, 300 (*Chem. Abstr.*, 1996, **124**, 87 629).
31   M. Murakami, K. Ikeda and K. Achiwa, *Carbohydr. Res.*, 1996, **280**, 101.
32   D.R.J. Palmer and J.A. Gerlt, *J. Am. Chem. Soc.*, 1996, **118**, 10323.
33   M.C. Shelton, I.C. Cotterill, S.T.A. Novak, R.M. Poonawala, S. Sudarshan and E.J. Toone, *J. Am. Chem. Soc.*, 1996, **118**, 2117.
34   B.R. Knappmann, A. Steigel and M.-R. Kula, *Biotechnol. Appl. Biochem.*, 1995, **22**, 107 (*Chem. Abstr.*, 1996, **124**, 30 129).
35   G. Limberg and J. Thiem, *Aust. J. Chem.*, 1996, **49**, 349.
36   C. Broennimann, Z. Bodnar, R. Aeschimann, T.Mallat and A. Baiker, *J. Catal.*, 1996, **161**, 720 (*Chem. Abstr.*, 1996, **125**, 276 385).
37   S.V. Turik, I.I. Bicherova, V.I. Kornilov and Y.A. Zhdanov, *Zh. Obshch. Khim.*, 1995, **65**, 1191 (*Chem. Abstr.*, 1996, **124**, 176 737).
38   B. Foessel, M. Stenzel, R. Baudouy, G. Condemine, J. Robert-Baudouy and B. Fenet, *Bull. Soc. Chim. Fr.*, 1995, **132**, 829 (*Chem. Abstr.*, 1996, **124**, 117 812).
39   S. Du, D. Plat and T. Baasov, *Tetrahedron Lett.*, 1996, **37**, 3545.
40   F. Alessi, A. Doutheau, D. Anker, G. Condemine and J. Robert-Baudouy, *Tetrahedron*, 1996, **52**, 4625.
41   J. Mlynarski and A. Banaszek, *Carbohydr. Res.*, 1996, **295**, 69.
42   C.-J. Li, *Tetrahedron*, 1996, **52**, 5643.
43   S.-X. Choi, S. Lee and G.M. Whitesides, *J. Org. Chem.*, 1996, **61**, 8739.
44   J. Gao, V. Martichonok and G.M. Whitesides, *J. Org. Chem.*, 1996, **61**, 9538.
45   R. Csuk, U. Hörning and M. Schaade, *Tetrahedron*, 1996, **52**, 9759.
46   M. Shimizu, A. Yoshida and K. Mikami, *Synlett*, 1996, 1112.
47   Y. Hatanaka, M. Hashimoto, K.I.-P. Jwa Hidari, Y. Sanai, Y. Nagai and Y. Kanaoka, *Heterocycles*, 1996, **43**, 531.
48   M.J. Kiefel, B. Beisner, S. Bennett, I.D. Holmes and M. von Itzstein, *J. Med. Chem.*, 1996, **39**, 1314.

49   G.B. Kok, D.R. Groves and M. von Itzstein, *J. Chem. Soc., Chem. Commun.*, 1996, 2017.

50   T.W. Brandstetter, C. de la Fuente, Y.-h. Kim, L.N. Johnson, S. Crook, P.M. de Q. Lilley, D.J. Watkin, K.E. Tsitsanou, S.E. Zographos, E.D. Chrysina, N.G. Oikonomakos and G.W.J. Fleet, *Tetrahedron*, 1996, **52**, 10721.

51   J.C. Estevez, M.D. Smith, A.L. Lane, S. Crook, D.J. Watkin, G.S. Besra, P.J. Brennan, R.J. Nash and G.W.J. Fleet, *Tetrahedron: Asymm.*, 1996, **7**, 387.

52   J. Scheigetz, R. Zamboni, M.A. Bernstein and B. Roy, *Org. Prep. Proceed. Int.*, 1995, **27**, 637 (*Chem. Abstr.*, 1996, **124**, 202 859).

53   P.W. Smith, I.D. Starkey, P.D. Howes, S.L. Sollis, S.P. Keeling, P.C. Cherry, M. von Itzstein, W.Y. Wu and B. Jui, *Eur. J. Med. Chem.*, 1996, **31**, 143 (*Chem. Abstr.*, 1996, **124**, 343 829).

54   G.B. Kok, M. Campbell, B. Mackey and M. von Itzstein, *J. Chem. Soc., Perkin Trans. 1*, 1996, 2811.

55   S.L. Sollis, P.W. Smith, P.D. Howes, P.C. Cherry and R.C. Bethell, *Bioorg. Med. Chem. Lett.*, 1996, **6**, 1805.

56   G. Westerberg, M. Bamford, M.J. Daniel, B. Langstron and D.R. Sutherland, *J. Labelled Compd. Radiopharm.*, 1996, **38**, 585 (*Chem. Abstr.*, 1996, **125**, 196 190).

57   P. Kosma, H. Sekljic and G. Balint, *J. Carbohydr. Chem.*, 1995, **15**, 701.

58   Z. Györgydeák, L. Szyágyi, Z. Dinya and J. Jekö, *Carbohydr. Res.*, 1996, **291**, 183.

59   T.W. Brandstetter, C. de la Fuente, Y.-h. Kim, R.I. Cooper, D.J. Watkin, N.G. Oikonomakos, L.N. Johnson and G.W.J. Fleet, *Tetrahedron*, 1996, **52**, 10711.

60   T. Kai, X.-L. Sun, M. Tanaka, H. Takayanagi and K. Furuhata, *Chem. Pharm. Bull.*, 1996, **44**, 208.

61   J.J. Potter and M. von Itzstein, *Carbohydr. Res.*, 1996, **282**, 181.

62   W. Fitz, P.B. Rosenthal and C.-H. Wong, *Bioorg. Med. Chem.*, 1996, **4**, 1349 (*Chem. Abstr.*, 1996, **125**, 301 362).

63   M.J. Kiefel and M. von Itzstein, *Tetrahedron Lett.*, 1996, **37**, 7307.

64   D.R. Groves and M. von Itzstein, *J. Chem. Soc., Perkin Trans. 1*, 1996, 2817.

65   M.J. Kiefel, S. Bennett and M. von Itzstein, *J. Chem. Soc., Perkin Trans.1*, 1996, 439.

66   V.A. Timoshchuk, *Usp. Khim.*, 1995, **64**, 721 (*Chem. Abstr.*, 1996, **124**, 146 592).

67   A.E.M. Boelrijk, A.M.J. Jorna and J. Reedijk, *J. Mol. Catal., A: Chem.*, 1995, **103**, 73 (*Chem. Abstr.*, 1996, **124**, 146 670).

68   M. Klozlowski and M. Jezowska-Bojczuk, *Handb. Met.-Ligand Interact. Biol. Fluids: Bioinorg. Chem.*, 1995, **1**, 679 (*Chem. Abstr.*, 1996, **124**, 343 808).

69   G. Dubreucq, B. Domon and B. Fournet, *Carbohydr. Res.*, 1996, **290**, 175.

70   C. Jabeur, F. Machetto, J.-M. Mallet, P. Duchaussoy, M. Petitou and P. Sinaÿ, *Carbohydr. Res.*, 1996, **281**, 253.

71   D. Joniak, *Chem. Pap.*, 1995, **49**, 198 (*Chem. Abstr.*, 1996, **124**, 261 541).

72   M.T. Barros, M.O. Januario-Charmier, C.D. Maycock and M. Pires, *Tetrahedron*, 1996, **52**, 7861.

73   E. Graf von Roedern, E. Lohof, G. Hessler, M. Hoffmann and H. Kessler, *J. Am. Chem. Soc.*, 1996, **118**, 10156.

74   C.G. Francisco, C.C. González and E. Suárez, *Tetrahedron Lett.*, 1996, **37**, 1687.

75   S. Czernecki, J.-M Valery and R. Wilkens, *Bull. Chem. Soc. Jpn.*, 1996, **69**, 1347.

76   P.D. Howes and P.W. Smith, *Tetrahedron Lett.*, 1996, **37**, 6595.

77   K. Krishnudu, P.R. Krishna and H.B. Mereyala, *Tetrahedron Lett.*, 1996, **37**, 6007.

78   A. Leydet, C. Jeantet-Segonds, P. Barthélémy, B. Boyer and J.P. Rogue, *Recl. Trav. Chim. Pays-Bas*, 1996, **115**, 421.

79   P. Ge and K.L. Kirk, *J. Org. Chem.*, 1996, **61**, 8671.

80   S. Arzik, A. Altunata and S.K. Timur, *J. Fac. Sci. Ege Univ., Ser A-B*, 1995, **18**, 43 (*Chem. Abstr.*, 1996, **125**, 87 069).

81   J. Schachtner and H.-D. Stachel, *Tetrahedron: Asymm.*, 1996, **7**, 3263.

82   K.N. Drew, T.J. Church, B. Basu, T. Vuorinen and A.S. Serianni, *Carbohydr. Res.*, 1996, **284**, 135.

83   K. Abdur-Rashid, T.P. Dasgupta and J. Burgess, *J. Chem. Soc., Dalton Trans.*, 1996, 1385 (*Chem. Abstr.*, 1996, **125**, 11 303).

84   K. Abdur-Rashid, T.P. Dasgupta and J. Burgess, *J. Chem. Soc., Dalton Trans.*, 1996, 1393 (*Chem. Abstr.*, 1996, **125**, 11 304).

85   I.A. Salem and A.H. Gemeav, *Transition Met. Chem. (London)*, 1996, **21**, 130 (*Chem. Abstr.*, 1996, **125**, 11 316).

86   M. Scarpa, F. Vianello, L. Signor, L. Zennaro and A. Rigo, *Inorg. Chem.*, 1996, **35**, 5201 (*Chem. Abstr.*, 1996, **125**, 115 023).

87   M.H. Pournaghi-Azar and R. Ojani, *Talanta*, 1995, **42**, 1839 (*Chem. Abstr.*, 1996, **124**, 202 862).

88   M.G. Kulkarni and S.R. Thopate, *Tetrahedron*, 1996, **52**, 1293.

89   K. Morisaki and S. Ozaki, *Chem. Pharm. Bull.*, 1996, **44**, 1647.

90   K. Morisaki and S. Ozaki, *Bull. Chem. Soc. Jpn.*, 1996, **69**, 725.

91   K. Morisaki and S. Ozaki, *Carbohydr. Res.*, 1996, **286**, 123.

92   C. Humeau, M. Girardin, D. Coulon and A. Miclo, *Biotechnol Lett.*, 1995, **17**, 1091 (*Chem. Abstr.*, 1996, **124**, 117 813).

93   Y. Lu, L. Gan and B. Chen, *Jingxi Huagong*, 1996, **13**, 17 (*Chem. Abstr.*, 1996, **125**, 196 192).

94   M.K. Kojic-Prodic, Z. Banic, M. Grdisa, V. Vela, B. Suskovic and K. Pavelic, *Eur. J. Med. Chem.*, 1996, **31**, 23 (*Chem. Abstr.*, 1996, **124**, 290 103).

95   H.-D. Stachel, K. Zeitler and S. Dick, *Liebigs Ann. Chem.*, 1996, 103.

96   M. Schmidt and K. Eger, *Pharmazie*, 1996, **51**, 11 (*Chem. Abstr.*, 1996, **124**, 317 715).

17
Inorganic Derivatives

1 Carbon-bonded Phosphorus Derivatives

Some new *C*-glycosidic difluoromethylenephosphonates have been prepared by phosphonyl radical addition to difluoroenol ethers (Scheme 1)[1], while glycosyl phosphinic acids (*e.g.* **1**) have been produced directly from free sugars.[2] Phosphonate analogues of α-D-*N*-acetylglucosamine- and α-D-*N*-acetylmannosamine-1-phosphate have been synthesized by way of an Arbuzov reaction, using triethyl phosphite, applied to an iodomethyl α-*C*-glycoside derivative,[3] and the phosphonate carbocyclic compounds **2** and **3** have been made by way of an Arbuzov reaction applied to suitable bromomethyl-substituted cyclitols.[4] 6-Deoxy-6-phosphono-D-glucopyranose and -D-galactopyranose have been synthesized and shown to be weak inhibitors of 1L-*myo*-inositol 1-phosphatase.[5]

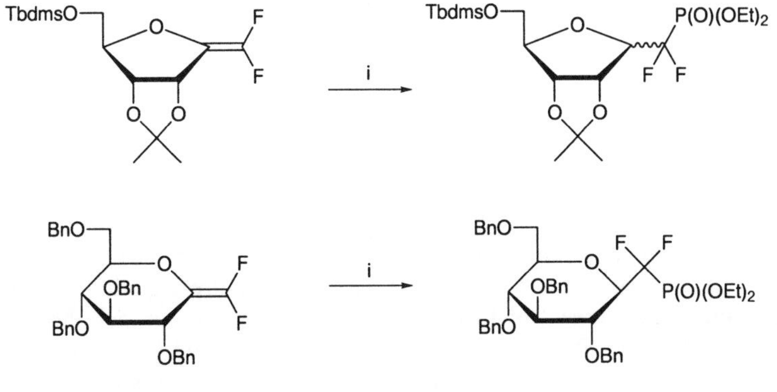

Reagents: i, $(EtO)_3P$, Bu^{t}OOBu^t or $(EtO)_2P(O)SePh$, AIBN, Bu_3SnH

Scheme 1

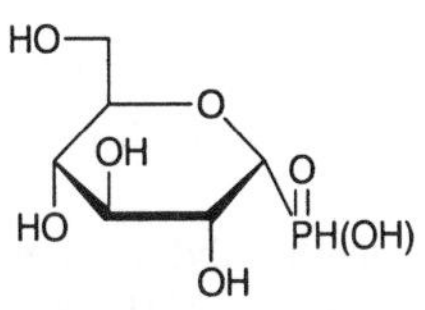

1

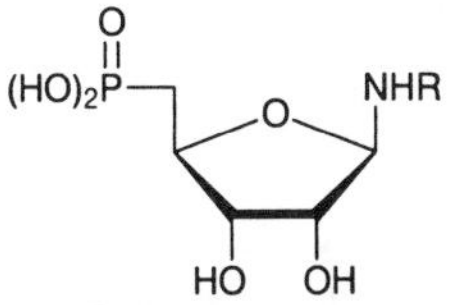

2 R = H
3 R = $COCH_2NH_2$

Carbohydrate Chemistry, Volume 30

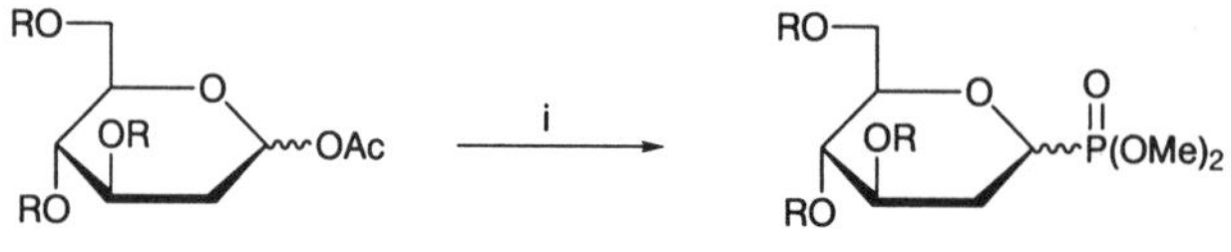

4 R = Ac, Bn

Reagents: i, P(OMe)₃, TmsOTf

Scheme 2

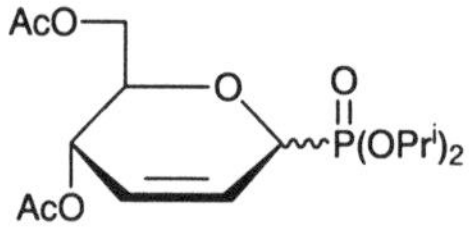

5

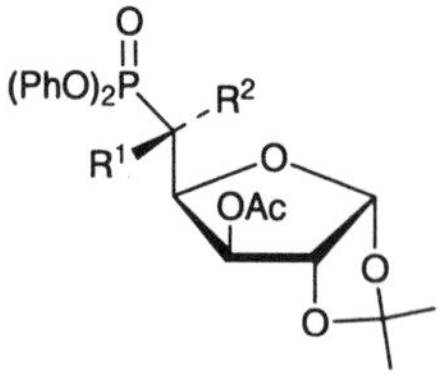

6 R¹ = H, R² = OH
or R¹ = OH, R² = H

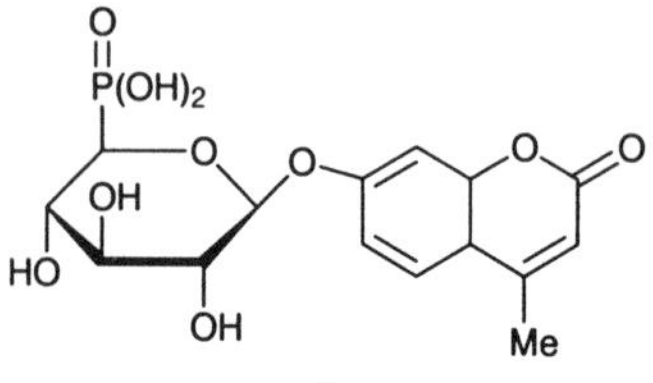

7

Glycosyl phosphonates **4** have been prepared from the corresponding glycosyl acetates (Scheme 2), whereas the diethylphosphonate analogue of the β anomer of **4** (R=Bn) was formed when a C-1 carbanion was treated with chlorodiethylphosphate.[6] Treatment of tri-*O*-acetyl-D-glucal with triisopropyl phosphite and boron trifluoride afforded the glycosyl phosphonates **5** which were converted into nucleotide analogues with a base moiety attached at C-4. Similar chemistry was performed on tri-*O*-acetyl-L-glucal and di-*O*-acetyl-L-rhamnal.[7] 3-*O*-Acetyl-1,2-*O*-isopropylidene-α-D-*xylo*-pento-1,5-dialdose in the presence of diphenylphosphinate has produced phosphonates **6**, and one epimer was transformed into **7** which was not an inhibitor (or substrate) of bovine liver glucuronidase.[8] Similarly the phenylphosphinates **8** were prepared from the corresponding 5-aldehydo derivative and were converted into **9**, from which all four isomers were separated.[9]

Treatment of methyl 6-*O*-tosyl-2,3,4-tri-*O*-trimethylsilyl-α-D-glucopyranoside with lithium diphenylphosphide at reflux temperatures (in THF or ether) afforded methyl 6-deoxy-6-diphenylphosphino-α-D-glucopyranoside, whereas at lower temperatures S-O bond cleavage predominated.[10] Similar treatment applied to methyl 2,3-*O*-isopropylidene-5,6-di-*O*-methanesulfonyl-α-D-mannofuranoside gave, after subsequent air oxidation, a mixture of **10** and **11**,[11] and a new

8   **9**   **10** X = OMs, Y = H
                             11 X = H, Y = OH   **12**

carbohydrate diphenylphosphine **12** has been prepared from D-glucitol.[12] The new analogues **13** of 2-deoxy-D-*ribo*- and *arabino*-hexose have been prepared,[13] and all eight diastereomers of the 3-phosphapentopyranoses **14** have been synthesized and separated.[14] Preparation of the phosphine **15** has been reported and subsequent oxidation by oxygen or hydrogen peroxide gave the corresponding phosphine oxide and phosphinic acid respectively.[15]

13   **14**   **15**

16 R = Me, Pr^i **17**

The phosphonates **16** have been prepared as 4'-branched analogues of nucleoside 5'-monophosphates[16] and 2-diethylphosphonyl-1,3-dithiane has been lithiated and used in the preparation of the phosphonic acid analogue **17** of 3-deoxy-D-*manno*-2-octulonic acid[17] while the phosphonate analogue **18** of *N*-acetylneuraminic acid has been prepared in excellent yield from *N*-acetyl-mannosamine (Scheme 3).[18]

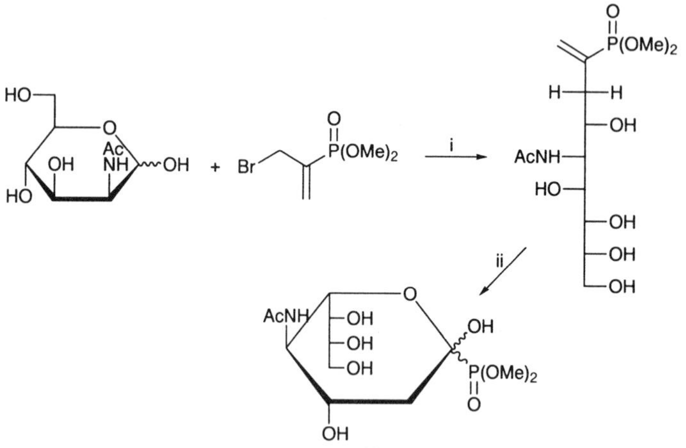

Reagents: i, Indium; ii, O₃ then Me₂S

Scheme 3

2   Other Carbon-bonded Derivatives

Treatment of 1,2-*O*-isopropylidene-3,5-di-*O*-tosyl-α-D-xylofuranose with lithium diphenylarsinide has afforded 5-deoxy-1,2-*O*-isopropylidene-5-*C*-diphenylarsino-3-*O*-tosyl-α-D-xylofuranose,[19] and the arsenic containing ribosides **19** have been prepared using standard procedures.[20,21]

The displacement of primary iodides in some carbohydrate derivatives by lithiated 2-trimethylsilyl-1,3-dithiane followed by hydrolysis of the dithioacetal has allowed the synthesis of some acylsilanes (*e.g.* **21** from **20**),[22] and a review on the oxidation of the carbon-silicon bond has included a number of carbohydrate examples.[23]

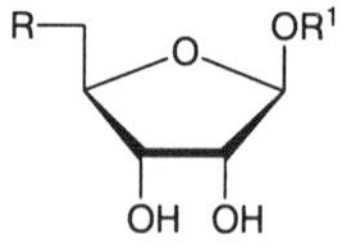

19 R = Me₂As, Me₂As(O), Me₂As(S),
      Me₃As⁺
    R¹ = (*R*)-2,3-dihydroxypropyl

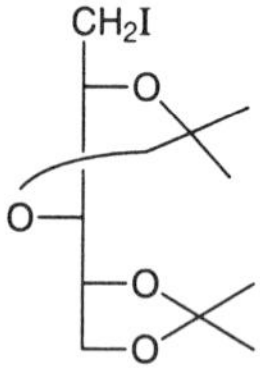

20

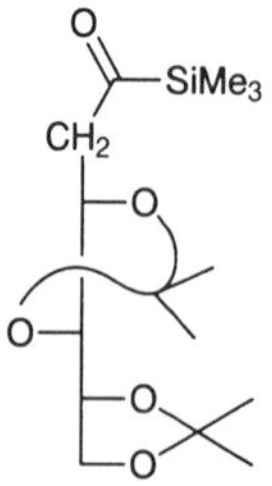

21

Lithium bis(dimethylphenylsilyl)cuprate and lithium methyl(dimethylphenyl-silyl)cuprate have been used to effect a conjugate addition of the dimethylphenyl-silyl moiety to some carbohydrate enones. Use of trimethylstannyl lithium and tributylstannyl lithium also afforded conjugate addition products.[24] In a new approach to carbohydrate organotin compounds, tributylstannylcopper has been used as a source of nucleophilic tributyltin to effect displacements of a primary tosylate, an allylic bromide and addition to an aldehyde.[25]

Syntheses of 1,2:3,4-di-O-isopropylidene-6-O-3-(triphenylstannyl)propyl-α-D-galactopyranose and 1,6-anhydro-3,4-O-isopropylidene-2-O-triphenylstannyl methyl-β-D-galactopyranose have been described[26] as well as a family of [6-O-(1,2:3,4-di-O-isopropylidene-α-D-galactopyranosyl)methyl]tin species Ph_nSn-$(CH_2OR)_{4-n}$ (n = 1-3; ROH = 1,2:3,4-di-O-isopropylidene-α-D-galacto-pyranose),[27] and 1,2:3,4-di-O-isopropylidene-6-O-triphenylstannylmethyl-α-D-galactopyranose.[28]

Some carbohydrate acetylenic alcohols have been converted to metal carbene spirocyclic derivatives, *e.g.* **22** (Scheme 4).[29] Acetobromoglucose has been converted into tri-O-acetyl-D-glucal in 90% yield by way of a glycosyl chromium-(III) complex,[30] and an organomercury intermediate featured in a synthesis of tri-O-benzyl-2-C-methyl-D-glucal covered in Chapter 14.

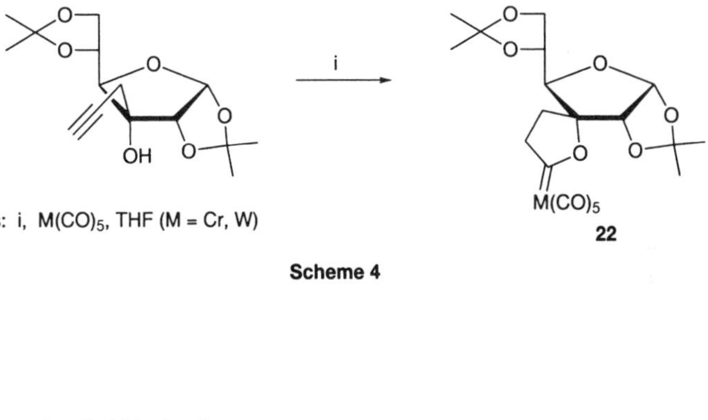

Reagents: i, M(CO)$_5$, THF (M = Cr, W)

22

Scheme 4

3   Oxygen-bonded Derivatives

A series of 3,4-di-O-diarylphosphinite derivatives of fructose have been prepared as potential chiral ligands for Ni(0) complexes,[31] and the reaction of 1,2:5,6-di-O-isopropylidene-α-D-glucofuranose with chlorophosphines (ClPPr^{i}R, R = Ph, Bu^t) gave 3-O-phosphinite diastereomers, the ratio of which depended on the base used in their formation.[32] Treatment of an anomerically unprotected aldo-pyranose with N,N-diisopropylchlorophosphoramidite has generated the four diastereomers of the glycosyl phosphoramidite.[33]

Organotin ether derivatives continue to be utilized in the regioselective acylation and alkylation of carbohydrates. The 3-O-acylation of benzyl α-L-lyxopyranoside used the dibutylstannylene method[34] while a polymer supported aryl(butyl)tin dichloride reagent has been used catalytically in the selective 6-O-

acylation of sucrose.[35] A stannylene-mediated selective silylation of primary hydroxy groups has been outlined,[36] while the use of different stannylenes in a study of regioselective benzylations is covered in Chapter 5. The synthesis of 4-methoxyphenyl 3,6-di-*O*-benzyl-2-deoxy-2-*N*-phthalimido-β-D-glucopyranoside used dibutylstannylene-mediated selective benzylation.[37] A β-D-galactopyrano-side moiety in an otherwise fully protected trisaccharide has been 3-*O*- and 3,6-di-*O*-sulfated using a stannylene intermediate.[38]

Semiempirical MO calculations have been used to predict low energy con-formations for α-D-galacturonate ion, and thus the preferred sites for metal ion binding which were at the carboxyl group, the pyranose ring oxygen atom, and the C-4 hydroxy group.[39] Complexes formed between W(VI) ions and D-galacturonic acid or D-glucuronic acid were shown to exist as 1:2 (metal:sugar) complexes with the furanose forms of both acids, whereas with Mo(VI) ions, complexes with the α-pyranoses were also observed.[40,41] D-*glycero*-L-*manno*-Heptose has been found to form tungstate and molybdate complexes in aqueous solution in a tetradentate fashion and always with the sugar in a furanose or hydrated acyclic form.[42] Similarly D-fructose and D-sorbose form tungstate and molybdate complexes in aqueous solution only in their acyclic forms.[43]

A crystalline sodium salt of bis(D-glucuronato)oxovanadium(IV) has been reported,[44] and other saccharide complexes of oxovanadium(IV) have been synthesized and characterized[45] as have a number of complexes of Mn(II) with sugars. They were of the general form $Na[Mn_2(Sugar)_2X_3]$ or $Na[Mn_3(Sugar)_3X_4]$ where X_3 or X_4 could be combinations of Cl and OH.[46] Similar complexes with Zn(II) ions have been prepared. They were of the general formula $Na[Zn_2(Sugar)_2Cl_3].nH_2O$ for fructose, galactose and glucose, whereas for xylose and ribose, hydroxide ions displaced one or more of the chloride ions in the complex.[47]

The use of ^{11}B NMR has unambiguously demonstrated that phenylboronic acid and methyl glycopyranosides in water form two types of cyclic boronate esters, one involving the vicinal *cis* 3-OH and 4-OH groups (galactose and arabinose) and the other involving the 4-OH and 6-OH groups (galactose and glucose).[48] The cooperative binding of boronic acid with saccharides in basic aqueous media which creates a sugar-diboronic acid macrocycle has been utilized in the design of molecular receptors. Cooperative binding together with a photoinduced electron transfer mechanism has been utilized in the design of sensitive and selective saccharide sensors.[49] The complexing of *p*-tolylboronic acid with D-fructose has been studied by ^{13}C NMR[50] and complexes of a porphyrin-substituted phenylboronic acid with D-fucose, D-arabinose, methyl α-D-manno-pyranoside and D-threitol have been prepared. Two phenylboronic acid moieties bind to each sugar unit with different angles between the porphyrin pairs.[51]

4   Nitrogen-bonded Derivatives

Some oxorhenium(V) complexes of 2′,3′-diamino-2′,3′-dideoxy-nucleosides have been prepared as inhibitors of purine specific ribonuclease[52] and other oxorhenium(V) complexes of 3′,5′-diamino-3′,5′-dideoxy-nucleosides have also been made.[53] The coordination of 1-(2-aminoethylamino)-1-deoxy-D-alditols and 1-(2-aminopropylamino)-1-deoxy-D-alditols with Cd(II) ions at neutral pH involved both the primary and secondary amines whereas at high pH (12) an additional OH group coordination occurs.[54] A review on the stability constants of copper(II) complexes of amino-sugars has appeared[55] and 1,6-anhydro-2-deoxy-2-*N*-methylamino-β-D-mannopyranose has been shown to be very effective in coordinating to copper(II) ions, but contrary to the analogous glucosamine derivative, it does not form dimeric species.[56]

References

1   T.F. Herpin, J.S. Houlton, W.B. Motherwell, B.P. Roberts and J.-M. Weibel, *J. Chem. Soc., Chem. Commun.*, 1996, 613.

2   S.B. Tzokov, I.T. Devedjiev and D.D. Petkov, *J. Org. Chem.*, 1996, **61**, 12.

3   F. Casero, L. Cipolla, L. Lay, F. Nicotra, L. Panza and G. Russo, *J. Org. Chem.*, 1996, **61**, 3428.

4   R Vince, M. Hua and C.A. Caperelli, *Nucleosides Nucleotides*, 1996, **15**, 1711.

5   N.S. Padyukova, S. Afsar, M.B.F. Dixon and M. Ya. Karpeiski, *Bioorg. Khim.*, 1995, **21**, 382 (*Chem. Abstr.*, 1996, **124**, 9 106).

6   N.J. Barnes, M.A. Probert and R.H. Wightman, *J. Chem. Soc., Perkin Trans. 1*, 1996, 431.

7   P. Alexander, V.V. Krishnamurthy and E.J. Prisbe, *J. Med. Chem.*, 1996, **39**, 1321.

8   R. Hoos, J. Huixin, A. Vasella and P. Weiss, *Helv. Chim. Acta,* 1996, **79**, 1757.

9   K. Seo, M. Yamashita, T. Oshikawa and J. Kobayashi, *Carbohydr. Res.*, 1996, **281**, 307.

10   W.V. Dahloff and K. Radkowski, *Z. Naturforsch., B : Chem. Sci.*, 1996, **51**, 891 (*Chem. Abstr.*, 1996, **125**, 196 134).

11   M.A. Brown, P.J. Cox, R.A. Howie, O.A. Melvin and J.L. Wardell, *J. Chem. Soc., Perkin Trans. 1*, 1996, 809.

12   A.L. Wang, X.D. Wang, S.J. Lu, H.X. Fu and H.Q. Wang, *Chin. Chem. Lett.*, 1996, **7**, 299 (*Chem. Abstr.*, 1996, **125**, 33 990).

13   M.J. Gallagher, M.G. Ranasinghe and I.D. Jenkins, *J. Org. Chem.*, 1996, **61**, 436.

14   C.J.R. Fookes and M.J. Gallagher, *Heteroat. Chem.*, 1996, **7**, 391 (*Chem. Abstr.*, 1996, **125**, 329 145).

15   K. Schürrle and W. Piepersberg, *J. Carbohydr. Chem.*, 1996, **15**, 435.

16   Z. Tocik, I. Kavenova and I. Rosenberg, *Collect. Czech. Chem. Commun.*, 1996, **61**, S76 (*Chem. Abstr.*, 1996, **125**, 329 231).

17   P. Coutrot, C. Grison and M. Lecouvey, *Tetrahedron Lett.*, 1996, **37**, 1595.

18   J. Gao, V. Martichonok and G.M. Whitesides, *J. Org. Chem.*, 1996, **61**, 9538.

19   M.A. Brown, P.J. Cox, O.A. Melvin and J.L. Wardell, *Main Group Met. Chem.*, 1995, **18**, 175 (*Chem. Abstr.*, 1996, **124**, 117 700).

20   J. Liu, D.H. O'Brien and K.J. Irgolic, *Appl. Organomet. Chem.*, 1996, **10**, 1 (*Chem. Abstr.*, 1996, **124**, 290 029).

21   J. Liu, D.H. O'Brien and K.J. Irgolic, *Appl. Organomet. Chem.*, 1996, **10**, 13 (*Chem. Abstr.*, 1996, **124**, 317 626).

22   R. Plantier-Royon and C. Portella, *Tetrahedron Lett.*, 1996, **37**, 6113.

23   G.R. Jones and Y. Landais, *Tetrahedron*, 1995, **52**, 7633.

24   A. Kirschning and J. Harders, *Synlett*, 1996, 772.
25   S. Jarosz, *Tetrahedron Lett.*, 1996, **37**, 3063.
26   S.J. Garden, J.L. Wardell, O.A. Melvin and P.J. Cox, *Main Group Met. Chem.*, 1996, **19**, 251 (*Chem. Abstr.*, 1996, **125**, 58 887).
27   P.J. Cox, S.J. Garden, R.A. Howie, O.A. Melvin and J.L. Wardell, *J. Organomet. Chem.*, 1996, **516**, 213 (*Chem. Abstr.*, 1996, **125**, 143 145).
28   J.P. Cox, O.A. Melvin, S.J. Garden and J.L. Wardell, *J. Chem. Crystallogr.*, 1995, **25**, 469 (*Chem. Abstr.*, 1996, **124**, 56 447).
29   K.H. Dötz, O. Neuß and M. Nieger, *Synlett*, 1996, 995.
30   G. Kovács, J. Gyarmati, L. Somák and K. Micskei, *Tetrahedron Lett.*, 1996, **37**, 1293.
31   T.V. Rajanbabu and A.L. Casalnuovo, *J. Am. Chem. Soc.*, 1996, **118**, 6325.
32   O.I. Kolodyazhnyi, *Zh. Obshch. Khim.*, 1995, **65**, 1926 (*Chem. Abstr.*, 1996, **125**, 11 235).
33   S.J. Freese and W.F. Vann, *Carbohydr. Res.*, 1996, **281**, 313.
34   A.K.M.S. Kabir, M.A. Sattar, A.Z.M.S. Chowdhury, M.A. Sattar and M.M. Rahman, *J. Bangladesh Chem. Soc.*, 1995, **8**, 95 (*Chem. Abstr.*, 1996, **125**, 301 356).
35   W.M. Macindoe, A. Williams and R. Khan, *Carbohydr. Res.*, 1996, **283**, 17.
36   M.W. Bredenkamp, *S. Afr. J. Chem.*, 1995, **48**, 154 (*Chem. Abstr.*, 1996, **125**, 222 275).
37   I. Robina, F. López-Barba and J. Fuentes, *Synth. Commun.*, 1996, **26**, 2847.
38   Y.-M. Zhang, A. Brodzki, P. Sinay, F. Uzabiaga and C. Picard, *Carbohydr. Lett.*, 1996, **2**, 67 (*Chem. Abstr.*, 1996, **125**, 222 301).
39   Y. Nakamura, *Kagoshima Daigaku Kyoikugakubu Kenkyu Kiyo, Shizen Kagaku Hen*, 1995, **47**, 135 (*Chem. Abstr.*, 1996, **125**, 11 238).
40   M.L.D. Ramos, M.M.M. Caldeira and V.M.S. Gil, *Carbohydr. Res.*, 1996, **286**, 1.
41   M. Hlaibi, M. Benaissa, S. Chapelle and J.-F. Verchere, *Carbohydr. Lett.*, 1996, **2**, 9 (*Chem. Abstr.*, 1996, **125**, 196 187).
42   Matuleva, J.-F. Verchere and S. Chapelle, *Carbohydr. Res.*, 1996, **287**, 37.
43   J.P. Sauvage, J.-F. Verchere and S. Chapelle, *Carbohydr. Res.*, 1996, **286**, 67.
44   S.B. Etcheverry, P.A.M. Williams and E.J. Baran, *J. Inorg. Biochem.*, 1996, **63**, 285 (*Chem. Abstr.*, 1996, **125**, 276 388).
45   A. Seedhara, C.P. Rao and B.J. Rao, *Carbohydr. Res.*, 1996, **289**, 39.
46   R.P. Bandwar and C.P. Rao, *Carbohydr. Res.*, 1996, **287**, 157.
47   R.P. Bandwar, M. Giralt, J. Hidalgo and C.P. Rao, *Carbohydr. Res.*, 1996, **284**, 73.
48   K. Oshima, H. Toi and Y. Aoyama, *Carbohydr. Lett.*, 1995, **1**, 223 (*Chem. Abstr.*, 1996, **125**, 114 962).
49   K.R.A.S. Sandanayake, T.D. James and S. Shinkai, *Pure Appl. Chem.*, 1996, **68**, 1207, (*Chem. Abstr.*, 1996, **125**, 248 248).
50   J.C. Norrild and H. Eggert, *J. Chem. Soc., Perkin Trans. 2*, 1996, 2583.
51   M. Takeuchi, Y. Chin, T. Imara and S. Shinkai, *J. Chem. Soc., Chem. Commun.*, 1996, 1867.
52   P. Wentworth, Jr., T. Wiemann and K.D. Janda, *J. Am. Chem. Soc.*, 1996, **118**, 12521.
53   P. Wentworth, Jr. and K. D. Janda, *J. Chem. Soc., Chem. Commun.*, 1996, 2097.
54   H. Lammers, H. van Bekkum and J.A. Peters, *Carbohydr. Res.*, 1996, **284**, 159.
55   G. Micera and M. Kozlowski, *Handb. Met.-Ligand Interact. Biol. Fluids: Bioinorg. Chem.* 1995, **1**, 707 (*Chem. Abstr.*, 1996, **124**, 343 809).
56   M. Jezowska-Bojczuk, S. Lamotte and T. Trnka, *J. Inorg. Biochem.*, 1996, **61**, 213 (*Chem. Abstr.*, 1996, **124**, 317 689).

18
Alditols and Cyclitols

1 Alditols and Derivatives

1.1 Alditols – A review on homogenous catalysts used in the hydrogenation of aldoses with emphasis on mechanistic aspects of their activity and modes of their deactivation has appeared,[1] and the hydrogenation of D-fructose to D-mannitol has been described using ruthenium molecular sieves modified with chiral ligands as catalysts.[2]

Density measurements of mannitol in potassium or sodium acetate solutions have revealed the nature of solute-solvent interactions in these mixtures[3] and the kinetics of oxidation of several hexitols (and cyclitols) by molecular oxygen, catalysed by copper compounds, has been studied.[4]

The syntheses of all eight 4,6,7,8,9-pentahydroxy-1,2,3,5-tetradeoxy-D-non-1-enitols, their conversion into 9-O-anthroyl-4,6,7,8-tetra-O-p-methoxycinnamoate derivatives and their subsequent use in determining the absolute stereochemistry of polyols by bichromic exciton coupled circular dichroism have been described. It was found that extension of lower homologues bearing four contiguous OH groups gave a major difference in spectra when there was a 1,3-*anti*-extension compared with a 1,3-*syn*-extension.[5]

A study has been conducted on the effects that sugar alditols and aminoalditols have on the rate of phosphate ester hydrolysis in the substrates bis(4-nitrophenyl)phosphate and supercoiled DNA, when catalysed by lanthanide cations.[6] See Chapter 23 for the effect of solvents on the relative stabilities of the complexes formed between alditols and lanthanide cations.

A hydrogenase from *Alcaligenes entrophus* coupled with a xylulose reductase from a yeast, have been used to produce xylitol from D-xylulose.[7]

The preparation of the D-*threo* compound **1**, together with the corresponding D-*erythro* derivative, from tartaric acid has been described. These compounds represent the chiral aryloxy C(8)-O-C(4′) ether linkages present in a major class of neolignans and are structurally related to the major interunit linkage of lignin. The syntheses of such compounds may help shed some light on why small molecule neolignans are optically active, whilst polymeric lignin is optically inactive.[8]

The reaction of 1,2:5,6-di-O-isopropylidene-D-mannitol with the metal oxo complex, $(C_5Me_5)ReO_3$, in the presence of triphenylphosphine effected a deoxydehydration reaction, and gave (E)-**2**. The same reaction applied to erythritol gave only 1,3-butadiene.[9]

Carbohydrate Chemistry, Volume 30
© The Royal Society of Chemistry, 1998

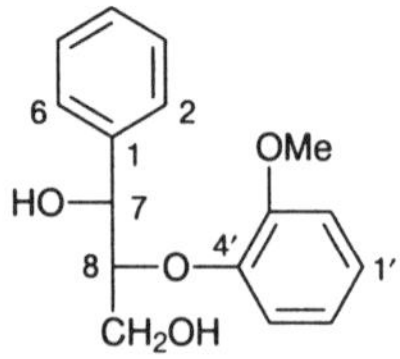

1 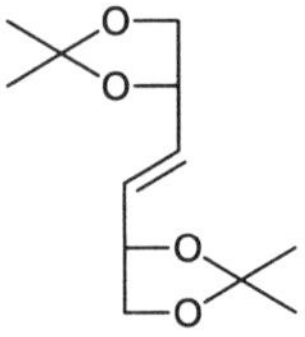   **2** 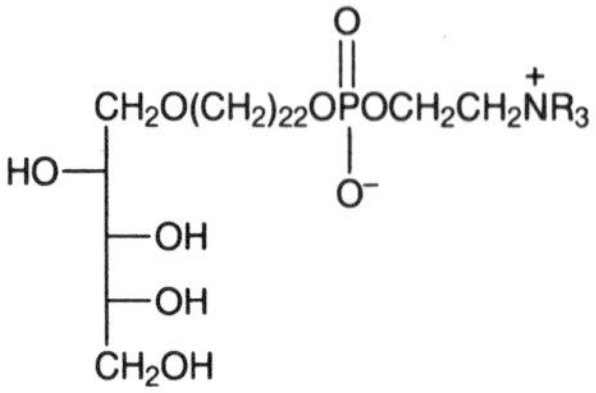   **3** R = Me or H

The preparations of the phosphate derivatives **3** and several analogues as bipolar phospholipids have been reported.[10]

A 'one pot' transformation of a tosylated 2,3-epoxy alcohol into an allylic alcohol is illustrated in Scheme 1. Several similar examples were also described. In cases involving substrates containing acid sensitive groups, 4-(dimethylamino-phenyl)diphenylphosphine was used in place of triphenylphosphine.[11]

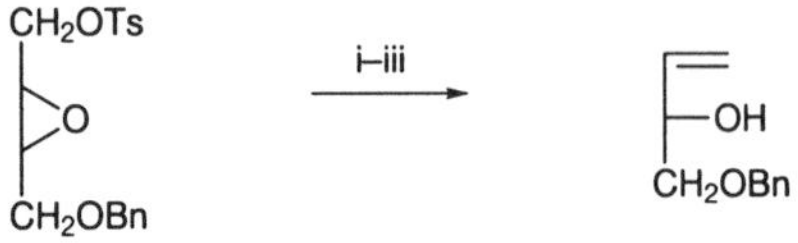

Reagents: i, KI (3 eq.), DMF; ii, Ph$_3$P (1 eq.); iii, I$_2$, 10 mol %

Scheme 1

In their reaction with benzyl chloride-potassium hydroxide-dimethylsulfoxide, 1,3:2,4-di-*O*-ethylidene-D-glucitol and 1,2-*O*-isopropylidene-3-*O*-methyl-α-D-glu-cofuranose showed a marked preference for alkylation at the secondary alcohol group relative to the primary one.[12]

Esterification of xylitol with oleic acid has been reported to take place at a lower temperature and with the formation of fewer side products when zinc oxide instead of sodium hydroxide was used as catalyst.[13]

The addition of methyllithium or methylmagnesium chloride to 2,3:4,6-di-*O*-isopropylidene-D-mannopyranose yielded a single D-*glycero*-D-*manno*-heptitol, whereas the same reaction with 2,3:5,6-di-*O*-isopropylidene-D-mannofuranose produced two diastereomers. The diastereoselectivity in the former case was attributed to a large ring chelate between the O-5, the aldehyde group and the metal cation.[14]

All D-pentoses and six D-hexoses have been transformed into their corresponding 1-deoxy-1-nitroalditols in a 'one pot' reaction involving a strongly basic anion exchange resin in the bicarbonate form as catalyst.[15]

Amidothiophosphorylation of 2,5-*O*-methylene-D-mannitol afforded **4** as a mixture of three diastereomers, one of which was characterized by X-ray crystallography.[16]

Reaction of 5-*O*-benzyl-2,3-*O*-isopropylidene ribonolactone with diiodomethane-samarium diiodide gave a low yield of cyclopropane derivative **5**.[17]

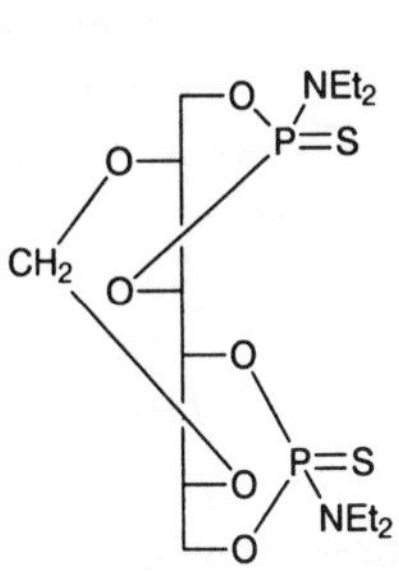

4

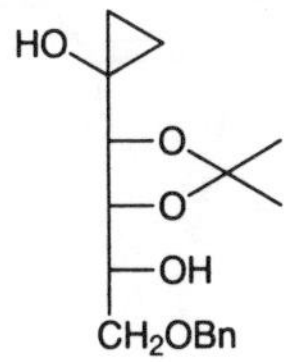

5

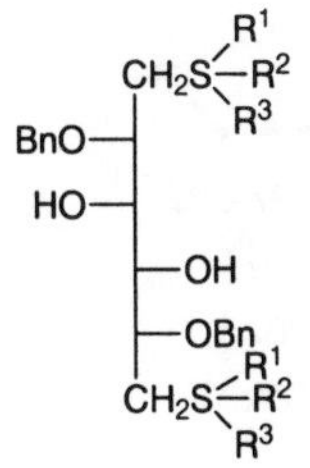

6 R¹—R² = O, R³ = Ph or Bn
7 R¹ = R² = O, R³ = Ph or Bn

Sulfoxide **6** has been prepared by treating 1,2:5,6-dianhydro-3,4-*O*-isopropylidene-D-mannitol successively with the appropriate thiol-potassium carbonate, benzyl bromide-sodium hydride, MCPBA then aqueous acetic acid. Substitution of MCPBA by meta-periodic acid led to the formation of sulfone **7**. The compounds were tested as potential inhibitors of HIV-1 protease.[18]

The diastereoselective reduction of hemiacetals derived from isopropylidene derivatives of carbohydrate lactones is depicted in Scheme 2. Diastereomeric ratios ranged from 86:14 to 100:1 (major epimer only shown in the Scheme).[19]

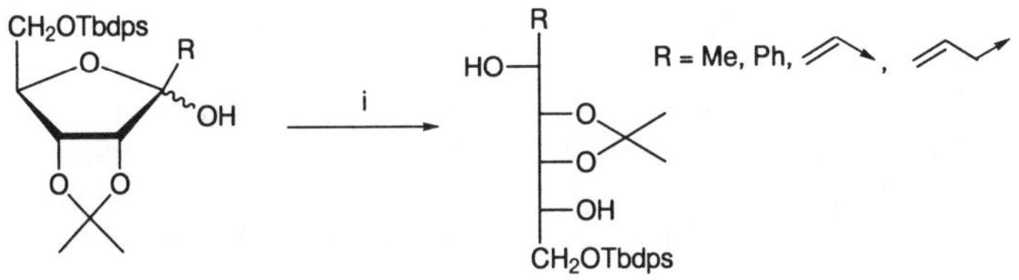

R = Me, Ph,

Reagents: i, DiBAL

Scheme 2

A route to 3,5-dideoxy heptitols is depicted in Scheme 3 with the preparation of 3,5-dideoxy-*meso*-xyloheptitol which represents the C-16 to C-22 and C-18 to C-24 fragments found in the macrolide antibiotics roxaticin and mycoticin A (and B), respectively. Other related studies were also performed.[20]

L-Mannitol has been prepared in high yield by reduction of L-*manno*-1,4-lactone using a new Cu-Cr oxide catalyst. It was further converted into the enantiomer of the pyrrolidine described in Vol. 25, p. 334, ref. 108 and used as a chiral auxiliary.[21]

The synthesis of 3,4-*O*-isopropylidene-L- and -D-erythrulose from L-ascorbic- and D-isoascorbic acids, respectively, proceeding through the corresponding protected tetritols, is covered in Chapter 2. The syntheses of natural and modified trapoxins and (+)-lanomycin by way of suitably protected tetritol derivatives, and

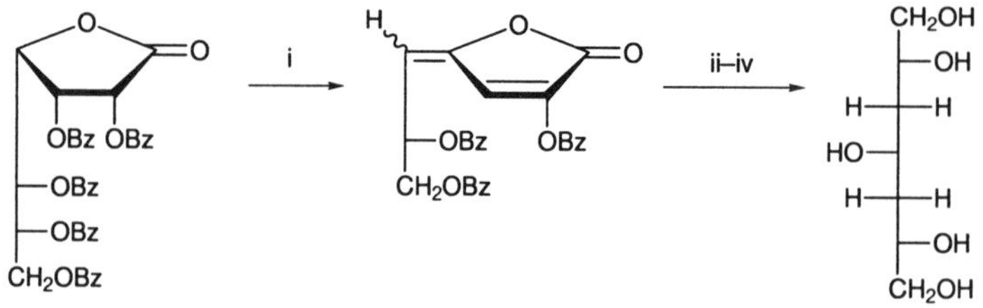

Reagents: i, Et$_3$N, CHCl$_3$, 20 °C, 45 min; ii, H$_2$, Pd/C; iii, NaOMe/MeOH; iv, NaBH$_4$

Scheme 3

the use of 1,2:5,6-di-*O*-cyclohexylidene-D-mannitol as a covalently binding aux-illiary for the low valent McMurry reaction of acetophenone are mentioned in Chapter 24.

1.2   Anhydro-alditols – The preparations of polyesters based on 1,4:3,6-di-anhydro-D-glucitol and -D-mannitol with succinyl-, glutaryl- and adipoyl-dichlor-ides have been reported.[22] Also reported have been the syntheses of poly(1→6)-2,5-anhydro-D-glucitol and its 3,4-di-*O*-alkylated derivatives by; Lewis acid induced polymerization of 3,4-di-*O*-alkyl-1,2:5,6-dianhydro-D-mannitol;[23] base induced polymerization of 1,2:5,6-dianhydro-3,4-di-*O*-methyl- or 3,4-di-*O*-iso-propylidene-D-mannitol;[24] base induced polymerization of 1,2:5,6-dianhydro-3,4-di-*O*-pentyl- or di-*O*-decyl-D-mannitol;[25] and base induced polymerization of 1,2:5,6-dianhydro-3,4-di-*O*-methyl-L-iditol.[26] The solubilities of these 3,4-di-*O*-alkylated and deprotected polymers in aqueous and organic media were studied. Further, it has been reported that 2,5-anhydro-3,4-di-*O*-methyl-D-glucitol linked 1,6 to silica gel was useful for the optical resolution of amines and amino acid salts.[27]

The synthesis of the anhydroglucitol **8** and its 1,2[^{3}H]-labelled analogue by addition of hydrogen or tritium to a protected terminal alkene followed by treatment with MCPBA has been described as an affinity label for *E. coli* derived glucosamine 6-phosphate synthase.[28]

Several *O*-benzylated glycal derivatives have been hydroborated and oxida-tively worked-up to afford anhydroalditols in which the newly introduced 2-OH group is formed *anti-* to the 3-*O*-benzyl substituent. For example, tri-*O*-benzyl-D-glucal gave **9**.[29]

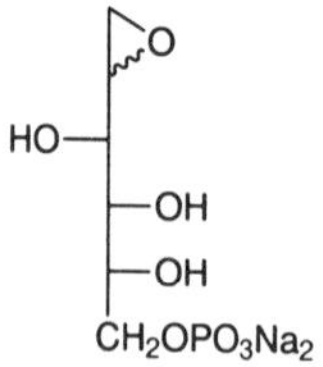

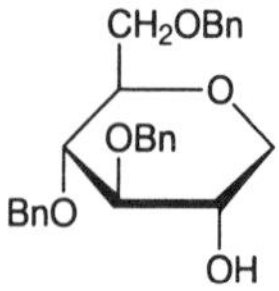

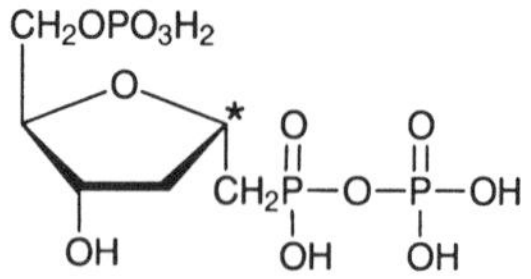

The anhydropentitol phosphonate derivative **10** has been synthesized, together with the epimer at the indicated carbon as a 2'-deoxy analogue of 5-phosphorylribose 1-α-diphosphate and found to be a competitive inhibitor of yeast orotate phophoribosyltransferase.[30]

Per-acylated pyranoid and furanoid glycosyl bromides were easily reduced in high yield to the corresponding anhydroalditols by catalytic quantities of titanocene borohydride (prepared *in situ* from Cp_2TiCl_2 and $NaBH_4$).[31]

The reaction of methyl 2,3,4,6-tetra-*O*-benzyl-α-D-glucopyranoside or -manno-pyranoside with triethylsilane-Tms triflate gave the corresponding 1,5-anhydro-hexitols, which can be viewed as hydroxymethyl-*C*-glycosides of β-L-xylopyranose and α-D-arabinopyranose, respectively.[32]

Wittig methylenation of 3,5-di-*O*-benzyl-2-deoxy-D-erythrofuranose then reaction with MCPBA-boron trifluoride etherate afforded 2,5-anhydro-4,6-di-*O*-benzyl-3-deoxy-D-ribohexitol. Mesylation of the latter followed by addition of the anion derived from adenine, furnished derivative **11**. Cytidine and thymidine could be treated in a similar manner and de-*O*-benzylated by standard methods. Structures were verified by X-ray analysis.[33]

Addition of the anion derived from uracil to 1,5:2,3-dianhydro-4,6-*O*-benzylidene-D-allohexitol, followed by inversion of configuration of the resulting 3-OH group by displace,ment of its mesylate derivative with sodium hydroxide, afforded **12**. Compound **12** was a useful intermediate for making 1,5-anhydro-2-deoxy-D-mannitol-containing pyrimidine nucleoside analogues.[34]

The anhydroalditol **13** has been prepared in several steps from L-glucose, the anhydride being formed by tributyltin hydride reduction of the corresponding glycosyl bromide. Compound **13** was further transformed into the nucleoside analogues **14** (nucleobase = Ura, Thy, Gua, Ade). The enantiomer of **13**, prepared from D-glucose by an improved procedure to that previously reported (see Vol. 27, p. 269, ref. 250 and 251), was similarly transformed into its 7- and 8-deazaguanine analogues.[35]

The reaction of 1-amino-1-deoxy-D-arabinitol with anhydrous hydrofluoric acid-formic acid afforded the anhydroalditol **15**. Likewise, 1-amino-1-deoxy-D-glucitol was converted into **16**. Other 1-amino-1-deoxy-pentitols and -hexitols reacted similarly.[36]

It has previously been reported that 2,5-anhydro-3,6-di-*O*-mesyl-L-idose ethylene acetal (see Vol. 29, p. 229, ref. 40), prepared from D-glucose, was converted into (+)-*epiallo*-muscarine. It has now been disclosed that the same intermediate has been converted into (−)-*allo*-muscarine which is the C-4 epimer of (+)-*epiallo*-muscarine.[37]

Pentitols, for example D-arabinitol, reacted with thionyl chloride in pyridine at −30 °C and gave the corresponding 1,2:4,5-di-*O*-cyclic sulfites which reacted further with sodium azide in dimethyformamide affording a small quantity of anhydro derivative **17** together with the expected 1,5-diazido-dideoxy-D-arabinitol as major product.[38]

The use of dianhydro sugars derived from D-glucose as novel chiral auxilliaries is covered in Chapter 24.

1.3   Amino- and Imino-alditols – Reviews on the syntheses of azasugars (iminoalditols) from non-carbohydrate sources;[39] on the syntheses of azasugars, sugar amidines and related structures as inhibitors of carbohydrate-processing enzymes;[40] on the syntheses of glycosidase inhibitors with reference to nagstatin;[41] and on the use of aldolases in the chemo-enzymatic syntheses of iminoalditols[42,43] have appeared.

1.3.1 Acyclic Derivatives – 1,5-Diazido-1,5-dideoxy-D-pentitols are readily reduced to the corresponding 1,5-amino-1,5-dideoxy derivatives. (See above for the preparation of the diazido compounds).[38]

The reductive amination of monosaccharides with 8-aminopyrene-1,3,6-trisulfate in a variety of organic acids, which act as catalysts, has been studied by capillary electrophoresis with laser-induced fluorescence detection.[44]

Scheme 4 shows the preparation of a pyrrolidine-based alditol by addition of a

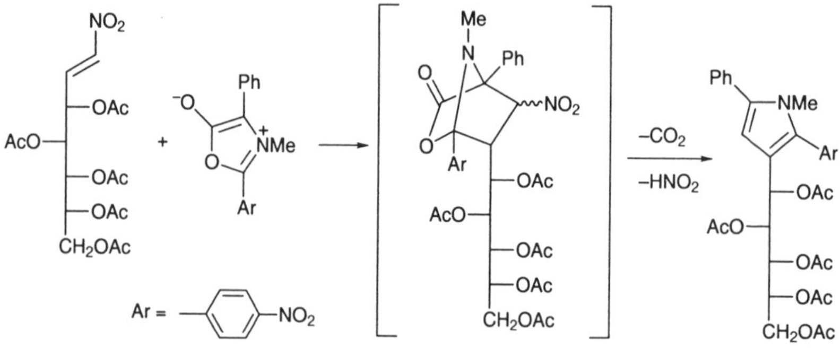

Scheme 4

münchnone to an unsaturated nitro derivative. The regioselectivity shown in the reaction is opposite to that expected from the FMO calculations.[45]

The thiazole-alditol derivative **18** has been synthesized by an asymmetric dihydroxylation of a diene-substituted thiazole derivative.[46]

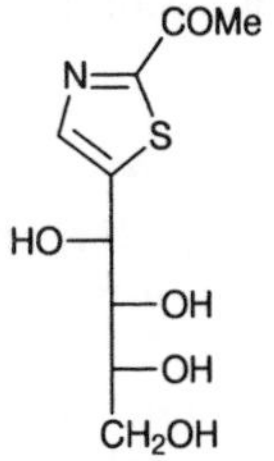

18 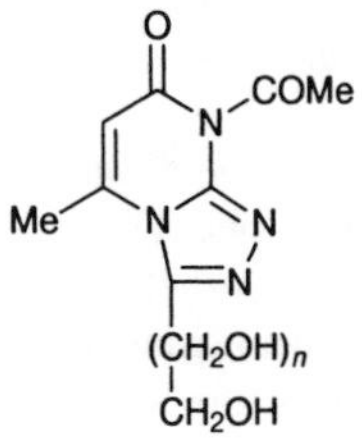   **19**

Condensation of D- and L-arabinose, D-ribose, D-xylose, D-galactose, D-glucose and D-mannose with 2-hydrazino-6-methyl-4-oxopyrimidine afforded intermediate sugar (6-methyl-4-oxo-2-pyrimidinyl)hydrazones. Treatment of the latter in 'one pot' with bromine-acetic acid-sodium acetate then acetic anhydride (to effect a cyclization-acetylation reaction) gave, after purification and de-*O*-acetylation, derivatives **19** (n = 3 or 4, stereochemistry as defined by the starting aldose).[47]

The sodium borohydride reduction of 2-amino-2,3 (and 2,6)-dideoxy-D-glucopyranose-3 (and 6)-sulfonic acids afforded 2-amino-2,3-dideoxy-D-glucitol-3-sulfonic acid and 2-amino-2,6-dideoxy-D-glucitol-6-sulfonic acid, respectively.[48]

Differentially *O*-benzylated derivatives of 1,6-diamino-1,6-dideoxy-D-mannitol or -D-talitol have been prepared from a common precursor, 1,6-diazido-1,6-dideoxy-3,4-*O*-isopropylidene-D-mannitol as potential HIV-1 protease inhibitors.[18]

The aminotetritol compound **20** has been synthesized from ethyl 3,4-*O*-isopropylidene-D-erythronate (available from D-isoascorbic acid according to Vol. 26, p. 190, ref. 22) by azide displacement of the 2-*O*-triflate followed by a reduction-de-*O*-isopropylidenation-reprotection sequence, as an intermediate for the preparation of D-*threo*-C-18-sphingosine.[49]

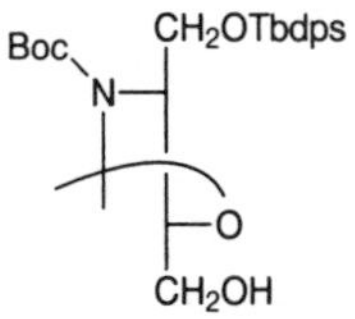

20

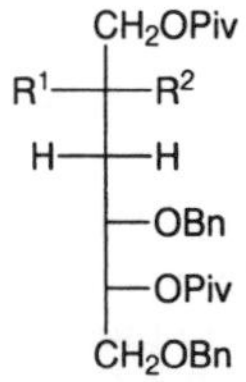

21 R^1 = H, R^2 = Ura, Ade, Thy.
22 R^1 = OH, R^2 = H

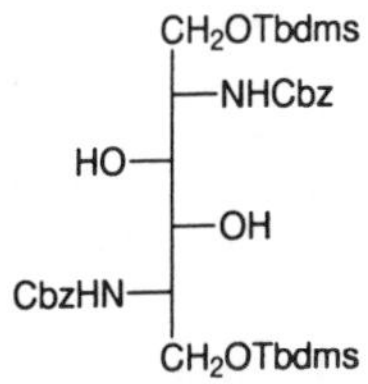

23

The nucleoside analogues **21** have been prepared by Mitsunobu reaction of alcohol **22** followed by deprotection. Several related examples were also described; no antiviral activity was noted.[50]

The aminoalditol **23** was prepared by a pinacol-like homocoupling of an appropriately protected L-serine-derived aldehyde and promoted by a vanadium-zinc complex $[V_2Cl_3(THF)_6)]_2[Zn_2Cl_6]$.[51] (See Section 2.2 for the further transformation of compound **23**).

1.3.2 Cyclic Imino Compounds – The phosphoramidite derivative **24** of a previously described phenyliminoribitol (see Vol. 27, p. 211, ref. 68) has been incorporated into a synthetic oligonucleotide chain (10 mer) and the inhibitory activity of the product against ricin toxin A-chain assessed.[52]

The preparation of the 'C-azanucleosides' **25** in a straightforward manner by the addition of aryllithium to 5-O-Tbdms-2,3-isopropylidene-D-ribofuranose followed by Swern oxidation, reductive amination and acid-catalysed hydrolysis has been reported.[53]

Acid-catalysed hydrolysis, followed by hydrogenation of the known methyl 4-azido-4,6-dideoxy-2,3-O-isopropylidene-α-L-mannopyranoside gave the 1,4-imino alditol **26**, which had L-rhamnosidase inhibitory activity. The corresponding lactam was also made but was devoid of glycosidase activity.[54]

The branched-chain compound **27** has been synthesized from **28** (available from D-mannose) by reductive amination then acid-catalysed hydrolysis. Compounds **27** thus prepared together with related compounds were tested for their inhibition against human blood purine nucleoside phosphorylase and α- and β-glucosidases.[55]

The aza-thymidine derivative **29** has been prepared and incorporated into oligodeoxyribonucleotides in an effort to resist nuclease activity,[56] and racemic nucleoside analogues **30**[57] and **31**[58] have been prepared as potential anti-AIDS agents.

The use of D-ribonolactone derived pyrrolidines as chiral catalysts for the addition of diethylzinc to aldehydes is mentioned in Chapter 24 and the synthesis of the pyrrolidine based antibiotic (+)-preussin is covered in Chapter 19.

Heating   N-benzyl-1,4-dideoxy-2,3-O-isopropylidene-5-O-mesyl-1,4-imino-L-lyxitol, prepared in several steps from a known D-ribofuranose derivative, produced intermediate **32** which reacted with various nucleophiles giving 1,5-imino-D-

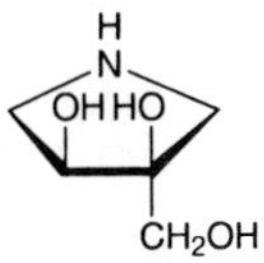

27 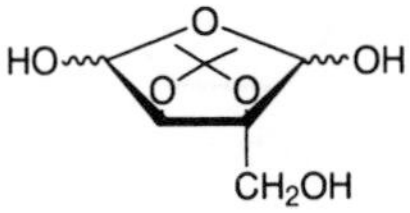   28 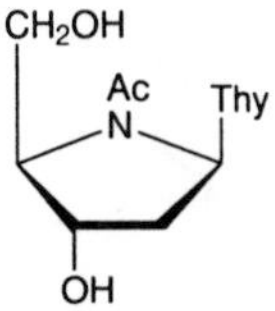   29

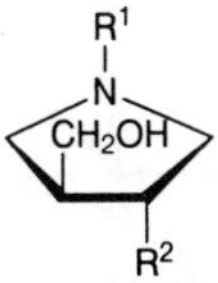

30 R^1 = various 5-substituted Ura, R^2 = OH,F,N_3
31 R^1 = various 5-substituted Cyt, R^2 = OH, F, N_3

ribitol derivatives **33** as major products, after deprotection.[59] The syntheses of 1,2,5-trideoxy-2-fluoro-1,5-imino-D-glucitol, as an analogue of 1-deoxynojirimycin (DNJ), and of the 2,5-imino-D-mannitol derivative **34** have been described and their glucosidase activity compared with that of the parent compounds.[60]

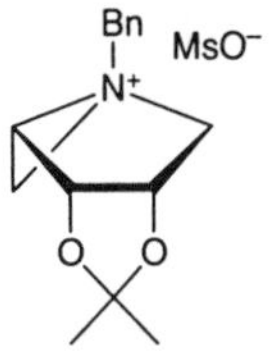

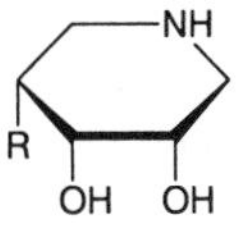

33 R = NH$_2$, F, OH

32

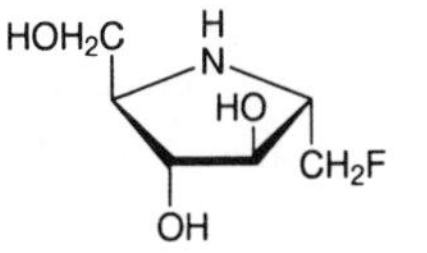

34 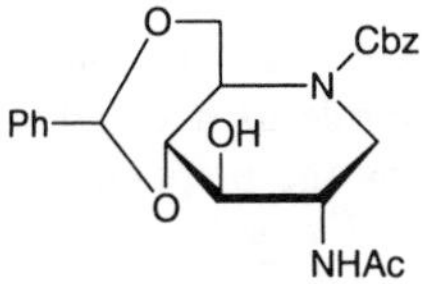   35

The 2-acetamido derivative of DNJ (**35**) has been prepared by reaction of azide followed by reduction and acetylation of the corresponding 2,3-anhydro-D-mannose derivative. Compound **35** was incorporated into oligosaccharide mimics of sialyl Le^a and sialyl Le^x.[61]

1,5-Dideoxy-1,5-imino-2,3-O-isopropylidene-D-ribitol has been prepared from 5-azido-1-O-benzoyl-5-deoxy-2,3-O-isopropylidene-D-ribofuranose by reaction with sodium methoxide then hydrogen. It was further transformed into derivatives **36** from which it was found that the compound with R^1 = H, R^2 = Me was a good α-fucosidase inhibitor.[62] The same compound has also been prepared independently from benzyl 2-deoxy-3,4-O-isopropylidene-2-C-methylene-β-L-*erythro*-pyranoside in six straightforward steps.[63]

36 R^1 = H, R^2 = Me
    R^1 = H, R^2 = OMe
    R^1 = alkyl, R^2 = Me or OH

37

38

39 R^1 = OH, R^2 = H, R^3 = NH—C(=NH)—NH$_2$
40 R^1 = OH, R^2 = H, R^3 = NH$_2$
41 R^1 = H, R^2 = OH, R^3 = NH$_2$

Reductive amination of 5-azido-2,3-di-*O*-benzyl-5-deoxy-D-arabinofuranose afforded 2,3-di-*O*-benzyl-1,5-dideoxy-1,5-imino-D-arabinitol from which the hydroxymethyl derivative **37** has been prepared (*via* hydroboration of the 4-*C*-methylene analogue) together with its C-4 epimer as a minor constituent. The hydroxymethyl group in **37** was further oxidized to a carboxylic acid function and the resulting product was found to be a good inhibitor of β-glucosidase.[64]

The homo proline mimic **38** has been synthesized by standard means and incorporated into peptides. Conformational aspects of the products were determined by ^{1}H NMR and molecular dynamics simulation.[65]

The L-iduronic acid-like 1-*N*-imino sugars **39-41** have been prepared from a siastatin B derivative (See Vol. 27, p. 237, ref. 96) as metastasis inhibitors.[66] The synthesis and glycosidase activity of other siastatin B analogues are mentioned in Chapter 19.

Further examples on the use of 2,3:5,6-di-*O*-isopropylidene-1-*C*-nitroso-D-mannofuranosyl chloride as a chiral dienophile for Diels-Alder reactions has been demonstrated by the preparation of 1,5-imino-1,5,6-trideoxy-D-allitol, -D-gulitol and -D-glucitol.[67] (See Vol. 29, p. 236, ref. 86 for other examples).

A conventional route has been reported from diacetone glucose to 3-*O*-benzyl-1,5-dideoxy-1,5-imino-D-glucitol and subsequently to 2-amino-4,6-di-*O*-benzoyl-3-*O*-benzyl-1,2,5-trideoxy-1,5-imino-D-mannitol which proceeded by way of an azide substitution reaction at the C-2 position.[68] A straightforward synthesis of 1,5-dideoxy-1,5-imino-L-glucitol together with the corresponding trihydroxypipecolic acid has been reported in which the piperidine ring was formed by an intramolecular displacement of a 6-*O*-mesyl group by the 2-amino group of a protected D-glucosamic acid derivative.[69]

A short synthesis of 3,4,6-tri-*O*-benzyl-1,5-dideoxy-1,5-imino-D-mannitol has been achieved by acid-catalysed hydrolysis, followed by controlled reductive amination of a per-benzylated 1,6′-diazido-1,6′-dideoxy-sucrose derivative. Further transformation by way of an azide displacement of a 2-*O*-triflyl derivative afforded ultimately 2-acetamido-1,2,5-trideoxy-1,5-imino-D-glucitol (2-acetamido-1,2-dideoxynojirimycin).[70]

The syntheses of 1,5,6-trideoxy-1,5-diimino-D-gulitol (5-*epi*-L-rhamnojiri-mycin), 2,6,7-tri-deoxy-2,6-imino-L-*glycero*-L-*galacto*-heptitol (β-homo-L-rham-nojirimycin), 2,6,7-trideoxy-2,6-imino-L-*glycero*-L-*manno*-heptitol (α-L-rham-nojirimycin) and of the tetrazole **42** have all been achieved from known 1,4- or 1,5-lactone derivatives as potential inhibitors of L-rhamnosidases.[71]

The DNJ analogue **43** has been prepared conveniently from a castanospermine derivative.[72]

Reduction of 6-azido-6-deoxy-D-galactose *N,N'*-diphenylformazan with triphe-nylphosphine afforded the 1,5-imino-D-lyxitol compound **44**.[73]

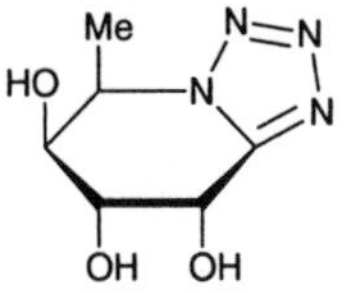

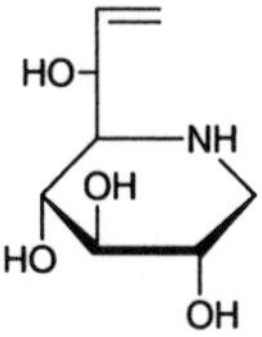

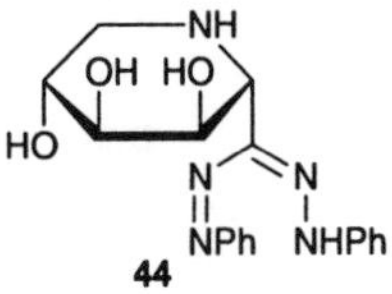

An interesting synthesis of iminoheptitols by a tandem Wittig [1+3] cycloaddi-tion reaction is depicted in Scheme 5 with the preparation of a D-*talo* derivative. The paper also described the preparation of the L-*allo*-isomer in a similar way starting from the diazo-ester intermediate shown.[74]

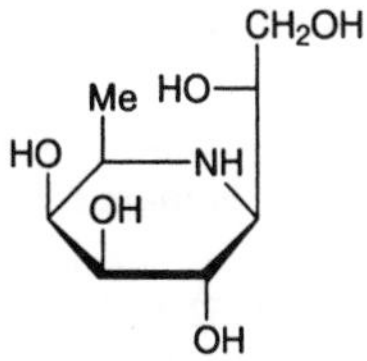

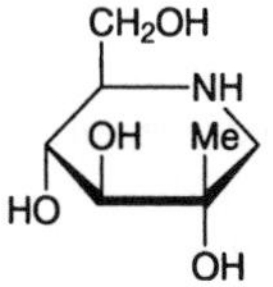

The readily available 2,3:6,7-di-*O*-isopropylidene-D-*glycero*-D-*gulo*-heptano-1,4-lactone has been converted into the iminooctitol **45** as a potential galacto-sidase inhibitor, but in practice was found to be a good β-glucosidase inhibitor.[75]

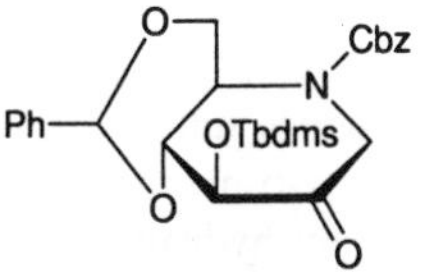

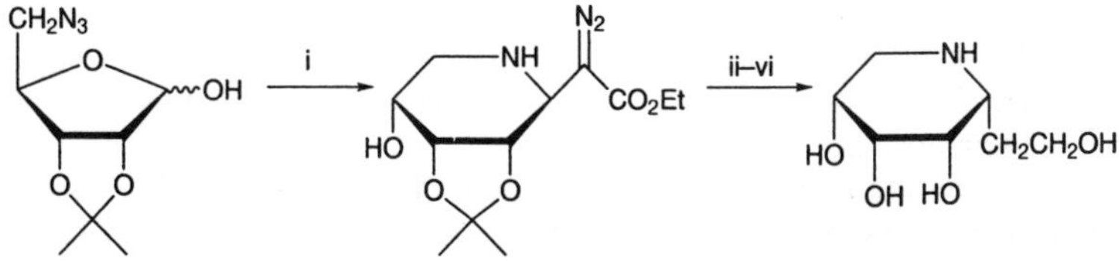

Reagents: i, Ph₃P═CHCO₂Et; ii, 90–100 °C; iii, H₂, Pd/C; iv, (Boc)₂O; v, LAH; vi, H₃O⁺

Scheme 5

The 2-*C*-methyl-branched iminoalditol **46** (together with the C-2 epimer) has been prepared by addition of methylmagnesium bromide to ketone **47**, followed by hydrolysis. The stereoselectivity of the Grignard addition was very dependent on the nature of the ether group adjacent to the ketone.[76]

Further examples have been reported of the addition of nucleophiles to the 1,2:5,6-diimino-L-iditol derivative **48** resulting in 6-amino-2,5-imino-D-glucitol or 6-amino-1,5-imino-D-mannitol derivatives depending on whether the primary or secondary carbon of the 1,2- or 5,6-imino group in **48** was attacked. The same compound, **48**, also served as starting material for 2,5-dideoxy-2,5-imino-D-glucitol.[77] (See Vol. 23, p. 231, refs. 48 and 49).

48

49 R^1 = H, R^2 = OH
50 R^1 = OH, R^2 = H

51 R^1 = OH, R^2 = H
52 R^1 = H, R^2 = OH

The additions of excess allylamine in the presence of perchloric acid to 1,2:5,6-dianhydro-3,4-di-*O*-allyl-L-iditol or -D-mannitol afforded the seven membered ring compounds **49** and **50**, respectively. The reaction proceeded *via* a preferred 7-*endo*-tet cyclization. Minor amounts of the piperidines formed by the 6-*exo*-tet ring closure were also produced. Some of the derivatives were very good glycosidase inhibitors.[78] The related azepane **51** has also been prepared from benzyl α-D-mannopyranoside by 6-*O*-tosylation, displacement with azide followed by hydrogenation-hydrogenolysis, and the preparation of **52** utilized an aldolase-catalysed reaction between ($\pm$)-3-azido-2-hydroxypropanaldehyde and dihydroxyacetonephosphate to produce 6-azido-6-deoxy-D-glucopyranose which was hydrogenated.[79]

Addition of tryptamine, catalysed by perchloric acid, to 1,2:5,6-dianhydro-3,4-di-*O*-benzyl-D-mannitol produced a mixture of the six-membered ring derivative **53** together with the corresponding seven-membered ring compound. Further sequential reaction of **53** with (Boc)$_2$O, TbdmsCl, MSCl-Et$_3$N, H$_2$N(CH$_2$)$_6$NHBoc followed by HCl gave the ring-contracted pyrrolidine **54**. A similar ring contraction occurred to give six-membered rings from seven-membered ones. The products were designed as non-peptide mimics of the cycloneuropeptide somatostatin.[80]

Some interesting interconversions have been observed between five-, six- and seven-membered rings. For example, conversion of the L-*gulo*-piperidine derivative **55** to its dimesylate followed by reaction with caesium acetate gave the ring contracted pyrrolidine **56** as the main product together with the ring-expanded derivative **57**. On the other hand the expected inversion at C-2 of **55** takes place under Mitsunobu conditions to afford the corresponding L-*ido* form of **55**.[81] Similarly, reaction of the seven-membered ring derivative **57** with mesyl chloride-triethylamine gave the piperidine chloride **59**, whilst reaction of **57** under Mitsunobu conditions with benzoic acid gave a mixture of **58** and **60**.[82]

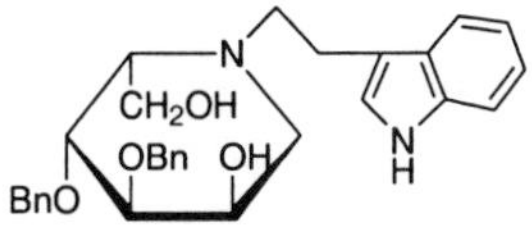

53

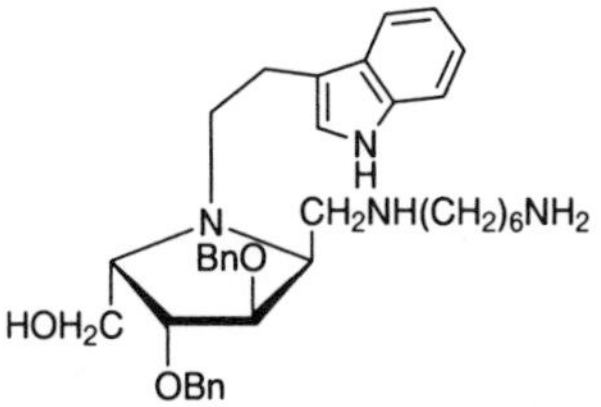

54

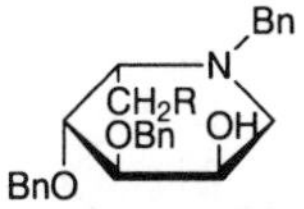

55 R = OH
59 R = Cl
60 R = OBz

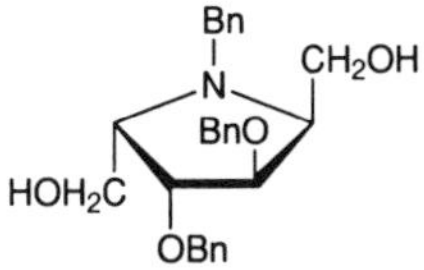

56

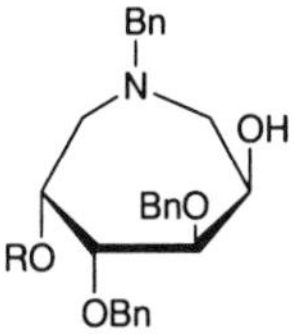

57 R = H
58 R = Bz

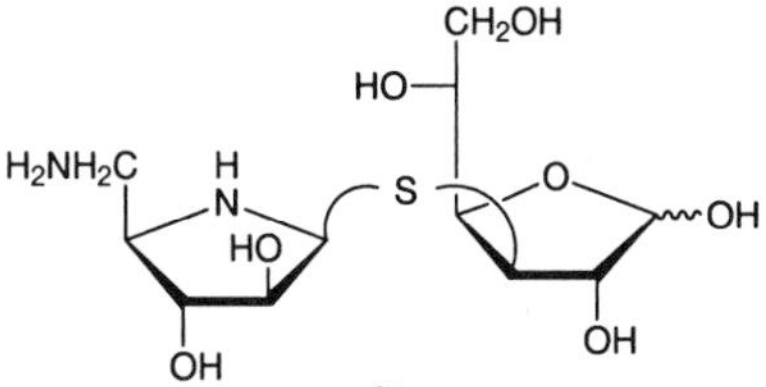

61

In the area of di- or tri-saccharides containing an iminoalditol group as pyranose or furanose mimics, it has been reported that 3-*O*- and 4-*O*-β-D-galactopyranosyl-DNJ can be formed by a β-galactosidase-catalysed transglycosylation reaction from lactose to DNJ.[83] Reaction of the anion derived from 1,2:5,6-di-*O*-isopropylidene-3-thio-α-D-glucofuranose with the diiminoalditol **48**, followed by deprotection, gave the pseudo disaccharide **61** as a potential glycosidase inhibitor.[84]

N,N'-Diacetylchitobiose has been converted into the pseudo disaccharide, β-D-GlcNAc*p*-(1→4)-2-acetamido-1,2,5-trideoxy-1,5-imino-D-glucitol by way of intermediate **62**. This was oxidized and reduced to give the 5-OH epimer of **62**, converted to a mesylate, reduced to the 1-amino compound (which effects cyclization) with displacement of mesic acid and finally de-benzylated.[85,86]

The β-homonojirimycin derivative **63**, has been made by reductive amination of the corresponding 2,6-diketone. Compound **63**, after deprotection, was converted into the 'homoaza methylcellobioside' **64**.[87]

The aza *C*-glycosyl compound **65** was the product from the samarium diiodide coupling of 1,2:3,4-di-*O*-isopropylidene-α-D-galactodialdose with the iodo derivative **66**.[88]

The pseudo trisaccharide **67** has been prepared by coupling of 1,3,4,5-tetra-*O*-

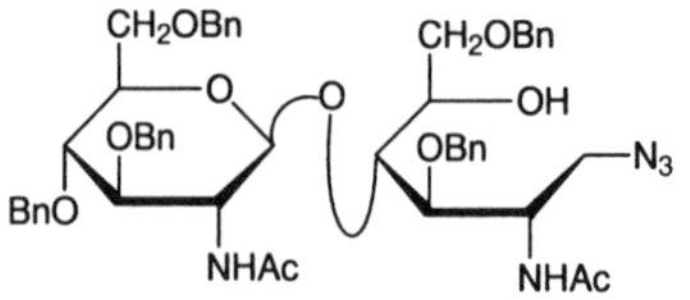

62

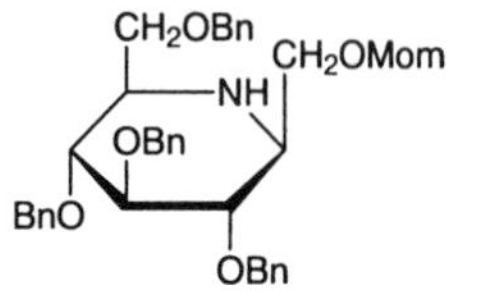

63

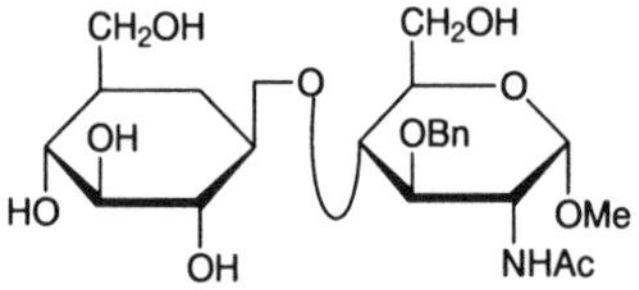

64

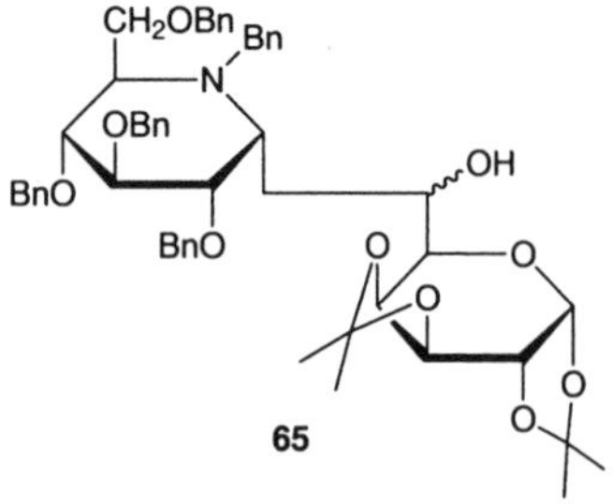

65

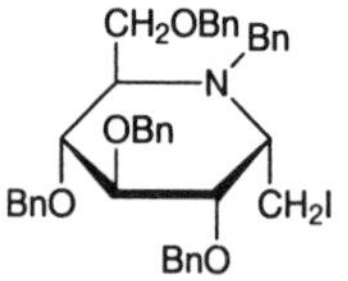

66

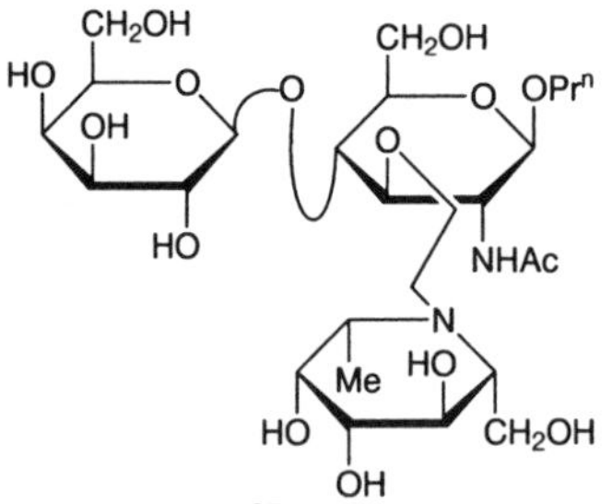

67

benzyl-2,6,7-trideoxy-2,6-imino-L-galactitol (a β-L-homofuconojirimycin derivative) with a 3′-hydroxyethyl ether lactosamine derivative, followed by deprotection.[89]

Several iminoalditols have been prepared from non-carbohydrate sources.
Thus 1,4-dideoxy-1,4-imino-D-lyxitol and 1,4,5-trideoxy-1,5-imino-D-lyxitol have
been made *via* a 'Meyers bicyclic lactam' derived from cyclocondensation of (*S*)-
phenylglycinol with the keto-acid derived from Jones oxidation of 2-benzyloxy-
methyl-4,5-dihydrofuran,[90] and a protected L-serine derivative,[91] respectively. 1-
Deoxymannojirimycin has been prepared from a chiral α-furfurylamine derivative[92] and 1,3,4,5-tetradeoxy-3-fluoro-1,5-imino-D-xylitol, together with its C-5
epimer, have been prepared from the fluorosulfinylhexenol, **68**.[93] Periodate
oxidation of 1-D-*myo*-inositol 1,2,6-trisphosphate (α-trinositol) followed by reductive amination with a large range of amines afforded *N*-alkylated D-*arabino*-
iminoalditols **69**.[94] A synthesis of calystegin B2 from a reduced tropanone-iron
carbonyl complex,[95] and the isolation of seven calystegins, including two novel
ones, calystegin A₆ (**70**) and calystegin N₁ (**71**), from the plant *Hyoscyamus niger*,
have been reported.[96]

68 **69** R = various alkyl groups

70 $R^1 = R^2 = H$, $R^3 = R^4 = OH$
71 $R^1 = R^2 = OH$, $R^3 = NH_2$, $R^4 = H$

2 Cyclitols and Derivatives

Reviews have appeared on the synthesis of inositols, carba sugars, conduritols and amino conduritols utilizing the non-carbohydrate sources of benzene *cis*-diols, quinic acid and Vogel's 'naked sugar' methodology.[39,97] Other reviews on the preparation of cyclophellitol and *epi*-cyclophellitol from glycals[98] and the use of D-glyceraldehyde as a chiral precursor in Diels-Alder and 1,3-dipolar cycloaddition approaches to carbocyclic derivatives[99] have also been reported.

2.1 Cyclopentane Derivatives – The isoxazolidinocarbocyclic derivative **72** was produced (together with a six-membered compound (**89**)-see Section 2.2) from cyclization of the nitrone **73**, itself generated *in situ* from the *N*-benzylhydroxylamine derivative of 3-*C*-allyl-1,2-*O*-isopropylidene-α-D-*ribo*-pentodialdofuranose. The formation of the cyclopentane adduct predominated in non polar and polar aprotic solvents while the cyclohexane adduct predominated in polar protic solvents.[100]

72 **73** **74**

Caryose, a carbocyclic monosaccharide-like compound from the lipopolysaccharide fraction of *Pseudomonas caryophylli* has been characterized as its glycoside with the L-*erythro*-D-*ido*-spiro-bicyclic structure **74**.[101] (See Vol. 29, p. 193, ref. 10 and 11 for the isolation-structure of a related sugar).

Several examples of the stereoselective formation of cyclopentane rings promoted by samarium diiodide have been reported. Thus a pinacol-like coupling of a 1,5-dialdehyde derivative produced **75** in which the newly formed hydroxyl groups R^1, R^5 were in a *syn*-arrangement.[102] (See Vol. 29, p.238, ref. 95 for a related reaction). In contrast, reaction of a 5-hexenal, produced by zinc-assisted Grob fragmentation of a methyl 6-deoxy-6-iodopyranoside, produced **76** and **77** in which the exocyclic groups R^1, R^4 (or R^2, R^5) were *anti* to each other.[103]

However, it appears unnecessary to perform the Grob fragmentation. For example, direct reaction of phenyl 2,6-dideoxy-6-iodopyranosides or the corresponding glycosyl acetates (these compounds being more reactive than the *O*-methyl glycosides) with samarium diiodide afforded (*via* a presumed 5-hexenal intermediate) compound **78**.[104] Similarly and independently, it was also found that reaction of various protected 6-deoxy-6-iodo-pyranosides with two equivalents of samarium diiodide gave cyclopentane rings directly, again *via* a 5-hexenal intermediate.[105]

75 $R^1 = R^5 = OH$, $R^2 = R^4 = H$, $R^3 = OBn$
76 $R^1 = Me$, $R^2 = R^5 = H$, $R^3 = OBn$, $R^4 = OH$
77 $R^1 = R^4 = H$, $R^2 = Me$, $R^3 = OBn$, $R^5 = OH$
78 $R^1 = Me$, $R^2 = R^3 = R^5 = H$, $R^4 = OH$

79

80

81

The cyclopentane-based phosphate **79**, designed as a mimic for *myo*-inositol 1,4,5-trisphosphate, has been synthesized by two different routes. The cyclitol ring was formed firstly by a 'Cp$_2$Zr' induced ring contraction of a hept-6-enose derivative, essentially as previously reported (Vol. 29, p. 238, ref. 94); secondly by a samarium diiodide promoted pinacol-like coupling of a 1,6-dialdehyde derived from methyl 2-*O*-benzyl-3,4-bis-(*p*-methoxybenzyl)-α-D-*gluco*-hexodialdopyrano-side.[106]

Reaction of the D-*arabino*-derived anisyl telluride **80** with 1-acetoxypyridin-2(1*H*)-thione under the influence of light, afforded **81** in which the initial cyclized radical was trapped as the thiopyridone adduct.[107] (See next Section for a different outcome when the D-*galacto* analogue of **80** was used).

In the area of carbocyclic nucleosides, (−)-2-azabicyclo[2.2.1]hept-5-en-3-one (see Vol. 29, p. 239, refs. 98 and 99 for applications of this material) has been transformed into the phosphoribosylamine analogue **82** and the glycinimide derivative **83**,[108] norbornadiene has served as starting material for *cis*-[4-(amino-methyl)-2-cyclopentenyl]methanol,[109] and compound **84** has been made by an initial Diels-Alder reaction of a carbomethoxy imine with cyclopentadiene, and enzymatically resolved.[110] (±)-Cyclopent-3-enol has been converted in several steps into **85** and resolved using vinyl acetate and a lipase and further transformed into a range of nucleotide analogues by displacement, with inversion of configuration at the indicated centre, with nucleobases.[111] (See Vol. 28, p. 235, ref.91 for an alternative synthesis).

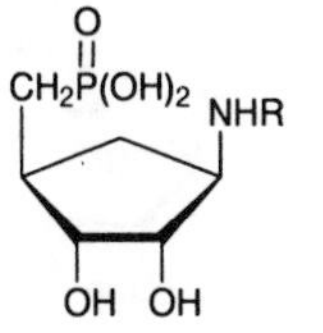

82 R = H
83 R = C(O)CH₂NH₂

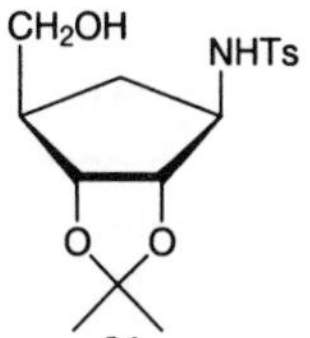

84

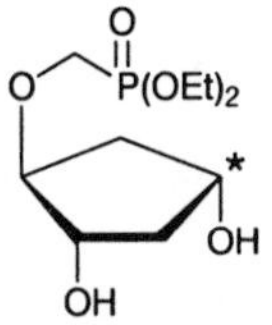

85

The C-1 branched ribofuranose analogues **86** have been prepared by the addition of the appropriate Grignard or alkyllithium to the corresponding ketone, itself prepared from D-ribono-1,4-lactone by known methods.[112]

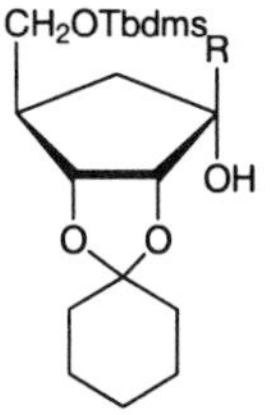

86 R = Me, CH=CH₂, CH₂CH=CH₂

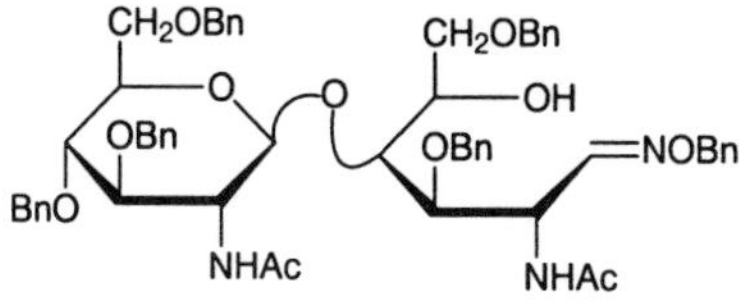

87

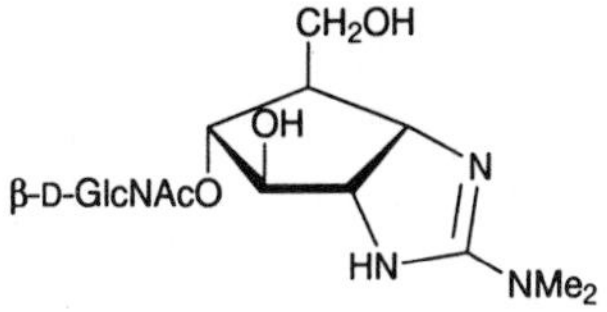

88

Standard conversion of the alcohol of **87** into a phenylthionocarbonate derivative, then reaction with tributyltin hydride-AIBN, produced a radical centred at C-5 which added to the oxime ether double bond to produce the corresponding cyclitol. Further standard conversions gave **88**.[113]

The carbocyclic analogue of the trehalase inhibitor, trehazolin, has been prepared and reported to have potency indistinguishable from the natural product.[114] The synthesis of other antibiotic-related cyclopentane derivatives such as analogues of mannostatin A, allosamidin (and allosamizolin) and a 4-membered carbocyclic thiazole analogue of oxetanocin, are covered in Chapter 19.

2.2   Inositols and Other Cyclohexane Derivatives – The isoxazolidinocarbocyclic derivative **89** has been produced together with a five-membered carbocyclic compound (diagram **72** in Section 2.1) by cycloaddition from nitrone **73**. The formation of **89** predominated in polar protic solvents whilst the reverse was observed in non polar and polar aprotic solvents.[100]

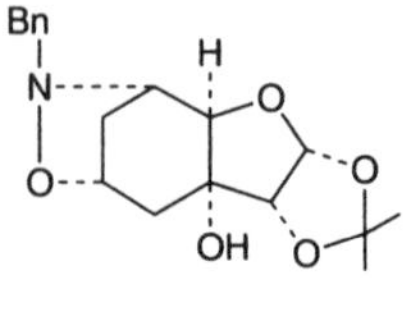

Access to new cyclitol thiirane derivatives has been reported by reaction of cyclitol epoxides with dimethylthioformamide in trifluoroacetic acid. For example the episulfide analogue **90** was produced from the conduritol B epoxide compound **91**. Several similar examples were described starting from cyclitol epoxides prepared by known methods.[115]

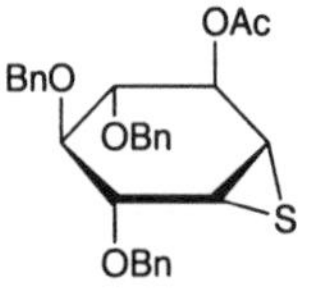

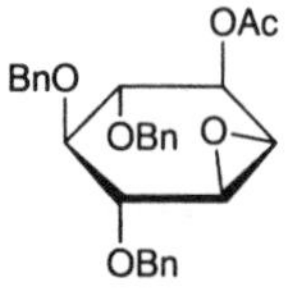

A systematic study on various salts used to catalyse the well known conversion of hex-5-enose derivatives into cyclohexanone compounds has revealed palladium(II) chloride to be the best, even working without the assistance of acid.[116]

A new, related synthesis of cyclohexanes from pyranose derivatives is illustrated in Scheme 6.[117]

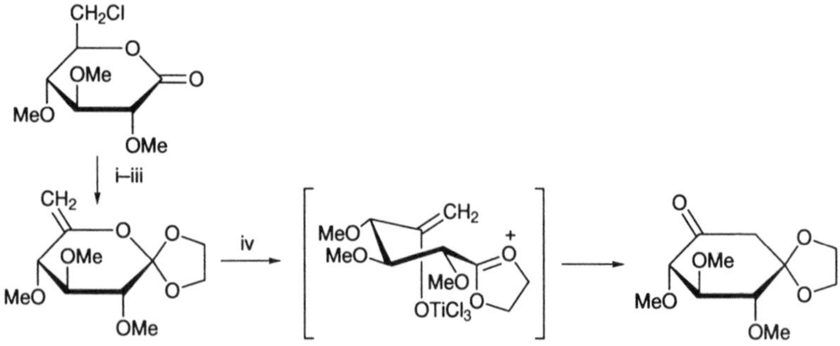

Reagents: i, TmsOCH$_2$CH$_2$OTms, TmsOTf; ii, NaI; iii, DBH; iv, TiCl$_4$, –78 °C

Scheme 6

Reaction of 1,4,5,6-tetra-*O*-benzyl-3-*O*-tosyl-D,L-*myo*-inositol with triethylborane afforded a high yield of the deoxygenated derivative **92**.[118]

2,3-Anhydro-1,5,6-tri-*O*-mesyl-*epi*-inositol reacted with pyridine or nicotinamide by opening at the 3-position of the epoxide to produce 6-membered *muco*-carbocyclic nucleoside analogues containing quaternary ammonium centres. Hydrogenation of these compounds led to saturation of the aromatic groups giving the piperidinyl and 3-carboxamidopiperidinyl derivatives.[119]

The known (S.M. Kupchan *et. al.*, *J. Org. Chem.*, 1969, **34**, 3898) natural product, crotepoxide A has been isolated from the rhizomes of *Kaempferia rotunda* and identified as the source of the plants' moderate insecticidal activity.[120] It has also been synthesized along with its isomer **93** from a quinic acid derivative as a potential glycosidase inhibitor. (See Vol. 28, p. 239, ref. 123 and 124).[121]

The syntheses of carba-β-D,L-fucopyranose and carba-β-D,L-galactopyranose have been reported from 3-*O*-benzyl-1,2:4,5-di-*O*-cyclohexylidene-*myo*-inositol. They involved the deoxygenation of the 6-OH, the conversion of the 1-OH into a ketone, and the addition of methylmagnesium bromide (for producing the *fuco* derivative) or benzyloxymethyllithium (for producing the *galacto* derivative) followed by deoxygenation and deprotection as the key step.[122]

Compound **23** (see Section 1.3.1) has been transformed into *N*-demethyl-*epi*-fortamine by a process involving a pinacol ring closure promoted with the complex $[V_2Cl_3(THF)_6]_2[Zn_2Cl_6]$.[51]

Benzene and its derivatives continue to be useful starting materials for cyclitol syntheses owing to their easy conversion into *cis*-diol derivatives by microbial oxidation. Products made from this kind of starting material include; a cyclophellitol analogue and a metabololite of chorismic acid;[123] (+)- and (−)- conduritols F involving an enzymatic asymmetrization step;[124] carba-α-L-fucopyranose and carba-6-deoxy-β-D-altropyranose from benzonitrile,[125] (−)-6(*R*)-hydroxyshikimic acid from benzonitrile;[126,127] cyanocyclitol **94**;[127] a conduritol D per-acetate which was kinetically resolved with a lipase (e.e. > 95 %) and the intermediate transformed into a variety of products including aminocyclitols and (+)-conduramine C;[128] and bis-carba disaccharide **95** together with some related carba tri- and tetra-saccharides.[129]

Reaction of the D-*galacto* homologue of **80** with 1-acetoxypyridin-2(1*H*)-thione under the influence of light resulted in cyclohexanoids **96**.[107]

A 1-*O*-silyl-2,5-cyclohexylidene derivative has been transformed into conduritol E and 2-deoxy-*allo*-inositol in which chirality was introduced using a Sharpless asymmetric epoxidation reaction,[130] and 1,4- and 1,3-cyclohexadienes have been bis-epoxidized and the epoxide rings cleaved with azide to afford di-azido diol compounds which were asymmetrized enzymatically.[131]

Of several lipases tested, porcine pancreatic lipase resulted in the highest enantioselectivity (> 97% e.e.) when used to resolve 1,2-*O*-cyclohexylidene-*myo*-inositol through esterification at O-5, although overall yield was only 10%.[132]

The controlled catalysed acylation of *myo*-inositol afforded 1,3,4,6-tetra- and 1,3,4,5,6-penta-acyl-*myo*-inositols These products were converted to the 2-keto derivatives by oxidation and were observed to undergo a stepwise deacylation under electron impact conditions.[133]

The regioselective etherification of 4,6-di-*O*-benzyl-1,2-*O*-isopropylidene-*myo*-inositol in the 3- or 5-position has been found to be strongly dependent on the nature of the alkylating agent used.[134]

myo-Inositol has been cyanoethylated with acrylonitrile and the product converted by ethanolysis and hydride reduction to the per-hydroxypropyl ether derivative.[135]

Carba disaccharide **97** has been prepared by insertion of a sugar glycosylidene carbene (made *in situ* from a glycosyl diazirine derivative) into OH groups of protected deoxy *myo*-inositol orthoformate derivatives.[136]

Other carba disaccharides to have been prepared include carba-α-D-Glc*p*-, carba-α-D-Man*p*-, carba-β-D-Man*p*- and carba-β-D-GlcNAc*p*- (1→4)-1,6-anhydro-2,3-*O*-isopropylidene-β-D-mannopyranoses. The α-linked compounds were prepared by adding the anion of 1,6-anhydro-2,3-*O*-isopropylidene-β-D-mannopyranose to epoxide **98** and the β-linked derivatives were obtained from the α-linked compounds by epimerization after oxidation of the free cyclitol OH group. Further standard transformations of the free cyclitol OH also gave the cyclitol configurations noted above.[137] In a similar way, opening of a cyclitol

97 X = F or H   **98**

99 X = NH, O, S

epoxide with N, O and S nucleophilic groups contained in methyl glucoside analogues afforded the carba maltose derivatives **99**.[138]

(+)-Pinitol has been prepared from a non-carbohydrate source by a Diels-Alder condensation of furan and *trans*-1,2-bis-phenylsulfonyl ethylene and resolved using an intermediate camphanic acid ester.[139] ($\pm$)-Cyclophellitol and its (1*R*, 6*S*)-enantiomer have been prepared from a 7-oxabicyclo[2.2.1]hept-5-ene-2-*endo*-carboxylic acid.[140]

Valiolamine and its C-1 and C-2 epimers, together with other analogues, have been prepared from $(-)$-quinic acid[141] and racemic validamine, together with its C-1 and C-2 isomers, have been synthesized from (phenylsulfonyl)-7-oxa-bicyclo[2.2.1]heptane derivatives.[142] Six *N*-alkyl derivatives of β-valienamine have been made and shown to be potent and specific inhibitors of β-glucocerebrosidase; the best was the *n*-octyl derivative.[143]

A GLC-MS approach to identifying the isomers of coumaroyl- and caffeoyl-D-quinic acid,[144] and syntheses of 3,4,5-tri- and 1,3,4,5-tetra-galloylquinic acids involving a description of their chirotopical properties using the benzoate chirality rule,[145] have been reported.

Stereospecific syntheses of $(-)$-3-*epi*-, 3(*R*)-fluoro- and 3(*S*)-fluoro-shikimic acids[146] and 3(*R*)-amino-4(*R*),5(*R*)-dihydroxy-1-cyclohexene-1-carboxylic acid (3(*R*)-aminoshikimic acid) which used a method for obtaining gram quantities of shikimic acid from the seeds of *Illicium anistum* (star aniseed)[147,148] have been described.

The synthesis of a lincosaminide derivative (a compound found in the antibiotic lincomycin) from a *myo*-inositol derivative is mentioned in Chapter 19.

2.3 Inositol Phophates and Derivatives – Reviews on the preparation and separation of inositol phosphates[149] and on the preparation of inositol polyphosphates from glucose[150] have appeared.

The total syntheses of all regioisomers of *myo*-inositol pentakisphosphate (4 isomers),[151] of *myo*-inositol tetrakisphosphate (9 isomers)[152] and of *myo*-inositol trisphosphate (12 isomers)[152,153] using base-catalysed acyl migration of mono-, di- and tri-benzoate derivatives, respectively, in a similar way to that described in Vol. 29, p. 246, ref. 156 and 157 have been reported. The separated, individual benzoates were phosphorylated and deprotected to afford products. The same group has also transformed the enol-acetates **100**, by way of a Ferrier reaction, into L-*chiro*-inositol 1,2,3-trisphosphate and L-*chiro*-inositol 1,2,3,5-tetrakisphosphate using standard chemical transformations.[154,155] Other inositols prepared by this type of carbocyclic ring forming reaction include D-*myo*-inositol 1,2,4,5-tetrakisphosphate and its *P*-2-(*O*-aminopropyl) analogue,[156] 2-deoxy-D-*myo*-inositol 1,3,4,5-tetrakisphosphate,[157] 1-L-phophatidyl-D-*myo*-inositol 4,5-bisphosphate and -3,4,5-trisphosphate containing benzophenone or 4-amino-7-nitro-2-oxa-1,3-diazobenzene photoactivatable groups[158] and a phosphatidyl-D-*myo*-inositol 3,4-bisphosphate incorporating a benzophenone photoactivatable group.[159]

Both enantiomers of 1,2:5,6-di-*O*-cyclohexylidene-*myo*-inositol have been prepared by standard kinetic resolution involving a lipase and each converted into 3-

100 R = Me or Ac   **101**

102 *n* = 6 or 14   **103**

104 R = Me, Pr, All,

etc.

mono-, 3,4-bis- and 3,4,5-tris-phosphate derivatives of L-α-phophatidyl-*myo*-inositol[160] and both enantiomers of 1,2:4,5-di-*O*-cyclohexylidene-3-*O*-allyl-*myo*-inositol as versatile starting materials for the synthesis of phospahtidyl derivatives in the 1-D-*myo*-inositol series have also been described.[161]

Glycosylation using 3,4,6-tri-*O*-acetyl-2-deoxy-2-phthalimido-D-glucal of a protected deoxy D-*chiro*-inositol derivative in the presence of boron trifluoride etherate afforded the expected *O*-glycoside with allylic rearrangement which was transformed into **101** for incorporation into GPI-anchor analogues.[162] The GPI membrane anchor analogue **102** has been prepared from a chiral *myo*-inositol derivative using standard trichloracetimidate methodology for glycosylating the inositol ring.[163] The glycosylation of a substituted *myo*-inositol has also been performed and led to the preparations of 1(3)-*O*-((±)-1,2-dipalmitoylglycerophospho)-4,6-*O*-β-D-glucopyranosyl-*sn*-*myo*-inositol[164] and the phosphatidyl *myo*-inositol **103** containing a polyene has also been reported.[165]

Twenty inositol phosphodiester derivatives **104** have been prepared by reaction of inositol 1,2-cyclic phosphates with the corresponding alcohol catalysed by phosphatidylinositol-specific phospholipase C.[166]

The heterocycle, quercetin, has been rendered more water soluble by conjugation through a 1-*O*-succinyl-*myo*-inositol-L-phosphate linkage.[167]

A full paper has appeared on the tin-mediated allylation and benzylation of 1,2-*O*-isopropylidene-*myo*-inositol (see Vol. 27, p. 219, ref. 126 and Vol. 25, p. 215, ref. 108 for preliminary report) which resulted in the preparation of various protected *myo*-inositol 1,6-bis- and 1,5,6-tris-phosphates.[168]

A wide range of *myo*- (both chiral and (±)) and *scyllo*-tetrakisphosphate derivatives have been prepared by standard means and the importance of their H-bonding with their target proteins probed. Several of the compounds were converted into acetoxymethyl esters in order to increase membrane permeability.[169]

Syntheses of (±) *myo*-inositol 1,2-bisphosphate and *myo*-inositol 1,2,3-trisphosphate by phosphorylation of (±) 3,4,5,6-tetra-*O*-benzyl-*myo*-inositol and 4,5,6-tri-*O*-benzoyl-*myo*-inositol, respectively, using dibenzyl *N*,*N*'-diisopropylphosphoramidite then oxidation have been reported,[170] and D- and L-*myo*-inositol 1,4,6-trisphosphate has been made from a *myo*-inositol derivative and resolved with (*S*)-(+)-*O*-acetylmandelic acid to obtain the required pure enantiomers.[171]

Full details of an earlier report (see Vol. 27, p. 222, ref. 146) on the syntheses of L-*scyllo*-inositol 1,2,4-trisphosphate, *scyllo*-inositol 1,2,4,5-tetrakis-phosphate and -phosphorothioate and 2-deoxy-2-fluoro-D,L-*myo*-inositol have been reported.[172]

myo-Inositol orthoformate has been transformed *via* an intermediate 4,6-di-*O*-benzoyl-1-*O*-butyryl derivative into 1-D-*myo*-inositol 1,2,6-trisphosphate (α-trinositol), as described in Vol. 27, p. 219, ref. 124.[173] The product, which has been reported to have antiinflammatory and analgesic properties, has also been converted into several 3,4,5-triester analogues with the aim of obtaining orally active compounds.[174,175] *myo*-Inositol orthoformate has also been used to prepare (±)-6-deoxy-6-hydroxymethyl *scyllo*-inositol 1,2,4-trisphosphate and found to be a potent agonist at the platelet D-*myo*-inositol 1,4,5-trisphosphate receptor.[176]

The syntheses from *myo*-inositol of 1-deoxy-1-(phosphonomethyl)-*myo*-2-inosose and 2-deoxy-*myo*-inositol trisphosphate as potential inhibitors of *myo*-inositol 1-monophosphate synthase have been described.[177]

(±)-1,4,5-Tri-*O*-benzyl-2,3-*O*-cyclohexylidene-*myo*-inositol has been transformed into (±)-6-deoxy-6-fluoro-*myo*-inositol 1,4,5-trisphosphate and information obtained about relative acceptor-donor effects of either the lone pair electrons or hydroxyl hydrogen in their affinities for the *endoplasmic reticulum* receptor.[178]

1,2-Dideoxy-1,2-difluoro-D,L-*myo*- and -*scyllo*-inositols have been prepared as analogues of D,L-*myo*-inositol 3,4,5,6-tetrakisphosphate.[179]

The syntheses of the 4,5-bisphosphorofluoridate **105** and the 4- and 5-monophosphorofluoridates have been described with the first reported to have twenty fold the inhibitory potency of *myo*-inositol 1,4,5-trisphosphate against a 5-phosphatase. The fluorines were introduced by reaction of a phosphate with 2-fluoro-1-methylpyridinium tosylate.[180]

105   **106**

The thioinositol derivative **106** has been made as an inhibitory analogue of *myo*-inositol 1-monophosphate, but only displayed poor activity against *myo*-inositol 1-phosphatase.[181]

References

1   S. Kolaric and V. Sunjic, *J. Mol. Catal. A: Chem.*, 1996, **111**, 239 (*Chem. Abstr.*, 1996, **125**, 301 322).

2   V.I. Parvulescu, V. Parvulescu, S. Coman, C. Radu, D. Macovei, E.M. Angelescu and R. Russu, *Stud. Surf. Sci. Catal*, 1995, **91**, 561 (*Chem. Abstr.*, 1996, **124**, 30 176).

3   U.N. Dash, E.R. Pattanaik, *Indian J. Chem., Sect. A: Inorg., Phys., Theor. Anal. Chem.*, 1995, **34A**, 834 (*Chem. Abstr.*, 1996, **124**, 9 189).

4   A.M. Sakharov, N.T. Silakhtaryan and I.P. Skibida, *Kinet. Catal.* (*Transl. of Kinet. Katal.*), 1996, **37**, 368 (*Chem. Abstr.*, 1996, **125**, 222 294).

5   D. Rele, N. Zhao, K. Nakanishi and N. Berova, *Tetrahedron*, 1996, **52**, 2759.

6   J. Rammo and H.-J. Schneider, *Liebigs Ann. Chem.*, 1996, 1757.

7   F. Hasumi, C. Teshima and I. Okura, *Chem. Lett.*, 1996, 597.

8   K. Li and R.F. Helm, *J. Chem. Soc., Perkin Trans. 1*, 1996, 2425.

9   G.K. Cook and M.A. Andrews, *J. Am. Chem. Soc.*, 1996, **118**, 9448.

10   U.F. Heiser and B. Dobner, *Chem. Commun.*, 1996, 2025.

11   N. Fujii, H. Habashita, M. Akaji, K. Nakai, T. Ibuka, M. Fujiwara, H. Tamamura and Y. Yamamoto, *J. Chem. Soc., Perkin Trans. 1*, 1996, 865.

12   E.A. El'Perina, M.I. Struchkova, M.I. Serkebaev and E.P. Serebryakov, *Izv. Akad. Nauk, Ser. Khim.*, 1993, 776 (*Chem. Abstr.*, 1996, **124**, 117 701).

13   Y. Wei, M. Lu and S. Bai, *Nanjing Ligong Daxue Xuebao*, 1994, 94 (*Chem. Abstr.*, 1996, **124**, 56 481).

14   G. Carchon, F. Chretien and Y. Chapleur, *Carbohydr. Lett.*, 1996, **2**, 17 (*Chem. Abstr.*, 1996, **125**, 196 158).

15   E. Lattova, M. Petrusova, A. Gaplovsky and L. Petrus, *Chem. Pap.*, 1996, **50**, 97 (*Chem. Abstr.*, 1996, **125**, 87 044).

16   I.A. Litvinov, V.A. Naumov, L.I. Gurarii and E.T. Mukmenev, *Izv. Acad. Nauk, Ser. Khim.* 1993, 406, (*Chem. Abstr.*, 1996, **124**, 87 503).

17   R. Csuk, U. Hörning and M. Schaade, *Tetrahedron*, 1996, **52**, 9759.

18   G. Niklasson, I. Kvarnström, B. Classon, B. Samuelsson, U. Nillroth, H. Danielson,
      A. Karlén and A. Hallberg, *J. Carbohydr. Chem.*, 1996, **15**, 555.
19   S. Jiang, G. Singh and R.H. Wightman, *Chem. Lett.*, 1996, 67.
20   C. Di Nardo, L.O. Joroncic, R.M. De Lederkremer and O. Varela, *J. Org. Chem.*,
      1996, **61**, 4007.
21   B. Giese, S.N. Müller, C. Wyss and H. Steiner, *Tetrahedron: Asymm.*, 1996, **7**, 1261.
22   M. Okada, Y. Okada and K. Aoi, *J. Polym. Sci., Part A: Polym. Chem.*, 1995, **33**,
      2813 (*Chem. Abstr.*, 1996, **124**, 117 772).
23   T. Kakuchi, S. Umeda, T. Satoh, H. Hashimoto and K. Yokota, *Macromol. Rep.*,
      1995, **A32**, 1007 (*Chem. Abstr.*, 1996, **124**, 117 752).
24   T. Satoh, T. Hatakeyama, S. Umeda, H. Hashimoto, K. Yokota and T. Kakuchi,
      Macromolecules, 1996, **29**, 3447 (*Chem. Abstr.*, 1996, **124**, 317 673).
25   T. Kakuchi, T. Satoh, J. Mata, S. Umeda, H. Hashimoto and K. Yokota,
      J. Macromol. Sci., Pure Appl. Chem., 1996, **A33**, 325 (*Chem. Abstr.*, 1996. **124**,
      290 064).
26   T. Satoh, T. Hatakeyama, S. Umeda, M. Kamada, K. Yokota and T. Kakuchi,
      Macromolecules, 1996, **29**, 6681 (*Chem. Abstr.*, 1996, **125**, 276 349).
27   T. Kakuchi, T. Satoh and K. Yokota, *Polym. Prepr. (Am. Chem. Soc., Div. Polym.
      Chem.)* 1996, **37**, 487 (*Chem. Abstr.*, 1996, **125**, 276 351).
28   C. Leriche, L. Rene, F. Derouet, B. Rousseau and B. Badet, *J. Labelled Compd.
      Radiopharm.*, 1995, **36**, 1115 (*Chem. Abstr.*, 1996, **124**, 117 778).
29   R. Murali and M. Nagarajan, *Carbohydr. Res.*, 1996, **280**, 351.
30   J.F. Witte and R.W. McClard, *Bioorg Chem.*, 1996, **24**, 29 (*Chem. Abstr.*, 1996, **124**,
      343 882).
31   C.L. Cavallaro and J. Schwartz, *J. Org. Chem.*, 1996, **61**, 3863.
32   Z.-W Guo and Y.-Z. Hui, *Synth. Commun.*, 1996, **26**, 2067.
33   N. Hossain, N. Blaton, O. Peeters, J. Rozenski and P.A. Herdewijn, *Tetrahedron*,
      1996, **52**, 5563.
34   M.-J. Pérez-Pérez, E. De Clercq and P. Herdewijn, *Bioorg. Med. Chem. Lett.*, 1996,
      6, 1457.
35   M.W. Andersen, S.M. Daluge, L. Kerremans and P. Herdewijn, *Tetrahedron Lett.*,
      1996, **37**, 8147.
36   J.C. Norrild, C. Pedersen and J. Defay, *Carbohydr. Res.*, 1996, **291**, 85.
37   V. Popsavin, O. Beric, J. Casanadi, M. Posavin and D. Miljkovic, *J. Serb. Chem.
      Soc.*, 1995, **60**, 625 (*Chem. Abstr.*, 1996, **124**, 9 199).
38   V. Glaçon, A. El Meslouti, R. Uzan, G. Demailly and D. Beaupère, *Tetrahedron
      Lett.*, 1996, **37**, 3683.
39   T. Hudlicky, D.A. Entwistle, K.K. Pitzer and A.J. Thorpe, *Chem. Rev.*, 1996, **96**,
      1195 (*Chem. Abstr.*, 1996, **124**, 290 002).
40   B. Ganem, *Acc. Chem. Res.*, 1996, **29**, 340 (*Chem. Abstr.*, 1996, **124**, 58 858).
41   K. Tatsuta, *Pure Appl. Chem.*, 1996, **68**, 1341 (*Chem. Abstr.*, 1996, **125**, 222 299).
42   H.J.M. Gijsen, L. Qiao, W. Fitz and C.-H. Wong, *Chem. Rev.*, 1996, **96**, 443 (*Chem.
      Abstr.*, 1996, **124**, 87 497).
43   C.-H. Wong, *Pure Appl. Chem.*, 1996, **67**, 1609 (*Chem. Abstr.*, 1996, **124**, 87 494).
44   R.A. Evangelista, A. Guttman and F.-T.A. Chen, *Electrophoresis*, 1996, **17**, 347
      (*Chem. Abstr.*, 1996, **124**, 317 713).
45   M. Avalos, R. Babiano, A. Cabanillas, P. Cintas, J.L. Jiménez, J.C. Palacios, M.A.
      Aguilar, J.C. Corchadó and J. Espinosa-García, *J. Org. Chem.*, 1996, **61**, 7291.
46   A.T. Ung, S.G. Pyne, B.W. Skelton and A.H. White, *Tetrahedron*, 1996, **52**,
      14069.

47   M.A.E. Shaban, M.A.M. Taha, A.Z. Nasr and A.E.A. Morgan, *Pharmazie*, 1995, **50**, 784 (*Chem. Abstr.*, 1996, **124**, 202 884).
48   J.G. Fernández-Bolaños, S. García, J. Fernández-Bolaños, M.J. Diánez, M.D. Estrada, A. López-Castro and S. Pérez, *Carbohydr. Res.*, 1996, **282**, 137.
49   A. Tuch, M. Sanière, Y. Le Merrer and J.-C. Depezay, *Tetrahedron: Asymm.*, 1996, **7**, 897.
50   N. Hossain, J. Rozenski, E. De Clercq and P. Herdewijn, *Tetrahedron*, 1996, **52**, 13655.
51   M. Kang, J. Park, A.W. Konradi and S.F. Pedersen, *J. Org. Chem.*, 1996, **61**, 5528.
52   X.-Y. Chen, T.M. Link and V.L. Schramm, *J. Am. Chem. Soc.*, 1996, **118**, 3067.
53   M. Yokoyama, T. Akiba, Y. Ochiai, A. Momotake and H. Togo, *J. Org. Chem.*, 1996, **61**, 6079.
54   B. Davis, A.A. Bell, R.J. Nash, A.A. Watson, R.C. Griffiths, M.G. Jones, C. Smith and G.W.J. Fleet, *Tetrahedron Lett.*, 1996, **37**, 8565.
55   M. Bols, M.P. Persson, W.M. Butt, M. Jorgensen, P. Christensen and L.T. Hansen, *Tetrahedron Lett.*, 1996, **37**, 2097.
56   K.-H. Altmann, S.M. Freier, U. Pieles and T. Winkler, *Angew. Chem.*, 1994, **106**, 1735 (*Chem Abstr.*, 1996, **124**, 261 558).
57   H.K. Kim, Y. Lee, I.K. Young and Y.B. Chae, *Korean J. Med. Chem.*, 1995, **5**, 68 (*Chem. Abstr.*, 1996, **124**, 202 916).
58   Y. Lee, H.K. Kim, I.K. Young and Y.B. Chae, *Korean J. Med. Chem.*, 1995, **5**, 66 (*Chem. Abstr.*, 1996, **124**, 202 915).
59   D.-K. Kim, G. Kim and Y.-W. Kim, *J. Chem. Soc., Perkin Trans. 1*, 1996, 803.
60   S. Andersen, A. de Raadt, M. Ebner, C. Ekhart, G. Gradnig, G. Legler, I. Lundt, M. Schichl, A.E. Stutz and S. Withers, *Electron. Conf. Trends Org. Chem.*, 1995, (pub. 1996), paper 19 (*Chem. Abstr.*, 1996, **125**, 33 993).
61   M. Kiso, H. Furui, H. Ishida and A. Hasegawa, *J. Carbohydr. Chem.*, 1996, **15**, 1.
62   Y. Igarashi, M. Ichikawa and Y. Ichikawa, *Bioorg. Med. Chem. Lett.*, 1996, **6**, 553.
63   A. Hansen, T.M. Tagmose and M. Bols, *Chem. Commun.*, 1996, 2649.
64   Y. Igarashi, M. Ichikawa and Y. Ichikawa, *Tetrahedron Lett.*, 1996, **37**, 2707.
65   E. Graf von Roedern, E. Lohof, G. Hessler, M. Hoffmann and H. Kessler, *J. Am. Chem. Soc.*, 1996, **118**, 10156.
66   Y. Nishimura, T. Satoh, H. Adachi, S. Kondo, T. Takeuchi, M. Azetaka, H. Fukuyasu and Y. Iizuka, *J. Am. Chem. Soc.*, 1996, **118**, 3051.
67   A. Defoin, H. Sarazin and J. Streith, *Hel. Chim. Acta*, 1996, **79**, 560.
68   J.I. Cho, S. Yoon, K.H. Chun and J.E.N. Shin, *Bull Korean Chem. Soc.*, 1995, **16**, 805 (*Chem. Abstr.*, 1996, **124**, 117 781).
69   K.H. Park, *Bull. Korean. Chem. Soc.*, 1995, 985 (*Chem. Abstr.*, 1996, **124**, 146 654).
70   G. Gradnig, G. Legler and A.E. Stutz, *Carbohydr. Res.*, 1996, **287**, 49.
71   J.P. Shilvock, J.R. Wheatley, B. Davis, R.J. Nash, R.C. Griffiths, M.G. Jones, M. Muller, S. Crook, D.J. Watkin, C. Smith, G.S. Besra, P.J. Brennan and G.W.J. Fleet, *Tetrahedron Lett.*, 1996, **37**, 8569.
72   J.B. Bremner, B.W. Skelton, R.J. Smith, G.J. Tarrant and A.H. White, *Tetrahedron Lett.*, 1996, **37**, 8573.
73   V. Zsoldos-Mády, I. Pintér, P. Sándor and A. Messmer, *Carbohydr. Res.*, 1996, **281**, 321.
74   C. Herdeis and T. Schiffer, *Tetrahedron*, 1996, **52**, 14745.
75   A. Baudat, S. Picasso and P. Vogel, *Carbohydr. Res.*, 1996, **281**, 277.
76   I.K. Khanna, R.M. Weier, J. Julien, R.R. Mueller, D.C. Lankin and L. Swenton, *Tetrahedron Lett.*, 1996, **37**, 1355.

77   I. McCort, A. Duréault and J.-C. Depezay, *Tetrahedron Lett.*, 1996, **37**, 7717.
78   X. Qian, F. Morís-Varas and C.-H. Wong, *Bioorg. Med. Chem. Lett.*, 1996, **6**, 1117.
79   F Morís-Varas, X.-H. Qian and C.-H. Wong, *J. Am. Chem Soc.*, 1996, **118**, 7647.
80   D. Damour, M. Barream, J.-C. Blanchard, M.-C. Burgevin, A. Doble, F. Herman, G. Pantel, E. James-Surcouf, M. Vuilhorgne, S. Mignani, L. Poitout, Y. Le Merrer and J.-C. Depezay, *Bioorg. Med. Chem. Lett.*, 1996, **6**, 1667.
81   L. Poitout, Y. Le Merrer and J.-C. Depezay, *Tetrahedron Lett.*, 1996, **37**, 1609.
82   L. Poitout, Y. Le Merrer and J.-C. Depezay, *Tetrahedron Lett.*, 1996, **37**, 1613.
83   M. Kojima, T. Seto, Y. Kyotani, H. Ogawa, S. Kitazawa, K. Mori, S. Maruo, T. Ohgi and Y. Ezure, *Biosci. Biotech. Biochem.*, 1996, **60**, 694.
84   L. Campanini, A. Duréault and J.-C- Depezay, *Tetrahedron Lett.*, 1996, **37**, 5095.
85   S. Takahashi, H. Terayama and H. Kuzuhara, *Carbohydr. Lett.*, 1996, **1**, 407 (*Chem. Abstr.*, 1996, **125**, 11 280).
86   S. Takahashi, H. Terayama and H. Kuzuhara, *Tetrahedron*, 1996, **52**, 13315.
87   O.M. Saavendra and O.R. Martin, *J. Org. Chem.*, 1996, **61**, 6987.
88   O.R. Martin, L. Liu and F. Yang, *Tetrahedron Lett.*, 1996, **37**, 1991.
89   L. Qiao, B.W. Murray, M. Shimazaki, J. Schultz and C-H. Wong, *J. Am. Chem. Soc.*, 1996, **118**, 7653.
90   A.I. Meyers, C.J. Andres, J.E. Resek, M.A. McLaughlin, C.A. Woodall and P.H. Lee, *J. Org. Chem.*, 1996, *61*, 2586.
91   Y. Doi, M. Ishibashi and J. Kobayashi, *Tetrahedron*, 1996, **52**, 4573.
92   Y.-M. Xu and W.-S. Zhou, *Tetrahedron Lett.*, 1996, **37**, 1461.
93   A. Arone, P. Bravo, A. Donadelli and G. Resnati, *Tetrahedron*, 1996, **52**, 131.
94   M. Malberg and N. Rehnberg, *Synlett*, 1996, 361.
95   J. Soulie, T. Faitg, J.-F. Betzer and J.-Y. Lallemand, *Tetrahedron*, 1996, **52**, 15137.
96   N. Asano, A. Kato, Y. Yokoyama, M. Miyauchi, M. Yamamoto, H. Kizu and K. Matsui, *Carbohydr. Res.*, 1996, **284**, 169.
97   T. Hudlicky, K. Abbond, D.A. Entwistle, R. Fan, R. Maurya, A.J. Thorpe, J. Bolonick and B. Myers, *Synthesis*, 1996, 897.
98   B. Fraser-Reid, *Acc. Chem. Res.*, 1996, **29**, 57 (*Chem. Abstr.*, 1996, **124**, 176 622).
99   R. Casas, Z. Chen, M. Diaz, N. Hanafi, J. Ibarzo, J.M. Jimenez and R.M. Ortuno, *An. Quim.*, 1995, **91**, 42 (*Chem. Abstr.*, 1996, **125**, 196 105).
100   R. Patra, N.C. Bar, A. Roy, B. Achari, N. Ghoshal and S.B. Mandal, *Tetrahedron*, 1996, **52**, 11265.
101   M. Adinolfi, M.M. Corsaro, C. De Castro, A. Evidente, R. Lanzetta, A. Molinaro and M. Parrilla, *Carbohydr. Res.*, 1996, **284**, 111.
102   E. Perrin, J.-M. Mallet and P. Sinaÿ, *Carbohydr. Lett.*, 1995, **1**, 215 (*Chem. Abstr.*, 1996, **125**, 114 991).
103   J.J.C. Grové, C.W. Holzapfel and D.B.G. Williams, *Tetrahedron Lett.*, 1996, **37**, 1305.
104   J.J.C. Grové, C.W. Holzapfel and D.B.G. Williams, *Tetrahedron Lett.*, 1996, **37**, 5817.
105   J.L. Chiara, S. Martínez and M. Bernabé, *J. Org. Chem.*, 1996, **61**, 6488.
106   D.J. Jenkins, A.M. Riley and B.V.L. Potter, *J. Org. Chem.*, 1996, **61**, 7719.
107   D.H.R. Barton, S.D. Gero, P. Holliday and B. Quiclet-Sire, *Tetrahedron*, 1996, **52**, 8233.
108   R. Vince, M. Hua and C.A. Caperelli, *Nucleosides Nucleotides*, 1996, **15**, 1711.
109   C. Balo, B. Dominguez, F. Fernandez, E. Lens and C. Lopez, *Org. Proced. Int.*, 1996, **28**, 211 (*Chem. Abstr.*, 1996, **124**, 343 901).

110   M.L. Edwards, J.E. Matt, Jr., D.L. Westrup, C.A. Kemper, R.A. Persichetti and
      A.L. Margolin, *Org. Proced. Int.*, 1996, **28**, 193 (*Chem. Abstr.*, 1996, *125*, 11 281).
111   A.F. Drake, A Garofalo, J.M.L. Hillman, V. Merlo, R. McCague and S.M. Roberts,
      J. Chem. Soc., Perkin Trans. 1, 1996, 2739.
112   N. Nga and A. Grouiller, *Carbohydr. Lett.*, 1995, **1**, 381 (*Chem. Abstr.*, 1996, **124**,
      317 679).
113   S. Takahashi, H. Terayama, H. Koshino and H. Kuzuhara, *Chem. Lett.*, 1996, 97.
114   C. Uchida and S. Ogawa, *Carbohydr. Lett*, 1994, **1**, 77 (*Chem. Abstr.*, 1996, **125**, 114
      997).
115   P. Letellier, A. El Meslouti, D. Beaupere and R. Uzan, *Synthesis*, 1996, 1435.
116   T. Iimori, H. Takahashi and S. Ikegami, *Tetrahedron Lett.*, 1996, **37**, 649.
117   D.G. Bourke, D.J. Collins, A.I. Hibberd and M.D. Mcleod, *Aust. J. Chem.*, 1996,
      49, 425.
118   J. Yu and J.B. Spencer, *J. Org. Chem.*, 1996, **61**, 3234.
119   M.Y. Grass, M.E. Gelpi, C. Elias and R.A. Cadenas, *An. Asoc. Quim. Argent.*, 1995,
      83, 201 (*Chem. Abstr.*, 1996, **125**, 196 198).
120   B.W. Nugroho, B. Schwarz, V. Wray and P. Proksch, *Phytochemistry*, 1996, **41**,
      129.
121   T.K.M. Shing and E.K.W. Tam, *Tetrahedron: Asymm.*, 1996, **7**, 353.
122   R. Verduyn, S.H. van Leeuwen, G.A. van der Marel and J.H. van Boom, *Recl. Trav.
      Chim. Pays-Bas*, 1996, **115**, 67.
123   S.M. Roberts, P.W. Sutton and L. Wright, *J. Chem. Soc., Perkin Trans. 1*, 1996,
      1157.
124   A. Patti, C. Sanfilippo, M. Piattelli and G. Nicolosi, *J. Org. Chem.*, 1996, **61**, 6458.
125   C.H. Tran and D.H.G. Crout, *Tetrahedron: Asymm.*, 1996, **7**, 2403.
126   C.H. Tran, D.H.G. Crout and W. Errington, *Tetrhedron: Asymm.*, 1996, **7**, 691.
127   H.A.J. Carless and Y. Dove, *Tetrahedron: Asymm.*, 1996, **7**, 649.
128   A. Patti, C. Sanfilippo, M. Piattelli and G. Nicolosi, *Tetrahedron: Asymm.*, 1996, **7**,
      2665.
129   T. Hudlicky, K.A. Abboud, J. Bolonick, R. Maurya, M.L. Stanton and A.J. Thorpe,
      Chem. Commun., 1996, 1717.
130   R. Angelaud and Y. Landais, *J. Org. Chem.*, 1996, **61**, 5202.
131   M. Gruber-Khadjawi, H. Hönig and C. Illaszewicz, *Tetrahedron: Asymm.*, 1996, **7**,
      807.
132   M.T. Rudolf and C. Schultz, *Liebigs Ann. Chem.*, 1996, 533.
133   F. Kurzer and D. Chapman, *Z. Naturforsch., B: Chem. Sci.*, 1996, **51**, 68 (*Chem.
      Abstr.*, 1996, **124**, 317 674).
134   C.-Y. Yuan, H.-X. Zhai and S.-S. Li, *Chin. J. Chem.*, 1996, **14**, 271 (*Chem. Abstr.*,
      1996, **125**, 276 363).
135   L. Guo, L.-L. Weng and H. Zheng, *Youji Huaxue*, 1995, **15**, 308 (*Chem. Abstr.*, 1996,
      124, 9 187).
136   A. Zapata, B. Bernet and A. Vasella, *Hel. Chim. Acta*, 1996, **79**, 1169.
137   S. Ogawa, K. Hirai, M. Ohno, T. Furuya, S. Sasaki and H. Tsunoda, *Liebigs Ann.
      Chem.*, 1996, 673.
138   H. Tsunoda, S. Sasaki, T. Furuya and S. Ogawa, *Liebigs Ann. Chem.*, 1996, 159.
139   J.L. Aceña, O. Arjona and J. Plumet, *Tetrahedron: Asymm.*, 1996, **7**, 3535.
140   J.L. Aceña, E. da Alba, O. Arjona and J. Plumet, *Tetrahedron Lett.*, 1996, **37**, 3043.
141   T.K.M. Shing and L.H. Wan, *J. Org. Chem.*, 1996, **61**, 8468.
142   O. Arjona and J. Plumet, *Ars. Pharm.*, 1995, **36**, 445 (*Chem. Abstr.*, 1996, **125**,
      301 400).

143 S. Ogawa, M Ashiura, C. Uchida, S. Watanabe, C. Yamazaki, K. Yamagishi and J.-i. Inokuchi, *Bioorg. Med. Chem. Lett.*, 1996, **6**, 929.

144 C. Fuchs and G. Spiteller, *J. Mass Spectrom.*, 1996, **31**, 602 (*Chem. Abstr.*, 1996, **125**, 143 163).

145 R. Altmann and H. Falk, *Monatsh. Chem.*, 1995, **126**, 1225 (*Chem. Abstr.*, 1996, **124**, 202 851).

146 R. Brettle, R. Cross, M. Frederickson, E. Haslam, F.S. MacBeath and G.M. Davies, *Bioorg. Med. Chem. Lett.*, 1996, **6**, 1275.

147 R. Brettle, R. Cross, M. Frederickson, E. Haslam and G.M. Davies, *Bioorg. Med. Chem. Lett.*, 1996, **6**, 291.

148 H. Adams, N.A. Bailey, R. Brettle, R. Cross, M. Frederickson, E. Haslam, F.S. MacBeath and G.M. Davies, *Tetrahedron*, 1996, **52**, 8565.

149 C. Schultz, A. Burmester and C. Stadler, *Subcell. Biochem.*, 1996, **26**, 371 (*Chem. Abstr.*, 1996, **125**, 86 991).

150 G.D. Prestwich, *Acc. Chem. Res.*, 1996, **29**, 503 (*Chem. Abstr.*, 1996, **125**, 276 298).

151 S.-K. Chung and Y.-T. Chang, *Bioorg. Med. Chem. Lett.*, 1996, **6**, 2039.

152 S.-K. Chung, Y.-T. Chang and Y. Ryu, *Pure Appl. Chem.*, 1996, **68**, 931 (*Chem. Abstr.*, 1996, **125**, 143 168).

153 S.-K. Chung, Y.-T. Chang and K.-H. Sohn, *Chem. Commun.*, 1996, 163.

154 S.-K. Chung and S.-H. Yu, *Korean J. Med. Chem.*, 1996, **6**, 35 (*Chem. Abstr.*, 1996, **125**, 114 995).

155 S.-K. Chung and S.-H. Yu, *Bioorg. Med. Chem. Lett.*, 1996, **6**, 1461.

156 J. Chen, G. Dormán and G.D. Prestwich, *J. Org. Chem.*, 1996, **61**, 393.

157 D.J. Jenkins, D. Dubreuil and B.V.L. Potter, *J. Chem. Soc., Perkin Trans. 1*, 1996, 1365.

158 J. Chen, A.A. Profit and G.D. Prestwich, *J. Org. Chem.*, 1996, **61**, 6305.

159 O. Thum, J. Chen and G.D. Prestwich, *Tetrahedron Lett*, 1996, **37**, 9017.

160 D.-S. Wang and C.-S. Chen, *J. Org. Chem.*, 1996, **61**, 5905.

161 R. Aneja, S.G. Aneja and A. Parra, *Tetrahedron Lett.*, 1996, **37**, 5081.

162 M.M. Silva, J. Cleophax, A.A. Benicio, M.V. Almeida, J.-M. Delaumeny, A.S. Machado and S.D. Gero, *Synlett*, 1996, 764.

163 J. Ye, W.T. Doerrer, M.A. Lehrman and J.R. Falk, *Bioorg. Med. Chem. Lett.*, 1996, **6**, 1715.

164 N.S. Shastina, L.I. Ernisman, A.E. Stepanov and V.I. Shvets, *Bioorg. Khim.*, 1996, **22**, 446 (*Chem. Abstr.*, 1996, **125**, 301 360).

165 N. Baba, T. Kosugi, H. Daido, H. Umino, Y. Kishida, S. Nakajima and S. Shimizu, *Biosci. Biotech. Biochem.*, 1996, **60**, 1916.

166 K.S. Bruzik, Z. Guan, S. Riddle and M.-D. Tsai, *J. Am. Chem. Soc.*, 1996, **118**, 7679.

167 P. Calias, T. Galanopoulos, M. Maxwell, A. Khayat, D. Graves, H.N. Antoniades and H. d'Alacao, *Carbohydr. Res.*, 1996, **292**, 83.

168 T. Desai, J. Gigg, R. Gigg and E. Martín-Zamora, *Carbohydr. Res.*, 1996, **296**, 97.

169 S. Roemer, C. Stadler, M.T. Rudolf, B. Jastorff and C. Schultz, *J. Chem. Soc., Perkin Trans. 1*, 1996, 1683.

170 I.D. Spiers, C.J. Barker, S.-K. Chung, Y.-T. Chang, S. Freeman, J.M. Gardiner, P.H. Hirst, P.A. Lambert, R.H. Mitchell, D.R. Poyner, C.H. Schwalbe, A.W. Smith and K.R.H. Solomons, *Carbohydr. Res.*, 1996, **282**, 81.

171 S.J. Mills and B.V.L. Potter, *J. Org. Chem.*, 1996, **61**, 8980.

172 D. Lampe, C. Liu, M.F. Mahon and B.V.L. Potter, *J. Chem. Soc., Perkin Trans. 1*, 1996, 1717.

173   P. Andersch and M.P. Schneider, *Tetrahedron: Asymm.*, 1996, **7**, 349.
174   A. Lindahl, M. Malmberg and N. Rehnberg, *J. Carbohydr. Chem.*, 1996, **15**, 549.
175   M. Malmberg and N. Rehnberg, *J. Carbohydr. Chem.*, 1996, **15**, 459.
176   A.M. Riley, C.T. Murphy, C.J Lindley, J. Westwick and B.V.L. Potter, *Bioorg. Med. Chem. Lett*, 1996, **6**, 2197.
177   M.E. Migaud and J.W. Frost, *J. Am. Chem. Soc.*, 1996, **118**, 495.
178   P. Guédat, M. Poitras, B. Spiess, G. Guillemette and G. Schlewer, *Bioorg. Med. Chem. Lett.*, 1996, **6**, 1175.
179   K.R.H. Solomons, S. Freeman, D.R. Poyner and F. Yafai, *J. Chem. Soc., Perkin Trans. 1*, 1996, 1845.
180   Y. Watanabe, S. Sofue, S. Ozaki and M. Hirata, *Chem. Commun.*, 1996, 1815.
181   N. Schnetz, P. Guédat, B. Spiess and G. Schlewer, *Bull. Soc. Chim. Fr.*, 1996, **133**, 205 (*Chem. Abstr.*, 1996, **124**, 343 880).

19
Antibiotics

1 Aminoglycosides and Aminocyclitols

The analogue **2** of kanamycin A has been synthesized by condensation of the thioglycoside **1** with a protected deoxystreptamine-3-amino-3-deoxyglucose unit followed by deprotection. Similar chemistry was carried out with the L-enantiomer of **1**, in which case the β-glycoside was also isolated, as was a regioisomer where the oxasugar unit was attached to O-5.[1] The same workers have also described a series of analogues of dibekacin in which the 3-amino-3-deoxy-D-glucose unit is replaced with an acyclic fragment (as in **3**) or with a 3-oxa- or 3-aza-sugar moiety. Analogue **3** was made by selective tetra-N-tosylation of dibekacin, followed by periodate cleavage, reduction and deprotection, whilst the 3-oxa-analogue was made by removal of the 3-amino-3-deoxyglucose unit and then condensation with a thioglycoside similar to **1**.[2] The related semisynthetic aminoglycoside arbekacin could be acetylated selectively at N-2′ by an enzymic method.[3]

The aldehyde **4** can be prepared from neomycin B, and this has been used in a combinatorial synthesis of neomycin B mimetics **5** (R = Bu^t or CH$_2$CO$_2$Me, R′ = side-chains of 12 different aminoacids), using Ugi-type four-component condensations in which the glycine unit was linked to polyethylene glycol. The products were tested as inhibitors of the HIV Rev-RRA interaction, and several of them were found to be more effective than neomycin B itself.[4]

A new semisynthetic antibiotic, 89-07, has been prepared by *N*-ethylation of N-1 of the deoxystreptamine unit in gentamycin C_{1a}.[5] A study of the reaction of gentamycin with hydrated electrons, using pulse radiolysis, has been described.[6]

A cell-free extract of an acarbose-producing *Actinoplanes sp.* phosphorylates acarbose at O-7 (i.e. the primary hydroxyl of the aminocyclitol unit).[7] Some other papers of relevance to the chemistry of acarbose-like compounds are mentioned in Chapter 18.

In the area of aminocyclopentitol antibiotics, Griffith and Danishefsky have given a full account of their synthesis of allosamidin (Vol. 25, p. 235). They also describe the synthesis of the disaccharide analogue **6**, in which the allosamizoline unit is replaced by an azasugar, and which displays strong chitinase inhibition; the corresponding GlcNAc derivative was also made.[8] Various analogues of allosamidin have been reported, including ones in which glycosylation has occurred on the alternative secondary alcohol of allosamizoline, and one (**7**) in which the oxazoline ring is fused to the cyclopentane in the reversed sense.[9] Some studies on the biosynthesis of allosamizoline have been reported.[10] The synthesis of trehazolin and its derivatives, and their biological evaluation, has been reviewed in Japanese.[11] The tetrahydropyrano[2,3-*d*]oxazoline analogue **8** of trehazolin has been prepared, but the equivalent product with a *cis*-ring junction underwent rearrangement to the furano-fused compound **9**.[12] A paper on related glycosylamino-oxazolines is mentioned in Chapter 10. Mannostatins A and B, and their enantiomers and the diastereomer of mannostatin B with the (*S*)-sulfinyl function, have been prepared from myo-inositol, and their inhibitory activity against jack-bean α-mannosidase was assessed.[13] 2-*Epi*-mannostatin A (**10**), its enantiomer, and some positional isomers have been prepared from optically-pure 4-hydroxy-2-cyclopentenone.[14]

2 Macrolide Antibiotics

The new antitumour antibiotic pyrrolosporin A (**11**) is a macrolide with an unusual spiro-α-acyl tetronic acid moiety. It also contains a 4-amino-2,4,6-trideoxy-β-D-*arabino*-hexopyranose unit, a sugar not previously found in nature (see Vol. 24, p. 148-9 for the synthesis of this sugar in studies of antibiotic analogues), and methanolysis of pyrrolosporin A gave the methyl α-pyranoside, with the pyrrole unit still linked to the sugar.[15]

erythromycin A
12

11

Coupling of 2',4''-di-*O*-acetyl-4'-*O*-demycarosyltylosin ethylene acetal *N*-oxide with phenyl 2,3,4-tri-*O*-acetyl-1-thio-α-L-rhamnopyranoside in the presence of NIS and triflic acid gave the α-L-rhamnoside with complete regioselectivity for O-3 of the macrolide, and this product was then deprotected.[16] When the 9-oxime of erythromycin A was converted to its oximate anion, reaction with carbohydrate triflates and subsequent deprotection of the sugar gave 9-*O*-glycosyl oxime derivatives such as **12**. This work, which could be carried out with no protection of the erythromycin, was designed to make more hydrophilic analogues,[17,18] and similar work was done on 9-oximes of tylosin.[18]

Cladinose analogues of 16-membered macrolide antibiotics have been prepared. Ethyl β-L-cladinoside was produced by ethanolysis of erythromycin A, and, after alkylation at O-4, the alkylated products were linked to O-4 of the mycaminose residue to give 3''-*O*-methyl-4''-*O*-alkyl derivatives of 9-dehydro-3-*O*-propionyl-leucomycin V.[19] Microbial glycosylation by a strain of *Streptomyces hygroscopicus* has been used to make the 2'-*O*-β-D-glucopyranosyl derivatives of erythromycin B, erythromycin A oxime and azithromycin.[20]

3 Anthracyclines and Other Glycosylated Polycyclic Aromatics

The daunorubicin analogues **13** (X = N₃ and CF₃CONH) have been made by linking the sugar (Chapter 9) with daunomycinone,[21] and the doxorubicin analogue **14**, with a trifluoromethyl group, was similarly prepared (see Chapter 8 for the synthesis of the sugar).[22] Similar analogues in which the daunosamine unit has been replaced with either an α-L-rhamnopyranosyl or -4-amino-4-deoxyrhamnopyranosyl moiety have also been described.[23] Linking of the appropriate protected disaccharide to the anthracyclinones was used to make the anthracycline disaccharides **15** (R = H or OMe), and also the epimers at C-4'.[24] Microbial conversion of β-rhodomycinone and aclavinone using an aclarubicin-negative *Streptomyces* mutant gave the new anthracyclinone trisaccharides CG21-C (**16**, X=Y=OH) and CG1-C (**16**, X=H, Y=CO$_2$Me), with a rednosyl-2-deoxyfucosyl-rhodosaminyl trisaccharide at C-7.[25] The amino function of doxorubicin has been coupled to 3,3-bis(diethylphosphono)propanoic acid, to give after deprotection a *gem*-bisphosphonic acid - doxorubicin conjugate, and some other related bisphosphonates were also prepared.[26]

A variety of anthracyclines fluorinated in ring A have been reported. Glycosylation of fluorinated aglycones was used to make 8-fluorocompounds such as **17**, whilst 10-fluoro-derivatives were prepared by modification of daunorubicin or idarubicin. Some C-4' epimers were also described.[27] 8-Fluoroanthracyclines have also been the subject of a symposium report.[28]

Formaldehyde efficiently cross-links the 3'-amino function of daunorubicin to the 2-amino group of guanine in some synthetic deoxyoligonucleotides and in natural DNA. In the latter case, the cross-linked drug remains intercalated.[29] Calicheamicin-daunorubicin hybrids have been prepared, by linking daunomycinone to the carbohydrate domain of the calicheamicins via a spacer unit.[30]

The benz[*a*]anthraquinone-related antibiotic urdamycinone B (**21**), formed from urdamycin B by cleavage of two *O*-glycosyl moieties, has been prepared by condensation of **18** with unprotected olivose (**19**) to give in 27% yield the *C*-glycoside (Scheme 1), which on hydrogenation gave directly the *C*-glycosyl juglone **20**. This was then converted to urdamycinone B (**21**) and its C-3 epimer by a Diels-Alder protocol.[31] The juglone derivative **20**, and similar compounds with other 2,6-dideoxyhexoses, could also be made by condensation of the unprotected sugar with naphthalene-1,5-diol followed by photooxygenation, thus giving a direct access to the common structural feature of the angucycline antibiotics.[32] Galtamycinone (**22**), the common aglycone of the *C*-glycosyl naphthacenequinone antibiotics, has been made by condensation of a previously-described intermediate (Vol. 29, p. 47), with the enolate of a substituted homophthalic anhydride.[33]

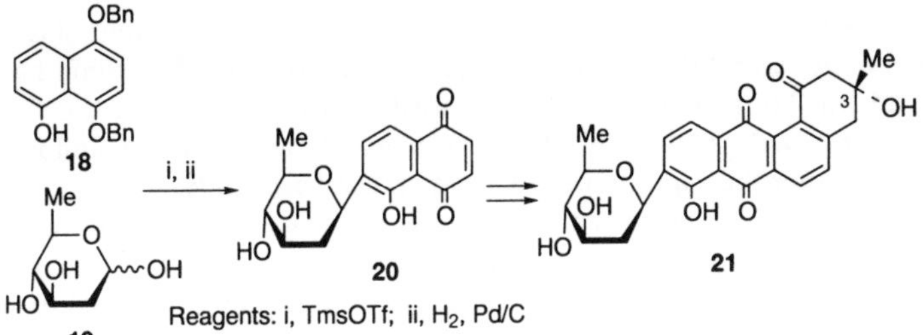

Reagents: i, TmsOTf; ii, H₂, Pd/C

Scheme 1

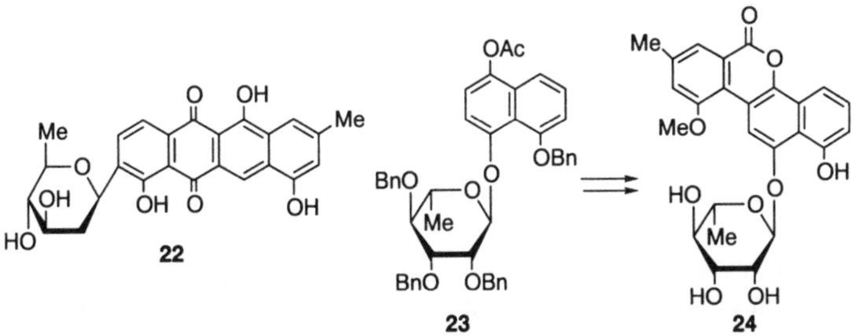

The total synthesis of the gilvocarcin class of aryl *C*-glycoside antibiotics has been reviewed.[34] The *O*-glycoside **23** was formed by condensation of the phenol with the α-glycosyl fluoride in the presence of Cp₂HfCl₂, AgClO₄ and 2,6-di-*t*-butyl-4-methylpyridine, conditions designed to suppress *C*-glycosylation, which was the major course of reaction in the absence of base. Subsequent manipulations led to the synthesis of BE-12406-A (**24**), which is related to the gilvocarcin-ravidomycin class of antibiotics, but is an *O*-glycoside rather than a *C*-glycoside.[35]

4 Nucleoside Antibiotics

A full account has been given of the synthesis of cordycepin via a furanoid glycal intermediate (Vol. 29, p. 258), and a similar approach has been adopted for the synthesis of the puromycin aminonucleoside (**25**), as outlined in Scheme 2.[36]

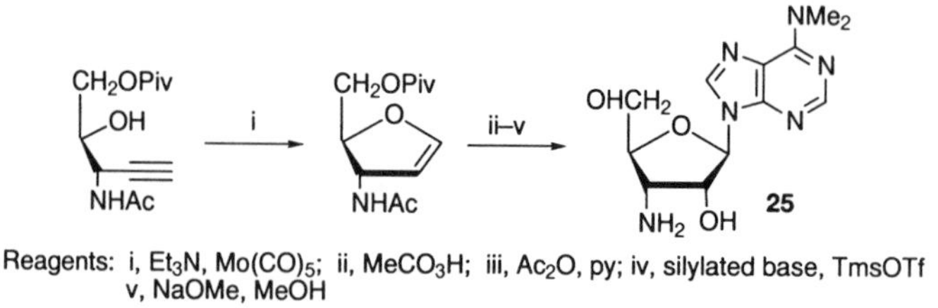
CH2OPiv
OH
NHAc
i
CH2OPiv
O
NHAc
ii–v
NMe2
N
N
OHCH2
O
N
N
25
NH2 OH
Reagents: i, Et3N, Mo(CO)5; ii, MeCO3H; iii, Ac2O, py; iv, silylated base, TmsOTf
v, NaOMe, MeOH

Scheme 2

The 2′-deoxy analogue **26** of the immunosuppressant bredinin has been prepared using radical deoxygenation, and the 5′-phosphate of bredinin has also been made; in both cases, bredinin was used as a precursor, but the reaction sequences involved photochemical ring opening of the imidazole, followed by recyclization after the appropriate modification had been made.[37]

Two syntheses have been reported of the antifungal polycyclopropane FR-900858 (**27**), both of which rely on Charette's asymmetric cyclopropanation to induce chirality, with the dihydrouridine nucleoside being introduced at a late stage.[38,39]

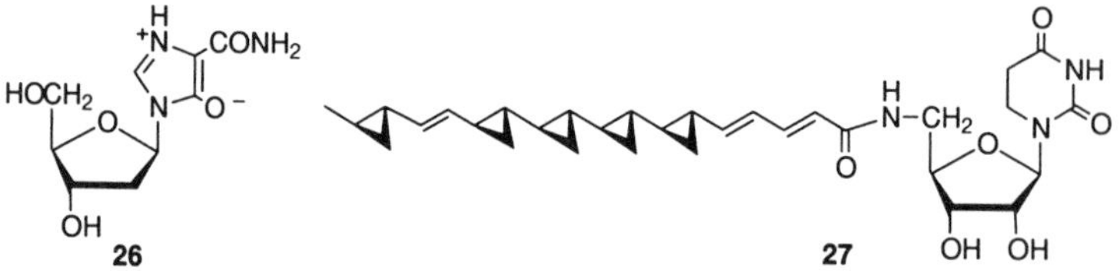
H
+N
CONH2
HOCH2
N
O-
O
OH
H
N-CH2
O
NH
N
O
O
OH OH

26 **27**

Another route to the IP_3 receptor agonist adenophostin A (**29**) has been reported. In this approach, in contrast to an earlier synthesis (Vol. 29, p. 258-9), the selectively-protected disaccharide **28** was prepared, and adenine was introduced at a late step.[40]

A review has been published, in Russian, on uronic acid nucleosides as a component of natural antibiotics.[41] The compound with the originally-proposed structure of diacetyl neosidomycin has been prepared, although the structure of neosidomycin has since been revised (see Vol. 28, p. 254).[42]

Two syntheses have been reported of the nucleoside disaccharide **30** of the cytotoxic tunicate metabolite shimofuridin. Both involve the reaction of 3′,5′-*O*-Tips-inosine, protected at N-2, with a fucosyl donor.[43,44] Shimofuridin has an unsaturated fatty acid moiety attached at O-4″, and in one of the syntheses a differentially-protected fucose unit was used, permitting selective access to O-4″, to which sorbic acid was linked as a simple analogue of the natural fatty acid.[44]

In studies directed towards the synthesis of the bacteriocin Agrocin 84, the phosphoramidate **31**, and the equivalent ribofuranosyl compound, have been prepared. The key step involved reactions of anomeric phosphoramidites with tri-*O*-benzoyladenosine, followed by oxidation at phosphorus.[45]

In the area of compounds based on higher sugars, synthesis of the tunicamycin system has been reviewed.[46] Conformational studies have been carried out on tunicamycin V and related model compounds, and on the natural substrate UDP-GlcNAc, to investigate the details of the inhibition of GlcNAc transferase by tunicamycin.[47]

A report from Gallagher's laboratory has described recent synthetic studies towards herbicidin,[48] and the first nucleoside analogue **32** with a close resemblance to herbicidin has been made by coupling adenine to a previously-described intermediate.[49] A new synthesis of sinefungin (**33**) has been reported in which both acyclic stereocentres were created with high diastereomeric purity, in the case of the centre at C-6′ by asymmetric alkylation of an enolate followed by Curtius degradation, whilst the centre at C-9′ was created by asymmetric homogeneous hydrogenation.[50] Various sinefungin analogues have also been prepared, including 6′-desaminosinefungin, and the decarboxylated compound.[51]

The analogue **34** of miharamycin, lacking the chain-branch at C-3′, has been prepared by extension of previous work (Vol. 29, p. 123-4) in which the sugar was prepared; in this new report both the 2-aminopurine and the aminoacid have been attached.[52] In an improved synthesis of uracil polyoxin C (**37**) which involves stereocontrol at C-5′, the mesylate **35** (Scheme 3) was made via asymmetric dihydroxylation of the 5′-ene; cyclization with base gave the anhydronucleoside **36**, convertible as indicated into **37**.[53] A new approach to the synthesis of nucleosides from furan using asymmetric organopalladium chemistry (see Chapter 20, Scheme 1) has also been applied to the synthesis of **37**, together with the 5′-epimer.[54]

Organopalladium chemistry was also used in the synthesis of the carbocyclic analogue of **37**, in order to link the base to an unsaturated bicyclic lactone.[55] The carbocyclic analogue **38** of oxanosine has also been prepared from a known

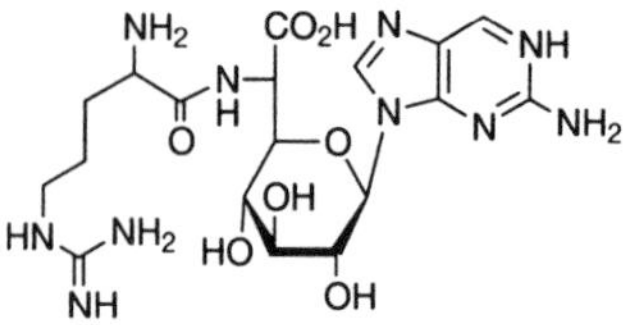

34

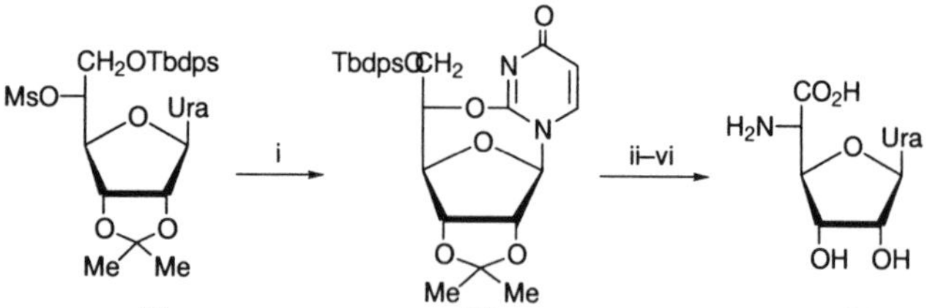

35   **36**   **37**

Reagents: i, DBU, DMF; ii, NaN$_3$; iii, TBAF; iv, PDC, DMF; v, H$_3$O$^+$; vi, H$_2$, Pd/C

Scheme 3

cyclopentane derivative by insertion of an amine function and subsequent elaboration of the heterocycle.[56]

Aristeromycin 5'-aldehyde has been prepared, and was shown to be a potent inhibitor of SAH hydrolase; this study indicated that the fluoromethylene compounds (Vol. 26, p. 247), which are also inhibitors, do not act as precursors of the aldehyde.[57] 2-Halo-derivatives **39** (X=F, Cl) of neplanocin A have been prepared, as have the equivalent structures lacking the CH$_2$OH group and the analogue **40**.[58] Neplanocin analogues with the primary alcohol oxidized to the carboxylic acid level have also been made by oxidation,[59] and aristeromycin and its cytosine analogue (carbodine) have been incorporated into a hammerhead ribozyme domain.[60]

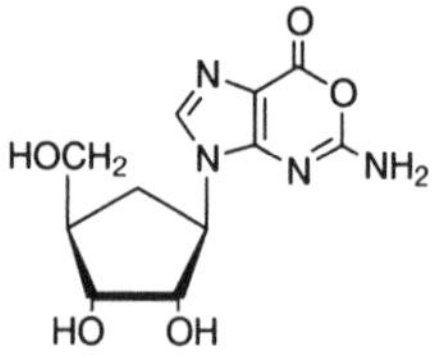

38 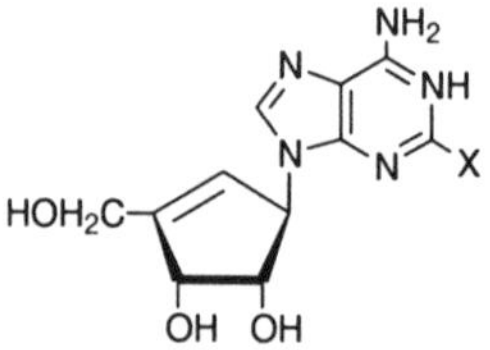   **39** 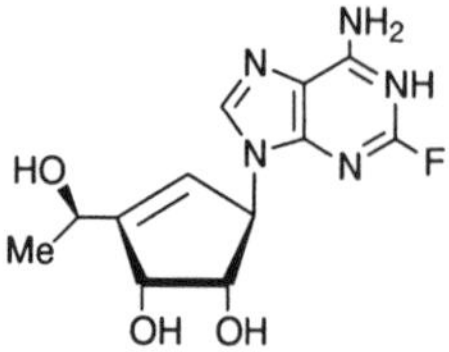   **40**

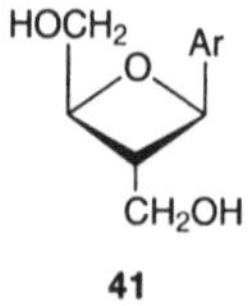

41 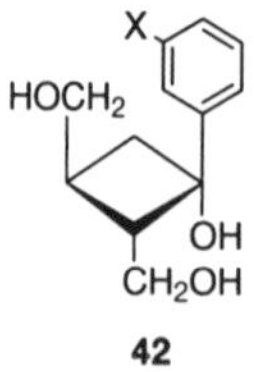   **42** 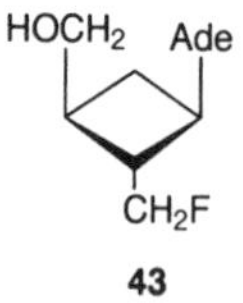   **43** 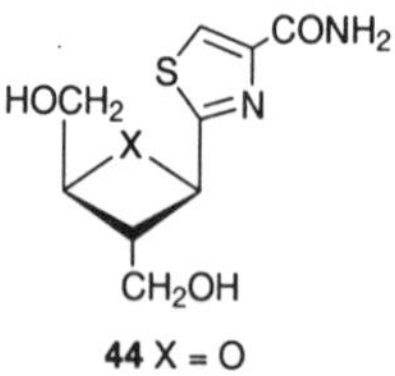   **44** X = O
                                                          45 X = CH$_2$

Further analogues of oxetanocin have been made, including the racemic aryl compounds **41** (Ar=Ph, *p*-MeO-Ph, *m*-tolyl),[61] and the phenyl carbocyclic oxetanocins **42** (X=H, CN, CONH$_2$).[62] The fluoromethyl derivative **43** of carbocyclic oxetanocin A has been prepared, as has the alternative regioisomer with F and OH interchanged.[63] The analogue **44** incorporating the heterocycle of tiazofurin has been prepared, the thiazole being elaborated from a known intermediate (Vol. 24, p. 176 and Vol. 26, p. 169) derived by ring contraction of a 2-*O*-triflyl-furanonolactone, and the carbocyclic system **45** was also made from a known cyclopropane.[64]

5   Miscellaneous Antibiotics

The desacetamido analogue of staurosporine (i.e. the compound lacking the γ-lactam ring), has been synthesized, along with some related structures.[65] A series of new antibiotics, the pyralomycins, containing a benzopyranopyrrole chromophore, have been isolated from *Microtetraspora spiralis*, and their structures were established by extensive NMR work. Pyralomycin 1a - 1d contain a cyclitol unit, as in pyralomycin 1a (**46**), whilst pyralomycin 2a (**47**) - 2c have a glucopyranose residue, although the absolute configurations of these compounds were not determined. The other pyralomycins have alternative chlorination patterns in the heterocycle.[66]

Full details have been given of the the synthesis of (+)-hydantocidin and 5-*epi*-hydantocidin from D-fructose carried out in Terashima's laboratory (see Vol. 27, p. 339-340).[67] The same group has also described a one-step synthesis of 5-*epi*-hydantocidin from D-isoascorbic acid and urea, by heating them together at 130 °C with no solvent, although the yield was very low. Studies on the acid-catalysed epimerization of hydantocidin to the 5-*epi*-compound were also reported, with the epimer being preferred at equilibrium.[68]

The α-L-arabinopyranosyl derivative of 9-hydroxyellipticine shows good anti-tumour activity. All three isomers with an *O*-methyl group specifically located on the sugar have now been made, the 2'-*O*-methyl compound **48** by alkylation of the 3',4'-*O*-isopropylidene derivative, and the other two isomers by glycosylation with an *O*-methylated glycosyl bromide. The antitumour activity suggested that all three hydroxy groups are important.[69]

The triazole **49** related to nagstatin has been prepared from tri-*O*-benzyl-L-ribofuranose by a route similar to that used for nagstatin itself (Vol. 29, p. 262), and the compound of D-*manno*-configuration was also made from tri-*O*-benzyl-L-xylofuranose.[70] The synthesis, chemistry and potential application in antiviral and antitumour chemotherapy of siastatin B (50) has been reviewed.[71] Siastatin B has been isolated from a strain of *Streptomyces nobilis*, along with the structurally-related novel heparanase inhibitors A-72363 A-1 (51), A-72363 A-2 (52) and A-72363 C (53), the absolute configurations of which were not determined.[72] Various analogues of siastatin B such as 54 and the corresponding 4-deoxy-compound have been prepared,[73] as have the analogues with the *N*-acetyl group replaced by NHCOCF$_3$ or a guanidino group.[74]

Methyl penta-*N,O*-acetyl-α-D-lincosaminide, related to lincomycin (55, X=OH), has been prepared from myo-inositol.[75] Lincomycin has been converted by a double inversion sequence via the chloride (clindamycin) into the analogues 55, X=N$_3$, imidazol-2-thiyl, etc.[76] and the lincosamine-related structure 56 has been made from 1,2:3,4-di*O*-isopropylidene-D-galactose.[77]

Calaporoside (57) (Vol. 28, p. 258) is a phospholipase C inhibitor. Deacetyl calaporoside, which is itself an inhibitor of the GABA$_A$ receptor ion channel, has been synthesized, in a process which had ~3:1 selectivity in favour of the β-linkage using 2-naphthyl tetra-*O*-benzyl-1-thio-D-mannopyranoside as glycosyl donor, and NIS-TfOH as activator.[78,79] The glycoside 58, lacking the mannonic acid unit, was also made along with its α-anomer. Both of these compounds, as well as both anomers of deacetyl calaporoside, have PLC inhibitory activity at similar levels.[79]

New glycopeptide antibiotics has been isolated which contain a 4-oxovancos-amine (dehydrovancosamine) unit, which is largely hydrated (see Vol. 28, p. 257 for previous occurrence of this sugar).[80] Reductive alkylation of the A 82846 family of glycopeptide antibiotics, which occurred selectively on the amino function of the disaccharide, gives increased antibiotic activity.[81]

A new enediyne antitumour antibiotic, namenamycin (59), has been isolated from the marine ascidian *Polysyncraton lithostrotum*. This structure has a significantly different mode of linkage between rings A and B in the trisaccharide, as compared with the hydroxylamino link in the calicheamicins.[82] A dimer of the

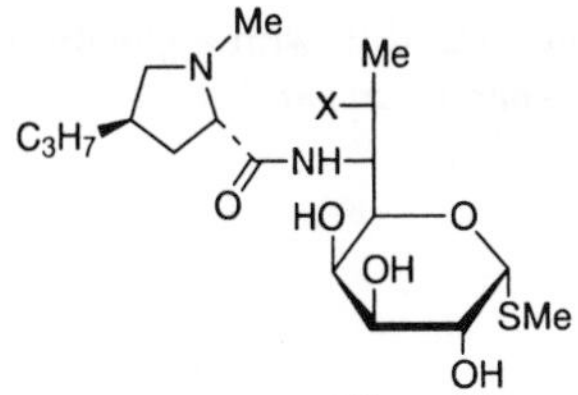

55 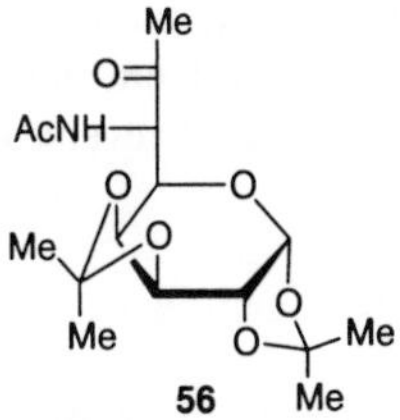   **56**

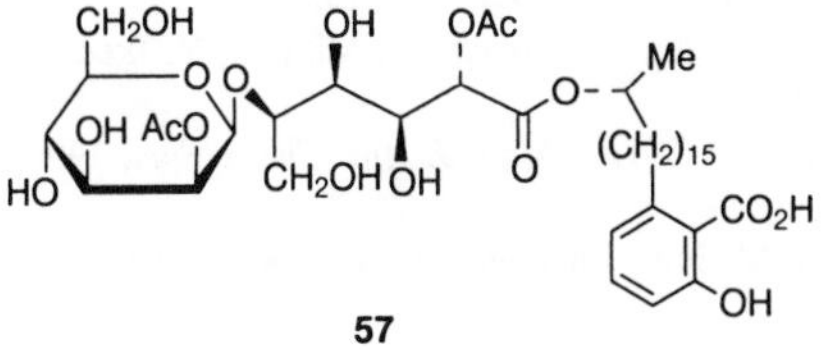

57 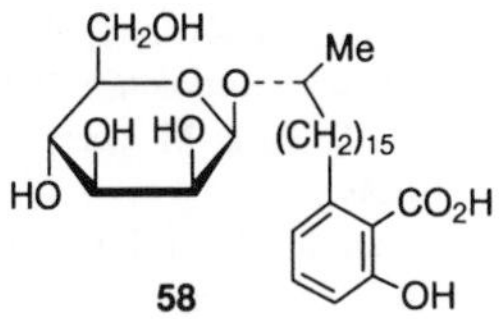   **58**

calicheamicin oligosaccharide has been synthesized, linked head-to-tail via a butane-1,4-diol spacer.[83]

Various glycosylated derivatives of rifamycins have bee prepared, in which the sugar unit is linked to the side-chain amino function, and some of these had activity against Gram-positive bacteria.[84]

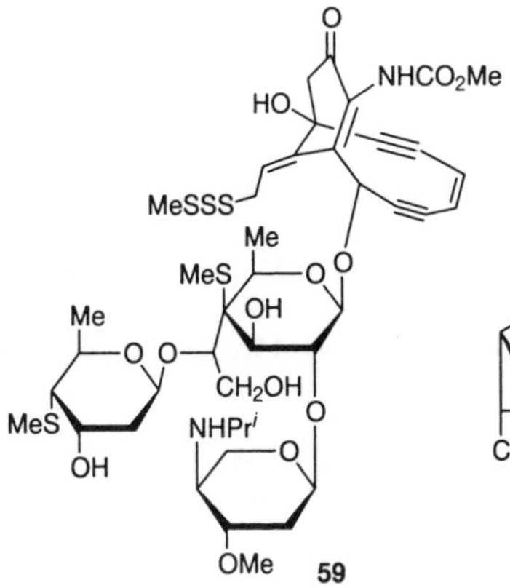

59 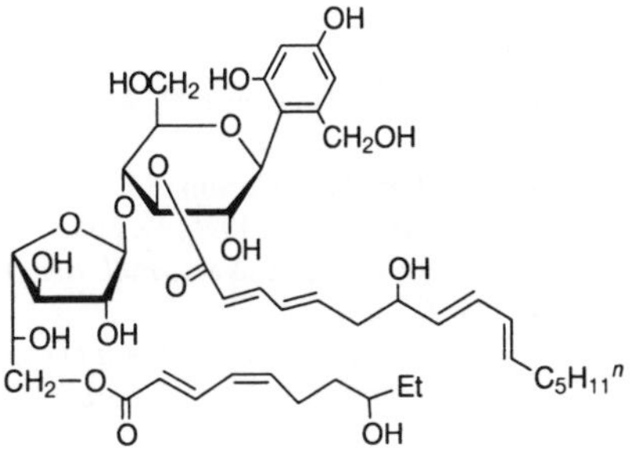   **60**

In the papulacandin area, new structures continue to emerge. Saricandin has the spirocyclic galactopyranosyl-glucopyranosyl core structure found in many of these compounds, but with a cinnamyl unit attached at O-6″,[85] whilst BE-29602 has a C_{10} side-chain at C-6″ which is identical to that of chaetiacandin (see Vol. 29, p. 262-3).[86] Furanocandin, from a *Tricothecium* species, has been assigned structure **60**, which does not have the spiro-system, and also contains the D-galactose unit in the furanose form (again see Vol. 29, p. 262-3).[87]

Extensive spectroscopic evidence supports the presence of a 1,2-linkage in the antifungal rhamnolipid produced by *Pseudomonas aeruginosa*.[88]

Two new hydrolyzable tannins, shephagenins A and B, with a glucuronic acid core, have been isolated from the leaves of *Sheperdia argentea*, and shown to be potent inhibitors of HIV-1 reverse transcriptase.[89]

References

1 R. Kuwahara and T. Tsuchiya, *Carbohydr. Res.*, 1996, **286**, 107.

2 R. Kuwahara and T. Tsuchiya, *Carbohydr. Res.*, 1996, **293**, 15.

3 K. Hotta, C.-B. Zhu, T. Ogata, A. Sunada, J. Ishikawa, S. Mizuno, Y. Ikeda and S. Kondo, *J. Antibiotics*, 1996, **49**, 458.

4 W.K.C. Park, M Auer, H. Jaksche and C.-H. Wong, *J. Am. Chem. Soc.*, 1996, **118**, 10150.

5 J. Fan, M. Zhao, J. Liu, X. Hu and M. Fan, *Zhongguo Kangshengsu Zazhi*, 1995, **20**, 401 (*Chem. Abstr.*, 1996, **125**, 115048).

6 K. Pietrucha, L. Gora and C.J. Doillon, *Radiat. Phys. Chem.*, 1996, **47**, 93 (*Chem. Abstr.*, 1996, **124**, 146 652).

7 K. Goeke, A. Drepper and H. Pape, *J. Antibiotics*, 1996, **49**, 661.

8 D.A. Griffith and S.J. Danishefsky, *J. Am. Chem. Soc.*, 1996, **118**, 9526.

9 R. Blattner, R.H. Furneaux and G.P. Lynch, *Carbohydr. Res.*, 1996, **294**, 29.

10 S. Sakuda, Z.-Y. Zhuo, H. Takao and Y. Yamada, *Tetrahedron Lett.*, 1996, **37**, 5711.

11 Y. Kobayashi and M. Shiozaki, *Yuki Gosei Kagaku Kyokaishi*, 1996, **54**, 300 (*Chem. Abstr.*, 1996, **124**, 317607).

12 M. Shiozaki, T. Mochizuki, H. Hanzawa and H. Haruyama, *Carbohydr. Res.*, 1996, **288**, 99.

13 S. Ogawa and Y. Yuming, *Bioorg. Med. Chem.*, 1995, **3**, 939 (*Chem. Abstr.*, 1996, **124**, 9207).

14 Y. Nishimura, Y. Umezawa, H. Adachi, S. Kondo and T. Takeuchi, *J. Org. Chem.*, 1996, **61**, 480.

15 D.R. Schroeder, K.L. Colson, S.E. Klohr, M.S. Lee, J.A. Matson, L.S. Brinen and J. Clardy, *J. Antibiotics*, 1996, **49**, 865.

16 C. Grandjean and G. Lukacs, *Bull. Soc. Chim. Fr.*, 1995, **132**, 1145 (*Chem. Abstr.*, 1996, **124**, 232919).

17 C. Grandjean and G. Lukacs, *J. Antibiotics*, 1996, **49**, 1036.

18 C. Grandjean and G. Lukacs, *J. Carbohydr. Chem.*, 1996, **15**, 831.

19 K. Kurihara, K. Ajito, S. Shibahara, T. Ishizuka, O. Hara, M. Araake and S. Omoto, *J. Antibiotics*, 1996, **49**, 582.

20 J. Sasaki, K. Mizoue, S. Morimoto and S. Omura, *J. Antibiotics*, 1996, **49**, 1110.

21 N. Aligiannis, N. Pouli, P. Marakos, A.-L. Skaltsounis, S. Leonce, A. Pierre and G. Atassi, *Bioorg. Med. Chem. Lett.*, 1996, **6**, 2473.

22 Y. Takagi, K. Nakai, T. Tsuchiya and T. Takeuchi, *J. Med. Chem.*, 1996, **39**, 1582.

23 K.-D. Ok, J.-B. Park, M.-S. Kim, D.-Y. Jung, S.-Y. An, C.-S. Bae and J. Yang, *Yakhak Hoechi*, 1996, **40**, 10 (*Chem. Abstr.*, 1996, **124**, 343900).

24 F. Animati, F. Arcamone, M. Berettoni, A. Cipollone, M. Franciotti and P. Lombardi, *J. Chem. Soc., Perkin Trans. 1*, 1996, 1327.

25 O. Johdo, T. Yoshioka, H. Naganawa, T. Takeuchi and A. Yoshimoto, *J. Antibiotics*, 1996, **49**, 669.

26 O. Fabulet and G. Sturtz, *Phosphorus, Sulfur, Silicon Relat. Elem.*, 1995, **101**, 225 (*Chem. Abstr.*, 1996, **124**, 87 608).

27   F. Pasqui, F. Canfarini, A. Giolitti, A. Guidi, V. Pestellini and F. Arcamone, *Tetrahedron*, 1996, **52**, 185.

28   P. Lombardi, F. Animati, A. Cipollone, G. Giannini and E. Monteagudo, *Acta Biochim. Pol.*, 1995, **42**, 433 (*Chem. Abstr.*, 1996, **125**, 11284).

29   F.F. Leng, R. Savkur, I. Fokt, T. Przewloka, W. Priebe and J.B. Chaires, *J. Am. Chem. Soc.*, 1996, **118**, 4731.

30   K.M. Depew, S.M. Zeman, S.H. Boyer, D.J. Denhart, N. Ikemoto, S.J. Danishefsky and D.M. Crothers, *Angew. Chem., Int. Ed. Engl.*, 1996, **35**, 2797.

31   G. Matsuo, Y. Miki, M. Nakata, S. Matsumuro and K. Toshima, *Chem. Commun.*, 1996, 225.

32   G. Matsuo, S. Matsumura and K. Toshima, *Chem. Commun.*, 1996, 2173.

33   T. Matsumoto, H. Yamaguchi and K. Suzuki, *Synlett*, 1996, 433.

34   K. Suzuki, T. Matsumoto and T. Hosoya, *Yuki Gosei Kagaku Kyokaishi*, 1995, **53**, 1045 (*Chem. Abstr.*, 1996, **124**, 117697).

35   T. Hosoya, E. Takashiro, Y. Yamamoto, T. Matsumoto and K. Suzuki, *Heterocycles*, 1996, **42**, 397.

36   F.E. McDonald and M.M. Gleason, *J. Am. Chem. Soc.*, 1996, **118**, 6648.

37   S. Shuto, K. Haramuishi and A. Matsuda, *Tetrahedron Lett.*, 1996, **37**, 187.

38   J.R. Falck, B. Mekonnen, J. Yu and J.-Y. Lai, *J. Am. Chem. Soc.*, 1996, **118**, 6096.

39   A.G.M. Barrett and K. Kasdorf, *Chem. Commun.*, 1996, 325.

40   N.C.R. van Straten, G.A. van der Marel, and J.H. van Boom, *Tetrahedron Lett.*, 1996, **37**, 3599.

41   V.A. Timoshchuk, *Khim.-Farm. Zh.*, 1995, **29**, 51 (*Chem. Abstr.*, 1996, **124**, 87491).

42   J.-L. Zhang and M.-S. Cai, *Huaxue Xuebao*, 1996, **54**, 84 (*Chem. Abstr.*, 1996, **124**, 290160).

43   S. Knapp and V.K. Gore, *J. Org. Chem.*, 1996, **61**, 6744.

44   H.I. Duynstee, E.R. Wijsman, G.A. van der Marel and J.H. van Boom, *Synlett*, 1996, 313.

45   T. Moriguchi, T. Wada and M. Sekine, *J. Org. Chem.*, 1996, **61**, 9223.

46   A. Zamojski and A. Banaszek, *Wiad. Chem.*, 1995, **49**, 683 (*Chem. Abstr.*, 1996, **125**, 58855).

47   M.B. Chen, Y. Liu, Z.W. Guo and G.Q. Cai, *Chin. J. Chem.*, 1996, **7**, 153 (*Chem. Abstr.*, 1996, **125**, 11326).

48   H.M. Binch, A.M. Griffin and T. Gallagher, *Pure Appl. Chem.*, 1996, **68**, 589.

49   H.M. Binch and T. Gallagher, *J. Chem. Soc., Perkin Trans. 1*, 1996, 401.

50   A.K. Ghosh and W. Liu, *J. Org. Chem.*, 1996, **61**, 6175.

51   P. Blanchard, M.S. El Kortbi, J.-L. Fourrey, F. Lawrence and M. Robert-Gero, *Nucleosides Nucleotides*, 1996, **15**, 1121.

52   S. Czernecki, S. Franco, S. Horns and J.-M. Valéry, *Tetrahedron Lett.*, 1996, **37**, 4003.

53   C.M. Evina and G. Guillerm, *Tetrahedron Lett.*, 1996, **37**, 163.

54   B.M. Trost and Z. Shi, *J. Am. Chem. Soc.*, 1996, **118**, 3037.

55   V.K. Aggarwal, N. Monteiro, G.J. Tarver and S.D. Lindell, *J. Org. Chem.*, 1996, **61**, 1192.

56   H. Kurata, S. Nishiyama, S. Yamamura, K. Kato, S. Fujiwara and K. Umezawa, *Bioorg. Med. Chem. Lett.*, 1996, **6**, 283.

57   S. Liu, C. Yuan and R.T. Borchardt, *J. Med. Chem.*, 1996, **39**, 2347.

58   T. Obara, S. Shuto, Y. Saito, R. Snoeck, G. Andrei, J. Balzarini, E. De Clercq and A. Matsuda, *J. Med. Chem.*, 1996, **39**, 3847.

59   T. Obara, S. Shuto, Y. Saito, M. Toriya, K. Ogawa, S. Yaginuma, S. Shigeta and A. Matsuda, *Nucleosides Nucleotides*, 1996, **15**, 1157.

60   F. Burlina, A. Favre, J.-L. Fourrey and M. Thomas, *Chem. Commun.*, 1996, 1623.

61   M.E. Jung and C.J. Nichols, *Tetrahedron Lett.*, 1996, **37**, 7667.

62   T. Maruyama and M. Fukuhara, *Chem. Pharm. Bull.*, 1996, **44**, 1407.

63   Y. Sato, K. Ueyama, T. Maruyama and D.D. Richman, *Nucleosides Nucleotides*, 1996, **15**, 109.

64   Y. Kikuchi, S. Nishiyama, S. Yamamura, K. Kato, S. Fjiwara, K. Umezawa and Y. Terada, *Bioorg. Med. Chem. Lett.*, 1996, **6**, 1897.

65   J.L. Wood, B.M. Stoltz, K. Onweme and S.N. Goodman, *Tetrahedron Lett.*, 1996, **37**, 7335.

66   N. Kawamura, R. Sawa, Y. Takahashi, K. Isshiki, T. Sawa, H. Naganawa and T. Takeuchi, *J. Antibiotics*, 1996, **49**, 651.

67   N. Nakajima, M. Matsumoto, M. Kirihara, M. Hashimoto, T. Katoh and S. Terashima, *Tetrahedron*, 1996, **52**, 1177.

68   N. Nakajima, M. Kirihara, M. Matsumoto, M. Hashimoto, T. Katoh and S. Terashima, *Heterocycles*, 1996, **42**, 503.

69   T. Shimamoto, S. Imajo, T. Honda, S. Yoshimura and M. Ishiguro, *Bioorg. Med. Chem. Lett.*, 1996, **6**, 1331.

70   K. Tatsuta, Y. Ikeda and S. Miura, *J. Antibiotics*, 1996, **49**, 836.

71   Y. Nishimura, *Stud. Nat. Prod. Chem.*, 1995, **16**, 75 (*Chem. Abstr.*, 1996, **124**, 117696).

72   T. Takatsu, M. Takahashi, Y. Kawase, R. Enokita, T. Okazaki, H. Matsukawa, K. Ogawa, Y. Sakaida, T. Kagasaki, T. Kinoshita, M. Nakajima and K. Tanzawa, *J. Antibiotics*, 1996, **49**, 54.

73   Y. Nishimura, T. Satoh, T. Kudo, S. Kondo and T. Takeuchi, *Bioorg. Med. Chem.*, 1996, **4**, 91.

74   T. Satoh, Y. Nishimura, S. Kondo, T. Takeuchi, M. Azetaka, H. Fukuyasu, Y. Iizuka, S. Ohuchi and S. Shibahara, *J. Antibiotics*, 1996, **49**, 321.

75   N. Chida, K. Nakazawa, S. Ninomiya, S. Amano, K. Koizumi, J. Inaba and S. Ogawa, *Carbohydr. Lett.*, 1995, **1**, 335 (*Chem. Abstr.*, 1996, **124**, 317686).

76   F. Sztaricskai, Z. Dinya, M.M. Puskas, G. Batta, R. Masuma and S. Omura, *J. Antibiotics*, 1996, **49**, 941.

77   M.A.F. Prado, R.J. Alves, A. Braga de Oliveira and J. Dias de Souza Filho, *Synth. Commun.*, 1996, **26**, 1015.

78   K. Tatsuta and S. Yasuda, *Tetrahedron Lett.*, 1996, **37**, 2453.

79   K. Tatsuta and S. Yasuda, *J. Antibiotics*, 1996, **49**, 713.

80   L. Vértezy, H.-W. Fehlhaber, H. Kogler and M. Limbert, *J. Antibiotics*, 1996, **49**, 115.

81   R.D.G. Cooper, N.J. Snyder, M.J. Zweifel, M.A. Staszek, S.C. Wilkie, T.I. Nicas, D.L. Mullen, T.F. Butler, M.J. Rodriguez, B.E. Huff and R.C. Thompson, *J. Antibiotics*, 1996, **49**, 575.

82   L.A. McDonald, T.L. Capson, G. Krishnamurthy, W.-D. Ding, G.A. Ellestad, V.S. Bernan, W.M. Maiese, P. Lassota, C. Discafini, R.A. Kramer and C.M. Ireland, *J. Am. Chem. Soc.*, 1996, **118**, 10898.

83   K.C. Nicolaou, K. Ajito, H. Konematsu, B.M. Smith, P. Bertinato and L. Gomez-Paloma, *Chem. Commun.*, 1996, 1495.

84   C. Bartolucci, L. Cellai, C. Martuccio, A. Rossi, A.L. Segre, S.R. Savu and L. Silvesto, *Helv. Chim. Acta*, 1996, **79**, 1611.

85   R.H. Chen. S. Tennant, D. Frost, M.J. O'Bierne, J.P. Karwowski, P.E. Humphrey, L.-H. Malmborg, W. Choi, K.D. Brandt, P. West, S.K. Kadam, J.J. Clement and J.B. McAlpine, *J. Antibiotics*, 1996, **49**, 596.

86 H. Okada, M. Nagashima, H. Suzuki, S. Nakajima, K. Kojiri and H. Suda, *J. Antibiotics,* 1996, **49**, 103.

87 E. Magome, K. Harimaya, S. Gomi, M. Koyama, N. Chiba, K. Ota and T. Mikawa, *J. Antibiotics*, 1996, **49**, 599.

88 S.-S. Moon, P.M. Kang, B.S. Kim and B.K. Hwang, *Bull. Korean Chem. Soc.*, 1996, **17**, 291 (*Chem. Abstr.*, 1996, **124**, 343869).

89 T. Yoshida, H. Ito, T. Hatano, M. Kurata, T. Nakanishi, A. Inada, H. Murata, Y. Inatomi, N. Matsuura, K. Ono, H. Nakane, M. Noda, F.A. Lang and J. Murata, *Chem. Pharm. Bull.,* 1996, **44**, 1436.

20
Nucleosides

1 Synthesis

Vorbrüggen has reviewed recent trends in nucleoside synthesis, including the new method developed recently in his laboratory involving the condensation of persilylated sugars and persilylated bases (see Vol. 29, p. 268-269).[1] Reports from Mukaiyama's group have elaborated further on the use of methyl 2,3,5-tri-*O*-benzoyl-β-D-ribofuranosyl carbonate as a glycosyl donor (see Vol. 28, p. 213); various iodides (e.g. SbI_3, TeI_4) can function as Lewis acids in such condensations to form protected nucleosides,[2] and use of catalytic quantities of either SbI_3 or TeI_4 with sulfated *N*-benzoyladenine gives high yields of tetrabenzoyl adenosine.[3] The use of pyridine-containing leaving groups in the synthesis of α-ribonucleosides reported last year (Vol. 29, p. 269) has now been extended to the formation of D-galactopyranosyl and L-fucopyranosyl nucleosides; in these pyranose cases, *O*-benzylated glycosyl donors gave predominantly 1,2-*trans*-nucleosides.[4]

A novel *de novo* synthesis of nucleosides (Scheme 1) involves the conversion of the *meso*-compound **1**, derived from furan, into the pure enantiomer **2** by two successive allylations using organopalladium intermediates. The first reaction uses 6-chloropurine as nucleophile, in the presence of a chiral bisphosphine. The subsequent elaboration of **2** indicates the use of the substituted malonate as an ingenious CO_2H synthon. Use of the enantiomeric catalyst in the first step gives the D-nucleoside.[5]

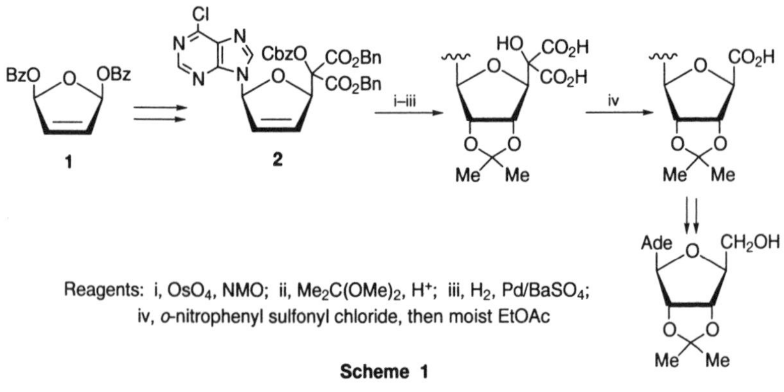

Reagents: i, OsO_4, NMO; ii, $Me_2C(OMe)_2$, H^+; iii, H_2, $Pd/BaSO_4$; iv, *o*-nitrophenyl sulfonyl chloride, then moist EtOAc

Scheme 1

Carbohydrate Chemistry, Volume 30

The first chemical synthesis of SAICAR (**3**), an intermediate in purine nucleotide biosynthesis, has been reported. The imidazole ring was elaborated around the amino group of 2,3-*O*-isopropylidene-β-D-ribofuranosylamine.[6]

When per-*O*-Tms-2-*N*-acylguanines and *O*-acetylated pentofuranoses are coupled in the presence of TiCl₄, the reaction gives mainly the 7-glycosylguanines, and the 7- and 9-glycosylated products are fortuitously separable by simple partitioning. Use of persilylated 2-*N*-acyl-6-*O*-diphenylcarbamoylguanine gives the 9-linked nucleosides. This finding, applicable to the formation of β-D-ribo-, α-D-arabino- and β-D-xylo-furanosylguanines, provides a solution to the long-standing problem of regiochemical control in glycosylation of guanine.[7]

Conventional base-sugar coupling procedures have been used to make the β-D-ribofuranosyl derivatives of 2- and 4-pyridone,[8] and to prepare uridine and cytidine multiply-labelled with ^{17}O and ^{15}N in the bases,[9] [5'-^{13}C]-ribonucleosides,[10] S^2-alkyl-2-thiouridines,[11] and 5-thioalkyluridines.[12] 6-Alkyluridines have been prepared by couplings where the 3-position of the uracil unit is blocked in one of two ways to avoid N^3-ribosylation.[13] Also reported have been β-D-ribofuranosyl derivatives of 5-amino-4-sulfonamidoimidazoles (see Vol. 28, p. 264),[14] 5-amino-3-pyrazolone,[15] pyrazone-2-ones **4** (R=H, Me, C₁₀H₂₁) from which 3-alkylpiperazin-2-ones, also accessible by direct condensation, could be made by reduction,[16] triazindiones **5**, reducible to their 2,3-dihydroderivatives, the 2-ribosylated analogues of which were made by ribosylation at the reduced level,[17] sulfahydantoins **6**,[18] 1,2,4-triazole-5(4H)-thiones,[19] quinolones such as **7**,[20] 2,5- and 2,6-disubstituted benzimidazoles (e.g. **8**),[21] and 2-thioxo-3,5,7-trisubstituted pyrido[2,3-*d*]pyrimidine-4(1H)-ones.[22] Either the betaine structures **9** (X=CH or N) or the corresponding N^1-ribosylated compounds could be obtained by condensation depending on the conditions.[23]

Pyrazolo[3,4-*d*]pyrimidine nucleosides **10** (R=H, Me) have been made in a new way by reaction between 2,3,6-tri-*O*-benzyl-β-D-ribofuranosylhydrazine and 6-chloro-pyrimidinone-5-carboxaldehydes.[24]

Either **11** or the N-1 ribosylated compound could be prepared selectively depending on the coupling conditions, the products being stretched analogues of the antiviral 1-β-D-ribofuranosyl-2,5,6-trichlorobenzimidazole.[25] The pyrazolo [3,4-*b*]quinoline nucleoside **12** has been prepared by silyl coupling and some 5′-chain-extended compounds were also described.[26]

A study has been made of the further ribosylation of inosine derivatives. Use of tri-*O* benzoylribofuranosyl bromide as glycosyl donor gave some of the N^1-β-product **13**, required for the synthesis of cADPR analogues, but other products were produced as well.[27] Bis(β-D-ribofuranosyl)-imidazolium species have also been described.[28]

A high-yielding and direct route to (β-D-lyxofuranosyl)uracil has been reported,[29] and 5-phenyl-1-(β-D-xylofuranosyl)-2-pyrimidinone and some related nucleosides have been prepared.[30]

Condensation methods have been used to make β-D-gluco- and β-D-galacto-pyranosyl nucleosides of some condensed pyridinethiones[31] and hydantoins with naphthylmethylene substituents at C-5,[32] and glucopyranosyl derivatives of 4-amino-1,2,4-triazin-5-on-3-thiones have been described.[33] When the acyclic amine **14** was treated with (EtO)₂CHOAc, 7-(β-D-glucopyranosyl)theophylline (**15**) was obtained, and the analogous α-L-arabinopyranosyl compound was similarly prepared.[34]

2   Anhydro- and Cyclo-nucleosides

2′,3′-Anhydronucleosides of pyrazolo[3,4-*d*]pyrimidines have been prepared by 2′,3′-*trans*-elimination of the corresponding xylofuranosyl systems.[35] When the α-hydroxyphosphonate **16** (see Vol. 29, p. 291 and earlier) was desilylated and then treated with triflyl chloride, the epoxyphosphonate **17** was obtained, and a 2′,3′-epoxy-2′-phosphonate was made similarly.[36]

There have been reports extending earlier work (Vol. 28, p. 281) concerning the formation of spirocyclic systems at C-1′ using radical chemistry. When the 1′,2′-

ene **18** of the earlier work was treated with tributylstannane-AIBN in the presence of oxygen, a moderate yield of the arabinofuranosyl compound **19** was obtained. The stereochemistry at C-2' could subsequently be inverted.[37] Such a cyclization, in the purine series, and in the absence of oxygen, gave some of the spiro-compound but a greater amount of the product **20**, together with the isomer with the alternative *cis*-ring fusion.[38] In the absence of a 1'-ene as is present in **18**, spiro-compounds are formed by a sequence involving 1,5-radical translocation, subsequent 5-*endo*-trig cyclization, and loss of a bromine atom; the example of the formation of **22** from **21** by action of Bu₃SnH-AIBN is illustrative, and similar sequences have been carried out on purine[39] and 2'-deoxypurine compounds.[39,40]

Treatment of compounds **23** with DBU gave the 2,2'-anhydronucleosides **24**.[41] When the 1'-cyanocompound **25** (Vol. 29, p. 275-76) was treated with Bu₄NF, the anhydrosystem **26** was obtained; it was speculated that the nitrile hydrolysis occurs by the participation of the 3'-OH as a general base. The amide **26** could be converted into 1'-cyano-β-D-arabinofuranosylcytosine (**27**).[42]

2,3'-Anhydrothymidine (**28**) has been prepared in 81% yield by a Mitsunobu reaction on unprotected thymidine,[43] whilst **29** can be obtained by treatment of 5'-*O*-tritylthymidine with perfluorobutane sulfonyl fluoride. Use of this reagent on thymidine itself gives the 5'-deoxy-5'-fluoro compound **30**.[44] 2',3'-Anhydroquinazoline nucleosides **31** (R=H, Me) have been made by cyclization of 3'-*O*-tosylates.[45] Intramolecular glycosylation was used during the synthesis of the methylene-bridged analogues **32** and **33** of nicotinamide ribotide; these were converted into NAD analogues in which the central oxygen of the diphosphate was replaced by a methylene unit, and these products were found to be unstable with regard to hydrolysis of the nicotinamide-sugar link.[46]

The bridged compounds **34** (R=H, allyl) have been made from the 5'-phenylselenouridine derivative, produced by a Pummerer reaction (Vol. 26,

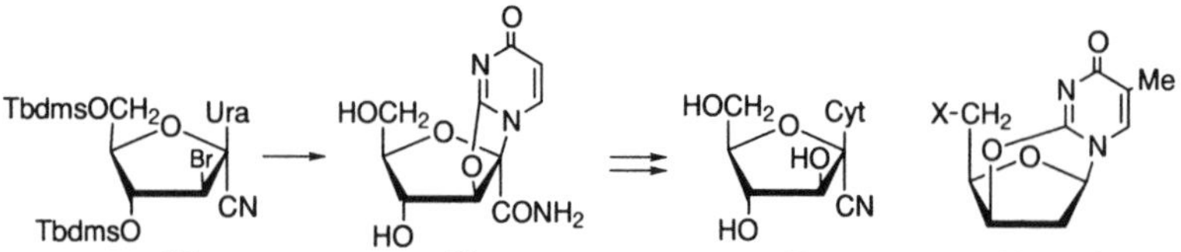

25   **26**   **27**

28 X = OH
29 X = OTr
30 X = F

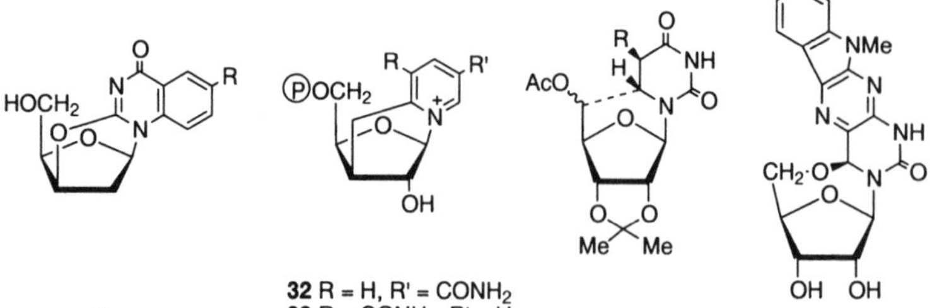

31

32 R = H, R' = CONH₂
33 R = CONH₂, R' = H

34

35

p. 239), with either tributylstannane or allyltributylstannane and AIBN.[47] When 5-aminocytosine was treated with *N*-methylisatin in aqueous ethanol, a high yield of **35** was obtained. Seemingly condensation of the diaminopyrimidine with the isatin makes C-6 of the pyrimidine more liable to nucleophilic attack.[48]

There has been a further report on the intramolecular hemiacetals formed from uridine-6-carboxaldehyde and related compounds (see Vol. 27, p. 244 and Vol. 26, p. 227-8), including evidence for the configuration at the hemiacetal carbons.[49]

A number of uses of anhydronucleosides in the synthesis of other types of nucleoside are mentioned elsewhere in this Chapter.

3   Deoxynucleosides

A new method for the stereocontrolled synthesis of pyrimidine β-D-2'-deoxyribonucleosides involves the use of a photolabile directing group at C-2', and is illustrated by the case in Scheme 2.[50] New glycosyl donors that have been employed with good to excellent stereoselectivity in reactions with silylated pyrimidines are the deoxyribofuranosyl phosphoramidate **36**[51] and the thiocarbamate **37**; in this latter case, intramolecular participation by the sulfur is thought to be involved in ensuring β-selectivity.[52]

2'-Deoxy-2-thiouridine has been prepared using immobilized *N*-deoxyribosyl transferase from *Lactobacillus leichmanii*,[53] whilst radical deoxygenation was used to prepare 2'-deoxy-3-isoadenosine (**38**), which was converted into the two regioisomeric dinucleoside monophosphates with thymidine.[54]

Base-sugar coupling procedures have been used to prepare 1-(2'-deoxy-β-D-

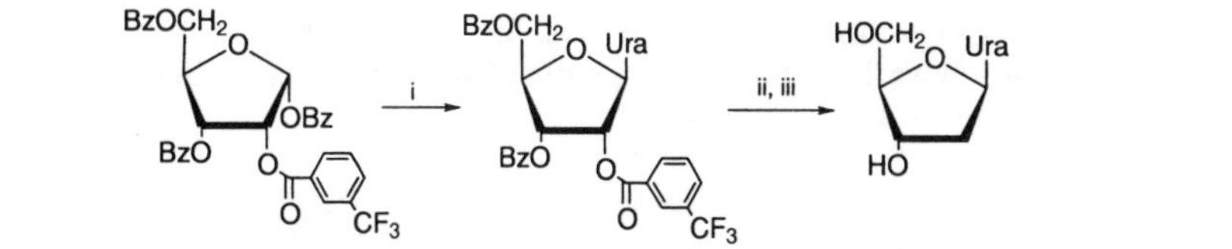

Reagents: i, bis(Tms)uracil, $SnCl_4$; ii, hv, *N*-methylcarbazole, $MgClO_4$, Pr^iOH (70%); iii, NH_3, MeOH

Scheme 2

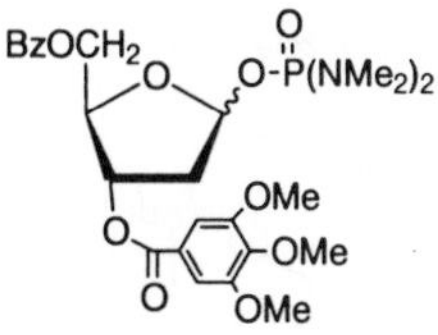

36 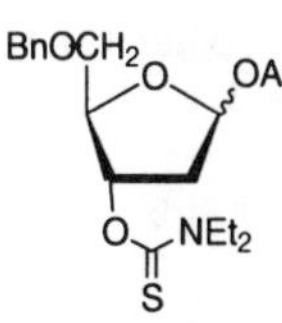 37 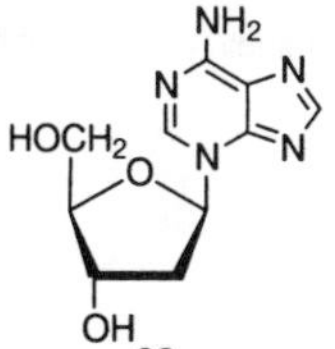 38 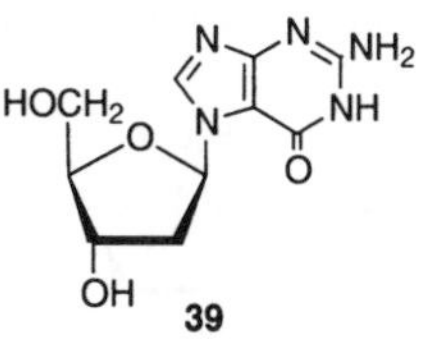 39

ribofuranosyl)-imidazole-4-carboxamide and -pyrazole-3-carboxamide,[55] 2'-deoxydiazepinone ribonucleosides,[56] 6-methyl-9-(2'-deoxy-β-D-ribofuranosyl) purine,[57] N^2-(*p*-*n*-octylphenyl)-deoxyguanosine (converted to the dGTP analogue),[58] the N^7-regioisomer of deoxyguanosine (**39**), the precursor of which was produced in significant amounts when 2-amino-6-methoxypurine was condensed with 2-deoxy-3,5-di-*O*-toluoyl-α-D-*erythro*-pentofuranosyl chloride,[59] 1-deazapurine 2'-deoxyribonucleosides **40** (R=NO_2, H, Cl), where the major regioisomer was as shown,[60] benzimidazoles **41** (R=NO_2, F, Me) and the regioisomers formed in the coupling,[61] the 7-deaza-analogue **42** of deoxyadenosine, to which various alkynes were subsequently coupled by palladium-catalysed processes,[62] 6- and 7-aryllumazine 2'-deoxy-β-D-ribofuranosides such as **43**,[63] and N^8-(2'-deoxy-β-D-ribofuranosyl)pteridines (e.g. **44**).[64,65]

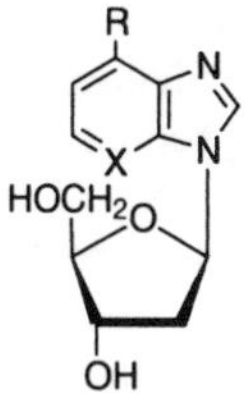

40 X = N
41 X = CH
42 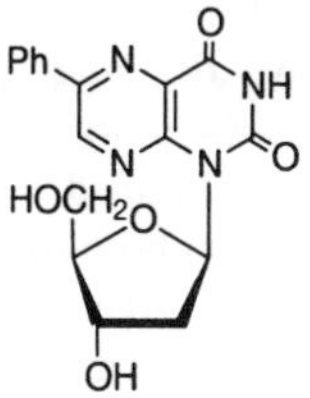 43 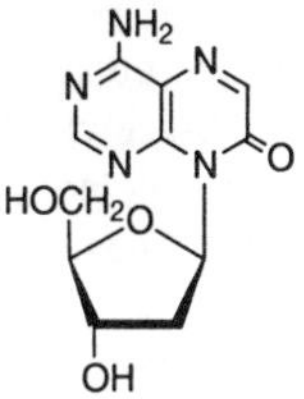 44 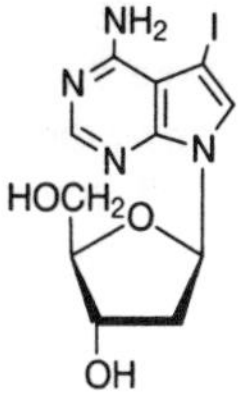

A stereospecific synthesis of α-2'-deoxycytidine (**45**) is outlined in Scheme 3,[66] and α-N^7-deoxyinosine (**46**) has been synthesized by a coupling procedure,[67] both α-nucleosides being incorporated into oligonucleotides.

There has been a report concerning the synthesis of 2'-deoxy-β-D-xylofuranosylpurines **47** (B=Gua, isoguanine, xanthine) and their assembly into oligodeoxynucleotides; in this work, the sugar configuration was established by a means previously reported (Vol. 23, p. 215).[68]

A previously described intermediate (Vol. 29, p. 275-6) has been reductively debrominated to give 1'-cyano-2'-deoxyuridine (48), and the thymidine analogue has been made by an analogous method.[42]

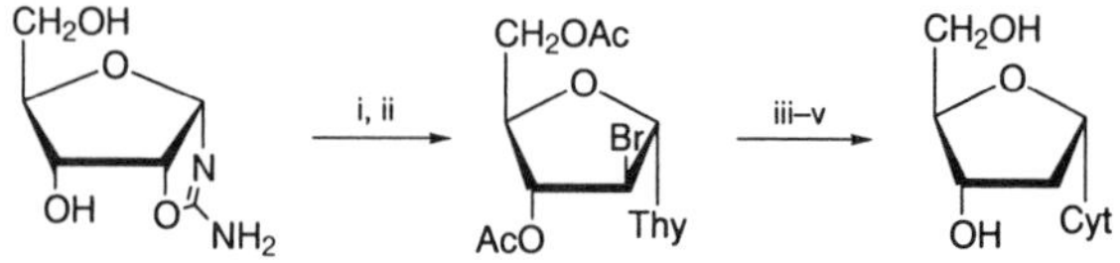

Reagents: i, methyl propiolate; ii, MeCOBr; iii, Bu₃SnH; iv, POCl₃, triazole; v, NH₄OH

Scheme 3

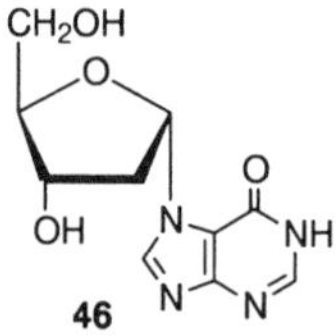

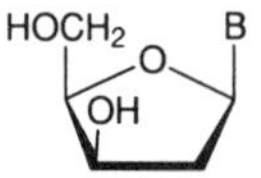

There have been reports on the synthesis of 2'-deoxynucleosides chain-extended at C-5'. The homologue 49 of deoxyadenosine has been prepared from 1,2:5,6-di-O-isopropylidene-α-D-allofuranose, with radical deoxygenations at C-2' and C-5',[69] whilst various compounds of type 50 (R=CN, N₃, OMe, -CH=CH₂, NO₂) have been made either by opening of a 5',6'-epoxide or by addition of nucleophiles to the 5'-aldehyde.[70] Other workers have shown that reaction of the same thymidine 5'-aldehyde with the silyl ketene acetal 1-t-butoxy-1-(Tbdms)oxyethene gives the product 50 (R=CO₂Buᵗ) stereoselectively.[71]

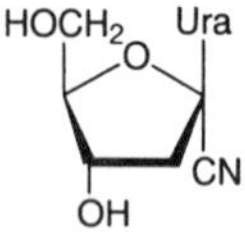

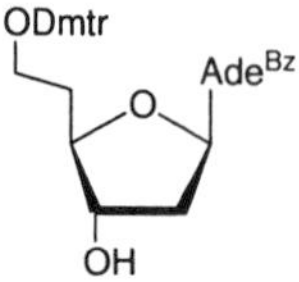

With regard to 2',3'-dideoxynucleosides, a photochemical method previously applied to the synthesis of branched-chain nucleosides (Vol. 28, p. 278) has now been carried out with optically-pure cyclobutanone 51; photolysis of this in the presence of purines gives the dideoxynucleosides 52 (X=H, Cl, OMe).[72] Dide-oxypurine nucleosides have also been made by the Pd-catalysed reduction of 9-(2,5-di-O-acetyl-3-bromo-β-D-xylofuranosyl)purines, the dideoxy-product, rather than the 3'-deoxy-, being obtained with the pH controlled at 9.5.[73] The 6-benzyluridine derivative 53 has been made by condensation using an intermediate similar to 37, followed by deoxygenation at C-3',[74] and both 2',3'-dideoxy- and 2',3'-didehydro-2',3'-dideoxy-derivatives of isoguanosine have been made using Corey-Winter deoxygenation.[75] A new coenzyme B₁₂ analogue containing 2',3'-dideoxyuridine has been prepared and characterized.[76]

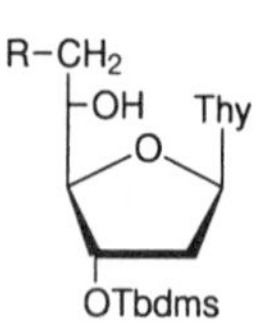

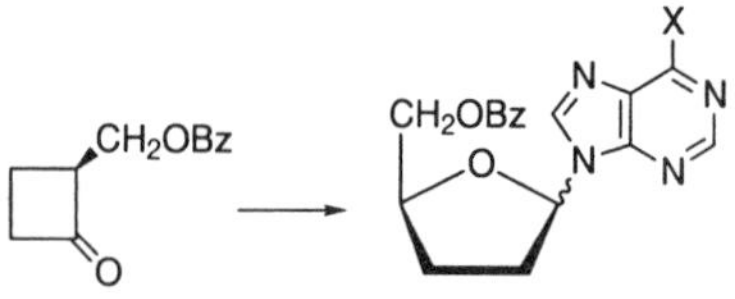

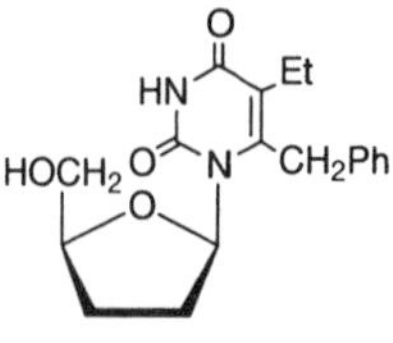

In the area of 2',3'-didehydro-2',3'-dideoxynucleosides, a new route to compounds of this type in the pyrimidine series is outlined in Scheme 4. The thioglycoside **54** was produced directly from deoxyribose and thiophenol in acidic conditions, and the condensations to form the nucleoside derivatives were β-selective by about 2:1.[77] A full account has been given of the formation of 2',3'-didehydro-2',3'-dideoxy systems from 2',3'-dimesylates, protected at O-5', by treatment with telluride anion (see Vol. 27, p. 247).[78] Treatment of the furanoid glycal **55**, made by cyclization of an acetylenic alcohol (Chapter 13), with silylated thymine in the presence of iodine, followed by sodium methoxide, provides a new route to d4T (**56**).[79] A new synthesis of d4T (**56**) from 5-methyluridine has also been described,[80] as has a route to d4T labelled with ^{14}C at C-1', which starts from [1-^{14}C]-ribose and proceeds via [1'-^{14}C]-5-methyluridine, convertible in very high yield to [1'-^{14}C]-d4T.[81]

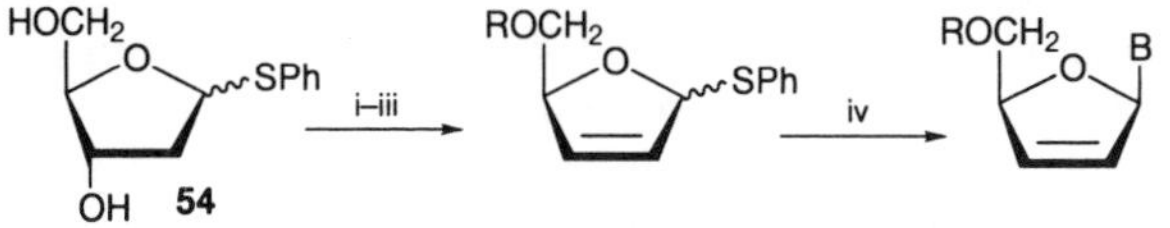

Reagents: i, RCl, py (R = Tbdms, Tbdps, Tr); ii, Tf₂O, py; iii, DBU; iv, silylated pyrimidine base, NBS

Scheme 4

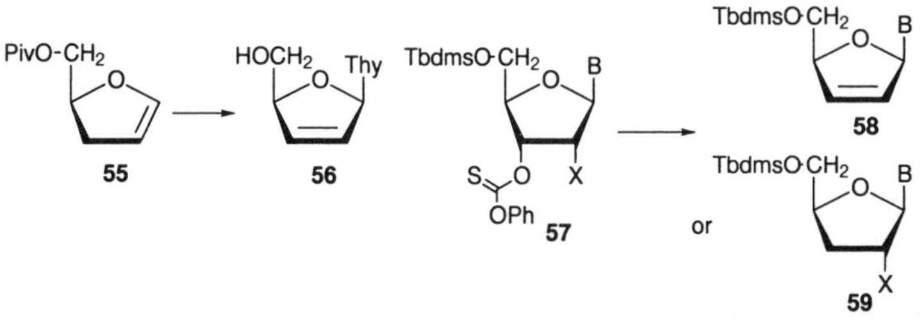

In a study related to the mechanism of inactivation of ribonucleotide diphosphate reductase (RDPR) by 2'-deoxy-2'-substituted nucleotide analogues, it was found that treatment of nucleosides of type **57** (B=Ura, Ade, X=I, Br, Cl, SMe, N₃) with tributylstannane and AIBN gave the d4 products **58**, whilst with X=F, OMs or OTs, the 3'-deoxysystem **59** was the product. The results pointed to the loss of radicals rather than anions from C-2' during mechanism-based inactivation of RDPR, and the authors suggest some modifications to Stubbe's mechanistic proposals.[82]

Routes have been developed for the synthesis of β-L-didehydrodideoxynucleosides in the purine series (**60**, B=Ade, Gua, Hypoxanthine), and the corresponding hydrogenated compounds, starting from either L-xylose or from D-glutamic acid.[83] The L-nucleoside derivative **61** has been prepared by known methods from L-arabinose, and was converted to β-L-d4C (**60**, B=Cyt), via **62** as an intermediate. Transglycosylation of **61** with 5-fluorouracil was used to make the L-enantiomer of 5-fluoro-d4C (**60**, B=5-fluoro-Ura).[84] Some of these L-enantiomers have potent anti-HIV and anti-HBV activity.

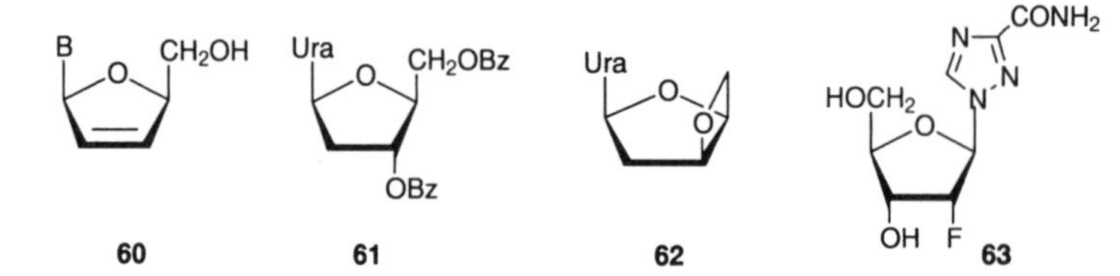

60   61   62   63

4   Halogenonucleosides

An enzymic synthesis of 2′-deoxy-2′-fluororibavirin (**63**) has been reported, involving the use of nucleoside transferase to exchange the triazole of ribavirin with the base of 2′-deoxy-2′-fluorouridine.[85]

Fluorination using DAST was employed in the synthesis of the 2′-deoxy-2′-fluoro-β-D-arabinofuranoside of 6-chloropurine (**64**),[86] and a number of pyrimidine nucleosides with a 2′-deoxy-2′-fluoro-β-L-arabinofuranose unit have been prepared using base-sugar coupling procedures; the thymine compound **65** (L-FMAU) had particular potency against HBV.[87]

The synthesis and antitumour activity of 2′,2′-difluoro-deoxycytidine (gemcytabine) has been reviewed by workers at Eli Lilly.[88]

Nucleophilic attack on 2′,3′-cyclic sulfates of inosine and adenosine occurs predominantly at C-3′, and in this way the 3′-deoxy-3′-fluoroxylofuranosides **66** (X=F) were prepared.[89] 5-Chloro-2′,3′-dideoxy-3′-fluorouridine labelled with tritium at C-5′ has been prepared by an oxidation-reduction sequence,[90] and the 2-*O*-ethyl analogues of 3′-fluoro-2′,3′-dideoxyuridine and 3′-fluorothymidine have been made by silyl base-sugar coupling procedures.[91]

The 2′-bromo-2′-deoxynucleosides **67** (B=Ura, Thy) have been made efficiently by treatment of the 2,3′-anhydronucleoside analogues with LiBr in the presence of BF₃.Et₂O,[92] whilst **67** (B=Ade) has been prepared by displacement of a 2′-'up' triflate. The arabinofuranosyl epimer of **67** (B=Ade) was similarly made, and

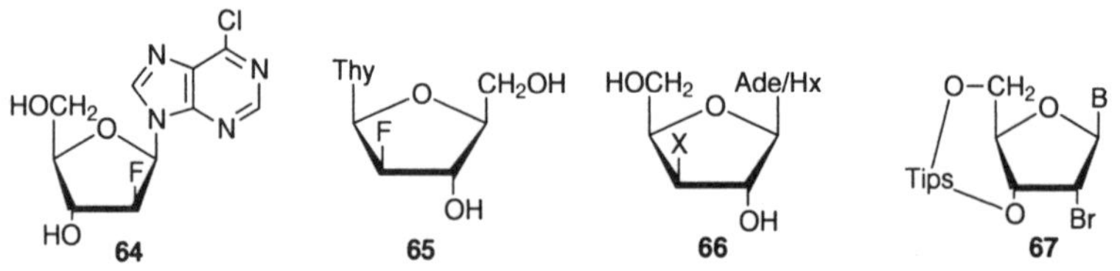

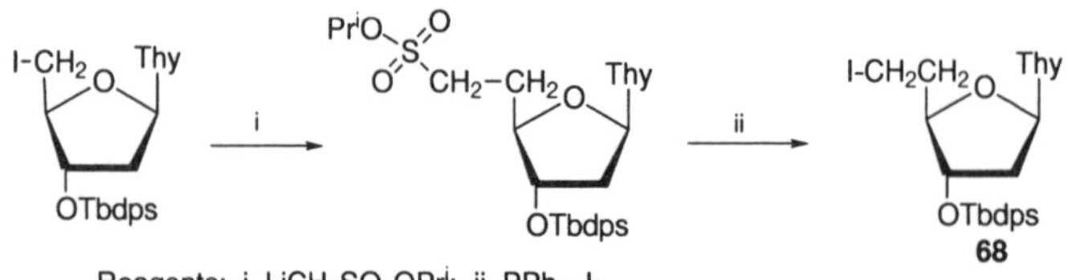

64   65   66   67

Reagents: i, LiCH₂SO₂OPrⁱ; ii, PPh₃, I₂

Scheme 5

both these compounds were incorporated into oligonucleotides using phosphor-amidite methods.[93]

The homologated iodide **68**, required for making sulfone-linked oligonucleo-tide analogues, has been prepared by the interesting sequence of reagents shown in Scheme 5.[94]

5   Nucleosides with Nitrogen-substituted Sugars

A report from Pfleiderer's laboratory has described propected 2'-amino-2'-deoxyarabinofuranosyl nucleosides of the five main nucleobases and their conver-sion into phosphoramidites.[95]

In a new approach to 2'-amino-2'-deoxyuridine (**70**) and related compounds, involving intramolecular delivery of a nitrogen nucleophile, the anhydronucleo-side **69** was treated as outlined in Scheme 6. The cytidine analogue was also prepared.[96] In a similar way (Scheme 7), 2'-alkoxyamino derivatives could be made from *N*-alkoxyurethanes,[97,98] and the same type of cyclization was possible with *N*-alkylurethanes, but required more vigorous conditions.[98]

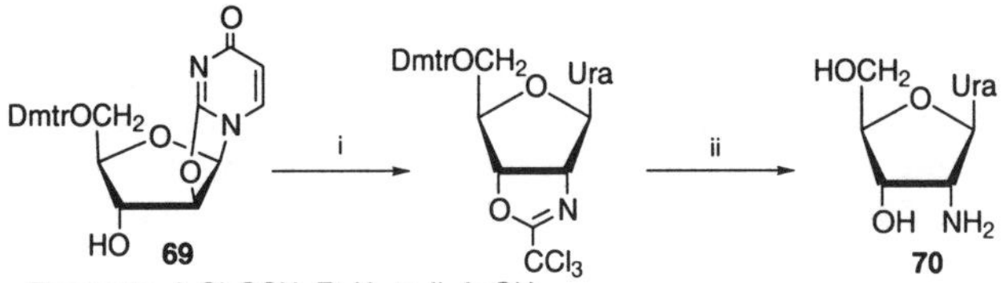

Reagents: i, Cl$_3$CCN, Et$_3$N, Δ; ii, AcOH

Scheme 6

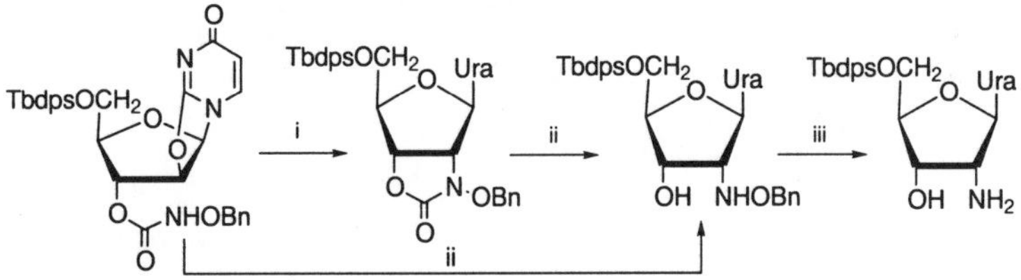

Reagents: i, DBU; ii, Cs$_2$CO$_3$, MeOH; iii, Pd(OH)$_2$, cyclohexene

Scheme 7

Other workers have used a similar approach to deliver a nitrogen nucleophile at C-3' via attachment at O-5'. An example is the case in Scheme 8; similar chemistry was used to make the uracil analogue **72**, and in this case a 2,2'-anhydronucleoside **71** was used as precursor, with the epoxide ring being generated *in situ* on treatment with NaH. 2'-Deoxyprecursors also could be used, with a 3'-mesylate acting as leaving group. In this work, the urethanes were generated by use of lipase-catalysed reactions (see Vol. 28, p. 295).[99]

A further report on the synthesis of AZT describes the reaction of anhydro-nucleoside **28**, prepared in a new way, with azide ion.[43] 3'-Azido-2-O-ethyl-2',3'-

dideoxyuridine and the 5-methyl compound (2-O-ethyl-AZT) have been made by condensation routes,[91] as has the 3′-azido-2′,3′-dideoxynucleoside of 5-phenyl-2-pyrimidinone,[100] whilst opening of a 2′,3′-cyclic sulfate by azide ion was used to make the 3′-azido-3′-deoxy-xylofuranosides **66** (X=N₃).[89]

The tetrazole **73** was made by the opening of a 2,3′-anhydronucleoside.[101]

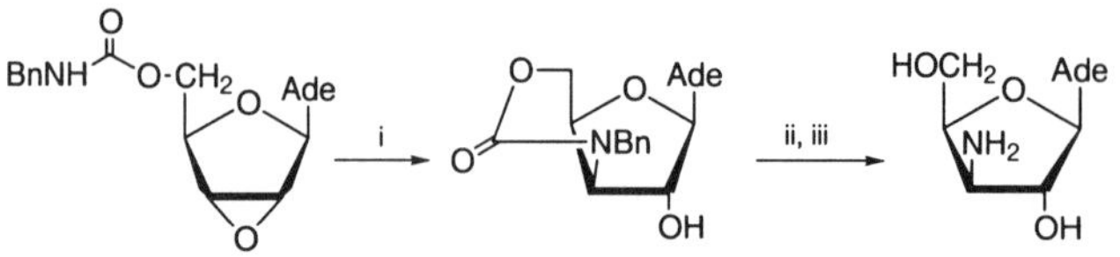

Reagents: i, NaH, DMF; ii, LiOH, EtOH; iii, Pd, HCO₂H, MeOH

Scheme 8

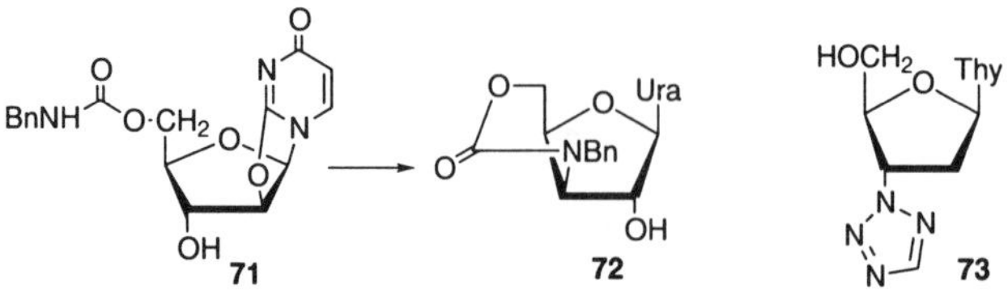

Two groups have independently described chemistry of the nitrocompound **74**, prepared via a seconucleoside, some of which is summarized in Scheme 9.[102,103] Similar chemistry was carried out in the D-*galacto*-series,[102] and some partial reductions of **75** to a hydroxylamine were carried out, followed by formation of a nitrone.[103]

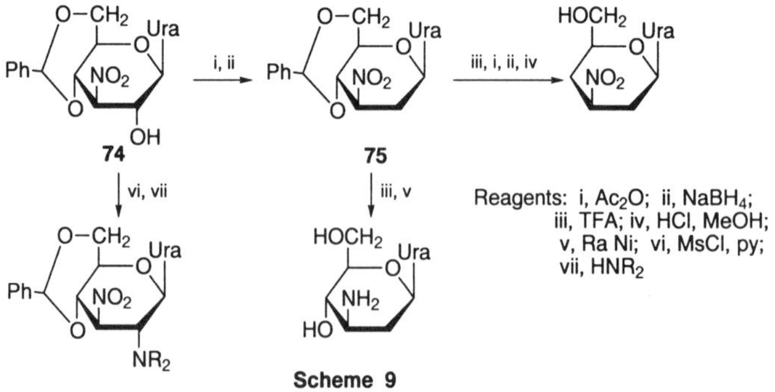

Reagents: i, Ac₂O; ii, NaBH₄; iii, TFA; iv, HCl, MeOH; v, Ra Ni; vi, MsCl, py; vii, HNR₂

Scheme 9

4′-Azido-2′-deoxy-2′-fluoro-arabinofuranosylpyrimidines of type **76** (R=H, Cl, Me) have been prepared from 2-deoxy-2-fluoro-1,3,5-tri-O-benzoyl-α-D-arabino-furanose in six-step syntheses; the products were not active against HIV.[104]

The interesting analogue **77** of S-adenosylmethionine has been made by a process in which 5′-deoxy-2′,3′-O-isopropylidene-5′-methylaminoadenosine was first prepared, and then linked to a side-chain unit derived from L-glutamic acid.[105] Thymidine analogues with a hydroxyalkylammonium moiety, such as **78**,[106] and 5′-amino- and 5′-azido-2′,5′-dideoxynucleosides of thieno[2,3-*d*]pyrimidine-2,4-dione have been made by base-sugar condensations.[107] The *O*-(1-benzo-triazolyl)sulfate leaving group has been used at O-5′ of an adenosine derivative, and gave 5′-deoxy-5′-dialkylaminoadenosines in high yield on reaction with secondary amines.[108] Nucleophilic substitution of a 5′-tosylate with azole salts gave access to products such as **79**,[109] and a series of amides based on 5′-amino-5-(2-thienyl)-2′,3′,5′-trideoxyuridine has been prepared.[110]

The oxorhenium(V) complex **80** has been prepared from 2′,3′-diamino-2′,3′-dideoxyadenosine, and exists as a 2:1 mixture of *syn*- and *anti*-isomers. Both were inhibitors of purine-specific ribonuclease, with the *syn*-isomer being more effective.[111] The same group has described a route to 3′,5′-diamino-3′,5′-dideoxy-adenosine (**82**) from the *lyxo*-epoxide **81** (Vol. 25, p. 251-2), as outlined in Scheme 10. The Mitsunobu inversion using benzyl alcohol as nucleophile is noteworthy, and proved superior to other strategies. An oxorhenium(V) complex was also formed from **82**.[112] Thymidine can be converted into the anhydronucleoside **83** by two successive Mitsunobu reactions, and **83** was converted into the aminoderivative **84** of AZT, and some phosphoramidates were produced from **84**.[113] Some 5′-deoxy-5′-sulfonylamido derivatives of AZT have also been produced by successive displacements at O-5′ and O-3′ by nitrogen nucleophiles.[114]

6   Thionucleosides

The carbaboranyluridine **85**, with a 2′-thiophenyl substituent, has been prepared by base-sugar coupling.[115]

There have been further reports (see Vol. 28, p. 287) on the preparation of nucleosidyl aryl disulfides such as **86**, and 2′-deoxyanalogues, by interaction of a 3′-thiol and either bis(*o*-nitrophenyl)-disulfide[116] or 2,2′-dithiobis-(5-nitro-pyridine).[117] The disulfides react with phosphites to give 3′-*S*-phosphorothiolate triesters,[116] and **86** was treated with a protected uridyl bis(Tms) phosphite to give

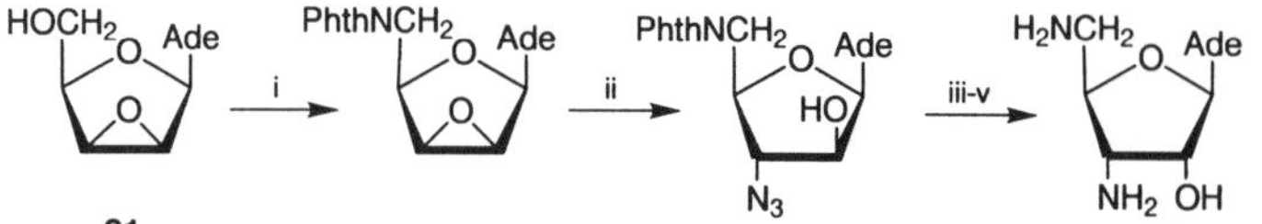

81 → **82**

Reagents: i, Ph₃P, DEAD, phthalimide; ii, LiN₃, DMF; iii, Ph₃P, DEAD, BnOH; iv, H₂, Pd/C; v, NH₂NH₂

Scheme 10

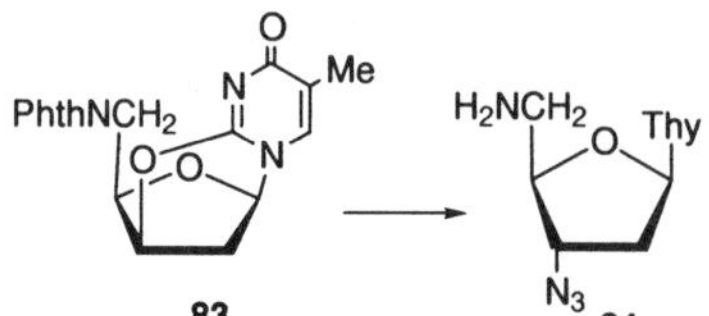

83 → **84** 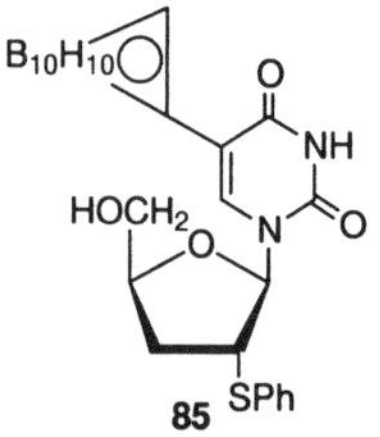   **85**

the phosphorothiolate **87** after deprotection. Base-catalysed cleavage of this was found to be 2000 times faster than for the equivalent phosphate.[117] Other workers have also described a similar route to the bis(uridylyl) analogue of **87**.[118]

In an approach to 2',3'-dideoxynucleoside libraries, reaction of **88** with thiols, followed by acetylation, gave the furanose mixture **89** and the equivalent pyranoses, which could be coupled to silylated thymine. With R=Bu^n, seven of the eight possible isomers could be detected in the product mixture.[119]

There is continued interest in 4'-thionucleosides. Reaction of **90** with silylated 5-ethyluracil in the presence of NBS gave a route to 2'-deoxy-5-ethyl-4'-thiouridine (**91**) with reasonable stereocontrol (β:α, 3.7:1), although the effect of the same directing group had been more pronounced in making normal 2'-deoxynucleosides (Vol. 28, p. 267-8).[120] The β-D-compound **91** has high antiherpetic activity; the β-L-, α-L- and α-D-isomers have been made, along with **91**, by a *de novo* synthesis. The isomers were separated by chiral HPLC, and only **91** had anti-HSV activity.[121] Various other 5-alkyl-2'-deoxy-4'-thiouridines (Alkyl = Pr^i, Bu^t, cyclopropyl, adamantyl) have also been made by similar coupling procedures, and some had significant anti-HSV and anti-VZV activity. NMR analysis supported a C-2'-*endo*-conformation for both the α- and β-anomers.[122]

A report from Walker's laboratory discusses some interesting reactions of 2'-deoxy-3'-thionucleosides and their sulfones, presenting evidence for the involvement of a bicyclic episulfonium ion in reactions of 4'-thiothymidine when a leaving group is present at C-5', and some reactions of sulfones which can be ascribed to the ease of anion formation at C-4'.[123]

Various NMR and physical data have been tabulated for about a hundred 2'-deoxy-4'-thio-pyrimidine nucleosides and examined as an indicator of anomeric configuration. Although there was not a single criterion which permitted an unequivocal assignment for all the structural variants examined, the authors suggest that if a set of criteria are examined and each considered in the light of their tabulation, then the anomeric configuration can be assigned with confidence, even if only one anomer is available.[124]

2'-Deoxy-4'-thio-purine nucleosides **92** (X=OMe, SMe, Cl, etc.) can be prepared from 2'-deoxy-4'-thiouridine, used as an anomeric mixture, by base exchange catalysed by trans-N-deoxy-ribosylase. This process effects a high conversion of the β-anomer of the precursor, and contrasts with chemical routes which tend to be α-selective. In the case of **92** (X=OMe), treatment with adenosine deaminase gave the guanosine analogue.[125] The 4'-thio-purine nucleoside **93** has been made by Vorbrüggen-type coupling, and converted to the thioanalogue of N⁴-cyclopentyl-adenosine, a potent adenosine A₁ agonist.[126]

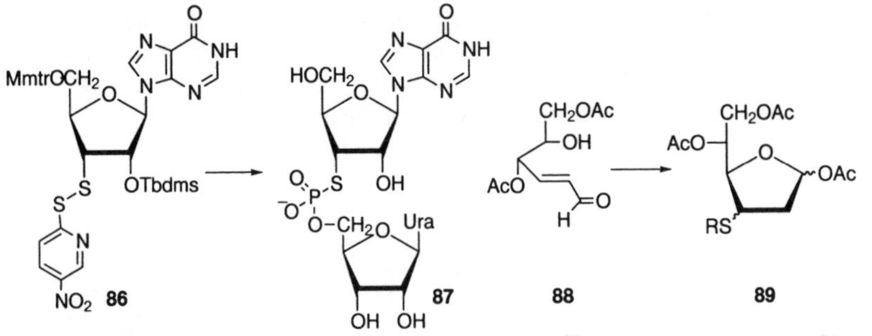

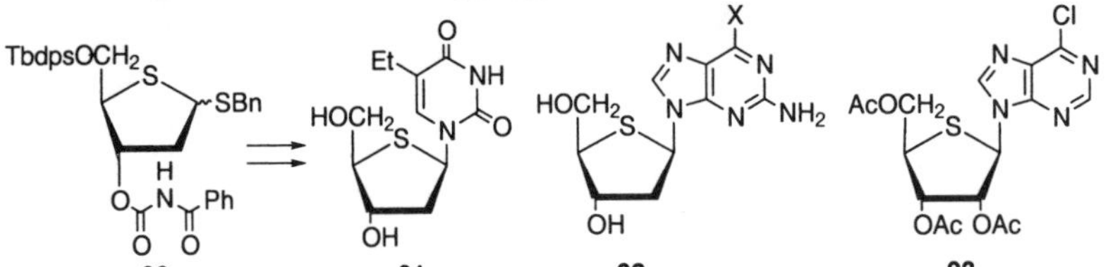

The 'sulfur-in-ring' analogues of two important antitumour agents have been prepared (Scheme 11). The tetrahydrothiophene **94** was prepared from D-glucose (sugar carbons indicated) in a way reminiscent of previous work by others (Vol. 29, p. 161), and converted as indicated into the 4'-thioanalogues of 2'-deoxy-2'-methylenecytosine (**95**) and gemcytabine (**96**); the use of Pummerer chemistry to attach the base to the thiosugar is noteworthy, and in both cases the α-anomer was also produced.[127]

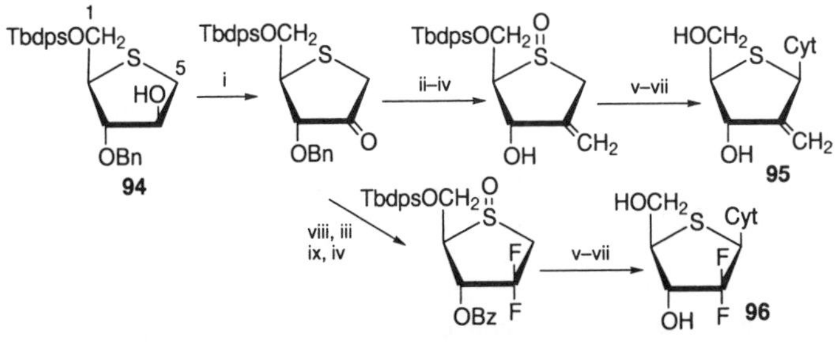

Reagents: i, DMSO, Ac₂O; ii, Ph₃P=CH₂; iii, BCl₃; iv, MCPBA; v, silylated N-AcCyt; vi, TBAF; vii, NH₃, MeOH; viii, DAST; ix, Bz₂O

Scheme 11

3′,5′-Dithiothymidine (**98**) has been prepared as outlined in Scheme 12. The dimesylate **97** could be converted into the *threo*-bis(disulfide) **99**, but all attempts to isolate the *threo*-dithiol led to the formation of the cyclic disulfide **100**, even though the dithiol could be trapped as an isopropylidene derivative.[128]

A reference to thietan nucleosides is mentioned in Section 14.

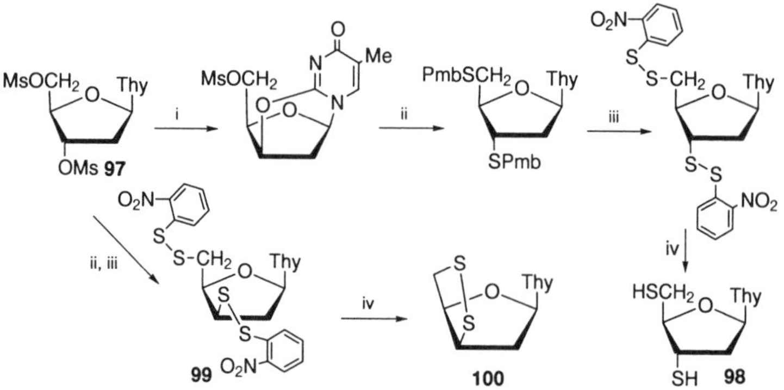

Reagents: i, Et₃N, MeOH; ii, PmbSNa; iii, *o*-nitrophenyl sulfenyl chloride, AcOH; iv, Zn, AcOH

Scheme 12

7   Nucleosides with Branched-chain Sugars

The lactone **101** was prepared diastereoselectively by *C*-methylation of the unsubstituted system, made from D-mannitol, and **101** could be converted to the nucleosides **102**, with all the principal nucleobases.[129] Deoxygenation of C-2′ tertiary alcohols via their methyl oxalyl esters was a key reaction in the synthesis of a range of 5-substituted uracil nucleosides **103**, and the equivalent cytidine analogues; the 5-iodouracil compound (**103**, R=I) showed the most potent anti-HSV activity.[130] Additions of acetylide anions to a 2′-ketonucleoside, followed by hydrogenation and deoxygenation, again via methyl oxalyl esters, were used to make a range of branched deoxyuridines **104**.[131] There has been a full account of the preparation of deoxyuridine analogues **105** (R=CO₂H, CONH₂, CH₂OH, CHOH.CH₂OH) from the corresponding 2′-*C*-allyl compound, and their incorporation into dinucleotides.[132]

The *Z*-alkene **106**, produced by a Wittig reaction, could be isomerized to the *E*-isomer **107** by Michael addition of thiophenol (selectively from the α-face), followed by oxidation to the sulfoxide and thermal elimination.[133]

The fused isoxazolidine **108** was made by regioselective and diastereoselective [3+2] cycloaddition, and has been converted to the nucleosides **109**.[134]

A full and extended account has been given of the synthesis of 2′,3′-dideoxy-3′-*C*-hydroxymethyl purine nucleosides by a photochemical method (see Vol. 28, p. 278).[135]

The branched glycosyl acetate **110** has been prepared from diisopropylidene-D-glucose and used to make the branched adenine nucleosides **111 - 114**.[136] Various fluorinated, branched nucleosides **115** (X=H, Me, Hal) have been described,[137] whilst **116** has been made by hydroboration-oxidation of the 3′-*C*-methylene compound and converted into the bicyclic species **117**.[138]

Acetolysis of the anhydrosugar **118** gave a mixture of pyranosyl and furanosyl glycosyl acetates, which could be converted to the 3′-hydroxymethyl nucleosides **119** and, after a deacetylation-periodate-borohydride sequence, **120** respectively.[139]

A new radical allylation procedure has been applied to the synthesis of a 3′-*C*-allyl-2′,3′-dideoxypyrimidine nucleoside,[140] and lactone **121**, made by photochemical addition of isopropanol to the corresponding 2,3-ene, has been converted to the nucleoside **122**.[141] Branched-chain sulfonates **123** have been made from 3′-ketonucleosides, and the isobutyl group was found to be superior to other possibilities, it being cleavable by iodide ion to give the sulfonic acid, an analogue of a 3′-phosphate.[142] An ingenious route to related phosphonates is outlined in

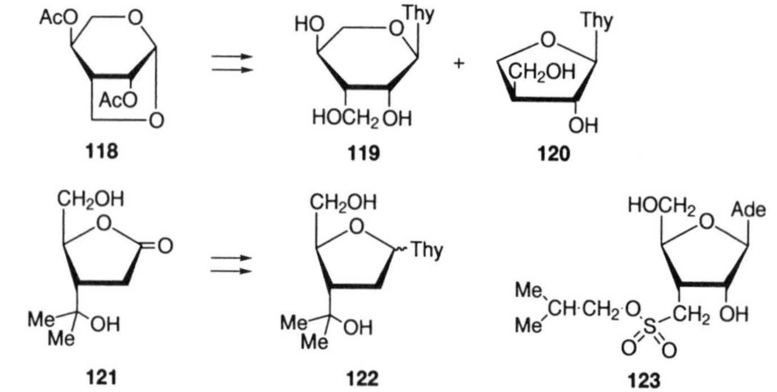

118   **119**   **120**

121   **122**   **123**

Scheme 13; although the case shown gave mostly α-nucleoside, the process was strongly β-selective when the 5-O-acetyl glycosyl acetate was used, in conjunction with $BF_3.Et_2O$, in the reaction with silylated thymine.[143] A related cyclization of **125** to give **126**, using Bu_3SnH and AIBN, has been reported, and the same workers also describe a similar process involving a 5'-propargyl ether to give **127**.[144]

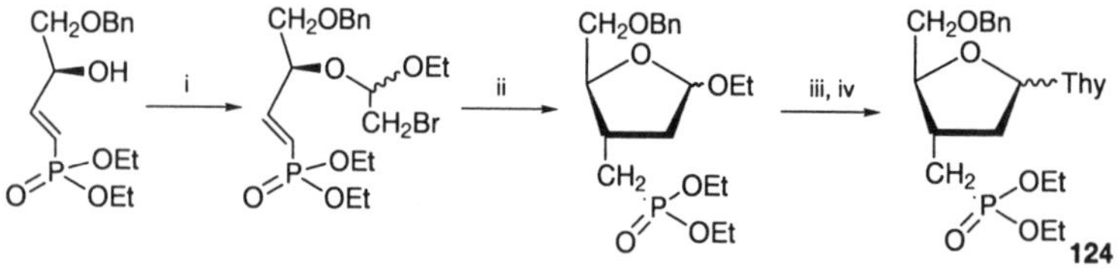

Reagents: i, $CH_2=CHOEt$, Br_2; ii, Bu_3SnH, AIBN; iii, Ac_2O, AcOH, H^+; iv, $(Tms)_2Thy$, BF_3, $SnCl_4$

Scheme 13

3'-Deoxy-3'-difluoromethyleneuridine (**128**), the first example of a 3'-difluoro-methylene nucleoside, has been made by use of the Wittig-type reagent derived from CF_2Br_2, HMPT and zinc metal. The 2'-difluoromethylene isomer could be similarly prepared; an earlier approach to 2'-difluoromethylene nucleosides (Vol. 26, p. 240) was not successful when applied to a 3'-keto-compound. 5'-O-Dmtr-3'-O-phosphoramidites of these difluoromethylene compounds were also made for assembly into oligonucleotides.[145]

The acetylenic nucleosides **130** (B=normal bases, 5-F-Ura, 5-F-Cyt) have been prepared by stereoselective addition of lithium Tms-acetylide to **129**, followed by nucleoside formation. The Ura and Cyt compounds had particular antitumour activity.[146,147] The allofuranosyl analogue **131** of TSAO-T has been prepared,[148] and 3'-ketonucleosides were treated with *N*-methylhydroxylamine, followed by reaction of the nitrones with lithioethyl acetate, to give the spirocyclic compounds **132**, also related to TSAO-T.[149] Some other compounds with branches at C-3', prepared in connection with making antisense oligonucleotides, are mentioned in Section 12.

As regards compounds branched at C-4', deoxygenations on the previously-described 4'-C-methylribonucleosides (Vol. 27, p. 256) has given rise to the 2'-deoxycompounds **133** (X=OH) and dideoxycompounds **133** (X=H), and also the d4 systems.[150] Modification of the 4'-hydroxymethyl compounds has led to the formation of the phosphoramidites **134** (X=OMe, NHCOCF$_3$) for incorporation into oligodeoxynucleotides.[151] Similarly, 4'-C-acylthymidines **135** (R=Me, Et, Bu^t, Ph) have been made either by regioselective modification of the 4'-hyd-roxymethyl compound or, in the cases of R=Ph or Bu^t, by initial conversion of the CH$_2$OH group of thymidine into an acyl group followed by aldol condensa-tion with formaldehyde, a procedure which also gave the 4'-β-acyl derivatives. The α-acyl compounds **135** were converted to derivatives for automated oligonu-cleotide synthesis.[152] The fluorinated compounds **136** (R=CF$_3$, CHF$_2$, CH$_2$F) have been made by base-'sugar' condensation.[153] Treatment of the 3'-ene **137** with allyl trimethylsilane in the presence of SnCl$_4$ gave **138** in good yield with high stereoselectivity. Similar processes occur with silyl enol ethers and TmsCN, but with somewhat reduced diastereoselectivity.[154] The pyranosyl nucleoside **139** has been made by condensation of silylated thymine with a sugar unit prepared from a ribopyranose derivative (Chapter 14).[155]

The bicyclic analogue **140** was prepared from the previously-described 3'-deoxy-3'-*C*-allylthymidine (Vol. 23, p. 218) by Cannizzaro reaction to establish a 4'-hydroxymethyl substituent followed by iodoetherification and hydrogenolysis.[156]

8   Nucleosides of Unsaturated Sugars, Ketosugars and Uronic Acids

As in previous volumes, 2',3'-didehydro-2',3'-dideoxyfuranosyl nucleosides (d4 systems) are discussed in Section 3, together with their saturated analogues.

When the 3'-mesylate **141** was treated with secondary amines, enamines of type **142** were obtained, as anomeric mixtures. The authors theorise that a 2'-ketonucleoside is involved, formed by a hydride shift under metal-free conditions;[157] for a similar reaction starting from a D-*lyxo*-precursor see Vol. 28, p. 273. Hydrolysis of the α-anomer of **142** (-NR$_2$ = morpholinyl) gave the α-2'-ketonucleoside **143**, the structure of which was secured by crystallography of a derivative, and some earlier structural assignments were revised.[158]

The 4'-ene **144** was obtained by elimination from the 5'-*O*-tosyl derivative.[114]

An improved route has been developed for the synthesis of adenosine receptor agonists of type **145**, involving the high-yield oxidation of 2',3'-*O*-isopropylideneadenosine with KMnO$_4$ at pH 12.[159] Esters and amides of adenosine-5'-carboxylic acid have also been prepared for investigation as inhibitors of *S*-adenosylhomocysteine hydrolase.[160]

9   *C*-Nucleosides

The deoxygenated analogues **146**[161] and **147**[162] of nicotinamide riboside have both been made by addition of a lithiated pyridine to an appropriate lactone, followed by further manipulation. Similarly, reaction of 2,3,5-tri-*O*-benzyl-D-ribonolactone with a lithiated pyridine, followed by ionic hydrogenation at the anomeric centre and further deoxygenation at C-2', gave access to the more basic analogue **148** of deoxycytidine, which was incorporated into oligodeoxynucleotides.[163] Reaction of lithiated pyrazines with 2,3,5-tri-*O*-benzyl-D-ribonolactone, followed again by treatment with Et$_3$SiH and BF$_3$.Et$_2$O. has led to the preparation of pyrazine *C*-nucleosides such as **149**.[164] The interesting 6-amino-pyrazine 2(1*H*)-one *C*-nucleoside **150** has been synthesized by elaboration of the heterocycle from a β-D-ribosylated α-aminonitrile; its 5'-triphosphate was also described, along with the base-pairing properties of oligonucleotides containing this base. The furanoside **150** rearranges on standing in solution to the β-pyranose isomer.[165]

Various 3-ribofuranosyl-indoles, -pyrroles and -pyrazoles have been made by reaction of *N*-blocked heterocycles with 2,3,5-tri-*O*-benzyl-β-D-ribofuranosyl fluoride.[166] Lithiation of 2,6-dichloroimidazo[1,2-*a*]pyridine occurs predominantly at C-5, and reaction with a ribono-γ-lactone derivative and anomeric deoxygenation gives the 5-ribofuranosyl system **151** with good β-selectivity.[167] The alternative 3-glycosylation pattern **152** could be obtained by palladium-catalysed coupling of 2,6-dichloro-3-iodoimidazo[1,2-*a*]pyridine with 2,3-dihydrofuran, followed by hydroxylation.[168] Various pyrazolo[4,3-*c*]pyridine *C*-nucleosides such as **153** have been made using an effective tetrazole-to-pyrazole transformation carried out on a *C*-ribofuranosyltetrazole.[169] A paper on the conformational properties of some purine-like *C*-nucleosides is mentioned in Chapter 21.

Condensation between 4,5-dichloro-*o*-phenylene diamine, D-*glycero*-D-*gulo*-heptose and an arylhydrazine gave acyclic systems, which, on acid-catalysed cyclization, gave *C*-nucleosides of type **154**, predominantly of β-configuration.[170]

The acetonitrile derivative **155** has been converted into a range of *C*-nucleosides with a dideoxy-motif, including the pyrazolotriazine **156** and a pyrrolo-[3,2-*d*]pyrimidine, the α-anomers also being obtained,[171] and the enantiomers of these compounds have been made from the enantiomer of **155**.[172]

Some *N*-alkyl 3-(4'-deoxy-4'-fluoro-α-L-arabinopyranosyl)-1,2,4-oxadiazole-5-carboxamides have been made by fluorination of the corresponding 3-β-D-xylofuranosyl compounds using DAST.[173]

10   Carbocyclic Nucleoside Analogues

The homologue **157** of neplanocin A, and several other analogues with different bases, have been synthesized from a known chiral cyclopentenone unit (Vol. 24,

p. 302).[174] Similar homologues of carbovir and its adenine analogue have also been made as racemates, using π-allyl palladium coupling procedures,[175] whilst structures of type **158** (B = various purines) and their enantiomers have been prepared from chiral bicyclic lactones; the inosine analogue had some anti-HIV activity.[176]

The *nor*-analogue **159** has been made by a Pd-catalysed reaction involving an allylic phosphate as reactant,[177] whilst others have also described routes to **159**, its hydrogenated analogue,[178] and the enantiomer and diastereoisomers of this,[179] using enzymic desymmetrization to introduce chirality; the products were evaluated for their ability to inhibit tumour necrosis factor-α.[179] A chemo-enzymatic approach has also been used to make the cyclopropane-fused analogue **160** and its enantiomer.[180]

There have been further reports on fused cyclopropanes of type **161**. Marquez and Siddiqui have prepared the adenine-containing compound from Altmann's previously-reported cyclopentene precursor **162** (Vol. 28, p. 285-6), using some improvements in methodology for introducing the cyclopropane ring with stereocontrol, and have shown it to have good activity against CMV and EBV.[181] The same laboratory has also made the compounds **161** with the other main nucleobases, the thymidine analogue having high anti-HIV activity, and carried out detailed conformational studies on these analogues and also the bicyclic compounds with the cyclopropane ring fused in the 1′,6′-position and which have a locked 'southern' conformation.[182] The cyclopentene **162** has also been used to make the cyclic sulfite **163**; regioselective reaction of this with azide ion was the key to the synthesis of the nucleoside analogue **164**.[183] Altmann's group have used their intermediate **165**, previously used to make 1′,6′-methylene-bridged species (Vol. 28, p. 286), to provide a route to the thymidine analogues **166** (for an alternative synthesis, see Vol. 24, p. 240) and **167**.[184]

Enzymic desymmetrization was used to develop an enantiospecific synthesis of (−)-BCA (**168**), and the same approach was also used to make (−)-carbovir.[185]

The carbocyclic nucleoside analogues **169**[186] and **170**[187] have been synthesized as racemates.

A number of reports on cyclohexenyl and cyclohexyl analogues have appeared. Compounds of type **171** (B=purines) have been prepared as racemates,[188] and the guanine member, a homologue of carbavir, has been produced in chiral form by the use of adenosine deaminase at high pressure on the racemic 2,6-diamino-purine analogue. Various other carbocyclic 6-aminopurine nucleosides were similarly resolved by this method, including carbovir.[189] Racemic compounds of type **172** (B=purine) have been prepared,[190] as have doubly-hydroxymethylated compounds such as **173**, related to (−)-BCA, and **174**, the guanine member of which can be regarded as a ring-expanded analogue of carbovir. Enzymic resolution was used during the syntheses.[191] Also reported have been compounds of type **175** (B=Ade, Thy, Cyt, X=OH), produced by opening of a racemic epoxide, and the deoxygenated systems, X=H.[192]

Some references to cyclopropyl analogues are mentioned in Section 14, and some phosphates and phosphonates of carbocyclic nucleosides are discussed in the next Section. Some papers concerned with carbocyclic nucleoside antibiotics and their close analogues are mentioned in Chapter 19.

11   Nucleoside Phosphates and Phosphonates

Triethylammonium salts of aryl H-phosphonates have been coupled to the 3′-OH of deoxynucleosides using pivaloyl chloride as condensing agent; subsequent hydrolysis by aqueous pyridine gives nucleoside H-phosphonate building blocks for oligonucleoside synthesis.[193] The 4-cyano-2-butenyl group, as in **176**, has been developed as a new type of protecting group for use during oligodeoxynucleotide synthesis by the phosphoramidite method. It is stable to acid, but can be removed with aqueous ammonia by a 1,4-elimination process.[194] The nitrotriazole **177** has been used as a new phosphitylating agent for 2'-deoxynucleosides, and can generate the 3'-phosphoramidite *in situ* during oligodeoxynucleotide synthesis.[195] Stereocontrolled transesterification has been used to make chiral bis(nucleosidyl) 2-cyanoethyl phosphate units.[196]

As regards modified internucleotidic links, bis(deoxynucleosidyl)-long-chain alkyl triphosphates have been reported, along with thymidine linked via either O-3' or O-5' to O-6 of methyl α-D-glucopyranoside through a hydrophobic phosphotriester.[197] A known derivative of 3'-deoxy-3'-*C*-formylthymidine (Vol. 28, p. 277-8) has been converted into a phosphoramidite which was used to make oligomers with a five-atom phosphate link, as in **178**.[198]

In an extension of a method previously used in the 2'-deoxy-series(Vol. 29, p. 288), oxathiaphospholanes of type **179** have been used to make 3',5'-phosphorothioates.[199] The efficiency of coupling in the synthesis of a bis(thymidylyl)-*o*-chlorophenyl phosphorothioate triester, when various coupling reagents are used, has been studied,[200] and the use of sulfur and Et_3N has been shown to convert internucleotidic phosphite triesters into phosphorothioate triesters with high efficiency; in this study, the diphenylmethylsilylethyl protecting group (Vol. 29, p. 288) was used as the third substituent on the dinucleosidyl phosphite.[201] In an ingenious route to pure diastereomers of phosphorothioates (Scheme 14), the phosphite triester **181** was prepared with 6:1 diastereoselectivity by sequential addition of thymidine units to the cyclic P(III) species **180** as indicated. Subsequent thionation and acid hydrolysis then gave the R_P-isomer **182**.[202]

Triesters **183** (R=Me, Bu^t, Ph) have been prepared by *S*-alkylation of the bis(thymidylyl) phosphorodithioate. The species **183**, particularly with R=Me, were rapidly converted back to the phosphorodithioate using various cell extracts, raising the possibility of using the *S*-acylthioethyl (SATE) blocking group in delivery of antisense oligonucleotides containing phosphorodithioate units.[203]

There has been interest in methylphosphonate analogues of internucleotidic linkages. Work in this area in Stawinski's laboratory has been reviewed, together

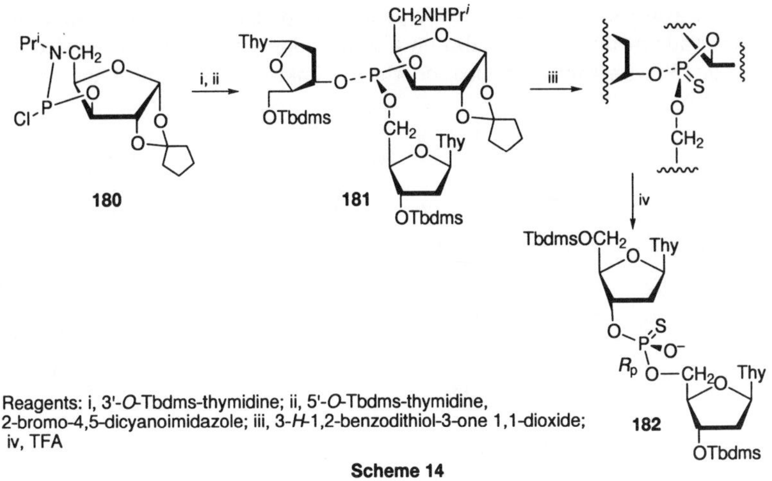

Reagents: i, 3'-O-Tbdms-thymidine; ii, 5'-O-Tbdms-thymidine, 2-bromo-4,5-dicyanoimidazole; iii, 3-H-1,2-benzodithiol-3-one 1,1-dioxide; iv, TFA

Scheme 14

with studies on making H-phosphonate and H-phosphonothiate esters and reagents for converting these into phosphate derivatives.[204] 3'O-Methanephosphonofluoridates (**184**) can be prepared non-stereoselectively from methylselenylphosphonates as indicated in Scheme 15, and used to make methylphosphonates (**185**). It is possible to generate and use **184** *in situ*.[205] Others have generated **184** (B=Thy) from the corresponding methylphosphonothioate, made in a new way, by treatment with 2,6-dinitro-4-trifluoromethylfluorobenzene; subsequent reaction of **184** (B=Thy) with 5'-O-silylated nucleosides in the presence of KF gave methylphosphonates of type **185**.[206] In related work, 3'-O-methylphosphonates could be activated as their mixed anhydrides such as **186**; methanolysis in the presence of Ag(I) then gave the methyl methylphosphonate **187** with inversion, in a procedure that should be applicable to dinucleotide analogues.[207] Both stereoisomers of dicytidylyl methylphosphonate have been prepared and used to make ribozyme substrates with a chiral methylphosphonate

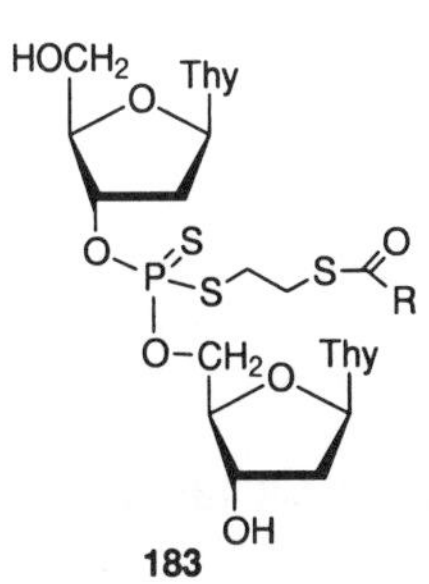

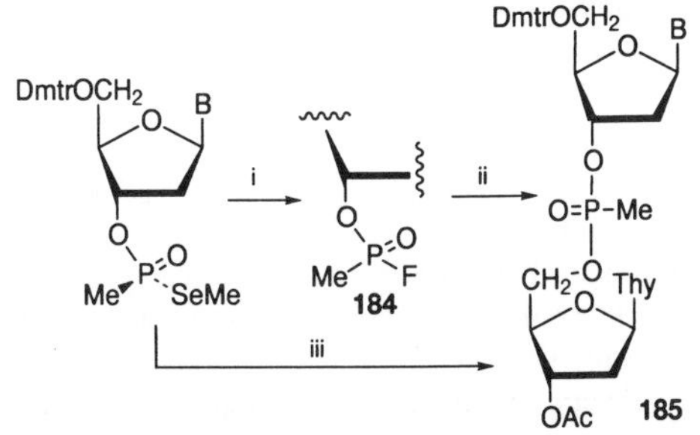

Reagents: i, AgF, H₂O, MeCN; ii, 3'-O-acetylthymidine; iii, Et₃N.3HF, 3'-O-acetylthymidine, DBU

Scheme 15

at the cleavage site.[208] It has proved possible to prepare diastereomerically-pure dinucleoside methylphosphonates on a solid support by Grignard-induced coupling of **188** with another nucleoside which is linked to the support at O-3′ via a spacer.[209]

186   **187**   **188**   **189**

The 1,1-dianisyl-2,2,2-trichloroethyl (DATE) moiety (Vol. 29, p. 295) has been used as a protecting group for O-5′ in the synthesis of dinucleoside trifluoromethylphosphonates,[210] and a phosphonoacetate unit has been introduced into a dinucleotide analogue **189**.[211]

As regards in-chain phosphonates, the novel type of analogue **190** has been made by coupling a 5′-phosphonate with a derivative of 3′-deoxy-3′-hydroxymethylthymidine, prepared from a known 3′-cyano-3′-deoxy-derivative (Vol. 22, p. 214). This isosteric replacement was incorporated into oligodeoxynucleotides.[212] A related branched-chain phosphonate was mentioned earlier in Section 7. A one-pot synthesis of the dinucleotide phosphonate GpCH$_2$U has been reported.[213]

A phosphoramidite and a methylphosphonamidite of 2′-*O*-methylisocytidine have been prepared, and used for incorporation into oligonucleotide analogues.[214] Di(2′-deoxy-2′-fluoro)nucleoside phosphates and phosphorothioates have been prepared, and their stability to snake venom phosphodiesterase was investigated.[215]

1-Deaza-analogues of (2′→5′)ApApA, with the replacement specifically in each position, have been described,[216] and an analogue has also been made with two 3′-deoxy- (cordycepin) units and with a 2-hydroxyethyl spacer attached to the 3′-OH at the 2′-terminal unit.[217]

A number of aryl 3′-uridylyl phosphorothioates have been prepared, using a new method involving the 3′-H-phosphonate as an intermediate; the kinetics of base catalysis and leaving group dependence in the intramolecular alcoholysis of the phosphorothioates was studied.[218] The 3′-phenylphosphonates of adenosine and uridine have been made and evaluated as substrates of 3′-nucleotidases.[219] Reaction of thymidine with an oxyphosphorane was used to make thymidin-3′-yl benzoin phosphate.[220]

In the area of 5′-phosphates, a chemical synthesis of SAICAR (3) has been described, along with a route to carbocyclic AMP and carbocyclic adenylosuccinate.[6] Ferrocene has been attached, via a 5′-phosphodiester and a spacer arm, to thymidine. Subsequent attachment of a thioethanol unit, via a 3′-phosphodiester, gave a compound used to prepare a redox-active nucleotide monolayer on a gold surface.[221] 5'-Fluorophosphates and 5'-H-phosphonates of various 2'-modified nucleosides have been prepared as potential antiviral agents,[222] and 1,2-cyclic monoalkyl-glycero-5'-thiophosphates of isopropylidene adenosine have been described.[223]

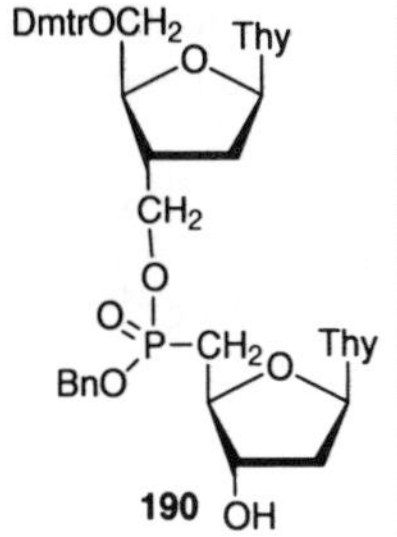

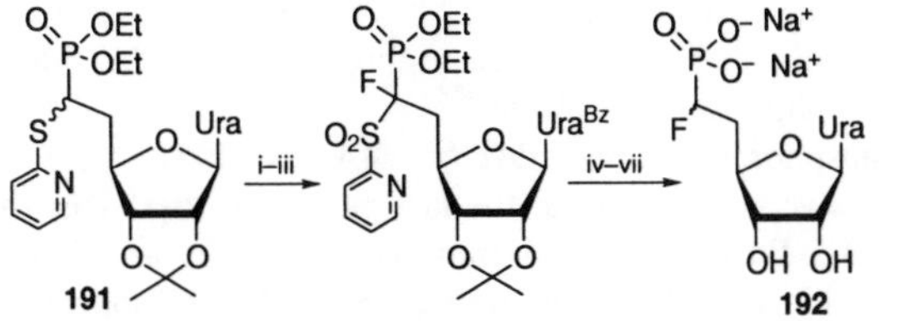

Reagents: i, MCPBA; ii, BzCl, py; iii, KH, selectafluor; iv, NH₃, MeOH; v, Bu₃SnH, AIBN; vi, TFA, vii, TmsBr

Scheme 16

Scheme 16 outlines the use of a known intermediate **191** (Vol. 26, p. 252 and Vol. 23, p. 224) to prepare the α-fluoromethylene phosphonate **192** using electrophilic fluorination and radical-mediated cleavage of a π-deficient sulfone.[224]

In the area of sugar nucleotides, various labelled analogues of CMP-NeuNAc have been prepared in order to study the mechanism of solvolysis,[225] and a chemical synthesis of CMP-*N*-glycolylneuraminic acid has been described.[226] In elegant work, Imamura and Hashimoto have prepared the phosphonate analogue **195** of CMP-NeuNAc, the key step (Scheme 17) being the conversion of phosphite **193** into phosphonate **194**.[227] The same workers have also reported the isosteric phosphonates **196** (X=OH, H), starting from a known *C*-allyl derivative of NeuNAc (Vol. 25, p. 52).[228] UDPG, and its analogues with adenine and cytosine as bases, have been prepared from the nucleoside monophosphates and sucrose by the combined use of nucleoside monophosphate kinase and sucrose synthase. A similar enzymic route was also used to make dUTP-6-deoxy-α-D-*xylo*-4-hexulose.[229]

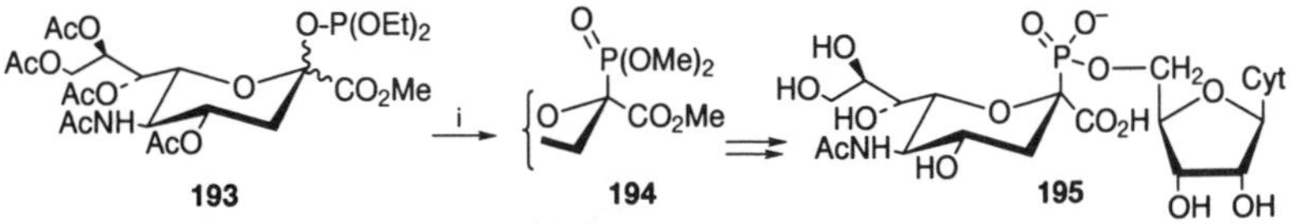

Reagents: i, TmsOP(OMe)₂, TmsOTf

Scheme 17

There have been a number of reports of 5'-phosphates of bioactive nucleosides being linked, via the 5'-triphosphate, to other entities as a means of devising prodrugs of the nucleoside monophosphate. New alkoxy- and aryloxy-phosphate derivatives of AZT have been described,[230] and specifically-deuterated alkoxyethyl phosphodiesters of AZT have been made to study interactions with membranes.[231] Acyl phosphates of AZT and d4T have been reported.[232] 5'-Phosphatidyl derivatives have been described for the antitumour agents 2'-cyano-2'-deoxy-ara-C (e.g. **197**),[233] 2'-deoxy-2'-methylenecytidine (DMDC), and its 5-fluoro-congener, these being made by enzymic transfer of the phosphatidyl group from phosphatidylcholine.[234] Similar phosphatidyl derivatives have been made chemically from AZT,[235,236] ddI[236] and ara-C.[237] Ether and thioether phospholipids have also been conjugated with AZT, ddC and related compounds via a 5'-diphosphate linkage.[238]

Phosphoramidates of d4T,[239] d4A (e.e. **198**),[240] FdU and ara-C,[241] and iso-ddA have been reported; the last paper also describes the bis(SATE) esters **199** (R=Me, Bu^t) of iso-ddA monophosphate as an alternative prodrug.[242] *N,N*-Bis(2-chloroethyl)phosphoramidates of 2',3'-*O*-isopropylidene nucleosides have been described.[243]

196   197   198

199   200   201

A new type of prodrug of d4T, **200**, has been prepared. Such compounds, and the analogous ddT species, degrade hydrolytically under physiological conditions without the need for an enzyme.[244] AZT and ddT derivatives **201** (R=Me, Ph) have also been prepared. Hydrolysis of the carboxylic ester leads to rapid release of the 5',5'-dinucleosidyl phosphate by a process of neighbouring group catalysis, with a mixed phosphoric-carboxylic anhydride as an intermediate.[245]

In the area of 5'-di- and tri-phosphates, ATP, GTP, UTP and CTP, with deuterium in all of the 3'-, 4'-, and 5'-positions, have been made enzymically from the labelled ribose (Chapter 2),[246] whilst chemical routes to the 5'-triphosphates of 4-thiouridine, 2-thiocytidine, 5-bromocytidine and [6-^{15}N]-adenosine have been described.[247] Azidonucleosides can be converted by treatment with a polystyryl diphenylphosphine into polymer-bound phosphine imines (Vol. 28, p. 272); it has now been reported that the nucleoside phosphine imine can be converted to its 5'-triphosphate before hydrolysis to give the amine. The method was used to convert AZT into the 5'-triphosphate of 3'-amino-3'-deoxythymidine, amongst other cases.[248] A method previously used to make the P$^\alpha$-methyl-phosphonyl analogue of TTP (Vol. 28, p. 289) has now been extended to the same analogues of dATP, dCTP and dGTP; in these cases, penicillin amidase was used to remove an acyl protecting group from each of the bases, the methylphosphonate being incompatible with chemical deprotection.[249] The 5'-α,β-imido-analogue of dUTP has been prepared by a combination of chemistry and enzymic phosphorylation, and was found to be a potent competitive inhibitor of dUTPase.[250] The mixed anhydride formed from a long-chain fatty acid and ethyl chloroformate interacts with ADP, GDP or ATP to form the appropriate acyl di- or triphosphate.[251]

The diphosphate-phosphinate **202** has been prepared, as a mixture of diastereoisomers, by Khorana coupling of uridine 5'-phosphomorpholidate with the 'lower' phosphinate-monophosphate in unprotected form. The product **202** is a good inhibitor of the ligase which adds D-glutamate during peptidoglycan biosynthesis, and the pentamethylene chain acts as a replacement for the *N*-acetyl muramyl unit in the natural substrate.[252]

The 3',5'-cyclic phosphorofluoridate of thymidine has been prepared,[253] whilst phosphorylation of AMP with cyclo-triphosphate gives a separable mixture of the 2',5'- and 3',5'-diphosphates.[254]

Cyclic ADP-ribose (cADPR) and related compounds continue to attract attention. Chemical cyclization of the appropriate etheno-bridged analogue of NAD$^+$ (NaBr, Et$_3$N, DMSO) gave the cADPR analogue **203**, with a different position of linkage to the product of enzymic cyclization (Vol. 29, p. 292); a similar cytidine analogue was also prepared either enzymically or chemically.[255] The cyclase enzyme will also convert nicotinamide adenine trinucleotide to cyclic ATP-ribose, a more potent inhibitor of Ca^{2+} release from rat-brain microsomes than is cADPR.[256] The carbocyclic analogue **204** of NAD$^+$, also containing a phosphonate replacement, has been prepared as a potential inhibitor of the cyclase.[257]

The *C*-glycoside *m*-(β-D-ribofuranosyl)benzamide (Vol. 26, p. 245-6) has been converted to benzamide adenine dinucleotide, which was an IMP dehydrogenase inhibitor, but without selectivity between type I and type II, which is upregulated in neoplastic cells.[258]

The metabolite **205**, produced during t-RNA splicing, has been synthesized by making the pyrophosphate bond through interaction of α-D-ribofuranose-1,2-cyclic phosphate and a mixed anhydride of AMP.[259] Cyclic dinucleotides of type **206** have been prepared by an H-phosphonate method in solution.[260]

The cyclic phosphate **207** has been prepared, in which the sugar ring is fixed in a C-3'-*endo*-conformation.[261]

Some carbocyclic nucleoside phosphonates have been described. There has been a further report on compounds of type **208** (see Vol. 28, p. 291), with both purine and pyrimidine bases. Although initially made as racemates, a precursor diol could be resolved by enantioselective acetylation using a lipase and vinyl acetate.[262] The triphosphate analogues **209** (B=Gua, Ade) have been made from the previously-described monophosphonates (Vol. 29, p. 285), and the cyclopropyl-fused diphosphate analogues **210** (X=CH$_2$ and O) were also reported, along with their enantiomers.[263] *Trans*-compounds of type **211** (n=1-3) have been made as racemates,[264] and so have the related *cis*-isomers (n=1 or 2).[265]

202   **203**   **204**

205   **206**   **207**

208   **209**   **210**   **211**

Some papers relating to mechanistic aspects of nucleoside phosphates are mentioned in Section 15, some 2',3'-epoxy-2'- and 3'-phosphates were discussed earlier,[36] and phosphonoalkylidene derivatives of nucleosides are referred to in Section 13.

12 Oligonucleotide Analogues with Phosphorus-free Linkages

The Ciba-Geigy group have reported further work on their promising amide-linked systems. *N*-Alkylated dimers of type **212** [Tttr = tris(*p*-*t*-butyl)trityl] have been prepared, and incorporated into oligonucleotides, to probe hydrophobic effects in duplexes with RNA complements.[266] The same team has also prepared building blocks of types **213** - **216** (B=Thy or 5-Me-Cyt) and investigated the effects of the methoxy/hydroxy substituents on hybridization with RNA sequences, and types **214** and **216** increased T_m by ~2-3 °C per replacement. In this work, a route was developed to the branched-chain unit **217**, in which an allyl group was introduced at C-3′ using radical chemistry, whilst for the synthesis of the hydroxy-compound **215**, the 'upper' base was introduced by coupling to a unit **218** already containing the internucleotidic replacement.[267] Meanwhile, other workers have synthesized the diribonucleotide analogue **221** by reaction between the lactone **219** and amine **220** in the presence of 2-hydroxypyridine as condensing agent.[268]

Compounds **222** (R=Me, Et) containing five-atom amide replacements have been prepared and incorporated into oligomers, as have others containing the related piperazine-type linkages (see Vol. 29, p. 293-4).[269] Other extended carbamate and urea-linked dimers, with a seven-atom link, have also been reported.[270]

The conformationally-constrained amide dimer **223** has been reported;[271] the cyclopropyl amine was made by reduction of a nitrocompound, which itself was prepared by chemistry similar to that described in Vol. 23, p. 218.

The group at Isis Pharmaceuticals have reported further on their methylene-imino-linked dimers (see Vol. 26, p. 242) and, significantly, have extended their work to the preparation of 2′-O-methylated diribonucleotide analogues **226** by linking units of type **224** and **225** under radical conditions. The compounds of type **226** could also be N-methylated.[272]

The acetylenic linkages of **227** (X=O or S) have been developed, as has the all-carbon species **228** and its epimer. When incorporated into oligomers, the (*R*)-isomer **228** of the all-carbon species proved most effective in giving good hybridization properties with complementary RNA when five replacements were present.[273]

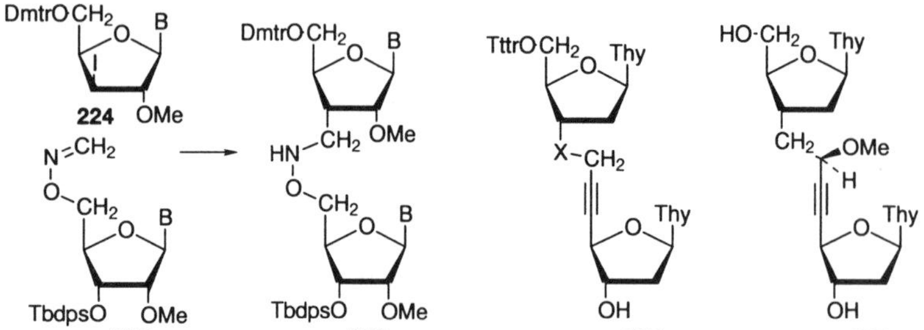

A report from Matteucci's laboratory has described the hexafluoroisopropyl-idene replacement, as in the unit **229**, but, when incorporated into two positions of a 15-mer, significant depression of T_m was observed.[274]

In elegant work, Benner has developed a route to oligoribonucleotide analogues with sulfone internucleotidic links. Reaction of **230** and **231**, both accessible from diacetone glucose, as outlined in Scheme 18, gave the dimer **232**. By selective deprotection at the 'top' and 'bottom', **232** and related dimers containing other bases can be converted, by chemistry as in the synthesis of **230** and **231**, into units with either a bromide at the 'top' or a thioacetyl group at the 'bottom', which can then be linked to give tetramers using the reagents of Scheme 18. This iterative process was used to produce several tetramers, and also octamers with seven sulfone links.[275]

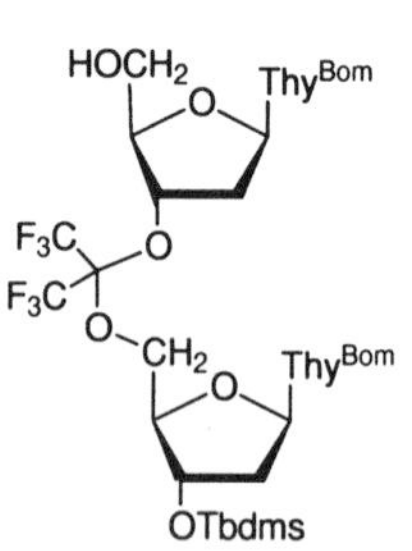

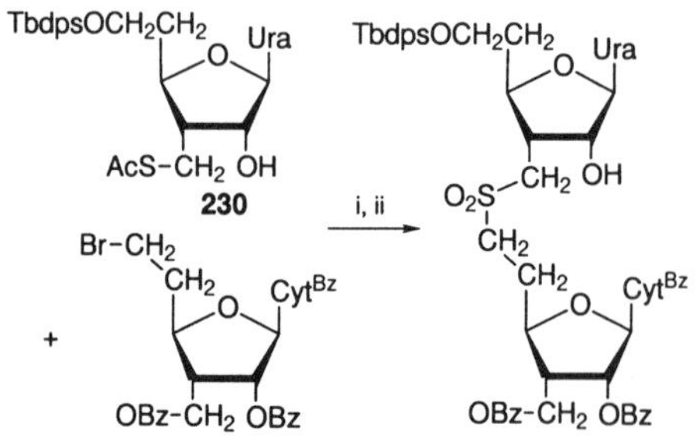

Reagents: i, Cs_2CO_3; ii, oxone

Scheme 18

Dimers with a novel three-atom $2' \rightarrow 5'$-linkage, as in **233**, have been synthesized, and their binding affinity to poly-U was reported.[276]

13 Ethers, Esters and Acetals of Nucleosides

An extensive study has been reported on the *O*-methylation of 5'-*O*-protected ribonucleosides by diazomethane in the presence of Lewis acids, with a view to optimization of the amount of 2'-*O*-methylated product.[277] Phase-transfer methylation has been used to obtain 2'-*O*-methyl-N⁶-cyclohexyladenosine,[278] whilst treatment of the anhydronucleoside **69** with magnesium or calcium alkoxides in DMF at 100°C led to the 2'-*O*-alkylated uridine derivatives. It was argued that the free hydroxy group at C-3' delivered the reagent from the α-face, since the 3'-*O*-methyl analogue of **69** underwent reaction at C-2 to give an *arabino*-product.[279] 2'-*O*-Methyl-β-D-arabinofuranosylthymine, and the α-anomer, have been prepared by base-sugar condensation, and incorporated into oligonucleotides by phosphoramidite chemistry.[280] 2'-*O*-Methoxyethyl ethers **234**, of use in making antisense constructs, have been prepared by base-sugar condensation with the α-methyl ribofuranoside, although prior conversion of the methyl glycoside into a glycosyl fluoride gave improved yields for B=Thy.[281] 2'-*O*-Aminopropyl ribonucleosides have also been prepared, by direct 2'-*O*-alkylation for purines and by the stannylene method for pyrimidines, and incorporated into oligonucleotides.[282]

3'- And 5'-*O*-alkyl derivatives of 2'-deoxy-5-fluorouridine, and some 5'-*O*-aminoacyl derivatives, have been described,[283] as have 3'- or 5'-*O*-arylmethyl derivatives of FdU which also have an aminoacyloxymethyl group at N-3 (e.g. **235**).[284]

The 1,1-dianisyl-2,2,2-trichloroethyl (DATE) protecting group (see Vol. 29, p. 295) has been used for procection at O-2' in ribonucleotide synthesis.[285] The dimethoxytrityl protecting group can be introduced at O-5' in highly electrophilic nucleosides using dimethoxytrityl borofluoride in the absence of any nucleophilic species,[286] whilst it has been shown that trityl and monomethoxytrityl groups can be removed in high yield using ceric ammonium nitrate in DMF-MeCN-H₂O.[287]

Selective enzymic acylation and deacylation of nucleosides have been reviewed.[288] Cordycepin (3'-deoxyadenosine) has been linked to various lipophilic vitamins and to diacylglycerols via dicarboxylic acid spacers at O-2' and O-5',[289] and similar conjugates have been made to the cordycepin trimer core, with linkages at either the 5'- or the 2'-end.[290] 3'-*O*-Aminoacyl-2'-deoxynucleosides have been made using Fmoc aminoacid fluorides as acylating agents.[291]

The diacetylated derivative **236** of (β-D-glucopyranosyl)thymine has been prepared as an intermediate for incorporation of this nucleoside into oligonucleotides by 4'-6'-links.[292]

Dicarboxylic acid 5'-monoesters of thymidine and of 5-(2-thienyl)-2'-deoxyuridine have been prepared as triphosphate mimics.[293] The ester of d4T with citric acid has also been made with the same objective,[294] and more lipophilic bioisosteres **237** and the AZT equivalent have been reported, these being designed to mimic the conformation of the nucleoside triphosphate when complexed to a metal ion.[295] Neither of these last two classes were effective, since all the activity against HIV could be shown to be due to hydrolysis to the parent nucleoside.

A tripyridyl unit has been linked to a known ketose nucleoside via a spacer arm to give **238**; this, and similar structures with the tripyridyl group linked to C-5 of the uracil, were incorporated into oligonucleotides, and the Cu(II) complexes were functional mimics of ribozymes.[296]

The *p*-nitrobenzyloxymethyl group has been used for the protection of the 2'-hydroxy function in solid-phase oligonucleotide synthesis, It is easily removed with fluoride ion and stable to basic and acidic conditions.[297]

Treatment of nucleoside 2',3'-orthoesters of type **239** (R=H, Me, Ph, B=normal bases) with (EtO)₂PCl in acetonitrile gives high yields of the phosphonoalkylidene compounds **240**, mostly as the *exo*-anomers, and similar 3',5'-*O*-phosphonoalkylidene derivatives were obtained from 2'-deoxy-D-*threo*-nucleosides.[298]

The ethylene-linked analogue **241** of CMP-NeuNAc has been prepared by Wittig chain extension of a cytosine 5'-aldehyde, followed by glycoside formation.[299]

237   238   239   240

14   Miscellaneous Nucleoside Analogues

There has been an extended report on the synthesis of 5'-epi-bicyclonucleosides of type **242** (see Vol. 29, p. 297) and their incorporation into oligodeoxynucleotides.[300]

The bicyclic nucleoside analogues **243** (B=Ade, Gua, Thy, Ura), related to isodideoxynucleosides, have been prepared starting from isosorbide,[301] and the 3'-deaza-analogue of (*S*,*S*)-iso-ddA has been reported, along with the 8-aza-analogue and some related 1,2,4-triazoles.[302] Some deaza-analogues of (*R*,*R*)-iso-ddA have also been prepared,[303] and there have been reports on the hydroxylated species **245** (B=Ade, Cyt, Ura),[304] and the branched azidocompounds **246** (B=Ade, Thy, Ura).[305]

1,5-Anhydrohexitol-based nucleosides continue to attract attention in Herdewijn's laboratory. The thymine compound **247** (Vol. 27, p. 269) has been converted into the azido-species **248** and into the d4-type compound **249**,[306] whilst **249** and related species with other bases have also been made starting from a 1,5-anhydro-D-glucitol derivative.[307] The enantiomer of **247**, and related species with other bases, have been made from L-glucose, whilst the 8-aza- and 7-deazaguanine analogues of **247** (D-series) have also been reported.[308] A route has been developed to make the nucleoside analogues **250** (B=Ura, Crt) based on 1,5-anhydro-D-mannitol.[309]

In the area of dioxolanyl and oxathiolanyl nucleoside analogues, a series of 5-substituted oxathiolanyl systems **251** [R=I, CO$_2$H, CONH(CH$_2$)$_2$NH$_2$] have been reported,[310] as has **251** (R=SePh) and its dioxolanyl and 6-aza-analogues,[311] and a species **251** with a carboranyl group at C-5.[115] Racemic thiazolidines **252**, and the regioisomers with sulfur and nitrogen interchanged, have been described.[312] An interesting series of compounds, namely the dioxolanyl species **253**,[313] the oxathiolanes **254** and **255**,[314] and the dithiolanyl compounds **256**[315] have been described by Swedish workers. The isoxazolidine systems **257** have been prepared by diastereoselective addition of *N*-methylhydroxylamine to a chiral α,β-unsaturated lactone, followed by base-'sugar' linkage.[316] Racemic isoxazolidine nucleosides have been made by condensation between bases and 5-acetoxy-isoxazolidines,[317] whilst dipolar cycloaddition between a nitrile oxide and an *N*-vinylpurine gave (±)-**258**.[318] Pyrrolidine analogues of type **259** (X=OH, N$_3$, F; B=5-substituted-Cyt or -Ura) have been reported.[319,320]

A novel type of nucleoside analogue reported for the first time is the thietane **261**. This was prepared from L-ascorbic acid, which was converted into the thietane **260** (sugar carbons numbered) by a process involving inversion of

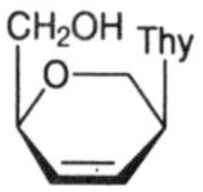

249 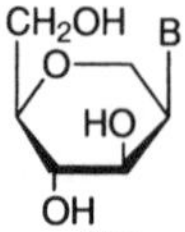   250 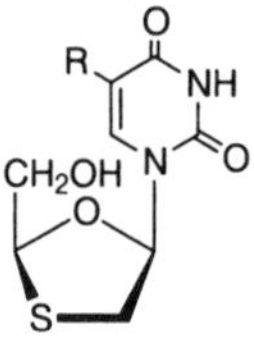   251   Boc 252

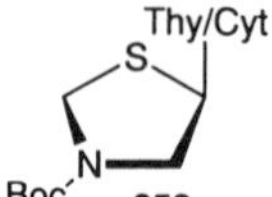

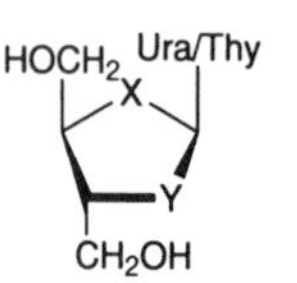

253 X = Y = O
254 X = O, Y = S
255 X = S, Y = O
256 X = Y = S

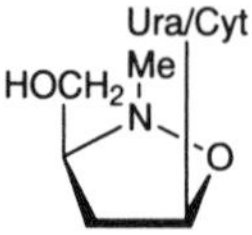

257 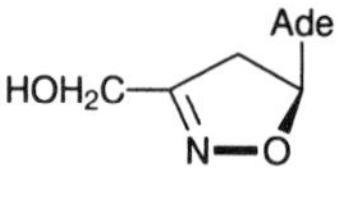   258 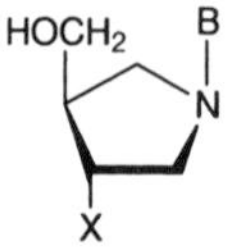   259

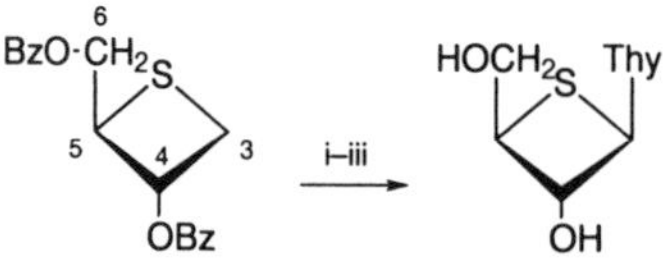

Reagents: i, MCPBA; ii, thymine, TmsOTf,
Et₃N, ZnI₂, toluene; iii, NaOMe, MeOH

Scheme 19

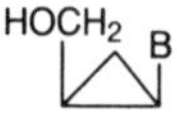

260   261

262 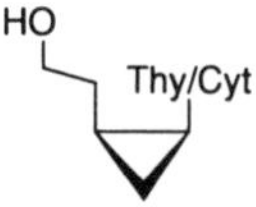   263

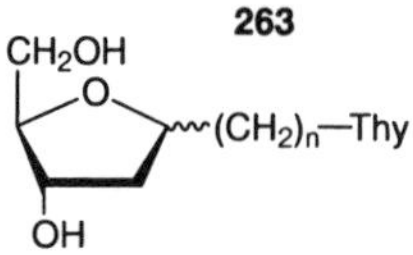

264, n = 1
265, n = 2

configuration at C-5. Formation of the nucleoside used Pummerer chemistry (Scheme 19), and was β-selective, presumably as a result of benzoyloxy participation, although the yield was only moderate.[321]

There have been reports on the synthesis of chiral cyclopropyl analogues of type 262,[322,323] and earlier work (Vol. 28, p. 298, and Vol. 29, p. 298-9) has been extended to the preparation of the homologues 263, designed to give a better fit of the hydroxy group with the corresponding OH of the substrate, but the compounds did not have anti-HIV activity.[324]

The extended thymidine analogues 264[325,326] and 265[326] (both anomers in both cases) have been prepared and incorporated into oligodeoxynucleosides by both 3′→3′ and 3′→5′ linkages to study conformational aspects of *alt*-DNA sequences. The synthesis of 264 involved reduction of a glycosyl cyanide, followed by elaboration of the heterocycle, whilst 265 was made by Wittig reaction-Michael cyclization on a 2-deoxyribose derivative, again followed by reduction and heterocycle synthesis. Some compounds with heterocyclic groups attached at C-5 of a ribose unit are mentioned in Chapter 10.

5′-Dimethylarsinyloxy nucleosides have been prepared for the first time, from dimethylarsinic acid and isopropylidene-protected nucleosides.[327]

15 Reactions

A review has been published on the hydrolysis of the *N*-glycosidic bond in nucleosides and nucleotides.[328] The electron distributions, proton affinities, and energies of the heterolytic fission of the *N*-glycosidic bond of adenosine, 3-methyladenosine and several related compounds have been calculated by semiempirical and *ab initio* methods. Preferred sites of protonation were calculated, together with the effect on the *N*-glycosidic bond.[329] Transglycosylation reactions of purine nucleosides have been reviewed, including both 3→9 and 7↔9 transglycosylations.[330]

Various nucleoside dialdehydes, obtained by periodate cleavage, have been tested for antitumour activity.[331] The periodate forms of strongly basic anion exchangers have been used for the selective oxidation of ribofuranosyl nucleosides in the presence of galacto- and gluco-pyranosides.[332] Routes have been developed for the regioselective synthesis of both adenosine monoaldehydes; the previously-unknown 3'-monoaldehyde **266** was made from the dialdehyde by selective protection of the 3'-aldehyde as the *N*,*N*'-diphenylimidazolidine derivative, followed by borohydride reduction.[332]

A detailed investigation has been reported into the cleavage of 3'→5'-uridyluridine to form the 2',3'-cyclic phosphate, and its isomerization to 2'→5'-uridyluridine.[334] The hydrolysis of uridine 2'-, 3'- and 5'-phosphoromonothioates under acidic and neutral conditions has been investigated; in mild acid only hydrolysis to uridine occurs, whilst at low pH desulfurization occurs in the cases of the 2'- and 3'-thioates.[335] The same workers have also studied the kinetics of hydrolysis and desulfurization of the diastereomeric monothio-analogues of uridine 2',3'-cyclic phosphate; under neutral or acidic conditions desulfurization competes with phosphoester hydrolysis.[336] The hydrolysis of the 2'-thionucleoside 3'-phosphate **267** (X=SH) has been studied; the predominant reaction pathway at pH 13 is the formation of the *S*-phosphate **268**, whilst at pH 7-10 mostly the 2',3'-cyclic monothiophosphate was produced.[337] The 2'-fluorocompound **267** (X=F), which has a C-3'-*endo*- conformation, underwent hydrolysis ten times faster than did the deoxycompound **267** (X=H).[338] The kinetics of hydrolysis of thymidine 5'-boranomonophosphate (**269**) have been studied by NMR. It was found that **269** hydrolyses slowly to thymidine and $[O_3P-BH_3^{3-}]$, with the latter hydrolysing even more slowly to phosphonate and boric acid.[339]

266   **267**   **268**   **269**

In studies on DNA cleavage by radical processes, the *t*-butyl ketone **270,** made from D-fructose, was photolysed to give the radical **271.** Under anaerobic conditions, with a hydrogen donor present, an α,β-mixture of deoxyuridines was formed, whilst aerobic photolysis in the absence of H-donors gave uracil and 2'-deoxyribonolactone. Experiments in the presence of oxygen and a thiol permitted an assessment of the relative importance of the two processes under physiological conditions.[340] The 4'-radical **272** has also been generated by photolysis of the 4'-*C*-pivaloyl compound (Vol. 28, p. 280), and, in the presence of methanol and Bu$_3$SnH, gives a mixture of **273** and **274.** Additional evidence from non-nucleosidic systems implicated a mesomeric radical cation, which reacts by an S$_N$1-type process.[341]

The reaction of 2'-deoxyguanosine with hydroxyl radicals gives two main products **275** and **276,** which have been the subject of a detailed study by NMR.[342]

When the nitrate ester **277** was treated with Bu$_3$SnD and AIBN, the deuteriated alcohol **278** was formed, by 1,5-radical translocation. However, the 2'-chloro-2'-deoxy-system **279** gave the furanone **280** under the same conditions. These results suggest that the inhibition of ribonucleoside diphosphate reductase by 2'-chloro-2'-deoxynucleosides involves loss of Cl from a radical at C-3'.[343] Some related studies from the same team, also supporting this conclusion, were mentioned earlier.[82]

References

1   H. Vorbrüggen, *Acta Biochim. Pol.*, 1996, **43**, 24 (*Chem. Abstr.*, 1996, **125**, 168 452).
2   T. Mukaiyama, M. Nagai, T. Matsutani and N. Shimomura, *Nucleosides Nucleotides*, 1996, **15**, 17.
3   M. Nagai, T. Matsutani and T. Mukaiyama, *Heterocycles*, 1996, **42**, 57.
4   S. Hanessian, J.J. Conde, H.H. Khai and B. Lou, *Tetrahedron*, 1996, **52**, 10827.
5   B.M. Trost and Z. Shi, *J. Am. Chem. Soc.*, 1996, **118**, 3037.
6   L. Schmitt and C.A. Caperelli, *Nucleosides Nucleotides*, 1996, **15**, 1905.
7   M.J. Robins, R. Zou, Z. Guo and S.F. Wnuk, *J. Org. Chem.*, 1996, **61**, 9207.
8   J. Matulic-Adamic, C. Gonzalez, N. Usman and L. Beigelman, *Bioorg. Med. Chem. Lett.*, 1996, **6**, 373.
9   A. Amantea, M. Henz and P. Strazewski, *Helv. Chim. Acta*, 1996, **79**, 244.
10   T. Sekine, E. Kawashima and Y. Ishido, *Tetrahedron Lett.*, 1996, **37**, 7757.
11   A. A.-H. Abdel-Rahman, M.A.Zahran, A. E.-S. Abdel-Megied, E.B. Pedersen and C. Neilsen, *Synthesis*, 1996, 237.
12   S. Sun, X.-Q. Tang, A. Merchant, P.S. Anjaneyulu and J.A. Piccirilli, *J. Org. Chem.*, 1996, **61**, 5708.
13   K. Felczak, A.K. Drabikowska, J.A. Vilpo, T. Kulikowski and D. Shugar, *J. Med. Chem.*, 1996, **39**, 1720.
14   A.S. Frame, R.H. Wightman and G. Mackenzie, *Tetrahedron*, 1996, **52**, 9219.
15   Y. Sha and M. Cai, *Huaxue Tongbao*, 1996, 38 (*Chem. Abstr.*, 1996, **125**, 301 423).
16   A. Benjahad, R. Granet, P. Krausz, C. Bosrigaud and S. Delebassée, *Nucleosides Nucleotides*, 1996, **15**, 1849.
17   J. Depelley, R. Granet, M. Kaouadji, P. Krausz, S. Piekarski, S. Delebassée and C. Bosrigaud, *Nucleosides Nucleotides*, 1996, **15**, 995.
18   G. Dewynter, N. Aouf, Z. Regainia and J.-L. Montero, *Tetrahedron*, 1996, **52**, 993.
19   H. Chen, J. Zhang, J. Mao and M. Cai, *Chem. Res. Chin. Univ.*, 1996, **12**, 258 (*Chem. Abstr.*, 1996, **125**, 329 263).
20   A.D. de Matta, A.M.R. Bernardino, G.A. Romeiro, M.R.P. de Oliveira, M.C.B.V. de Souza and V.F. Ferreira, *Nucleosides Nucleotides*, 1996, **15**, 889.
21   R. Zou, K.R. Ayres, J.C. Drach and L.B. Townsend, *J. Med. Chem.*, 1996, **39**, 3477.
22   A.K. Sharma, A.K. Yadav and L. Prakash, *Phosphorus, Sulfur Silicon Relat. Elem.*, 1996, **112**, 109 (*Chem. Abstr.*, 1996, **125**, 143 193).
23   J. Farràs, M. del Mar Lleó, J. Vilarrasa, S. Castillón, M. Matheu, X. Solans and M. Font-Bardia, *Tetrahedron Lett.*, 1996, **37**, 901.
24   P.J. Bhuyan and J.S. Sandhu, *J. Chem. Res. (S)*, 1996, 44.
25   Z. Zhu and L.B. Townsend, *Tetrahedron Lett.*, 1996, **37**, 3263.
26   R. Wolin, D. Wang, J. Kelly, A. Afonso, L. James, P. Kirchmeier and A.T. McPhail, *Bioorg. Med. Chem. Lett.*, 1996, **6**, 195.
27   K. Aritomo, C. Urashima, T. Wada and M. Sekine, *Nucleosides Nucleotides*, 1996, **15**, 1.
28   A. Al Mourabit, M. Beckmann, C. Poupat, A. Ahond and P. Potter, *Tetrahedron: Asymm.*, 1996, **7**, 3455.
29   V.A. Timoshchuk, *Zh. Org. Khim.*, 1995, **31**, 630 (*Chem. Abstr.*, 1996, **124**, 233 006).
30   M. Krecmerova, H. Hrebabecky, M. Masojidkova and A. Holy, *Collect. Czech. Chem. Commun.*, 1996, **61**, 458 (*Chem. Abstr.*, 1996, **125**, 58 953).
31   N.M. Fathy, *Phosphorus, Sulfur, Silicon Relat. Elem.*, 1995, **107**, 7 (*Chem. Abstr.*, 1996, **124**, 343 963).
32   A.I. Khodair and E.-S.E. Ibrahim, *Nucleosides Nucleotides*, 1996, **15**, 1927.

33   Y.A. Ibrahim, *Carbohydr. Lett.,* 1996, **1**, 425

34   R. Rico Gomez and J. Rios Ruiz, *Heterocycles,* 1996, **43**, 317.

35   R.P. Tripathi, A. Mishra, R. Pratap and D.S. Bhakuni, *Indian J. Chem., Sect. B:*
      Org. Chem. Incl. Med. Chem., 1996, **35B**, 441 (*Chem. Abstr.,* 1996, **125**, 34 025).

36   W.L. McEldoon and D.F. Wiemer, *Tetrahedron,* 1996, 52, 11695.

37   A. Kittaka, Y. Tsubaki, H. Tanaka, K.T. Nakamura and T. Miyasaka, *Nucleosides*
      Nucleotides, 1996, **15**, 97.

38   A. Kittaka, N. Yamada, H. Tanaka, K.T. Nakamura and T. Miyasaka, *Nucleosides*
      Nucleotides, 1996, **15**, 1447.

39   A. Kittaka, H. Tanaka, N. Yamada and T. Miyasaka, *Tetrahedron Lett.,* 1996, **37**,
      2801.

40   T. Gimisis and C. Chatgilialoglu, *J. Org. Chem.,* 1996, **61**, 1908.

41   J. Hiebl and E. Zbiral, *Nucleosides Nucleotides,* 1996, **15**, 1649.

42   Y. Yoshimura, F. Kano, S. Mirazaki, N. Ashida, S. Sakata, K. Haraguchi, Y. Itoh,
      H. Tanaka and T. Miyasaka, *Nucleosides Nucleotides,* 1996, **15**, 305.

43   M.I. Balagopala, A.P. Ollapally and H.J. Lee, *Nucleosides Nucleotides,* 1996, **15**, 899.

44   B. Bennua-Skalmowski and H. Vorbrüggen, *Nucleosides Nucleotides,* 1996, **15**, 739.

45   A.A. El-Barbary, N.R. El-Brollosy and E.B. Pedersen, *Bull. Soc. Chim. Fr.,* 1996,
      133, 51 (*Chem. Abstr.,* 1996, **124**, 317 760).

46   P. Lipka, A. Zatorski, K.A. Watanabe and K.W. Pankiewicz, *Nucleosides Nucleo-*
      tides, 1996, **15**, 149.

47   K. Haraguchi, H. Tanaka, S. Saito, S. Kinoshima, M. Hasoe, K. Kanmuri,
      K. Yamaguchi and T. Miyasaka, *Tetrahedron,* 1996, **52**, 9467.

48   P. Ge, G.O. Voronin and T.I. Kalman, *Nucleosides Nucleotides,* 1996, **15**, 1701.

49   M.P. Groziak, R. Lin, W.C. Stevens, L.L. Wotring, L.B. Townsend, J. Balzarini,
      M. Witvrouw and E. De Clercq, *Nucleosides Nucleotides,* 1996, **15**, 1041.

50   M. Park and C.J. Rizzo, *J. Org. Chem.,* 1996, **61**, 6092.

51   T. Iimori, H. Kobayashi, S. Hashimoto and S. Ikegami, *Heterocycles,* 1996, **42**, 485.

52   T. Mukaiyama, N. Hirano, M. Nishida and H. Uchiro, *Chem. Lett.,* 1996, 99.

53   N. Hicks and D.W. Hutchinson, *Biocatalysis,* 1994, **11**, 1 (*Chem. Abstr.,* 1996, **125**,
      11 337).

54   N.J. Leonard and Neelima, *Nucleosides Nucleotides,* 1996, **15**, 1369.

55   D.E. Bergstrom, P. Zhang and W.T. Johnson, *Nucleosides Nucleotides,* 1996, **15**, 59.

56   G. Cristalli, R. Volpini, S. Vittori, E. Camaioni, G. Rafaiani, S. Potenza and
      A. Vita, *Nucleosides Nucleotides,* 1996, **15**, 1567.

57   J.E. Anderson-McKay, G.W. Both and G.W. Simpson, *Nucleosides Nucleotides,*
      1996, **15**, 1307.

58   J. Jansons, J. Stattel, A. Verri, J. Gambino, N. Khan, F. Focher, S. Spadari and
      G. Wright, *Nucleosides Nucleotides,* 1996, **15**, 669.

59   F. Seela and P. Leonard, *Helv. Chim. Acta,* 1996, **79**, 477.

60   T. Wenzel and F. Seela, *Helv. Chim. Acta,* 1996, **79**, 169.

61   F. Seela, W. Bourgeois, H. Rosemeyer and T. Wenzel , *Helv. Chim. Acta,* 1996, **79**,
      488.

62   F. Seela and M. Zulauf, *Synthesis,* 1996, 726.

63   Y. Maurinsh and W. Pfleiderer, *Nucleosides Nucleotides,* 1996, **15**, 431.

64   O. Jungmann and W. Pfleiderer, *Tetrahedron Lett.,* 1996, **37**, 8355.

65   M. Melguiso, M. Gottlieb and W. Pfleiderer, *Pteridines,* 1995, **6**, 85 (*Chem. Abstr.,*
      1996, **124**, 261 581).

66   K. Shinozuka, N. Yamada, A. Nakamura, H. Ozaki and H. Sawai, *Bioorg. Med.*
      Chem. Lett., 1996, **6**, 1843.

67   J. Marfurt, J. Hunziker and C. Leumann, *Bioorg. Med. Chem. Lett.*, 1996, **6**, 3021.
68   F. Seela, M. Heckel and H. Rosemeyer, *Helv. Chim. Acta*, 1996, **79**, 1451.
69   K.A. Henningfeld, T. Arslan and S.M. Hecht, *J. Am. Chem. Soc.*, 1996, **118**, 11701.
70   G. Wang and P.J. Middleton, *Tetrahedron Lett.*, 1996, **37**, 2739.
71   J.-M. Escudier, I. Tworkowski, L. Bouziani and L. Gorrichon, *Tetrahedron Lett.*, 1996, **37**, 4689.
72   E. Lee-Ruff, F. Xi and J.H. Qie, *J. Org. Chem.*, 1996, **61**, 1547.
73   H. Shiragami, Y. Amino, Y. Honda, M. Arai, Y. Tanaka, H. Iwagami, T. Yukawa and K. Isawa, *Nucleosides Nucleotides*, 1996, **15**, 31.
74   K. Danel, E. Larsen, E.B. Pedersen, B.F. Vestergard and C. Nielsen, *J. Med. Chem.*, 1996, **39**, 2427.
75   C.-S. Chen and J.-W. Chen, *Nucleosides Nucleotides*, 1996, **15**, 1253.
76   B.H. Zhu, H. Yan, L.B. Luo, Z.H. Liu and H.L. Chen, *Chin. Chem. Lett.*, 1996, **7**, 503 (*Chem. Abstr.*, 1996, **125**, 222 315).
77   K. Sujino, T. Yoshida and H. Sugimura, *Tetrahedron Lett.*, 1996, **37**, 6133.
78   D.L.J. Clive, P.L. Wickens and P.W.M. Sgarbi, *J. Org. Chem.*, 1996, **61**, 7426.
79   F.E. McDonald and M.M. Gleason, *J. Am. Chem. Soc.*, 1996, **118**, 6648.
80   H. Shiragami, T. Ineyama, Y. Uchida and K. Isawa, *Nucleosides Nucleotides*, 1996, **15**, 47.
81   R.P. Discordia, *J. Labelled Compd. Radiopharm.*, 1996, **38**, 613 (*Chem. Abstr.*, 1996, **125**, 143 202).
82   M.J. Robins, S.F. Wnuk, A.E. Hernández-Thirring and M.C. Samano, *J. Am. Chem. Soc.*, 1996, **118**, 11341.
83   P.J. Bolon, P. Wang, C.K.Chu, G. Gosselin, V. Boudou, C. Pierra, C. Mathé, J.-L. Imbach, A. Faraj, M. A. el Alaoui, J.-P. Sommadossi, S.B. Pai, Y.-L. Zhu, J.-S. Lin, Y.-C. Cheng and R.F. Schinazi, *Bioorg. Med. Chem. Lett.*, 1996, **6**, 1657.
84   T.-S. Lin, M.-Z. Luo, M.-C. Liu, Y.-L. Zhu, E. Gullen, G.E. Dutschman and Y.-C. Cheng, *J. Med. Chem.*, 1996, **39**, 1757.
85   M.J. Slater, C. Gowrie, G.A. Freeman and S.A. Short, *Bioorg. Med. Chem. Lett.*, 1996, **6**, 2787.
86   T. Maruyama, Y. Sato, Y. Oto, Y. Takahashi, R. Snoeck, G. Andrei, M. Witvrouw and E De Clercq, *Chem. Pharm. Bull.*, 1996, **44**, 2331.
87   T. Ma, S.B. Pai, Y.L. Zhu, J.S. Lin, K, Shanmuganathan, J. Du, C. Wang, H. Kim, M.G. Newton, Y.C. Cheng and C.K. Chu, *J. Med. Chem.*, 1996, **39**, 2835.
88   L.W. Hertel, J.S. Kroin, C.S. Grossmann, G.B. Grindley, A.F. Dorr, A.M.V. Storiolo, W. Plunkett, V. Ghandi and P. Huang, *A.C.S. Symp. Ser.*, 1996, **639**, 265 (*Chem. Abstr.*, 1996, **125**, 301 455).
89   V.I. Kobylinskaya, A.S. Shalamay, V.A. Gladkaya, V.L. Makitruk and V.I. Kondratyuk, *Bioorg. Khim.*, 1994, **20**, 1226 (*Chem. Abstr.*, 1996, **124**, 30 239).
90   J.A. Hill and D.D. Bankston, *J. Labelled Compd. Radiopharm.*, 1995, **36**, 713 (*Chem. Abstr.*, 1996, **124**, 9281).
91   H.M. Abdel-Barg, A.A.-H. Abdel-Rahman, E.B. Pedersen and C. Nielsen, *Monatsh. Chem.*, 1995, **126**, 811 (*Chem. Abstr.*, 1996, **124**, 9302).
92   Y. Aoyama, T. Sekine, Y. Imamoto, E. Kawashima and Y. Ichido, *Nucleosides Nucleotides*, 1996, **15**, 733.
93   M. Aoyagi, Y. Ueno, A. Ono and A. Matsuda, *Bioorg. Med. Chem. Lett.*, 1996, **6**, 1573.
94   D.A. Baeschlin, M. Daube, M.O. Blättler, S.A. Benner and C. Richert, *Tetrahedron Lett.*, 1996, **37**, 1591.
95   G. Walcher and W. Pfleiderer, *Helv. Chim. Acta*, 1996, **79**, 1067.

96   D.P.C. McGee, A. Vaughn-Settle, C. Vargeese and Y. Zhai, *J. Org. Chem.*, 1996, **61**, 781.

97   D.P.C. McGee, D.P. Sebesta, S.S. O'Rourke, R.L. Martinez, M.E. Jung and W.A. Pieken, *Tetrahedron Lett.*, 1996, **37**, 1995.

98   D.P. Sebesta, S.S. O'Rourke, R.L. Martinez, W.G. Pieken and D.P.C. McGee, *Tetrahedron*, 1996, **52**, 14385.

99   L.F. García-Alles, J. Magdalena and V. Gotor, *J. Org. Chem.*, 1996, **61**, 6980

100   M. Krecmerova, H. Hrebebecky, M. Masojidkova and A. Holy, *Collect. Czech. Chem. Commun.*, 1996, **61**, 478 (*Chem. Abstr.*, 1996, **125**, 58 954).

101   A.A. Malin, V.A. Ostrovskii, M.V. Yas'ko and A.A. Kraevskii, *Zh. Org. Chem.*, 1995, **31**, 628 (*Chem. Abstr.*, 1996, **124**, 233 005).

102   A. Matsuda and K.A. Watanabe, *Nucleosides Nucleotides*, 1996, **15**, 205.

103   N. Ohta, K. Minamoto, T. Yamamoto, N. Koide and R. Sakoda, *Nucleosides Nucleotides*, 1996, **15**, 833.

104   Y.-H. Jin, M. Bae, Y.-J. Byun, J.H. Kim and M.W. Chun, *Arch. Pharmacol. Res.*, 1995, **18**, 364 (*Chem. Abstr.*, 1996, **124**, 146 711).

105   M.J. Thompson, A. Mekhalfia, D.C. Jakeman, S.E.V. Phillips, K. Phillips, J. Porter and G.M. Blackburn, *Chem. Commun.*, 1996, 791.

106   X. Chen, K. Bastow, B. Goz, L.S. Kucera, S.L. Morris-Natschke and K.S. Ishaq, *J. Med. Chem.*, 1996, **39**, 3412.

107   A.A. Barbary, N.R. El-Brollosy, E.B. Pedersen and C. Nielsen, *Monatsh. Chem.*, 1996, **124**, 593 (*Chem. Abstr.*, 1996, **124**, 9279).

108   Y.S. Zhang, L.X. Wang and L.H. Zhang, *Chin. Chem. Lett.*, 1995, **6**, 549 (*Chem. Abstr.*, 1996, **124**, 9294).

109   K. Walczak and J. Suwinski, *Pol. J. Chem.*, 1996, **70**, 861 (*Chem. Abstr.*, 1996, **125**, 196 212).

110   T. Persson, A.-B. Hoernfeldt and N.G. Johansson, *Antiviral Chem. Chemother.*, 1996, **7**, 101 (*Chem. Abstr.*, 1996, **124**, 343 950).

111   P. Wentworth, Jr., T. Wiemann and K.D. Janda, *J. Am. Chem. Soc.*, 1996, **118**, 12521.

112   P. Wentworth, Jr. and K.D. Janda, *Chem. Commun.*, 1996, 2097.

113   C. McGuigan, S. Turner, N. Mahmood and A.J. Hay, *Bioorg. Med. Chem. Lett.*, 1996, **6**, 2445.

114   W. Urjasz, L. Celewicz and K. Golankiewicz, *Nucleosides Nucleotides*, 1996, **15**, 1189.

115   K. Imamura and Y. Yamamoto, *Bioorg. Med. Chem. Lett.*, 1996, **6**, 1855.

116   A.P. Higson, G.K. Scott, D.J. Earnshaw, A.D. Baxter, R.A. Taylor and R. Cosstick, *Tetrahedron*, 1996, **52**, 1027.

117   L.B. Weinstein, D.J. Earnshaw, R. Cosstick and T.R. Cech, *J. Am. Chem. Soc.*, 1996, **118**, 10341.

118   X. Liu and C.B. Reese, *Tetrahedron Lett.*, 1996, **37**, 925.

119   S.K. Singh, V.S. Palmer and J. Wengel, *Tetrahedron Lett.*, 1996, **37**, 7617.

120   S. Shaw-Ponter, G. Mills, M. Robertson, R.D. Bostwick, G.W. Hardy and R.J. Young, *Tetrahedron Lett.*, 1996, **37**, 1867.

121   D.L. Selwood, K. Carter, R.J. Young and K.S. Jandu, *Bioorg. Med. Chem. Lett.*, 1996, **6**, 991.

122   I. Basnak, M. Sun, P.L. Coe and R.T. Walker, *Nucleosides Nucleotides*, 1996, **15**, 121.

123   E.L. Hancox and R.T. Walker, *Nucleosides Nucleotides*, 1996, **15**, 135.

124   D.F. Ewing and G. Mackenzie, *Nucleosides Nucleotides*, 1996, **15**, 809.

125 N.A. Van Draanen, G.A. Freeman, S.A. Short, R. Harvey, R. Jensen, G. Szczech and G.W. Koszalka, *J. Med. Chem.*, 1996, **39**, 538.

126 P. Barraclough and C.J. Wharton, *J. Chem. Res.*, 1996, (*S*) 514; (*M*) 2946.

127 Y. Yoshimura, K. Kitano, H. Satoh, M. Watanabe, S. Miura, S. Sakata, T. Sasaki and A. Matsuda, *J. Org. Chem.*, 1996, **61**, 822.

128 A. Eleuteri, C.B. Reese and Q. Song, *J. Chem. Soc., Perkin Trans. 1*, 1996, 2237.

129 I. Giri, P.J. Bolon and C.K. Chu, *Nucleosides Nucleotides*, 1996, **15**, 183.

130 H. Awano, S. Shuto, T. Miyashita, N. Ashida, H. Machida, T. Kira, S. Shigeta and A. Matsuda, *Arch. Pharm. (Weinheim, Ger.)*, 1996, **329**, 66.

131 T. Iino, S. Shuto and A. Matsuda, *Nucleosides Nucleotides*, 1996, **15**, 169.

132 A.J. Lawrence, J.B.J. Pavey, R. Cosstick and I.A. O'Neil, *J. Org. Chem.*, 1996, **61**, 9213.

133 A.E.A. Hassan, N. Nishizono, N. Minakawa, S. Shuto and A. Matsuda, *J. Org. Chem.*, 1996, **61**, 6261.

134 Y. Xiang, R.F. Schinazi and K. Zhao, *Bioorg. Med. Chem. Lett.*, 1996, **6**, 1475.

135 E. Lee-Ruff, M. Ostrowski, A. Ladha, D.V. Stynes, I. Vernik, J.-L. Jiang, W.-Q. Wan, S.-F. Ding and S. Joshi, *J. Med. Chem.*, 1996, **39**, 5276.

136 I.A. Mikhailopulo, N.E. Poopieko, T.M. Tsvetkova, A.P. Marochkin and J. Balzarini, *Carbohydr. Res.*, 1996, **285**, 17.

137 M.W. Chun, K. Lee, Y.S. Choi, J. Lee, J.H. Kim, C.K. Lee, B.G. Choi and Y.C. Xu, *Arch. Pharmacol. Res.*, 1996, **19**, 243 (*Chem. Abstr.*, 1996, **125**, 196 214).

138 J. Wengel, R.F. Schinazi and M.H. Caruthers, *Bioorg. Med. Chem.*, 1995, **3**, 1223 (*Chem. Abstr.*, 1996, **124**, 87 653).

139 B. Doboszewski and P. Herdewijn, *Tetrahedron*, 1996, **52**, 1651.

140 B. Quiclet-Sire and S.Z. Zard, *J. Am. Chem. Soc.*, 1996, **118**, 1209.

141 J. Wengel, K. Østergaard and A. Hager, *Nucleosides Nucleotides*, 1996, **15**, 1361.

142 M. Xie and T.S. Widlanski, *Tetrahedron Lett.*, 1996, **37**, 4443.

143 W.Y. Lau, L. Zhang, J. Wang, D. Cheng and K. Zhao, *Tetrahedron Lett.*, 1996, **37**, 4297.

144 K. Hisa, A. Hittaka, H. Tanaka, K. Yamaguchi and T. Miyasaki, *Nucleosides Nucleotides*, 1996, **15**, 85.

145 P.J. Serafinowski and C.L. Barnes, *Tetrahedron*, 1996, **52**, 7929.

146 A. Matsuda, H. Hattori, M. Tanaka and T. Sasaki, *Bioorg. Med. Chem. Lett.*, 1996, **6**, 1887.

147 H. Hattori, M. Tanaka, M. Fukushima, T. Sasaki and A. Matsuda, *J. Med. Chem.*, 1996, **39**, 5005.

148 R. Alvarez, A. San-Felix, E. De Clercq, J. Balzarini and M.J. Camarasa, *Nucleosides Nucleotides*, 1996, **15**, 349.

149 J.M.J. Tronchet, I. Kovacs, F. Barbalat-Rey and N. Dolatshahi, *Nucleosides Nucleotides*, 1996, **15**, 337.

150 T. Waga, H. Ohrui and H. Meguro, *Nucleosides Nucleotides*, 1996, **15**, 287.

151 G. Wang and W.E. Seifert, *Tetrahedron Lett.*, 1996, **37**, 6515.

152 A. Marx, P. Erdmann, M. Senn, S. Körner, T. Jungo, M. Petretta, P. Imwinkelried, A. Dussy, K.J. Kulicke, L. Macko, M. Zehnder and B. Giese, *Helv. Chim. Acta*, 1996, **79**, 1980.

153 P. Bravo, A. Mele, G. Salani, F. Viani and P. La Collo, *Gazz. Chim. Ital.*, 1995, **125**, 295 (*Chem. Abstr.*, 1996, **124**, 30 216).

154 K. Haraguchi, H. Tanaka, Y. Itoh, K. Yamaguchi and T. Miyasaka, *J. Org. Chem.*, 1996, **61**, 851.

155 B. Doboszewski and P.A.M. Herdewijn, *Nucleosides Nucleotides*, 1996, **15**, 1495.

Carbohydrate Chemistry

156   I. Bousquié, V. Madiot, J.-C. Florent and C. Monneret, *Bioorg. Med. Chem. Lett.*, 1996, **6**, 1815.
157   K. Sakthivel and T. Pathak, *Tetrahedron*, 1996, **52**, 4877.
158   K. Sakthivel, T. Pathak and C.G. Suresh, *Tetrahedron*, 1996, **52**, 1767.
159   S.B. Ha and V. Nair, *Tetrahedron Lett.*, 1996, **37**, 1567.
160   S.F. Wnuk, S. Liu, C.-S. Yuan, R.T. Borchardt and M.J. Robins, *J. Med. Chem.*, 1996, **39**, 4162.
161   P.E. Joos, A. De Groot, E.L. Esmans, F.C. Alderweireldt, A. De Bruyn, J. Balzarini and E. De Clercq, *Heterocycles*, 1996, **42**, 173.
162   P.E. Joos, E.L. Esmans, F.C. Alderweireldt, A. De Bruyn, J. Balzarini and E. De Clercq, *Heterocycles*, 1996, **43**, 287.
163   S. Hildbrand and C. Leumann, *Angew. Chem., Int. Ed. Engl.*, 1996, **35**, 1968.
164   W. Liu, J.A. Walker, J.J. Chen, D.S. Wise and L.B. Townsend, *Tetrahedron Lett.*, 1996, **37**, 5325.
165   J.J. Voegel and S.A. Benner, *Helv. Chim. Acta*, 1996, **79**, 1863.
166   M. Yokoyama, M. Nomura, H. Togo and H. Seki, *J. Chem. Soc., Perkin Trans. 1*, 1996, 2145.
167   K.S. Gudmundsson, J.C. Drach and L.B. Townsend, *Tetrahedron Lett.*, 1996, **37**, 2365.
168   K.S. Gudmundsson, J.C. Drach and L.B. Townsend, *Tetrahedron Lett.*, 1996, **37**, 6275.
169   M. Prhavc and J. Kobe, *Nucleosides Nucleotides*, 1996, **15**, 1779.
170   M.A.E. Sallam, H.M. El Nahas, S.M.E. Abdel Megid and T. Anthonsen, *Carbohydr. Res.*, 1996, **280**, 127.
171   Y. Xiang, J. Du and C.K. Chu, *Nucleosides Nucleotides*, 1996, **15**, 1821.
172   C.S. Lee, J. Du and C.K. Chu, *Nucleosides Nucleotides*, 1996, **15**, 1223.
173   H. Cheng, L. Ma and L. Zhang, *Gaodeng Xuexiao Huaxue Xuebao*, 1996, **17**, 1078 (*Chem. Abstr.*, 1996, **125**, 196 130).
174   S. Shuto, T. Obara, Y. Saito, G. Andrei, R. Snoeck, E. De Clercq and A. Matsuda, *J. Med. Chem.*, 1996, **39**, 2392.
175   L.B. Akella and R. Vince, *Tetrahedron*, 1996, **52**, 2789.
176   R. Vince and L.B. Akella, *Tetrahedron*, 1996, **52**, 8407.
177   T.T. Curran, *Synth. Commun.*, 1996, **26**, 1209.
178   D.R. Borcherding, B.T. Butler, M.D. Linnik, S. Mehdi, M.W. Dudley and C.K. Edwards, III, *Nucleosides Nucleotides*, 1996, **15**, 967.
179   D.R. Borcherding, N.P. Peet, H.R. Munson, H. Zhang, P.F. Hoffman, T.L. Bowlin, and C.K. Edwards, III, *J. Med. Chem.*, 1996, **39**, 2615.
180   F. Theil, S. Ballschuh, M. von Janta-Lipinski and R.A. Johnson, *J. Chem. Soc., Perkin Trans. 1*, 1996, 255.
181   M.A. Siddiqui, H. Ford, Jr., C. George and V.E. Marquez, *Nucleosides Nucleotides*, 1996, **15**, 235.
182   V.E. Marquez, M.A. Siddiqui, A. Ezzitouni, P. Russ, J. Wang, R.W. Wagner and M.D. Matteucci, *J. Med. Chem.*, 1996, **39**, 3739.
183   L.S. Jeong and V.E. Marquez, *Tetrahedron Lett.*, 1996, **37**, 2353.
184   K.-H. Altmann, R. Kesselring and U. Pieles, *Tetrahedron*, 1996, **52**, 12699.
185   M. Tanaka, Y. Norimine, T. Fujita, H. Suemune and K. Sakai, *J. Org. Chem.*, 1996, **61**, 6952.
186   M.C. Balo, F. Fernández, E. Lens, C. Lopez, E. De Clercq, G. Andrei, R. Snoeck and J. Balzarini, *Nucleosides Nucleotides*, 1996, **15**, 1335.
187   L. Santana, M. Teijeira, E. Uriarte, C. Teran, U. Castellato and R. Graziani, *Nucleosides Nucleotides*, 1996, **15**, 1179.

188   M.J. Konkel and R. Vince, *Tetrahedron*, 1996, **52**, 799.

189   N. Katagiri, Y. Ito, T. Shiraishi, T. Maruyama, Y. Sato and C. Kaneko, *Nucleosides Nucleotides*, 1996, **15**, 631.

190   M.J. Konkel and R. Vince, *Tetrahedron*, 1996, **52**, 8969.

191   A. Rosenquist, I. Kvarnström, B. Classon and B. Samuelsson, *J. Org. Chem.*, 1996, **61**, 6282.

192   S.N. Mikhailov, N. Blaton, J. Rozenski, J. Balzarini, E. De Clercq and P. Herdewijn, *Nucleosides Nucleotides*, 1996, **15**, 867.

193   V. Ozola, C.B. Reese and Q. Song, *Tetrahedron Lett.*, 1996, **37**, 8621.

194   V.T. Ravikumar, Z.S. Cheruvallath and D.L.Cole, *Tetrahedron Lett.*, 1996, **37**, 6643.

195   Z. Zhang and J.Y. Tang, *Tetrahedron Lett.*, 1996, **37**, 331.

196   M. Mizuguchi and K. Makino, *Nucleosides Nucleotides*, 1996, **15**, 407.

197   A. De Nino, A. Liguori, A. Procopio, E. Roberti and G. Sindona, *Carbohydr. Res.*, 1996, **286**, 77.

198   B. Haly, L. Bellon, V. Mohan and Y. Sanghvi, *Nucleosides Nucleotides*, 1996, **15**, 1383.

199   A. Sierzchala, A. Okruszek and W.J. Stec, *J. Org. Chem.*, 1996, **61**, 6713.

200   V.T. Ravikumar and Z.S. Cheruvallath, *Nucleosides Nucleotides*, 1996, **15**, 1149.

201   Z.S. Cheruvallath, D.L. Cole and V.T. Ravikumar, *Nucleosides Nucleotides*, 1996, **15**, 1441.

202   Y. Jin, G. Biancotto and G. Just, *Tetrahedron Lett.*, 1996, **37**, 973.

203   G. Tosquellas, I. Barber, F. Morvan, B. Rayner and J.-L. Imbach, *Bioorg. Mec Chem. Lett.*, 1996, **6**, 457.

204   A. Kers, I. Kers, A. Kraszewski, M. Sobowski, T. Szabó, M. Thelin, R. Zain and J. Stawinski, *Nucleosides Nucleotides*, 1996, **15**, 361.

205   L. Wozniak, J. Pyzowski and W.J. Stec, *J. Org. Chem.*, 1996, **61**, 879.

206   M.C. Pirrung and N. Chidambaram, *J. Org. Chem.*, 1996, **61**, 1540.

207   M.M. Vaghefi and K.A. Langley, *Tetrahedron Lett.*, 1996, **37**, 4853.

208   Y.-Z. Zhou, L.-X. Wang and L.-H. Zhang, *Guodeng Xuexiao Huaxue Xuebao*, 1995, **16**, 915 (*Chem. Abstr.*, 1996, **124**, 30 234).

209   C. Le Bec and E. Wickstrom, *J. Org. Chem.*, 1996, **61**, 510.

210   R.M. Karl, W. Richter, R. Klösel, M. Mayer and I. Ugi, *Nucleosides Nucleotides*, 1996, **15**, 379.

211   M.J. Rudolph, M.S. Reitman, E.W. MacMillan and A.F. Cook, *Nucleosides Nucleotides*, 1996, **15**, 1725.

212   T. Kofoed and M.H. Caruthers, *Tetrahedron Lett.*, 1996, **37**, 6457.

213   A. Haikal, J. Douman and L. Wyns, *Monatsh. Chem.*, 1995, **126**, 1031 (*Chem. Abstr.*, 1996, **124**, 146705).

214   D. Wang and P.O.P. Ts'o, *Nucleosides Nucleotides*, 1996, **15**, 387.

215   Z.-F. Wang, X.-H. Gu and Y.-Q. Chen, *Chin. J. Chem.*, 1996, **14**, 80 (*Chem. Abstr.*, 1996, **125**, 11 335).

216   I.A. Mikhailopulo, E.N. Kalinichenko, T.L. Podkopaeva, T. Wenzel, H. Rosemeyer and F. Seela, *Nucleosides Nucleotides*, 1996, **15**, 445.

217   C. Hörndler and W. Pfleiderer, *Helv. Chim. Acta*, 1996, **79**, 718.

218   H. Almer and R. Strömberg, *J. Am. Chem. Soc.*, 1996, **118**, 7921.

219   H. Ushida, Y.-D. Wu, S. Sonoda, S. Fukushima, J. Fukuda, I. Takahashi, S. Maeda and A. Nomura, *Nucleosides Nucleotides*, 1996, **15**, 649.

220   X. Chen, Y.P. Feng, N.J. Zhang and Y.F. Zhao, *Chin. Chem. Lett.*, 1995, **6**, 577 (*Chem. Abstr.*, 1996, **124**, 9296).

221   R.C. Mucic, M.K. Herrlein, C.A. Mirkin and R.L. Letsinger, *Chem. Commun.*, 1996, 555.

222   Y.A. Sharkin, M.V. Yas'ko, A.Y. Skoblov and L.A. Aleksandrova, *Bioorg. Khim.*, 1996, **22**, 297 (*Chem. Abstr.*, 1996, **125**, 276 428).

223   H.M. Chen, R.Y. Chen, P.Y. Li and C.X. Zhang, *Chin. Chem. Lett.*, 1996, **7**, 799 (*Chem. Abstr.*, 1996, **125**, 329 262).

224   S.F. Wnuk and M.J. Robins, *J. Am. Chem. Soc.*, 1996, **118**, 2519.

225   B.A. Horenstein and M. Bruner, *J. Am. Chem. Soc.*, 1996, **118**, 10371.

226   T. Yoshino and R.R. Schmidt, *Carbohydr. Lett.*, 1995, **1**, 329 (*Chem. Abstr.*, 1996, **124**, 317 743).

227   M. Imamura and H. Hashimoto, *Tetrahedron Lett.*, 1996, **37**, 1451.

228   M. Imamura and H. Hashimoto, *Chem. Lett.*, 1996, 1087.

229   A. Zervosen, A. Stein, H. Adrian and L. Elling, *Tetrahedron*, 1996, **52**, 2395.

230   A. Tsotinis, T. Calogeropoulou, M. Koufaki, C. Souli, J. Balzarini, E. De Clercq and A. Makriyannis, *J. Med. Chem.*, 1996, **39**, 3418.

231   M. Koufaki, T. Calogeropoulou, T. Mavromoustakos, E. Theodoropoulou, A. Tsotinis and A. Makriyannis, *J. Heterocycl. Chem.*, 1996, **33**, 619.

232   D. Bonaffé, B. Dupraz, J. Ughetto-Monfrin, A. Namane, Y. Henin and T.H. Dihn, *J. Org. Chem.*, 1996, **61**, 895.

233   S. Shuto, H. Awano, N. Shimazaki, K. Hanaoka and A. Matsuda, *Bioorg. Med. Chem. Lett.*, 1996, **6**, 1021.

234   S. Shuto, H. Awano, A. Fujii, K. Yamagami and A. Matsuda, *Bioorg. Med. Chem. Lett.*, 1996, **6**, 2177.

235   H. Sigmund and W. Pfleiderer, *Helv. Chim. Acta*, 1996, **79**, 426.

236   E.L. Vodovozova, Y.B. Pavlova, M.A. Polushkina, A.A. Rzhaninova, M.M. Garaev and Y.G. Molotovskii, *Bioorg. Khim.*, 1996, **22**, 451 (*Chem. Abstr.*, 1996, **125**, 329131).

237   H. Schott and R.A. Schwendener, *Liebigs Ann. Chem.*, 1996, 365.

238   C.I. Hong, A. Nechaev, A.J. Kirisits, R. Vig, C.L. Wegt, K.K. Manouilov and C.K. Chu, *J. Med. Chem.*, 1996, **39**, 1771.

239   C. McGuigan, D. Cahard, H.M. Sheeka, E. De Clercq and J. Balzarini, *Bioorg. Med. Chem. Lett.*, 1996, **6**, 1183; *J. Med. Chem.*, 1996, **39**, 1748.

240   C. McGuigan, O.M. Wedgwood, E. De Clercq and J. Balzarini, *Bioorg. Med. Chem. Lett.*, 1996, **6**, 2359.

241   T.W. Abraham, T.I. Kalman, E.J. McIntee and C.R. Wagner, *J. Med. Chem.*, 1996, **39**, 4569.

242   G. Valette, A. Pompon, J.-L. Girardet, L. Cappellacci, P. Franchetti, M. Grifantini, P. La Colla, A.G. Loi, C. Périgaud, G. Gosselin and J.-L. Imbach, *J. Med. Chem.*, 1996, **39**, 1981.

243   R.-Y. Chen, G.-C. Chi and X.-R. Chen, *Gaodeng Xuexiao Huaxue Xuabao*, 1996, **17**, 429 (*Chem. Abstr.*, 1996, **125**, 34 020).

244   C. Meier, *Angew. Chem., Int. Ed. Engl.*, 1996, **35**, 70.

245   S. Khamnei and P.F. Torrence, *J. Med. Chem.*, 1996, **39**, 4109.

246   T.J. Tolbert and J.R. Williamson, *J. Am. Chem. Soc.*, 1996, **118**, 7929.

247   B. Nawrot, P. Hoffmüller and M. Sprinzl, *Chem. Pap.*, 1996, **50**, 151 (*Chem. Abstr.*, 1996, **125**, 168 559).

248   T. Schoetzau, T. Holletz and D. Cech, *Chem. Commun.*, 1996, 387.

249   M.A. Dineva and D.D. Petkov, *Nucleosides Nucleotides*, 1996, **15**, 1459.

250   T. Persson, G. Larsson and P.O. Nyman, *Bioorg. Med. Chem.*, 1996, **4**, 553.

251   A. Kreimeyer, J. Ugletto-Monfrin, A. Namane and T. Huynh-Dinh, *Tetrahedron Lett.*, 1996, **37**, 8739.

252 M.E. Tanner, S. Vaganay, J. van Heijenoort and D. Blanot, *J. Org. Chem.*, 1996, **61**, 1756.

253 W. Dabkowski and J. Michalski, *Pol. J. Chem.*, 1995, **69**, 979 (*Chem. Abstr.*, 1996, **124**, 9313).

254 H. Inoue, Y. Baba and M. Tsuhako, *Phosphorus Res. Bull.*, 1995, **5**, 137 (*Chem. Abstr.*, 1996, **125**, 11 342).

255 F.-J. Zhang and C.J. Sih, *Bioorg. Med. Chem. Lett.*, 1996, **6**, 2311.

256 F.-J. Zhang, S. Yamada, Q.-M. Gu and C.J. Sih, *Bioorg. Med. Chem. Lett.*, 1996, **6**, 1203.

257 E.J. Hutchinson, B.F. Taylor and G.M. Blackburn, *Chem. Commun.*, 1996, 2765.

258 A. Zatorski, K.A. Watanabe, S.F. Carr, B.M. Goldstein and K.A. Pankiewicz, *J. Med. Chem.*, 1996, **39**, 2422.

259 J. Hall, P. Genschik and W. Filipowicz, *Helv. Chim. Acta*, 1996, **79**, 1005.

260 F. Zeng and R.A. Jones, *Nucleosides Nucleotides*, 1996, **15**, 1679.

261 K. Seio, T. Wada, K. Sakamoto, S. Yokoyama and M. Sekine, *J. Org. Chem.*, 1996, **61**, 1500.

262 A.F. Drake, A. Garofalo, J.M.L. Hillman, V. Merlo, R. McCague and S.M. Roberts, *J. Chem. Soc., Perkin Trans. 1*, 1996, 2739.

263 N. Dyatkina, E. Shirokova, F. Theil, S.M. Roberts and A. Krayevsky, *Bioorg. Med. Chem. Lett.*, 1996, **6**, 2639.

264 R. Liboska, M. Masojidkova and I. Rosenberg, *Collect. Czech. Chem. Commun.*, 1996, **61**, 313 (*Chem. Abstr.*, 1996, **125**, 11 347).

265 R. Liboska, M. Masojidkova and I. Rosenberg, *Collect. Czech. Chem. Commun.*, 1996, **61**, 778 (*Chem. Abstr.*, 1996, **125**, 168 549).

266 A. Waldner, A. De Mesmaeker and S. Wenderborn, *Bioorg. Med. Chem. Lett.*, 1996, **6**, 2363.

267 A. De Mesmaeker, C. Lesueur, M.-O. Bévièrre, A. Waldner, V. Fritsch and R.M. Wolf, *Angew. Chem., Int. Ed. Engl.*, 1996, **35**, 2790.

268 M.J. Robins, S. Sarker, M. Xie, W. Zhang and M.A. Peterson, *Tetrahedron Lett.*, 1996, **37**, 3921.

269 G. Viswanadham, G.V. Petersen and J. Wengel, *Bioorg. Med. Chem. Lett.*, 1996, **6**, 987.

270 S.M. Ali and P.K. Bridson, *Nucleosides Nucleotides*, 1996, **15**, 1531.

271 B. Haly, R. Bharadwaj and Y.S. Sanghvi, *Synlett*, 1996, 687.

272 B. Bhat, E.E. Swayze, P. Wheeler, S. Dimock, M. Perbost and Y.S. Sanghvi, *J. Org. Chem.*, 1996, **61**, 8186.

273 S. Wenderborn, C. Jouanno, R.M. Wolf and A. De Mesmaeker, *Tetrahedron Lett.*, 1996, **37**, 5511.

274 K.-Y. Lin and M.D. Matteucci, *Tetrahedron Lett.*, 1996, **37**, 8667.

275 C. Richert, A.L. Roughton and S.A. Benner, *J. Am. Chem. Soc.*, 1996, **118**, 4518.

276 C.R. Noe, N. Windhab and G. Haberhauer, *Arch. Pharm. (Weinheim, Ger)*, 1995, **328**, 743 (*Chem. Abstr.*, 1996, **124**, 202 905).

277 H. Cramer and W. Pfleiderer, *Helv. Chim. Acta,* 1996, **79**, 2114.

278 M. Prashad, K. Prasad and O. Repic, *Synth. Commun.*, 1996, **26**, 3967.

279 D.P.C. McGee and Y. Zhai, *Nucleosides Nucleotides*, 1996, **15**, 1797.

280 C.H. Gotfredsen, J.P. Jacobsen and J. Wengel, *Bioorg. Med. Chem.*, 1996, **4**, 1217.

281 P. Martin, *Helv. Chim. Acta,* 1996, **79**, 1930.

282 R.H. Griffey, B.P. Monia, L.L. Cummins, S. Freier, M.J. Greig, C.J. Guinosso, E. Lesnik, S.M. Manalili, V. Mohan, S. Owens, B.R. Ross, H. Sasmor, E. Wanciewicz, K. Weiler, P.D. Wheeler and P.D. Cook, *J. Med. Chem.*, 1996, **39**, 5100.

283   A. Uemura, Y. Tada, S. Takeda, J. Uchida and J. Yamashita, *Chem. Pharm. Bull.*,
       1996, **44**, 150.
284   N. Harada, M. Hongu, T. Kawaguchi, M. Ohohashi, K. Oda, T. Hashiyama and
       K. Tsujihara, *Chem. Pharm. Bull.*, 1996, **44**, 1196.
285   R. Klösel, S. König, S. Lehnhoff and R.M. Karl, *Tetrahedron*, 1996, **52**, 1493.
286   M.K. Lakshman and B. Zajc, *Nucleosides Nucleotides*, 1996, **15**, 1029.
287   J.R. Hwu, M.L. Jain, S.-C. Tsay and G.H. Hakimelahi, *Chem. Commun.*, 1996, 545.
288   A.K. Prasad and J. Wengel, *Nucleosides Nucleotides*, 1996, **15**, 1347.
289   M. Wasner, R.J. Suhadolnik, S.E. Horvath, M.E. Adelson, N. Kon, M.-X. Guan,
       E.E. Henderson and W. Pfleiderer, *Helv. Chim. Acta*, 1996, **79**, 609.
290   M. Wasner, R.J. Suhadolnik, S.E. Horvath, M.E. Adelson, N. Kon, M.-X. Guan,
       E.E. Henderson and W. Pfleiderer, *Helv. Chim. Acta*, 1996, **79**, 619.
291   J.S. Oliver and A. Oyelere, *J. Org. Chem.*, 1996, **61**, 4168.
292   B.S. Ermolensky, M.V. Fomitcheva, E.V. Efimtseva, S.V. Meshkov, S.N. Mikhailov,
       D.M. Esipov, E.F. Boldyreva and V.G. Korobko, *Nucleosides Nucleotides*, 1996, **15**,
       1619.
293   U. Wellmar, A.-B. Hörnfeldt, S. Gronowitz and N.G. Johansson, *Nucleosides
       Nucleotides*, 1996, **15**, 1059.
294   R. Weaver, I.H. Gilbert, N. Mahmood and J. Balzarini, *Bioorg. Med. Chem. Lett.*,
       1996, **6**, 2405.
295   A.O. Goldring, I.H. Gilbert, N. Mahmood and J. Balzarini, *Bioorg. Med. Chem.
       Lett.*, 1996, **6**, 2411.
296   J.K. Bashkin, J. Xie, A.D. Daniher, U. Sampath and J. L.-F. Kao, *J. Org. Chem.*,
       1996, **61**, 2314.
297   G.R. Gough, T.J. Miller and N.A. Mantick, *Tetrahedron Lett.*, 1996, **37**, 981.
298   M. Endová, M. Masojídková, M. Budesinsky and I. Rosenberg, *Tetrahedron Lett.*,
       1996, **37**, 3497.
299   Y. Hatanaka, M. Hashimoto, K.I.-P. Jwa Hidari, Y. Sanai, Y. Nagai and
       Y. Kanaoka, *Heterocycles*, 1996, **43**, 531.
300   J.C. Litten and C. Leumann, *Helv. Chim. Acta*, 1996, **79**, 1129.
301   L. Pickering and V. Nair, *Nucleosides Nucleotides*, 1996, **15**, 1751.
302   L.B. Zintek, T.S. Jahnke and V, Nair, *Nucleosides Nucleotides*, 1996, **15**, 69.
303   N. Navarre, P.N. Preston, A.V. Tsytovich and R.H. Wightman, *J. Chem. Res.*, 1996,
       (*S*) 444; (*M*) 2560.
304   H.-W. Yu, L.-R. Zhang, J.-C. Zhou, L.-T. Ma and L.H. Zhang, *Bioorg. Med.
       Chem.*, 1996, **4**, 609.
305   G.S. Jeon and V. Nair, *Tetrahedron*, 1996, **52**, 12643.
306   I. Verheggen, A. Van Aerschot, J. Rosenski, G. Janssen, E. De Clercq and
       P. Herdewijn, *Nucleosides Nucleotides*, 1996, **15**, 325.
307   I. Luyten and P. Herdewijn, *Tetrahedron*, 1996, **52**, 9249.
308   M.W. Andersen, S.M. Daluge, L. Kerremans and P. Herdewijn, *Tetrahedron Lett.*,
       1996, **37**, 8147.
309   M.-J. Pérez-Pérez, E. De. Clercq and P. Herdewijn, *Bioorg. Med. Chem. Lett.*, 1996,
       6, 1457.
310   N. Mourier, C. Trabaud, V. Niddam, J.-C. Graciet, M. Camplo, J.-C. Chermann
       and J.-L. Kraus, *Nucleosides Nucleotides*, 1996, **15**, 1397.
311   J.M. Yoo, H.J. Moon, B.H. Chung, B.G. Choi, J.H. Hong and M.W. Chun, *Yakhak
       Hoechi*, 1996, **40**, 46 (*Chem. Abstr.*, 1996, **124**, 343 952).
312   G. Rassu, F. Zanardi, M. Cornia and G. Casiraghi, *Nucleosides Nucleotides*, 1996,
       15, 1113.

313　J. Brånalt, I. Kvarnström, B. Classon and B. Samuelsson, *J. Org. Chem.*, 1996, **61**, 3599.

314　J. Brånalt, I. Kvarnström, B. Classon and B. Samuelsson, *J. Org. Chem.*, 1996, **61**, 3604.

315　J. Brånalt, I. Kvarnström, B. Classon and B. Samuelsson, *J. Org. Chem.*, 1996, **61**, 3611.

316　Y. Ziang, Y. Gong and K. Zhao, *Tetrahedron Lett.*, 1996, **37**, 4877.

317　U. Chiacchio, G. Gumina, A. Rescifina, R. Romeo, N. Uccella, F. Casuscelli, A. Piperno and G. Romeo, *Tetrahedron*, 1996, **52**, 8889.

318　Y. Ziang, J. Chen, R.F. Schinazi and K. Zhao, *Bioorg. Med. Chem. Lett.*, 1996, **6**, 1051.

319　Y. Lee, H.K. Kim, I.K. Young and Y.B. Chae, *Korean J. Med. Chem.*, 1995, **5**, 66 (*Chem. Abstr.*, 1996, **124**, 202 915).

320　H.K. Kim, Y. Lee, I.K. Young and Y.B. Chae, *Korean J. Med. Chem.*, 1995, **5**, 68 (*Chem. Abstr.*, 1996, **124**, 202 916).

321　N. Nishizono, N. Koike, Y. Yamagata, S. Fujii and A. Matsuda, *Tetrahedron Lett.*, 1996, **37**, 7569.

322　R. Czuk and Y. von Scholz, *Tetrahedron*, 1996, **52**, 6383.

323　M.W. Chun, *Korean J. Med. Chem.*, 1996, **6**, 141 (*Chem. Abstr.*, 1996, **125**, 168 547).

324　T.-F. Yang, H. Kim, L.P. Kotra and C.K. Chu, *Tetrahedron Lett.*, 1996, **37**, 8849.

325　C.L. Scremin, J.H. Boal, A. Wilk, L.R. Phillips and S.L. Beaucage, *Bioorg. Med. Chem. Lett.*, 1996, **6**, 207.

326　J.H. Boal, A. Wilk, C.L. Scremin, G.N. Gray, L.R. Phillips and S.L. Beaucage, *J. Org. Chem.*, 1996, **61**, 8617.

327　C. Lamberty, *Pharmazie*, 1995, **50**, 768 (*Chem. Abstr.*, 1996, **124**, 146 712).

328　M. Oivanen, J. Hovinen, P. Lehikoinen and H. Lönnberg, *Trends Org. Chem.*, 1993, **4**, 397 (*Chem. Abstr.*, 1996, **125**, 58 866).

329　M. Hotakka and H. Lönnberg, *THEOCHEM*, 1996, **363**, 191 (*Chem. Abstr.*, 1996, **125**, 87 075).

330　J. Boryski, *Nucleosides Nucleotides*, 1996, **15**, 771.

331　N.A. Brusentsov, Ya.V. Dobrynin, T.G. Nikolaeva, E.V. Sergeeva, N. Ya. Yurchenko and M.N. Preobrazhenskaya, *Khim.-Farm. Zh.*, 1995, **29**, 52 (*Chem. Abstr.*, 1996, **124**, 117 855).

332　N.A. Brusentsov and M.N. Preobrazhenskaya, *Bioorg. Khim.*, 1996, **22**, 215 (*Chem. Abstr.*, 1996, **125**, 248 299).

333　J.P. Neenan, S.M. Opitz, C.L. Cooke, M.A. Ussery, T.C. Morrill and L.M. Eckel, *Bioorg. Med. Chem. Lett.*, 1996, **6**, 1381.

334　R. Breslow, S.D. Dong, Y. Webb and R. Xu, *J. Am. Chem. Soc.*, 1996, **118**, 6588.

335　M. Ora, M. Oivanen and H. Lönnberg, *J. Chem. Soc., Perkin Trans. 2*, 1996, 771.

336　M. Ora, M. Oivanen and H. Lönnberg, *J. Org. Chem.*, 1996, **61**, 3951.

337　C.L. Dantzman and L.L. Kiessling, *J. Am. Chem. Soc.*, 1996, **118**, 11715.

338　L.Y. Kuo, K. Travers and J. Chock, *Nucleosides Nucleotides*, 1996, **15**, 1741.

339　H. Li, C. Hardin and B.R. Shaw, *J. Am. Chem. Soc.*, 1996, **118**, 6606.

340　B.K. Goodman and M.C. Greenberg, *J. Org. Chem.*, 1996, **61**, 2.

341　S. Peukert and B. Giese, *Tetrahedron Lett.*, 1996, **37**, 4365.

342　S. Raoul, M. Berger, G.W. Buchko, P.C. Joshi, B. Morin, M. Weinfeld and J. Cadet, *J. Chem. Soc., Perkin Trans 2*, 1996, 371.

343　M.J. Robins, Z. Gou, M.C. Samano and S.F. Wnuk, *J. Am. Chem. Soc.*, 1996, **118**, 11317.

21
NMR Spectroscopy and Conformational Features

1 General Aspects

Three major reviews with the following titles have been published: i, 'Molecular modelling of carbohydrates' (52 pp., 108 refs.);[1] ii, 'Carbon-proton coupling constants in the conformational analysis of sugar molecules' (46 pp., 156 refs.);[2] iii, 'Carbon-13 nuclear magnetic relaxation and motional behaviour of carbohydrate molecules in solution (68 pp., 160 refs.).[3] Two further reviews, one in English (77 refs.)[4] and one in Japanese (77 refs.),[5] deal with the determination of the absolute configurations of sugars based on a combination of novel chemical derivatizations with HPLC and NMR techniques, and a short (3 pp.) conference report discusses the internal motions of carbohydrates as probed by NMR spectroscopy and molecular modelling.[6]

By use of neural networks and a data set comprising 56 furanoid and 55 pyranoid compounds, a mathematical model for the computer-assisted simulation of ^{13}C-NMR spectra of monosaccharides has been developed.[7] A new computer simulation method of conformational movement, based on interconversion phenomena, has been proposed as an alternative to molecular dynamics. It has been illustrated for methyl α-D-glucopyranoside and α-D-GalpNAc-(1→3)-[α-L-Fucp-(1→2)]-D-Galp-OMe.[8] A ^{1}H-NMR data base aimed at assisiting carbohydrate chemists in the spectral identification of sugars of unknown structures has been established.[9] Application of the PASS-TOCSY pulse sequence to the 2D HETLOC (heteronuclear coupled hetero half filled proton-proton correlation) experiment allowed the accurate measurement of long-range coupling constants in crowded spectral regions from 1D subspectra; the technique has been applied to the conformational analysis of AZT.[10]

The ring [1]H- and ^{13}C-chemical shifts of 8 aldofuranoses have been recorded in DMSO-d_6 and those of 14 aldopyranoses in DMSO-d_6 as well as D$_2$O, to assess the solvent dependencies and the accuracy in calculated predictions of chemical shifts.[11] A new method for determining the absolute configuration of optically active secondary alcohols (menthol, cholesterol, etc.) involves measurement of the differences in the pyridine-induced chemical shifts of their β-D- and β-L-fucofuranosyl derivatives.[12]

Strong intramolecular hydrogen bonds were detected in DMSO solutions of sugars with *syn*-diaxial OH pairs by use of the SIMPLE (secondary isotope multiple NMR spectroscopy of partially labelled entities) technique; weaker H-bonds were observed when one of the two diaxial groups was a hydroxymethyl function.[13]

New molecular simulations on D-xylose support earlier findings on D-glucose that the observed anomeric equilibration of aldoses results from a competition between internal energy and solvation terms of opposite signs.[14] *Ab initio* molecular orbital calculations have been employed in an investigation of the anomeric, *exo*-anomeric and reverse anomeric effects in *C*-, *N*- and *S*-glycosides. The anomeric effect was found to decrease in the order Cl > OMe > F > SMe > NHMe > Et > NH$_2$Me for the anomeric substituents, the last three exhibiting reverse anomeric effects.[15] Extensive semiempirical SCF-MO calculations indicated that π-character along the anomeric bond contributes to restricted internal rotation and hence to the *exo*-anomeric effect in *O*-, *N*- and *S*-glycosides.[16] A reconsideration of the reverse anomeric effect in glycosylimidazoles, using *N*-tri-*O*-acetyl-α-D-xylopyranosylimidazole (1) as representative compound, suggests that stabilization of the $^1C_4$-form may be dependent on *N*-protonation rather than solution ionic strength.[17]

2 Acyclic Systems

A comparative study of MM2 and MM3 force fields for all 16 diastereomeric peracetylated 1-deoxy-1-nitroheptitols and peracetylated 2-deoxyaldooctoses has been presented.[18]

1   2   3

4   R^1 = NHC(S)NHX, R^2 = H
5   R^1 = H, R^2 = NHC(S)NHX
     X = Me or Ph

3 Furanose Systems

An analysis of the geometric parameters (bond lengths, bond angles, torsion angles) of over 100 ribose and deoxyribose derivatives using well-defined crystal structures from the Cambridge Crystallographic Database has been published.[19] In continuation of previous studies on the ribofuranosyl sugars present in RNA (see Vol. 24, Chapter 21, Ref. 34), $^1J_{CH}$-, $^2J_{CH}$- and $^3J_{CH}$-values for β-D-ribofuranose (2) as functions of the ring conformation were calculated from MO-derived structural data and confirmed experimentally through the measurement of J_{CH} values of conformationally rigid aldopyranosides, *e.g.*, compound 3, containing ^{13}C-^{1}H coupling pathways similar to those found in specific confor-mers of **2**.[20] The conformation of 4-thio-L-lyxono-1,4-lactone in aqueous solution has been investigated by NMR spectroscopy.[20a] On the basis of variable temperature NMR experiments with 3-deoxy-3-thioureido-furanose derivatives **4** and **5** in CDCl$_3$, hydrogen bonding between NH-3' and the ring oxygen atom in the gluco compounds **4** has been proposed.[21]

NMR spectroscopic experiments on L-ascorbic acid and on metal complexes of D-glucuronic and D-galacturonic acid in the furanose form are covered in Section 8 below and a conformational analysis of AZT is referred to in Section 1 above.

Quantum mechanical studies on 2α'-deoxycytosin and its 2',2'-difluoro ana-logue (gemcitabine),[22] and on cyclic 3',5'-AMP[23] have been reported. In an extensive new investigation, the pseudorotational equilibria of β-D-ribo- and 2-deoxy-β-D-ribo-furanosyl nucleosides have been correlated with the anomeric effects and pK$_a$ values of the nucleoside bases.[24]

^{1}H-NMR spectroscopic analysis of purine-like *C*-nucleosides revealed the presence of a hydrogen bond between N-1 and OH-5', an indicator of *syn*- over *anti*-conformational preference, in compound **6**, but not in **7**.[25] The bicyclic deoxynucleoside derivatives **8** have been subjected to conformational analysis by ^{1}H-NMR spectroscopy, to show that a 1'-*exo*/2'-*endo*-arrangement is preferred and that the acetoxy group at C-5' is axially disposed, *i.e.*, the torsion angle γ is in the unusual *syn*-clinal range.[26]

6   7   8

Base = Thy or Bz^6 Ade

D-*Glycero*-D-*ido*-2-octulose, the major sugar found in the resurrection plant *Craterostigma plantagineum*, has been identified by an NMR analysis of its 1,3:4,6:7,8-tri-*O*-isopropylidene-α-furanose derivative.[26a]

4 Pyranose and Related Systems

A united atom molecular mechanics parameter set for pyranoses has been developed from *ab initio* molecular orbital calculations and used to compute anomeric free energies for a series of simple monosaccharides in water.[27] An empirical method has been employed to predict the magnitudes and signs of two-bond ^{13}C-^{13}C spin coupling constants in aldopyranosyl rings. Although designed mainly for the prediction of $^2J_{CCC}$ values, application to the interpretation of

$^{2}J_{COC}$ values and conformational analysis of O-glycosidic linkages has not been excluded.[28]

In a density functional study, the relative stabilities of the $^{1}C_{4}$ and $^{4}C_{1}$ chair forms of β-D-glucopyranose have been calculated for two selected low-energy hydroxyl rotamers.[29] 1D and 2D NMR studies on D-galacturonic and D-glucuronic acid in aqueous solution showed these compounds to exist mainly in pyranose forms, in contrast to their complexes with W(VI) and Mo(IV) ions which prefer furanose forms (see Section 8 below).[30] Semiempirical MO calculations, on the other hand, suggested that the preferred sites for metal ion binding in D-galacturonic acid are the ring oxygen atom, the axial oxygen atom at C-4 of the pyranose form, and a carboxylate oxygen.[31]

| 9 | 10 X = F,Cl, Br, or I | 11 X = N₃ |

9   10 X = F,Cl, Br, or I   11 X = N$_3$
12 X = O(CH$_2$)$_3$SnPh$_3$
13 X = O(CH$_2$)$_3$SnIPh$_2$

14 X = S—(oxazole)

Conformational analyses by use of NMR spectroscopy and/or theoretical calculations have also been reported for the following compounds: methyl α-L-arabinopyranoside monoacetates;[32] 1-C-cyano sugar **9**;[33] 2-deoxy-2-sulfoamino-D-glucopyranoside anions;[34] several methyl 2,6-dideoxy-6-halo-α-L-hexopyrano-sides, *e.g.*, compounds **10**;[35] 6-azido-6-deoxy-1,2:3,4-di-O-α-D-galactopyranose **11**;[36] a number of stannylated α-D-galactopyranose derivatives, such as **12** and **13**;[37] the 6-S-benzoxazol-2-yl derivative **14**, as a representative azaheterocycle/thiosugar hybrid;[38] the S-glycoside **15** of 1,5-dithio-β-D-xylopyranose, a new antithrombotic agent;[39] all nine inositols;[40] the *myo*-inositol derivatives **16** and **17**,[41] and a number of *myo*-inositol mono- to hexakis-phosphates over the entire pH range.[42]

Special attention was given to the conformations about the C-5–C-6 bond in semiempirical quantum mechanical nuclear shielding calculations on a series of hexopyranoses,[43] and in molecular mechanics simulations of methyl α- and β-D-gluco-, -galacto-, and talo-pyranoside in aqueous solution,[44] as well as to the conformation of the acetamido group in 2-acetamido-2-deoxy-D-allo- and -D-gluco-pyranose derivatives, such as **18–20**.[45]

Examination of spin-lattice relaxation times and second moment measurements of crystalline monosaccharides revealed jump motions of a hydrogen-bonded hydroxyl proton in β-D-allopyranose and *trans/gauche* reorientations of the CH$_2$OH group in D-allopyranose and methyl α-D-mannopyranoside.[46] Structural and conformational studies on the mannose-derived amidine **21** and related

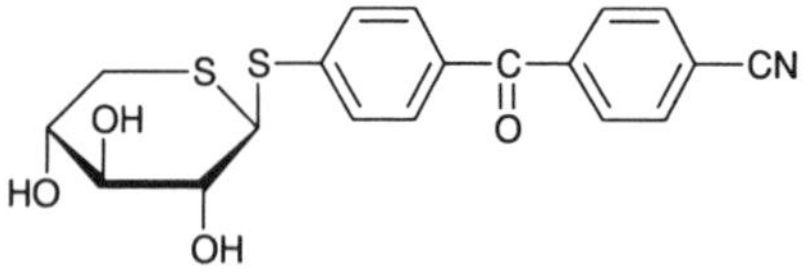

15

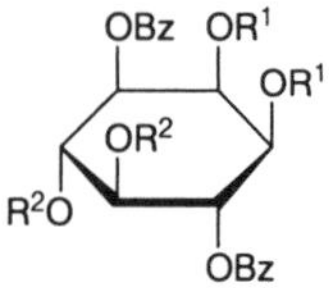

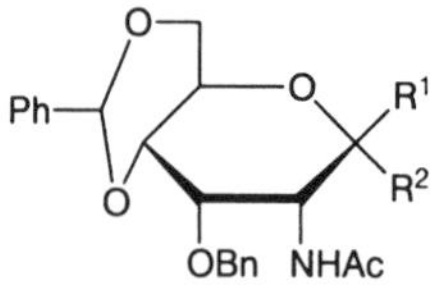

16 R^1, R^1 = Me / Me   R^2 = H

17 R^1 = H, R^2, R^2 = Me / Me

18 R^1 = H, R^2 = OMe

19 R^1 = OMe, R^2 = H

20 R^1, R^2 = NH / NH

compounds by NMR spectroscopy, as well as computational methods, showed that the orientation of its exocyclic NHBn-group differs significantly from that of the anomeric substituent of β-D-glucopyranosides, *i.e.,* amidine-sugars are not perfect mimics of the transition state of β-glycosidase hydrolysis.[47] The ^{13}C CP MAS solid-state NMR spectrum of ureido sugar **22** has been compared with its solution spectrum,[48] and a ^{1}H-NMR investigation of the corresponding *iso*-butyl and L-leucine derivatives **23** and **24**, respectively, showed that N-1′, protons are more acidic than those attached to N-3′.[49]

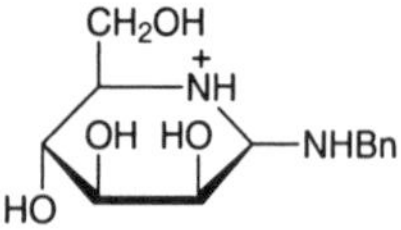

21

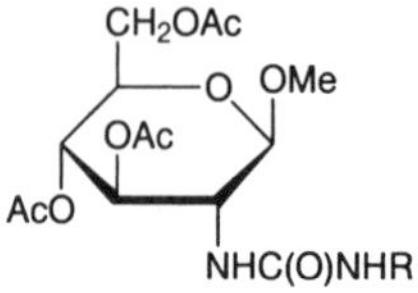

22 R = CH$_2$CH$_2$Ph
23 R = CH$_2$CHMe$_2$
24 R = CHCO$_2$Et
                    |
                    CH$_2$CHMe$_2$

5   Disaccharides

NMR spectroscopy played a vital role in the identification of disaccharides **25** and **26**, the former obtained in the chitosanase digestion of the chitinous component of a fungal cell wall,[50] the latter isolated as an aroma precursor of tea leaves.[51] In connection with the complete ^{1}H- and ^{13}C-NMR assignment of a natural flavonol rhamnosylrhamnoside, the NMR distinction of (1→2)-, (1→3)-, and (1→4)-linked rhamnopyranosylrhamnopyranosides has been discussed.[51a]

$$\beta\text{-D-Glc}p\text{NH}_2\text{-(1}\rightarrow\text{4)-D-Glc}p\text{NAc} \qquad \alpha\text{-L-Ara}p\text{-(1}\rightarrow\text{6)-}\beta\text{-D-Glc}p\text{-}O\text{-geranyl}$$

25   **26**

The structure, conformation and intramolecular hydrogen bonds of crystalline and aqueous sucrose have been discussed in detail.[52] Evidence for a transient inter-residue hydrogen bond in aqueous sucrose (O-2···H···O-1') has been discovered by ROESY spectroscopy under supercooled conditions.[53] Combined use of high resolution NMR techniques and dynamics simulations furnished a new model for the dynamical conformational behaviour of sucrose in water in which internal motions occur at the same rate as overall tumbling.[54] A GROMOS force field analysis, modified to include a potential energy term for the *exo*-anomeric effect, has been assessed for efficiency by application to a conformational analysis of α-maltose.[55,56]

Conformational analyses by use of NMR-spectroscopy and/or computational methods have been undertaken with lactose,[57] the methyl glycosides of disacharidess **27**[58] and **28**,[59] six different monosulfates of disaccharide **29**[60] and with uniformly ^{13}C-labelled digalactosyldiacylglycerols **30** in a membrane environment.[61] Addition of NaCl or CaCl$_2$ to the solution of methyl glycoside of disaccharide **31**, the repeating unit of hyaluronan, in D$_2$O caused only small, although measurable changes in the ^{1}H- and ^{13}C-chemical shifts and the ^{1}H-^{1}H coupling constants remained unchanged.[62]

$$\beta\text{-D-Glc}p\text{-(1}\rightarrow\text{4)-}\alpha\text{-D-Glc}p \quad \alpha\text{-D-Man}p\text{-(1}\rightarrow\text{3)-}\beta\text{-D-Glc}p \quad \alpha\text{-D-Gal}p\text{-(1}\rightarrow\text{3)-}\beta\text{-D-Gal}p$$

27   **28**   **29**

$$\alpha\text{-D-Gal}p\text{-(1}\rightarrow\text{6)-}\beta\text{-D-Gal}p\text{-OCH[CH}_2\text{OCO(CH}_2)_n\text{Me]}_2 \quad \beta\text{-D-Glc}p\text{UA-(1}\rightarrow\text{4)-}\beta\text{-D-Glc}p\text{NAc}$$

30   **31**

The solution conformations of the *C*-linked lactose analogues **32–34**, and in particular the flexibility of the interglycosidic linkages, have been examined by NMR spectroscopy, as well as computational methods. Compound **32** is very similar to natural lactose around the glycosidic bond but more flexible about the aglyconic bond.[63] Analogue **33** is similar to **32**, its isomer **34**, however, is more

32 X = Y = H
33 X = H, Y = OH
34 X = OH, Y = H

flexible than natural lactose about both, the aglyconic and glycosidic bond.[64] *C*-Lactose **32** adopts a different conformation to that of the native *O*-disaccharide when bound to the protein target ricin B, contrary to expectations based on Kishi's work.[63,65]

The complexation of octyl β-maltoside, -melibioside and -lactoside with a 1,1'-dinaphthyl-derived cyclophane which binds disaccharides but not monosaccharides has been investigated by ^{1}H NMR titrations.[66]

6   Oligosaccharides

NMR-spectroscopy was crucial in the identification of the *N*-linked oligosaccharide **35** from pokeweed lectin-B,[67] and of twelve novel, neutral galactofuranose-containing oligosaccharides obtained by degradation of an *N*-linked high-mannose oligosaccharide constituent of ascorbate oxidase.[68]

α-D-Man*p*

1

↓

3

α-D-Man*p*-(1→3)-α-D-Man*p*-(1→4)-α-D-Glc*p*NAc-(1→4)-D-Glc*p*NAc

2   3

↑   ↑

1   1

β-D-Xyl*p*   α-L-Fuc*p*

35

New insights into the three-dimensional structures of oligosaccharides gained by use of X-ray, NMR, and theoretical methods have been reviewed (94 refs.).[69] CHEAT95, a completely revised version of the CHEAT force field (carbohydrate hydroxyl groups represented by extended atoms) which allows simulation of oligosaccharides in aqueous solutions without explicit inclusion of water, has been applied to a complete conformational search for the exocyclic and inter-glycosidic dihedrals of a structural variant of Lewis X.[70] A new method for the measurement of long-range heteronuclear coupling constants from 2D HMQC spectra has been applied to the trisaccharide model **36**.[71] Long-range [4]*J* and [5]*J* values including interglycosidic correlations have been obtained by GCOSY

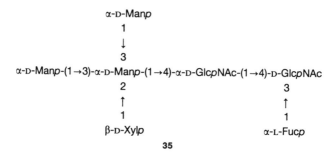

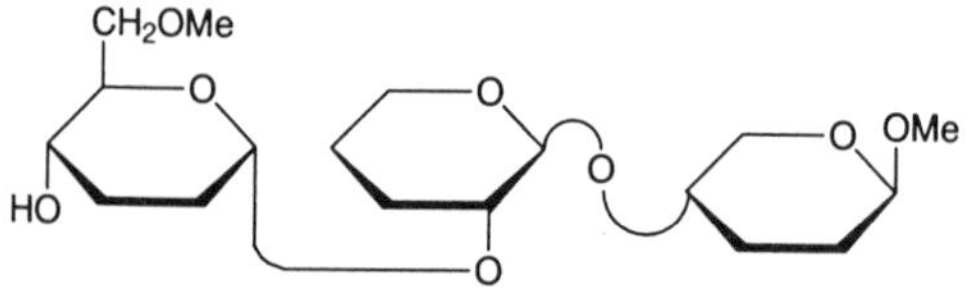

36

techniques; they are especially useful for the spectral assignments of pyranosyl ring protons of linear and branched oligosaccharides in the typically very crowded 3–4 ppm region.[72]

Complete assignment of ^{1}H- and ^{13}C-NMR spectra and conformational studies have been reported for six Lewis D-related trisaccharides,[73] nine protected α-(1→3)-linked digitoxoside trisaccharides,[74] the two trisaccharide moieties of the anthracycline antibiotic ditrisarubicin B,[75] the tetrasaccharide ganglioside GA1,[76,77] ganglioside GM1,[78] Lewis X, Lewis A, sialyl Lewis X and sialyl Lewis A,[79] 1,1,1-kestopentaose,[80] a hexasaccharide carrying a variety of protecting groups, that was a decisive synthetic Lewis X precursor,[81] a synthetic dimeric Lewis X octasaccharide,[82] and a decasaccharide representing the repeating unit of a *Streptococcus* capsular polysaccharide.[83]

The effects of the stereochemistry of the oligosaccharide head-groups on the physical properties of aqueous synthetic glycolipids, *e.g.*, 1,3-di-*O*-dodecylglycerol β-glycosidically linked to cello- or malto-oligosaccharides at O-2, have been studied.[84]

The structures of permethylated cyclodextrins in aqueous solutions, as derived from their unequivocally assigned ^{1}H- and ^{13}C-NMR solution spectra, have been compared with those of their parent oligosaccharides and with solid state data.[85] An investigation of α-, β- and γ-cyclodextrins and their per-*O*-methyl ethers by use of various molecular modelling methods furnished a method for determining the diameters of their cavities and observing the changes in the torus shapes of these molecules on permethylation.[86] The conformation of the inclusion complex of β-cyclodextrin with 2-naphthalenecarboxylate in aqueous solution has been examined.[87] Molecular modelling of the 1,3-bis(benzeneimidazol-2-yl)-benzene-capped αdextrin **37** suggests a very rigid, asymmetric structure, so that the ^{1}H- and ^{13}C-MNR spectra are highly resolved.[88]

37   **38**

7   Other Compounds

NOESY experiments and computerized molecular modelling have been employed to examine the configuration and conformation of the epoxypropyl side-chain of asperlin in benzene,[89] and the structure of fructosazine **38**, a selfcondensation product of 3,4,6-tri-*O*-benzyl-D-glucosamine (see Chapter 10) has been proved by use of 1D differential NOE experiments.[90]

Conformational analyses and related NMR investigations have been carried out on guanosine-5'-diphospho-fucose,[91] a tunicamycin model compound,[92] a coumaroylated flavonon glucorhamnoside,[93] a new ent-kaurane diterpenoid glucosyl ester, the main sweetening constituents of some traditional Chinese medicines,[94] and two new diterpene glycosides from *Nicotiana tabacum*.[95]

8 NMR of Nuclei Other than ¹H and ¹³C

^{13}C-, ^{15}N-, CP MAS- and high resolution multinuclear-NMR methods have been applied to a study of ureido sugars **39**.[96] The ^{17}O-NMR spectra of L-ascorbic acid recorded in aqueous solution as a function of pH have been presented under the title 'New spectroscopy of an old molecule',[97] and the ^{15}N- and ^{17}O-NMR spectra of doubly labelled pyrimidine nucleosides have been recorded for use as references in NMR analyses of labelled RNA fragments.[98] Information on the hydration behaviour of several monosaccharides (Glc, Gal, Man, All, and Tag) has been obtained by measurements of water ^{17}O-relaxation times.[99]

The relative proportions of furanose and pyranose forms of aqueous 2-deoxy-2-fluoro-D-ribose at equilibrium have been determined with the help of ^{1}H-, ^{13}C-, and ^{19}F-NMR spectroscopy.[100] Recordings of vicinal ^{13}C–^{31}P coupling constants were crucial for the determination of the configuration at phosphorus in the enantiomerically pure nucleoside phosphorothioates **40**.[101] After reassignment of the resonances for the phosphorus atoms in D-*myo*-inositol 1,2,6-trisphosphate, conclusions regarding the structure of its zinc complexes drawn previously on the basis of ^{31}P-NMR titrations have been revised.[102]

CH₂OAc
OMe
OAc
AcO
NHC(O)NH—⟨benzene⟩—X

X = H, Me, OMe, or Cl

39

(MeO)₂TrOCH₂
Thy
R¹O∿P=S
OR²

40

R¹ = ⟨CH₂—O—Thy⟩ OTMS

R² = CH₂CH₂CN

Evidence obtained by use of ^{1}H-, ^{13}C-, ^{17}O-, as well as ^{95}Mo- and ^{153}W-NMR spectroscopy, indicated that Mo(VI) and W(VI) ions form 1:2 (metal:ligand) complexes with the furanose forms of D-glucuronic- and D-galacturonic acid.[30] The complexes of L-ascorbic acid with dimethyl- and dibutyl-tin have been examined in the solid state by ^{13}C CP MAS and IR spectroscopy and in aqueous solution by ^{1}H-, ^{13}C-, and ^{119}Sn-NMR spectroscopy; the metal appears to interact with O-1, O-2, and O-3 of the ascorbate anion; similar spectroscopic experiments were carried out with an L-ascorbic acid/dimethyl thallium compound which is shown to behaves as a salt.[103]

References

1   B.R. Boswell, E.E. Coxon and J.M. Coxon, *Adv. Mol. Model.*, 1995, **3**, 145 (*Chem. Abstr.*, 1996, **124**, 317 610).
2   I. Tvaroska and F.R. Taravel, *Adv. Carbohydr. Chem. Biochem.*, 1995, **51**, 15 (*Chem. Abstr.*, 1996, **124**, 261 462).
3   P. Dais, *Adv. Carbohydr. Chem. Biochem.*, 1995, **51**, 63 (*Chem. Abstr.*, 1996, **124**, 232 892).
4   Y. Nishida, *Tohoku J. Agric. Res.*, 1995, **46**, 73 (*Chem. Abstr.*, 1996, **124**, 261 463).
5   Y. Nishida, *Nippon Nogei Kagaku Kaishi*, 1995, **69**, 1481 (*Chem. Abstr.*, 1996, **124**, 117 695).
6   S. Pérez, S.B. Engelsen and C. Herve du Penhoat, *Spectrosc. Biol. Mol., Eur. Conf., 6th*, 1995, 419 (*Chem. Abstr.*, 1996, **124**, 146 638).
7   B.E. Mitchell and P.C. Jurs, *J. Chem. Inf. Comput. Sci.*, 1996, **36**? 58 (*Chem. Abstr.*, 1996, **124**, 117 715).
8   J. Koca, *Theochem.*, 1995, **343**, 125 (*Chem. Abstr.*, 1996, **124**, 176 727).
9   T. Yoshino, *J. Chem. Software*, 1995, **2**, 196 (*Chem. Abstr.*, 1996, **124**, 261 469).
10 R. Bazzo, G. Barbato and D.O. Cicero, *J. Magn. Res. Ser. A*, 1995, **117**, 267 (*Chem. Abstr.*, 1996, **124**, 176 790).
11 P. Hobley, O. Howarth and R.N. Ibbett, *Magn. Reson. Chem.*, 1996, **34**, 755 (*Chem. Abstr.*, 1996, **125**, 301 344).
12 M. Kobayashi, *Tennen Yuki Kagobutsu Toronkai Koen Yoshishu*, 1995, **37th**, 301 (*Chem. Abstr.*, 1996, **124**, 232 910).
13 S.J. Angyal and J.C. Christofides, *J. Chem. Soc., Perkin Trans. 2*, 1996, 1485.
14 R.R. Schmidt, M. Karplus and J.W. Brady, *J. Am. Chem. Soc.*, 1996, **118**, 541.
15 I. Tvaroska and J.P. Carver, *J. Phys. Chem.*, 1996, **100**, 11305 (*Chem. Abstr.*, 1996, **125**, 86 994).
16 J. Llano and L.A. Montero, *J. Comput. Chem.*, 1996, **17** 1371 (*Chem. Abstr.*, 1996, **125**, 248 239).
17 A.R. Vaino, S.S.C. Chan, W.A. Sarek and G.R.J. Thatcher, *J. Org. Chem.*, 1996, **61**, 4514.
18 D. Valasco, C. Jaime and X. Sanchez-Ruiz, *J. Mol. Struct.*, 1995, **356**, 35 (*Chem. Abstr.*, 1996, **124**, 56 483).
19 A. Gelbin, B. Schneider, L. Clowney, S.-H. Hsieh, W.K. Olsen and H.M. Berman, *J. Am. Chem. Soc.*, 1996, **118**, 519.
20 C.A. Podlasek, W.A. Stripe, I. Carmichael, M. Shang, B. Basu and A. Serianni, *J. Am. Chem. Soc.*, 1996, **118**, 1413.
20a D. Varela, P.A. Zunszain, D.O. Cicero, R.F. Baggio, D.R. Vega and M.T. Garland, *Carbohydr. Res.*, 1996, **280**, 187.
21 J.M.G. Fernandez, C.O. Mellet, J.L.J. Blanco, J. Fuentes, M.J. Dianez, M.D. Estrada, A. Lopez-Castro and S. Perez-Garrido, *Carbohydr. Res.*, 1996, **286**, 55.
22 F.H. Hausheer, N.D. Jones, P. Seetharamulu, U.C. Singh, B.J. Deeter, W.L. Hertel and S.J. Kroin, *Comput. Chem.*, 1996, **20**, 459 (*Chem. Abstr.*, 1996, **125**, 301 452).
23 K.N. Kirschner and G.C. Shields, *Theochem.*, 1996, **362**, 297 (*Chem. Abstr.*, 1996, **124**, 233 007).
24 C. Thibaudeau, J. Plavec and J. Chattopadhyaya, *J. Org. Chem.*, 1996, **61**, 266.
25 B.A. Otter and R.S. Klein, *Nucleosides Nucleotides*, 1996, **15**, 793.
26 J.C. Litten and C. Leumann, *Helv. Chim. Acta*, 1996, **79**, 1129.
26a O.W. Howarth, N. Pozzi, G. Vlahov and D. Bartels, *Carbohydr. Res.*, 1996, **289**, 137.

27   H. Senderowitz, C. Parish and W. Clark Still, *J. Am. Chem. Soc.*, 1996, **118**, 2078.
28   T. Church, I. Carmichael and A.S. Serianni, *Carbohydr. Res.*, 1996, **280**, 177.
29   I. Csonka, K. Elias and I.B. Csizmadia, *Chem. Phys. Lett.*, 1996, **257**, 49 (*Chem. Abstr.*, 1996, **125**, 222 270).
30   M.L.D. Ramos, M.M.M. Caldeira and V.M.S. Gil, *Carbohydr. Res.*, 1996, **286**, 1.
31   Y. Nakmura, *Kagoshima Daigaku Kyoikugakubu Kenkyu Kiyo, Shizen Kagaku Hen*, 1995, **47**, 135 (*Chem. Abstr.*, 1996, **125**, 11 238).
32   R. Lanzetta, M. Parrilli, C. Garzillo, A. di Matteo and G. del Re, *J. Chem. Soc., Perkin Trans. 2*, 1996, 505.
33   L. Zhang and L. Zhang, *Gaodeng Xuexiao Huaxue Xuebao*, 1996, **17**, 1086 (*Chem. Abstr.*, 1996, **125**, 196 131).
34   D.M. Whitfield, D. Lamba, T.-H. Tang and I.G. Csizmadia, *Carbohydr. Res.*, 1996, **286**, 17.
35   R. El Bergmi and J. Molina Molina, *Bioorg. Med. Chem.*, 1996, **4**, 151 (*Chem. Abstr.*, 1996, **124**, 343 902).
36   Y. Du, F. Kong, Y. Wang and G. Xu, *Huanjing Huaxue*, 1996, **15**, 183 (*Chem. Abstr.*, 1996, **124**, 343 822).
37   S.J. Garden, J.L. Wardell, O.A. Melvin and P.J. Cox, *Main Group Met. Chem.*, 1996, **19**, 251 (*Chem. Abstr.*, 1996, **125**, 58 887).
38   M.T. Lakin, N. Mouhous-Riou, C. Lorin, P. Rollin, J. Kroon and S. Pérez, *Carbohydr. Res.*, 1996, **290**, 125.
39   J.-Y. Le Questel, N. Mouhous-Riou and S. Pérez, *Carbohydr. Res.*, 1996, **284**, 35.
40   M.K. Dowd, A.D. French and P.J. Reilly, *Aust. J. Chem.*, 1996, **49**, 327
41   S.-K. Chung, Y.T. Chang, D. Whang and K. Kim, *Carbohydr. Res.*, 1996, **295**, 1.
42   L.G. Barrientos and P.P.N. Murthy, *Carbohydr. Res.*, 1996, **296**, 39.
43   K. Mazeau, F.R. Taravel and I. Tvaroska, *Chem. Pap.*, 1996, **50**, 77 (*Chem. Abstr.*, 1996, **125**, 58 902).
44   N.W.H. Cheetham and K. Lam, *Carbohydr. Res.*, 1996, **283**, 13.
45   P. Fowler, B. Bernet, and A. Vasella, *Helv. Chim. Acta*, 1996, **79**, 269.
46   E.C. Reynhardt and L. Latanowicz, *Chem. Phys. Lett.*, 1996, **251**, 235 (*Chem. Abstr.*, 1996, **124**, 343 839).
47   V. Blériot, A. Genre-Grandpierre, A. Imberty and C. Tellier, *J. Carbohydr. Chem.*, 1996, **15**, 985.
48   R. Anulewicz, I. Waver, B. Piekarska-Bartoszewicz and A. Temeriusz, *Carbohydr. Res.*, 1996, **281**, 1.
49   A. Waver, I. Waver, B. Piekarska-Bartoszewicz and A. Temeriusz, *Spectrosc. Lett.*, 1996, **29**, 1079 (*Chem. Abstr.*, 1996, **125**, 301 347).
51   T. Fukamizo, Y. Honda, H. Toyoda, S. Ouchi and S. Goto, *Biosci. Biotech. Biochem.*, 1996, **60**, 1705.
51   M. Nishikitani, K. Kubota, A. Kobayashi and F. Sugawara, *Biosci. Biotech. Biochem.*, 1996, **60**, 929.
51a   W.-K. Li, J.-Q. Pan, M.-J. Lü, P.-G. Xiao and R.-Y. Zhang, *Phytochemistry*, 1996, **42**, 213.
52   S. Pérez, *Sucrose*, 1995, 11 (*Chem. Abstr.*, 1996, **124**, 290 052).
53   S. Sheng and H. van Halbeek, *Biochem. Biophys. Res. Commun.*, 1995, **215**, 504 (*Chem. Abstr.*, 1996, **124**, 30 166).
54   S.B. Engelsen, C. Herve du Penhoat and S. Pérez, *J. Phys. Chem.*, 1995, **99**, 13334 (*Chem. Abstr.*, 1996, **124**, 9 170).
55   K.-H. Ott and B. Meyer, *J. Comput. Chem.*, 1996, **17** 1068 (*Chem. Abstr.*, 1996, **125**, 58 898).

56   K.-H, Ott and B. Meyer, *Carbohydr. Res.*, 1996, **281**, 11.
57   J. Oh, Y. Kim and Y. Won, *Bull. Korean Chem. Soc.*, 1995, **16**, 1153 (*Chem. Abstr.*, 1996, **124**, 176 699).
58   B.J. Hardy, A. Gutierrez, K. Lesiak, E. Seidl and G. Widmalm, *J. Phys. Chem.*, 1996, **100**, 9187 (*Chem. Abstr.*, 1996, **124**, 317 668).
59   L. Maeler, G. Widmalm and J. Kowalewski, *J. Phys. Chem.*, 1996, **100**, 17103 (*Chem. Abstr.*, 1996, **125**, 301 373).
60   C.A. Stortz and A.S. Cerezo, *An. Asoc. Quim. Argent.*, 1995, **83** 171 (*Chem. Abstr.*, 1996, **125**, 196 183).
61   K.P. Howard and J.M. Prestegard, *J. Am. Chem. Soc.*, 1996, **118**, 3345.
62   W. Sicinska and L.E. Lerner, *Carbohydr. Res.*, 1996, **286**, 151.
63   J.-F. Espinosa, F.J. Cañada, J.L. Asensio, M. Martín-Pastor, H. Dietrich, M. Martín-Lomas, R.R. Schmidt and J. Jiménez-Barbero, *J. Am. Chem. Soc.*, 1996, **118**, 10862.
64   J.F. Espinosa, H. Dietrich, M. Martn-Lomas, R.R. Schmidt and J. Jiménez-Barbero, *Tetrahedron Lett.*, 1996, **37**, 1467.
65   J.-F. Espinosa, F.J. Cañada, J.L. Asensio, H. Dietrich, M. Martín-Lomas, R.R. Schmidt and J. Jiménez-Barbero, *Angew. Chem., Int. Ed. Engl.*, 1996, **35**, 303.
66   U. Neidlein and F. Diederich, *J. Chem. Soc., Chem. Commun.*, 1996, 1493.
67   Y. Kimura, K.-i. Yamaguchi and G. Funatsu, *Biosci. Biotech. Biochem.*, 1996, **60**, 537.
68   M. Otah, S. Emi, H. Iwamoto, J. Hirose, K. Hiromi, H. Hoh, M. Shin, S. Murao and F. Matsuura, *Biosci. Biotech. Biochem.*, 1996, **60**, 1123.
69   R.J. Woods, *Curr. Opin. Struct. Biol.*, 1995, **5**, 591 (*Chem. Abstr.*, 1996, **124**, 87 493).
70   M.L.C.E. Kouwizjer and P.D.G. Grootenhuis, *J. Phys. Chem.*, 1995, **99**, 13426 (*Chem. Abstr.*, 1996, **124**, 56 480).
71   D. Uhrin, V. Varma and J.-R. Brisson, *J. Magn. Res., Ser. A*, 1996, **119**, 267 (*Chem. Abstr.*, 1996, **124**, 317 728).
72   A. Otter and D.R. Bundle, *J. Magn. Res., Ser. B.*, 1995, **109**, 194 (*Chem. Abstr.*, 1996, **124**, 117 748).
73   J.Ø. Duus, N. Nifant'ev, A.S. Shashkov, E.A. Khatuntseva and K. Bock, *Carbohydr. Res.*, 1996, **288**, 25.
74   S. Köpper and B. Meyer, *Liebigs Ann. Chem.*, 1996, 1131.
75   J.P. Mackay, C.J. Shelton and M.M. Harding, *Tetrahedron*, 1996, **52**, 5617.
76   K. Lee, G. Jhon, G.Rhyu, E. Bang, B. Choi and Y. Kim, *Bull. Korean Chem. Soc.*, 1995, **16**, 864 (*Chem. Abstr.*, 1996, **124**, 56 501).
77   K. Lee and Y. Kim, *Bull. Korean Chem. Soc.*, 1996, **17**, 118 (*Chem. Abstr.*, 1996, **124**, 317 710).
78   A. Bernardi and L. Raimondi, *API Conf. Proc.*, 1995, **330**, 319 (*Chem. Abstr.*, 1996, **124**, 146 677).
79   F. Bizik and I. Tvaroska, *Chem. Pap.*, 1996, **50**, 84 (*Chem. Abstr.*, 1996, **125**, 87 057).
80   P. Wang and G.-Q. Song, *Huaxue Xuebao*, 1996, **54**, 96 (*Chem. Abstr.*, 1996, **124**, 261 507).
81   M. Hiegemann, *Z. Naturforsch., B: Chem. Sci.*, 1995, **50**, 1091 (*Chem. Abstr.*, 1996, **124**, 9 155.)
82   K.G.R. Pachler, *Magn. Reson. Chem.*, 1996, **34**, 711 (*Chem. Abstr.*, 1996, **125**, 301 402).
83   W. Zou, J.-B. Brisson, Q.-L. Yang, M. van der Zwan and H.J. Jennings, *Carbohydr. Res.*, 1996, **295**, 209.
84   M. Hato and H. Minamikawa, *Langmuir*, 1996, **12**, 1658 (*Chem. Abstr.*, 1995, **124**, 261 518).

85   A. Botsi, K. Yannakopoulou, E. Hadjoudis and B. Perly, *Magn. Reson. Chem.*, 1995, **34**, 419 (*Chem. Abstr.*, 1996, **125**, 114 985).
86   R. Reinhardt, M. Richter and P.P. Mager, *Carbohydr. Res.*, 1996, **291**, 1.
87   H. Hirai, Y. Shiraishi, H. Himori and T. Kawamura, *Polym. J.*, 1995, **27**, 1064 (*Chem. Abstr.*, 1996, **124**, 87 561).
88   D.-Q. Yuan, K. Koga, M. Yamaguchi and K. Fujita, *J. Chem. Soc., Chem. Commun.*, 1996, 1943.
89   E. Mikros, P. Dais and F Sauriol, *Carbohydr. Res.*, 1996, **294**, C5.
90   R.J. Kerns, T. Toida and R.J. Linhardt, *J. Carbohydr. Chem.*, 1996, **15**, 581.
91   M.B. Chen, Y. Liu, Z.W. Guo and C.Q. Cai, *Chin. Chem. Lett.*, 1996, **7**, 29 (*Chem. Abstr.*, 1996, **124**, 317 747).
92   M.B. Chen, Y. Liu, Z.W. Guo and C.Q. Cai, *Chin. Chem. Lett.*, 1996, **7**, 153 (*Chem. Abstr.*, 1996, **125**, 11 326).
93   J. Gao, G. Shi, G. Song, Y. Shao and R. Zhou, *Magn. Reson. Chem.*, 1996, **34**, 249 (*Chem. Abstr.*, 1996, **124**, 343 868).
94   R.X. Tan, W.Z. Wang, S.X. Wu and L. Yang, *Magn. Reson. Chem.*, 1995, **33**, 749 (*Chem. Abstr.*, 1996, **124**, 56 431).
95   Y. Shinozaki, T. Tobita, M. Mizutani and T. Matsuzaki, *Biosci. Biotech. Biochem.*, 1996, **60**, 902.
96   I. Waver, B. Piekarska-Bartoszewicz and A. Temeriusz, *Carbohydr. Res.*, 1996, **290**, 137.
97   A. Ruchmann, J. Lauterwein, T. Baecker and M. Klessinger, *Magn. Reson. Chem.*, 1996, **34**, 116 (*Chem. Abstr.*, 1996, **124**, 290 099).
98   A. Amantea, M. Henz and P. Strazewski, *Helv. Chim. Acta*, 1996, **79**, 244.
99   N.W.H. Cheetham and K. Lam, *Aust. J. Chem.*, 1996, **49**, 365.
100   P.N. Sandersen, B.C. Sweatman, R.D. Farrant and J.C. Linden, *Carbohydr. Res.*, 1996, **284**, 51.
101   E. Gács-Baitz, I. Tömösközi and M. Katjár-Peredy, *Tetrahedron: Asymm.*, 1996, **7**, 2447.
102   K. Mernissi-Arifi, C. Wehrer, G. Schlewer and B. Spiess, *J. Inorg. Biochem.*, 1996, **61**, 63 (*Chem. Abstr.*, 1996, **124**, 202 818).
103   J.S. Casas, M.V. Castano, M.S. Garcia-Tasende, T. Perez-Alvarez, A. Sanchez and J. Sordo, *J. Inorg. Biochem.*, 1996, **61**, 97 (*Chem. Abstr.*, 1996, **124**, 261 542).

22
Other Physical Methods

1 IR Spectroscopy

Mid-range IR attenuated total reflectance spectra of various sugars (both monosaccharides and oligosaccharides, as 10% aqueous solutions) have been evaluated by multidimensional statistical analysis, providing characteristic frequencies within a few seconds. Monosaccharides show a hollow at 998 cm^{-1} and a single unique major peak in the 1075-1030 cm^{-1} region, while oligosaccharides show three characteristic bands in the same region (main band shifted to 1033 cm^{-1}) as well as a band at 998 cm^{-1} due to the glycosidic link vibrational motion.[1]

IR data for crystalline β-L-arabinose over a temperature range from ambient to liquid N_2, including deuterium exchange experiments, and coupled with crystallographic data, indicate a somewhat stronger than previously suggested (intramolecular) H-bonding between the anomeric hydroxyl and the ring oxygen atom (which form an infinite crystalline chain).[2]

A new method for sucrose determination by flow injection analysis and difference FTIR detection is based on enzymatic cleavage of sucrose by invertase to give α-D-glucose and β-D-fructose. The method gives linear results in the 10–100 mmol/L range.[3] A series of papers on experimental and calculated IR and Raman spectra for a range of mono- and polysaccharides have appeared, covering α-D-glucose and α-D-galactose, glucitol and galactitol,[4] methyl β-D-glucopyranoside, methyl α-D-glucopyranoside and methyl β-D-xylopyranoside,[5] and methyl α-D-glucopyranoside, methyl α-D-galactopyranoside and methyl α-D-mannopyranoside.[6]

A study of the conformations of amylose and cellulose oligomers using vibrational spectroscopies has been reported.[7] At the same conference, the use of FTIR (near, mid and far IR), and EXAFS, for spectroscopic characterization of mono- and disaccharides and coordination transition metal complexes was described.[8] IR has been applied to characterize 1,2:3,4-di-O-isopropylidene-6-O-triphenylstannylmethyl-α-D-galactopyranose.[9]

Several metal derivatives of L-(+)-ascorbic acid have been studied in the solid state by various techniques, including IR. These compounds have formulae [TlMe$_2$(HAsc)]0.5C$_3$H$_8$O, [SnMe$_2$(Asc)] and [SnBu$_2$(Asc)] and the studies help to show that the thallium compound behaves as a salt, and that in the tin derivatives the metal interacts with the (Asc)$^{2-}$ ion through O-1, O-2 and O-3.[10]

An investigation of FT Raman spectra of selectively substituted nitrates of carbohydrates has appeared in a symposium report.[11] The depolarized low-frequency Raman spectra of aqueous solutions of L-*xylo*-ascorbic acid and of its epimer D-arabinoascorbic acid have been studied as a function of concentration (at 30 °C). This provides evidence that the effect on the dynamical structure of water is greater for the former carbohydrate.[12]

Vasella's group have published part 23 of their continuing series on the use of IR in evaluating the nature of intramolecular H-bonding in carbohydrates and related chemistry. In this case, H-bonding in orthoacetal protected inositols (2-OH, 2-F or 2-deoxy) is used to rationalize the favoured outcome of glycosylation with a diazirene glycosyl donor.[13]

The polarized Raman spectrum of a single crystal of AZT (using a Raman microscope with 488.0 nm argon-ion laser excitation) provides useful orientational parameters.[14] Adsorption of guanine, guanosine and derivatives onto silver film allows the application of near IR FT Surface Enhanced Raman Scattering (SERS). Assignments are discussed.[15] Polarized Raman spectroscopy of a single crystal of the barium salt of inosine monophosphoric acid hexahydrate has been observed with 488.0 nm excitation. Analysis provides a method for determining base orientation in duplex fibres.[16]

FTIR-ATR and laser-Raman spectroscopy of oligosaccharides in aqeuous solutions at various concentrations shows the influence of the glycosidic linkage position, the overall hydration and concentration on characteristic spectral ranges.[17] Raman spectra of guests (3-bromostyrene, 3-nitrostyrene and 4-methoxystyrene) in cyclodextrins, α, β and γ, indicate that the α-CD inclusion leads to conformational selection of the guest molecule.[18]

2   Mass Spectrometry

In the gas phase, both the α- and β-anomers of methyl 3-*O*-benzyl-2,6-dideoxy-D-*arabino*-hexopyranoside undergo ready deprotonation of the 4-hydroxyl, and collision activated dissociation (CAD) of these anions leads to E2 elimination, decarbonylation and ring opening fragmentations. Study of the trideuteromethyl α-glycoside, 2,2-dideutero- and dideuterobenzyl α-glycosides support the mechanism proposed.[19]

EI-MS of alkyl and phenyl *N*-alkyl- and *N,N*-dialkyl-2-amino-4,6-*O*-benzyl-idene-2-deoxy-D-hexopyranosides and benzyl and phenyl 2,3-di-*O*-alkyl-4,6-*O*-benzylidene-D-hexopyranosides leads to three different fragmentation pathways, and definitive chemical evidence for these different pathways is presented.[20]

FAB and EI spectra of some acenaphtho[1,2-e][1,2,4]triazines and acenaphtho[1,2,4]triazolo[4,3-*b*] and [3,4-*c*][1,2,4]triazines attached to acyclic monosaccharide derivatives have been reported.[21] The reaction of *N*-α-acetyl-L-lysinamide and glucose has been examined with the assistance of electrospray mass spectrometry (along with capillary zone electrophoresis), providing evidence for a number of products exhibiting different degrees of dehydration and oxidation and for species with two lysines per glucose (relevant to cross-linking of proteins exposed to high glucose concentrations).[22] Thermolysis of anhydrous acidified sucrose yields a range of disaccharide and higher oligomeric anhydrides, characterized by mass spectrometry of their per-*O*-Tms ethers.[23] Electrospray MS has been used to show that the iron(III) complexation of benzoxazinone derivatives **1** is not markedly influenced by the presence of the β-D-glucopyranosyloxy group; additionally, while the NH compound monoligates the NOH compounds di- or tri-ligate.[24] Collision-induced decomposition pathways and common fragmentation trends of the novel phenylpropanoid sucrose esters, vanicosides A and B and hydropiperoside (from *Polygonium pensylnvanicum L.*) have been established by negative ion FAB and tandem MS. These data allow characterization of mixtures of such compounds without the need for derivatization.[25] LSIMS of the potassium salts of eight glycobrassicins (eg. **2**) have been reported.[26] LSIMS and CI-MS have been used to characterize stereoisomeric α-methylene-γ-lactone furanosidic derivatives (**3**, R=H, $CONH_2$).[27] The structures of three new glycosidic jasmonoids, **4-6**, and glycoside **7** were established through a combination of HR-SIMS and 2D NMR.[28]

1 (a) X = OH; Y = H
(b) X = OH; Y = OMe
(c) X = H; Y = OMe
(d) X = OMe; Y = OMe

2

3

4 R_1 = β-D-Glu; R_2 = H
5 R_1 = β-D-Glu; R_2 = Me

6 R_1 = β-D-Glu; R_2 = H

7

EI-MS has been used in characterization of some acylated aldobionitriles.[29] Tandem mass spectrometry has been used in differentiating the glycosidic stereochemistry of diastereomeric cobalt-1→3-, 1→4-, and 1→6-glucobiose complexes. The $[Co^{3+}(acac)_2(disaccharide)]^+$ complexes were generated *in situ* by FAB ionization.[30] Novel unsaturated carbohydrate bolaforms, e.g. **8**, prepared by olefin catalytic metathesis reactions, have been characterized by CAD-electrospray MS.[31]

Geranyl 6-*O*-α-L-arabinopyranosyl-β-D-glucopyranoside has been identified in the leaves of a green tea cultivar with the assistance of HR-FAB-MS (along with COSY, HMQC and HMBC NMR analyses).[32] Both the linkage position and anomeric configuration have been deteriminied for underivatized glucopyranosyl disaccharides by negative ion electrospray MS.[33]

8   9

Cyclodextrins and analogues and their complexes have been the subject of a number of studies involving mass spectrometric techniques this year. MALDI-TOF-MS has provided a quantitative method of analysis of cyclodextrins using internal standard cyclodextrins (e.g. permethylated, β-CD and partly methylated derivatives).[34] ESIMS, MALDI-TOF-MS and FAB-MS have been employed for determination of molecular weight, degree of substitution and purity of charged derivatives of cyclomalto-heptaose and -octaose.[35] FAB-MS has been used to study the complexation of per-methylated α- and β-CDs with methyl esters of Trp and Ala, showing preferential binding of the D-amino acid enantiomers.[36] Cyclodextrins (α-, β- and γ-) are hosts for 1-anilinonaphthalene-8-sulfonate or 2-p-toluidinylnapthalene-6-sulfonate, with negative ion mode ionspray MS showing 2:1 CD:guest stoichiometry.[37]

Mass spectrometry has seen a considerable number of applications in the oligosaccharide arena this year. CIMS at low gas (ammonia) pressure has been used to determine sugar linkage positions in the triterpenoidal tetrasaccharide glycoside heteropapussaponin 2 **9**.[38] Sulfated and sialyl Lewis type glycosphingo-lipids have been characterized using FAB-MS, ESIMS and CID tandem MS, each technique providing different, but complementary structural information. This approach should be applicable to much larger mass glycoconjugates.[39] Molecular mass assignments of octa- to hepta-oligosaccharides (oligo-galacturonic acids) have been obtained using negative ion electrospray MS.[40] Both sensitivity and accuracy of the CI-MS methylation analysis of complex alditol acetates is enhanced through formation of protonated pyridine complexes.[41] Capillary electrophoresis electrospray MS and tandem MS have been employed for the structural characterization of *Pseudomonas aeruginosa* derived lipopoly-saccharides, and is the subject of an ACS Symposium publication.[42] MALDI of N-linked oligosaccharides was carried out using post-source decay on a reflection TOF instrument or CID on an instrument fitted with an orthogonal TOF analyser. The former provided glycosidic cleavage information and the latter major fragment ions from cross ring cleavages of most of the constituent monosaccharides, providing overall significant sequence and branching informa-tion.[43] Derivatized oligosaccharides have been analysed using post-source decay MALDI MS.[44] Cationic derivatization of oligosaccharides, by way of formation of hydrazones with Girard's T reagent, led to enhanced detection sensitivity in MALDI and electrospray MS analyses.[45] CAD-tandem MS assisted the structure determination of an *N*-linked oligosaccharide **10** from pokeweed lectin B (along with 500 MHz ^{1}H NMR analysis).[46]

Negative ion FAB and CID-tandem MS identified several sulfation sites in sulfated mucin oligosaccharides, including the apparently novel sulfation of C-3 of an amino sugar residue.[47] CID FAB MS has provided useful structural information about the fragmentation of various oligosaccharides, including evidence for internal monosaccharide loss.[48] High energy CID MS of synthetic mannose-6-phosphate oligosaccharides (6-phosphate on either or both non-reducing terminal or penultimate residues) shows that the positive ion mode can identify these phosphorylated residues.[49] Twelve major *N*-terminal linked oligosaccharides (novel β-D-galactofuranose-containing high-mannose oligosaccharides in ascorbate oxidase from *Acremonium sp.* HI-25) have been identified by FAB-MS.[50] FAB-MS has also been used to characterize the fully deacylated pentasaccharide from treatment of (GlcNAc)$_5$ with chitin deacylase from *C. lindemuthianum*.[51] MALDI-TOF allows identification of β-(1→3)xylo-β-oligosaccharides from cell wall glycans of *Caulerpa sp.*, showing the xylan to be composed of 25 β(1→3)-linked linear oligosaccharides.[52]

```
        α-D-Man
          1
          ↓
          6
α-D-Man(1→3)-Manβ1→4GlcNAcβ1→4GlcNAc
          2                    3
          ↑                    ↑
          1                    1
        β-D-Xyl              α-L-Fuc
                   10
```

Nucleosides and nucleotides have also seen a number of useful applications of mass spectrometric techniques this year. MALDI of dATP, dTTP, dCTP and dGTP gives clear molecular ions with minimal fragmentation (in contrast to FAB-MS which gives weak parent ions and more fragmentation).[53] FAB-MS of the glycosyl esters of nucleoside pyrophosphates and polyisoprenyl phosphates has been applied to the determination of anomeric configurations.[54] The following nucleoside-related reports have appeared: FAB-MS and FAB-MS/MS of Tms derivatives of 3'-*O*-methyladenonsine, guanosine and uridine;[55] MS of bisindole nucleosides 3-(1H-indol-3-yl)-4-[1-glycospyranosyl]-1H-indol-3-yl]-2,5-furandiones and 3-(1H-indol-3-yl)-4-[1-glycospyranosyl]-1H-indol-3-yl]-1H-pyrrole-2,5-diones;[56] and EI-MS of some potential anti-HIV carbocyclic nucleoside analogues **11** (R=Cl, OH, NH$_2$).[57]

3 X-Ray and Neutron Diffraction Crystallography

Specific X-ray crystal structures have been reported as follows (solvent molecules of crystallization are frequently not reported).

3.1   Free Sugars and Simple Derivatives Thereof – 1,2:3,4-Di-*O*-isopropylidene-6-*O*-methyacryloyl-α-D-galactose and 1,2:5,6-di-*O*-isopropylidene-3-*O*-methyl-acryloyl-α-D-glucose.[58] Dimethyl and diacetoxyacetals of 2,3,4,5,6-pentacetyl-D-glucose.[59]

3.2   Glycosides, Disaccharides and Derivatives Thereof – A study of the con-formation and intramolecular hydrogen bonding of crystalline sucrose.[60] The 1:1 complex of *N*-methylglycine and sucrose has been reported.[61] Methyl β-D-ribo-furanoside (as part of study involving *ab initio* calculations and NMR of ^{13}C-labelled materials),[62] methyl 3-*O*-benzoyl-4,6-*O*-benzylidene-2-*O*-trifluorometha-nesulfonyl-α-D-mannoside,[63] methyl α-D-lyxopyranoside,[64] methyl 6-*O*-[(*R*)-1-carboxyethyl]-α-D-galactopyranoside,[65] methyl 6-*O*-*n*-decanoyl-α-D-glucopyrano-side, methyl 6-*O*-*n*-dodecanoyl-α-D-glucopyranoside, methyl 6-*O*-*n*-dodecanoyl-β-D-glucopyranoside and methyl 6-*O*-*n*-dodecanoyl-α-D-galactopyranoside (the alkyl chains are all *trans*, and in the latter two cases the pyranoside rings adopt $^4C_1$ conformation, but disorded in the crystals of the other structures).[66]

The trichloroethylidene acetals **12**,[67] phenyl 2,3,4,6-tetra-*O*-acetyl-β-D-galacto-pyranoside,[68] benzyl β-L-arabinopyranoside and its 3,4-*O*-isopropylidene and 2-*O*-benzoyl-3,4-*O*-isopropylidene derivatives.[69] *p*-Nitrophenyl α-D-mannopyrano-side,[70] 3-deoxy-3-*C*-(ethoxycarbonyl)methyl-1,2:5,6-di-*O*-isopropylidene-α-D-allofuranose,[71] and bicyclic **13**.[72]

The tetracetate of sutherlandin epoxide, namely 4-β-D-glucopyranosyloxy-2*R*,3*R*-epoxy-3-hydroxymethylbutyronitrile, **14**,[73] methyl 1,1-*O*,*O*-ethylidene-4,5-dideoxy-β-D,L-glycerohex-4-eno-2-ulopyranoside,[74] methyl 1-*O*-(3,16α-dihydroxy-estra-1,3.5(10)-trien-17β-yl)-2,3,4-tri-*O*-acetyl-β-D-glucopyranosiduronate,[75] and methyl 5-*O*-acetyl-2-*O*-benzoyl-3,4-*O*-isopropylidene-β-L-idoseptanoside **15**.[76]

Glycolipids containing a Gal headgroup and a hexadecyl chain with or without different ethylene oxide spacers have been studied as 1:2 mixtures with 1,2-dipalmitoyl-sn-*glycero*-3-phosphocholine by X-ray analysis, as well as by DSC (see section 4).[77]

The disaccharide methyl 2-*O*-(α-L-fucopyranosyl)-β-D-galactopyranoside.[78] The 16-membered macrolide antibiotic chalcomycin, which contains chalcose (**16**) and mycinose (**17**) as monosaccharide subunits[79] and the macrocyclic lactone

11,12,4″-tri-*O*-methylazithromycin which contains L-cladinose and D-desosamine.[80]

3.3 Higher Oligosaccharides and *C*-Glycosides – A crystal structure of the methyl-β-glycoside of the Lewis^x trisaccharide has been reported.[81] The crystal structure of amylose V complexed with glycerol shows left-handed six-fold amylose helices.[82] Crystal structures of the lectin maltoporin bound to maltose, maltotriose and maltohexaose have been reported. These provide useful information regarding sugar specificity in the binding channel and on the possible mechanism of sugar translocation.[83]

A number of X-ray analyses of cyclic oligosaccharides, their derivatives or complexes have been reported this year. The structure of α-CD 5.5 hydrate with one glucose replaced by 3-amino-3-deoxy-α-D-altrose shows the amino sugar adopts the $^{3,0}B$ boat conformation.[84] 6′-(6-Aminohexyl)amino-6′-deoxycyclo-maltoheptaose,[85] 6′,6″-diamino-β-CD,[86] hexakis-(2,3,6-tri-*O*-methyl)-cyclomalto-hexaose (permethyl-α-CD).[87] Dimethyl-α-CD and complexed with acetone as a 1:1 complex (acetone is fully included in the cavity, and in the uncomplexed case a methoxy from a neighbouring CD occupies the cavity volume).[88] An analysis of literature data of complexes of *O*-methylated α- and β-CD evidencing that eclipsed conformations about the exocyclic bonds are common.[89] Various structural analyses of α-CD complexed with different fatty acid methyl or ethyl esters have been reported.[90] 6^A-Boc-L-phenylalanylamino-6^A-deoxy-β-CD shows the Ph group located above the parent macrocycle with the *t*-Bu located in the cavity of an *adjacent* molecule.[91]

The *C*-phenyl glycoside **18**,[92] fused *C*-phenyl glcoside **19**,[93] and methoxy-naphthyl *C*-glycoside **20**.[94] *C*-Glycosides **21**,[95] **22** (during synthesis of maitotoxin components by Kishi's group – see Chap. 24),[96] the acetylenic *C*-glycosidic monosaccharide **23**[104] and disaccharide **24**,[97] and the heterocycle-bridged disaccharide mimic **25**.[98]

3.4   Anhydro-sugars – The dianhydro sugar **26**,[99, 100] 1,6-anhydro-4-*O*-benzyl-
1-*C*-3-deoxy-isopropyl-2-*O*-methyl-D-*ribo*-hexitol and 1,6-anhydro-4-*O*-benzyl-3-
deoxy-2-*O*-4-*C*-dimethyl-D-*ribo*-hexitol,[101] 1,5-anhydro-2,3-dideoxy-2-(guanin-9-
yl)-D-*arabino*-hexitol,[102] and 1,4-anhydro-D-glucitol and 1,5-anhydro-D-glucitol
with a comparative discussion of the structures of these last two.[103]

**3.5   Arsenic, Halogen, Phosphorus, Sulfur, Selenium and Nitrogen Containing
Compounds** – The (acyclic) phenylhydrazone of D-mannose,[104] methyl 3,4-6-tri-
O-acetyl-2-deoxy-2-[3-(2-phenylethyl)-ureido]-β-D-glucospyranoside   **27**,[105]   1-
deoxy-1-amino-1-*N*-(*p*-tolyl)-D-fructose,[106] methyl 2-amino-2-deoxy-3,4-di-*O*-*t*-
butyldimethylsilyl-6,7-*O*-isopropylidene-β-D-glucohept-2-ulofuranosonate,[107] 2-
carboxymethylamino-2-deoxy-D-glucose   and   -galactose,[108]   and   2-amino-2,6-
dideoxy-D-glucitol-6-sulfonic acid.[109] *N*,*N*-Di(1-deoxy-D-fructos-1-yl)-glycine has
a spiro-bicyclic hemiketal structure with one carbohydrate unit in the normal $^2C_5$
β-pyranose form and the other in an acyclic chain (this tautomer is more than
half of the solution structure, in which there are at least 12 isomeric forms).[110]
Compound **28** was prepared by combinatorial means during work aimed at
sialidase inhibitor discovery.[111]

2-Methylthio-5-nitroso-6-*N*-(2,3,4-tri-*O*-acetyl-β-D-xylopyranosyl)amino-4-(3*H*)-
pyrimidinone,[112] the aminopyridyl *N*-glycosides **29** and **30**,[113] thiazole **31**,[114]
1-(*N*-benzyl-*N*-hydroxyamino)-1-deoxy-1-(2-furyl)-2,3:4,5-di-*O*-isopropylidene-L-
manno-pentitol,[115] the 6*S*-benzoxazol-2-yl galactose derivative **32**,[116] and the C-6
substituted amino compounds **33-35**.[117]

The fused β-lactams **36** and **37**,[118] the fused oxazolidinones **38** and **39**,
[intermediates in Danishefsky's synthesis of staurosporine (see Chapter 24)],[119] **41**
and a precursor to this, namely **40**,[120] and **42**.[121] The X-ray structure of D-
mannose derived azepine **43** shows this to adopt a *pseudo*-chair conformation.[122]
The glucose-based tetrazole **44**[123] and fused heterocycle **45**.[124]

The thio-analogues **46** and **47** related to castanospermine and australine,[125] 3-*O*-benzoyl-5,6-*S,O*-carbonyl-1,2-*O*-isopropylidene-5-thio-α-D-glucofuranose,[126] 4-thio-L-lyxono-1,4-lactone,[127] and the antithrombotic *S*-glycoside **48**.[128]

The D-mannitol-derived bisamido thiophosphate **49**,[129] phosphonates **50** and **51**,[130] diphenylphosphonate **52**[131] and 5-deoxy-5-*C*-diphenylarsino-1,2-*O*-isopropylidene-3-*O*-*p*-tosyl-α-D-xylofuranose.[132] Trichloroethylidene acetal fluoro glycoside **53**,[67] and 2,4,6-tri-*O*-benzoyl-3-*O*-benzyl-β-L-idopyranosyl fluoride.[133]

55   **56**   **57**   **58**

3.6   Branched-chain Sugars – Ketone **54**,[134] dihydropyran derivatives **55** and **56**,[135] and mannoside **57**, an oxanthrone derivative from the roots of *Picramnia hirsuta*.[136] The 3-deoxy-3-cyclopentadienyl **58** derived from diacetone-D-glucose.[137]

3.7   Sugar Acids and Their Derivatives – The epimers of 2,3-*O*-benzylidene-D-erythronolactone.[145]

3.8   Inorganic Derivatives – The calcium complex of inositol 'disaccharide' **59** shows an extended helical structure, with an ordered array of pentagonal bipyrimidal Ca^{2+} ions bridging two amino groups, and with an OH group from each amino compound, an equatorial and two axial waters also bound.[138] Calcium, sodium and potassium gluconate and lactobionate,[139] Ca^{2+}(1,6-anhydro-β-D-glucopyranose)$_2$, in which both O-2 and O-3 diols are involved in complexing.[140] 1,2:3,4-di-*O*-isopropylidene-6-*O*-triphenylstannylmethyl-α-D-galactopyranose in which the triphenyltin and carbohydrate moieties are linked by a *trans* methylene-oxygen-methylene arrangement, and the pyranose ring adopts a twist-boat conformation.[9] The carbohydrate-derived chiral P,S ligand-palladium complex **60**.[141]

59   **60**

3.9   Alditols and Cyclitols and Derivatives Thereof – *cis*-Inositol monohydrate,[142] 2,4,6-tri-*O*-benzyl-*myo*-inositol 1,3,5-tris-dibenzylphosphate,[143] 3,4,5,6-tetra-*O*-benzyl-1,2-*O*-cyclohexylidene-*myo*-inositol,[144] (±)-1,4-di-*O*-benzoyl-2,3-*O*-isopropylidene-*myo*-inositol and (±)-1,4-di-*O*-benzoyl-5,6-*O*-isopropylidene-*myo*-inositol,[145] *myo*-inositol derivatives **61** and **62**,[146] (D)-1-*O*-allyl-3,6-di-*O*-benzyl-2-*O*-[(-)-camphanoyl]-4,5-*O*-isopropylidene-*scyllo*-inositol **63**,[147] and fluoro-inositol derivative **64** (also Chapter 18, and section 1 for IR).[13]

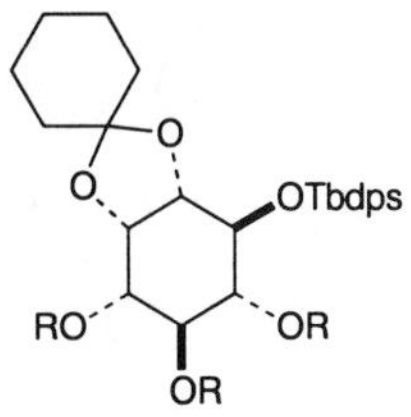

61 R = H
62 R = Bz

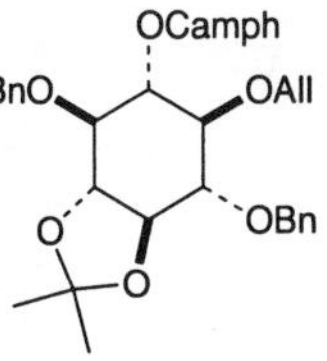

63

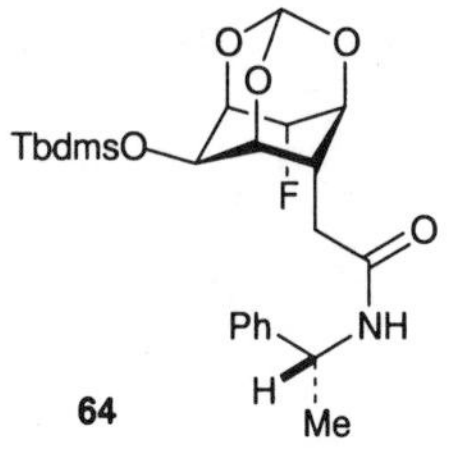

64

3.10 Nucleosides and Their Analogues and Derivatives Thereof – The disodium salt of AMP undergoes a four-step phase transition through 0-90% humidity.[148] 5-*O*-Benzoyl-3-deoxy-α-D-*glycero*-pent-2-ulofuranosyl)uracil,[149] 2'-deoxycytidine and gemcitabin (2',2'-difluoro-2'-deoxycytidine),[150] the N-1 sodium salt of inosine,[151] 3'-deoxy-2',3'-difluorothymidine,[152] 5-fluorocytidine,[153] 2',3'-dideoxy-3'-nitrothymidine and 3'-deoxy-2'-*O*-propyl-3'-nitrothymidine,[154] 2'-deoxycytidine-N-3-cyanoborane,[155,156] 5-cyclohexyl-2'-deoxyuridine (*syn*-glycosidic bond with ₄E ring),[157] 7-iodo-2'-deoxytubericidin **65**,[158] 7-(2-deoxy-α-D-ribofuranosyl)hypoxanthine,[159] 1-(2-deoxy-α-D-*erythro*-pentopyranosyl)4-triazolidine-3,5-dione,[160] 2-Bromo-5-(2-deoxy-α-D-ribofuranosyl)pyridine[157] and 1-(2'-deoxy-α-D-ribofuranosyl)4-triazolidine-3,5-dione **66**.[161]

4-Amino-2-(5-*O*-benzoyl-β-D-ribofuranosyl)-6-methyl-[1,3]-oxazolo[5,4-*d*]pyrimidine **67**,[162] 4-amino-1-(2-deoxy-β-D-ribofuranosyl)-6,7-dihydro-1*H*,5*H*-cyclopentapyrimidine-2-one **68**,[163] and 2-(2,3-anhydro-β-D-allofuranosyl)-4,6-bis(methylthio)-2*H*-pyrazolo[3,4-*d*]pyrimidine.[164] The polycyclic spiro-fused analogues **69**[165] and **70**.[166]

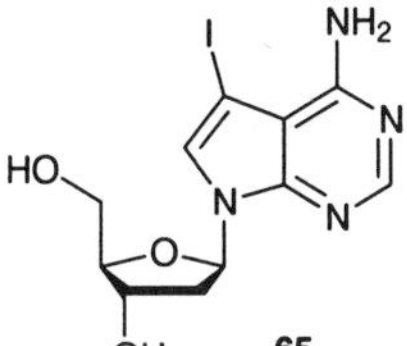

65 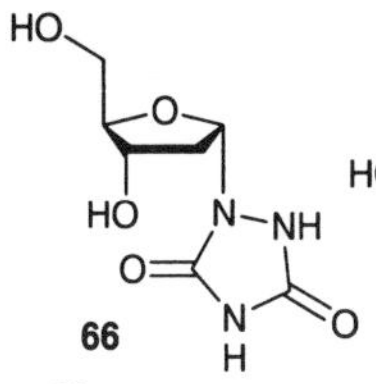 **66** 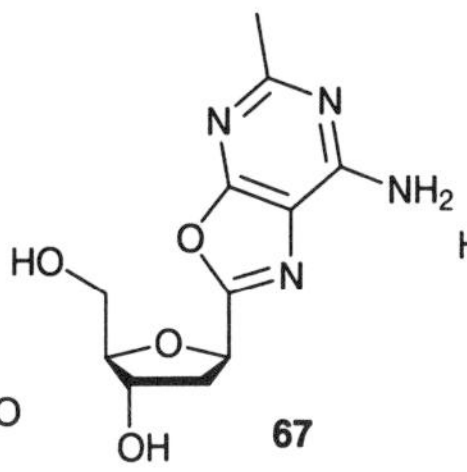 **67** 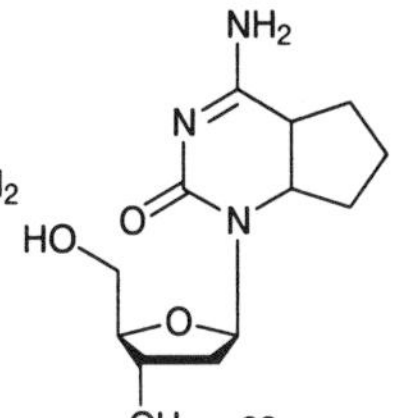 **68**

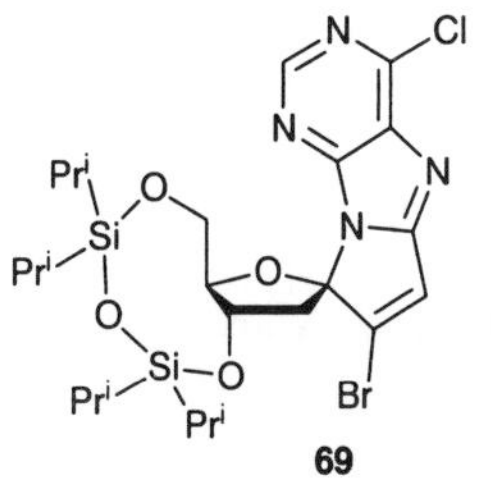

69 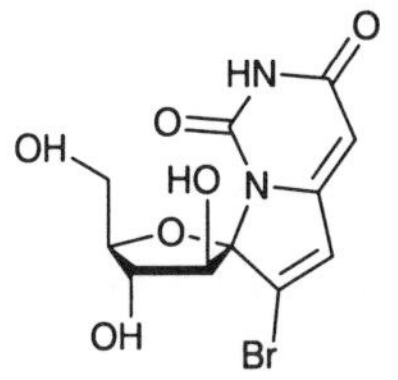 **70** 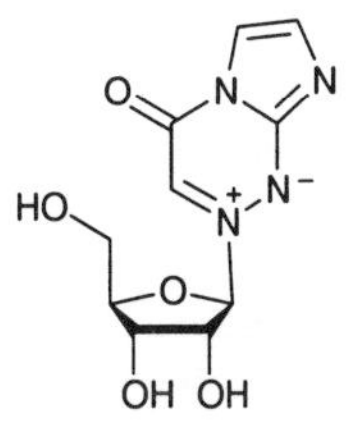 **71**

Structures of a range of base and/or sugar-modified nucleoside analogues have been described. The 6-azacytosine 71,[167] pyrazolo[3,4-*b*]quinoline β-ribofuranoside 72[168] 5′-*O*-benzoyl-3′-deoxy-2′-keto-α-uridine,[169] 3′-deoxy-3′-*S*-(2-nitrophenyldisulfanyl)thymidine,[170] anhydronucleoside 73,[171] 5-adamantyl-2′-deoxy-4′-thio-α-uridine,[172] the bicyclic nucleoside analogue 74,[173] and bicyclic analogues 75 (both epimers) and 76.[174]

The 4′-acyl nucleosides 77, 78, 4′-*C*-branched nucleosides such as 79,[175] and 1-*N*-[(2′-deoxy-β-D-*ribo*-pentofuranosyl)methyl] thymine, cytosine and adenosine.[176] A number of carbocyclic nucleosides; *cis*-1-[(2-hydroxymethyl)cyclopentyl]uridine,[177] 4,4-dihydroxymethyl-(1-thymin-1-yl)cyclohexane,[178] and the cyclopropyl-fused carbocyclic analogue 81.[179]

5′-Amino-5′-deoxyadenosine[180] and azaadomet bound to the MetJ repressor protein (showing the purine to lie in a hydrophobic pocket and the glycosidic bond to adopt an unusual *syn*-conformation).[181]

3.11 Neutron Diffraction – The one example reported this year involved an exploration of the structural features of the liquid-air interface affected by adsorbed sugar-based ionic surfactant 7,7-bis[(1,2,3,4,5-pentahydroxyhexanamido)-methyl]-*n*-tridecane.[182]

4   Polarimetry, Circular Dichroism, Calorimetry and Related Studies

The chirality induced in styrene-containing co-polymers (post-glycosidic hydrolysis) as a consequence of using carbohydrate-bearing monomers (L-threitol, α-D-gluco- and D-mannitol derivatives) has been assessed using the exciton chirality method.[183] Quantitative polarographic analysis of 6-β-D-glucopyranosyloxy-7-hydroxycoumarin has been described.[184]

The similarity between monosaccharide and methyl 3-*O*-(α-D-mannopyranosyl)-α-D-mannopyranoside vacuum UV CD suggests linkage flexibility in the disaccharide.[185] CD data for all stereoisomers of free 1,2,4-triols, 1,2,4,6-tetraols and 1,2,4,6,8-pentaols provides a method for configurational assignment from this data set.[186] Extension of this work to further homologues and spectral consequences has been described by the same group.[187] Solvent effects on the CD spectra of a range of 4-nitrophenyl glycosides (some peracylated) have been reported.[188] Induced Cotton effect in the CD spectra of sodium hyaluronate-acridine orange complex has been discussed.[189] The effect of sugars on the 1:1 complex of an anionic and cationic porphyrin was to produce specific exciton coupling bands, but only in the selective cases of glucose and xylose.[190] Shinkai's work on synthetic saccharide receptors has continued this year (see recent SPR volumes). A dendritic (PAMAM-based) boronic acid functions as a low concentration receptor for D-galactose, D-fructose or D-glucose.[191] A review on Shinkai's work has also appeared [82 refs].[192] CD has been used for conformational analysis of 7-methyl-7-deazaguanine-containing octanucleotides (adopting B-DNA duplex) and 8-methyl-7-deazaguanine-containing octanucleotides (inducing a switch to Z-DNA duplex). The base analogues are incorporated alternating with natural cytidine.[193]

A study of the relaxational transitions and related heat capacity anomalies for galactose and fructose has been described which employs calorimetric methods.[194] The kinetics of solution oxidation of L-ascorbic acid have been studied using an isothermal microcalorimeter.[195] Differential scanning calorimetry (DSC) has been used to measure solid state co-crystallization of sugar alcohols (xylitol, D-sorbitol and D-mannitol),[196] and the thermal behaviour of anticoagulant heparins.[197] Thermal measurements indicate a role for the structural transition from hydrated β-CD to dehydrated β-CD.[198] Calorimetry was used to establish thermodynamic parameters for (1:1) complexation equilibrium of citric acid and β-CD in water.[199] Several thermal techniques were used to study the decomposition of β-CD inclusion complexes of ferrocene and derivatives.[200] DSC and derivative thermogravimetric measurements have been reported for crystalline cytidine and deoxycytidine.[201] Heats of formation have been determined for α-D-glucose esters and compared with semiempirical quantum mechanical calculations.[202]

Spectrophotometric methods have been applied to inclusion complexes of α- and β-CDs with methyl orange (in the presence of alcohols).[203] A report on a symposium on spectroscopic and thermodynamical studies of metal-amino sugar complexes was published.[204] Electronic absorption and fluorescence spectra of L-ascrobic acid have been studied at a range of pH values.[205]

Atomic Force Microscopy (AFM) has been used to characterize the differing supramolecular structures of *N*-alkyl-D-gluconamide adsorbates[206, 207] and *N*-(*n*-alkyl)-*N*′-D-maltosylsemicarbazone adsorbates[208] (on silica, graphite). An analysis of a β-CD fibre-optic chemical sensor (pyrene binding) has been discussed.[209]

A variety of spectroscopies have been employed in a study of the electron transfer photophysics of 5-(1-carboxypyrenyl)- and 5-(1-pyrenyl)-2′-deoxyuridine.[210] The inclusion complexes of α-, β- and γ-CDs and 2,6-di-*O*-methyl and 2,3,6-tri-*O*-methyl β-CD have been investigated using UV (as well as NMR) spectroscopy.[211] The electrochemical oxidation of 4-*O*-glucopyranosyl-D-glucose in the presence of bromide ions has been examined using cyclic chronovoltammetry.[212] Underivatized oligosaccharides and other carbohydrate types were analysed by semimicrocolumn LC/pulsed amperometric detection.[213]

References

1   F. Cadet and B. Offman, *Spectrosc. Lett.*, 1996, **29**, 523 (*Chem. Abstr.*, 1996, **124**, 343 871).

2   E. T. G. Lutz, M. J. Luinge and J. H. van der Maas, *Bull. Pol. Acad. Sci., Chem.*, 1994, **42**, 513 (*Chem. Abstr.*, 1996, **124**, 9113).

3   B. Lendl and R. Kellner, *Mikrochim. Acta*, 1995, **119**, 73 (*Chem. Abstr.*, 1996, **124**, 30 164).

4   R. G. Zhbankov, V. M. Andrianov and H. Ratajczak, *Zh. Strukt. Chim.*, 1995, **36**, 430 (*Chem. Abstr.*, 1996, **124**, 30 137).

5   R. G. Zhbankov, V. M. Andrianov H. Ratajczak, and M. Marchewka, *Zh. Strukt. Chim.*, 1995, **36**, 443 (*Chem. Abstr.*, 1996, **124**, 30 138).

6   R. G. Zhbankov, V. M. Andrianov, H. Ratajczak and M. Marchewka, *Zh. Strukt. Chim.*, 1995, **36**, 456 (*Chem. Abstr.*, 1996, **124**, 30 138).

7   M. Sekkal, M. L. Benhafsa, J. P. Huvenne and G. Vergoten, *Spectrosc. Biol. Mol.*, *Eur. Conf., 6th*, 1995, 431 (*Chem. Abstr.*, 1996, **124**, 146 640).

8   J. G. Wu, L.-Q. Wang, X. Ju, Q. Zhou, Y.-M. Yang and D.-F. Xu, *Spectrosc. Biol. Mol., Eur. Conf., 6th*, 1995, 427 (*Chem. Abstr.*, 1996, **124**, 146 639).

9   P. J. Cox, O. A. Melvin, S. J. Garden and J. L. Wardell, *J. Chem. Crystallogr.*, 1995, **25**, 469 (*Chem. Abstr.*, 1996, **124**, 56 447).

10   J. S. Casas, M. V. Castano, M. S. Garcia-Tasende, T. Perez-Alvarez, A. Sanchez and J. Sordo, *J. Inorg. Biochem.*, 1996, **61**, 97 (*Chem. Abstr.*, 1996, **124**, 261 542).

11   R. G. Zhbankov, A. I.Usov, T. E. Kolosova, L. K. Prihodchenko, M. Marchewka and H. Ratajczak, *Spectrosc. Biol. Mol., Eur. Conf., 6th*, 1995, 429 (*Chem. Abstr.*, 1996, **124**, 146 615).

12   Y. Wang and Y. Tominaga, *J. Chem. Phys.*, 1996, **104**, 1.

13   A. Zapata, B. Bernet and A. Vasella, *Helv. Chim. Acta*, 1996, **79**, 1169.

14   A. Kumakwa, M. Tsuboi, K. Ushizawa and T. Ueda, *Biospectroscopy*, 1996, **2**, 233 (*Chem. Abstr.*, 1996, **124**, 301 456).

15   Y. Long and L. Zhou, *Fenzin Huaxue*, 1996, **24**, 258 (*Chem. Abstr.*, 1996, **124**, 317 769).

16   K. Ushizawa, T. Ueda, and M. Tsuboi, *Nucleosides Nucleotides*, 1996, **15**, 569.

17   M. Kacurakova and M. Mathlouthi, *Carbohydr. Res.*, 1996, **284**, 145.

18   P. J. A. Ribiero-Claro, A. M. Amado and J. J. C. Teixeira-Dias, *J. Raman Spectrosc.*, 1996, **27**, 155 (*Chem. Abstr.*, 1996, **124**, 290 048).

19   R. W. Binkley, E. R. Binkley, S. Duan, M. J. S. Tevesz, and W. Winnik, *J. Carbohydr. Chem.*, 1996, **15**, 879.

20   J. M. Vega-Perez, J. I. Candela, M. Vega, F. Alcudia and F. Iglesias-Guerra, *J. Mass Spectrom.*, 1996, **31**, 493 (*Chem. Abstr.*, 1996, **125**, 33 992c).

21 M. E. Mahamoud, H. A. Hamid, E. H. E. Ashry, and J. Cunniff, *Egypt J. Chem.*, 1996, **39**, 105 (*Chem. Abstr.*, 1996, **125**, 115 011w).

22 A. Lapolla, D. Federle, R. Aronica, O. Curcuruto, M. Hamdan, S. Catinella, R. Seraglia and P. Traldi, *J. Mass Spectrom.*, 1995, 569 (*Chem. Abstr.*, 1996, **125**, 11 299g).

23 M. Manley-Harris and G. N. Richards, *Carbohydr. Res.*, 1996, **287**, 183.

24 L. Bigler, A. Baumeler, C. Werner, and M. Hesse, *Helv. Chim. Acta*, 1996, **79**, 1701.

25 A. T. Sneden, M. L. Zimmermann and T. L. Sumpter, *J. Mass Spectrom.*, 1995, **30**, 1628 (*Chem. Abstr.*, 1996, **124**, 146 629).

26 C. Gardarat, B. Joseph, P. Rollin, C. Vitry and G. Bourgeois, *Analysis*, 1995, **23**, 222 (*Chem. Abstr.*, 1996, **124**, 117 718).

27 C. Borges, M. A. A. Ferreira and M. Claeys, *Rapid Commun, Mass Spectrom.*, 1996, **10**, 757.

28 T. Fujita, K. Terato and M. Nakayama, *Biosci. Biotech. Biochem.*, 1996, **60**, 732.

29 N. D'Accorso, I. M. E. Thiel and M. E. Gelpi, *Rev. Latinoam. Quim.*, 1994, **23**, 111 (*Chem. Abstr.*, 1996, **125**, 115 022a).

30 G. Smith and J. A. Leary, *J. Am. Soc. Mass Spectrom.*, 1996, **7**, 953.

31 G. Descotes, E. Gentil, J. Ramzal and J. Banoub, *J. Braz. Chem. Soc.*, 1996, **7**, 379 (*Chem. Abstr.*, 1996, **125**, 329 148).

32 M. Nishikatani, K. Kubota and A. Kobayashi, *Biosci. Biotech. Biochem.*, 1996, **60**, 929.

33 B. Mulroney, J. C. Traeger and B. A. Stone, *J. Mass Spectrom.*, 1995, **30**, 1277 (*Chem. Abstr.*, 1996, **124**, 56 457).

34 H. Bartsch, W. A. Konig, M. Strabner and U. Hintze, *Carbohydr. Res.*, 1996, **286**, 41.

35 B. Chankvetadze, G. Endresz, G. Blaschke, M. Juza, H. Jakubetz and V. Schurig, *Carbohydr. Res.*, 1996, **287**, 139.

36 S. N. Davey, D. A. Leigh, J. P. Smart, L. W. Tetler and A. M. Truscello, *Carbohydr. Res.*, 1996, **290**, 117.

37 P. Cescutti, D. Garazzo and R. Rizzo, *Carbohydr. Res.*, 1996, **290**, 105.

38 G. Resnicek and M. Mathà, *Carbohydr. Res.*, 1996, **293**, 133.

39 T. Ii, Y. Ohashi, T. Ogawa and Y. Nagai, *J. Mass Spectrom. Soc. Jpn.*, 1996, **44**, 183 (*Chem. Abstr.*, 1996, **125**, 87 068z).

40 M. Xie, D. Giraud, Y. Bertheau, B. Casetta and B. Arpuio, *Rapid Commun. Mass Spectrom.*, 1995, **9**, 1572.

41 P. Patoprsty, V. Kovacik and S. Karacsonyi, *Rapid Commun. Mass Spectrom.*, 1996, **9**, 840 (*Chem. Abstr.*, 1996, **124**, 181).

42 S. Auriola, P. Thibault, I. Sadovskaya and E. Altman, *ACS Symp. Ser.*, 1996, **619**, 149 (*Chem. Abstr.*, 1996, **124**, 261 510).

43 D. J. Harvey, T. J. P. Naven, R. Kuster, R. H. Bateman and M. R. Green, *Rapid Commun. Mass Spectrom.*, 1995, **9**, 1556.

44 J. Lemoine, F. Chivat and B. Domon, *J. Mass Spectrom.*, 1996, **31**, 908.

45 T. J. P. Naven and D. J. Harvey, *Rapid Commun. Mass Spectrom.*, 1996, **10**, 829.

46 Y. Kimura, K.-i. Yamaguchi and G. Funatsu, *Biosci. Biotech. Biochem.*, 1996, **60**, 538.

47 N. G. Karlsson, H. Karlsson and G. C. Hansson, *J. Mass Spectrom.*, 1996, **31**, 560.

48 V. Kovacik, J. Hirsch, P. Kovac, W. Heerma and J. Thomas-Oates, *J. Mass Spectrom.*, 1995, **30**, 949.

49 R. R. Townsend, A. L. Burlingame and O. Hindsgaul, *J. Am. Soc. Mass Spectrom.*, 1996, **7**, 182.

50   M. Ohta, S. Emi, H. Iwamoto, J. Hirose, K. Hiromi, H. Itoh, M. Shin, S. Murao and S. Matsuura, *Biosci. Biotech. Biochem.*, 1996, **60**, 1123.

51   K. Tokuyasu, M. Ohnishi-Kameyama and K. Hayashi, *Biosci. Biotech. Biochem.*, 1996, **60**, 1598.

52   T. Yamagaki, M. Maeda, K. Kanezawa, Y. Ishizuka and H. Nakanishi, *Biosci. Biotech. Biochem.*, 1996, **60**, 1222.

53   K. Burgess, D. H. Russell, A. Shitangkoon and A. J. Zhang, *Nucleosides Nucleotides*, 1996, **15**, 1719.

54   B. A. Woluncka, E. de Hoffmann, J. S. Rush and C. J. Waechter, *J. Am. Soc. Mass Spectrom.*, 1996, **7**, 541.

55   T. Kinoshita, K. Torii, T. Ishikawa, Y. Maruyama, T. Takazawa and T. Kosaka, *Bunseki Kagaku*, 1996, **45**, 457 (*Chem. Abstr.*, 1996, **125**, 58 974m).

56   I. V. Yartseva and S. Y. Melnik, *Collect. Czech. Chem. Commun.*, 1996, **61**, S150 (*Chem. Abstr.*, 1996, **125**, 329 149).

57   L. Santana, M. Teijeira, E. Uriarte, A. Fadda and G. Podda, *Rapid Commun. Mass Spectrom.*, 1996, **10**, 1316.

58   W. H. Ojala, W. G. Gleason, M. P. E. Connelly, R. R. Wallis and J. J. Kremer, *Acta Crystallogr.*, 1996, **C52**, 155.

59   P. Koll and J. Kopf, *Aust. J. Chem.*, 1996, **49**, 391.

60   S. Perez, *Sucrose*, 1995, 11 (*Chem. Abstr.*, 1996, **124**, 290 052).

61   R. V. Krishakumar and S. Natarjan, *Carbohydr. Res.*, 1996, **287**, 117.

62   C. A. Podlasek, W. A. Stripe, I. Carmichael, M. Shang, B. Basu and A. S. Serainni, *J. Am. Chem. Soc.*, 1996, **118**, 1413.

63   J. C. Barnes, J. S. Brimacombe and A. K. M. S. Kabir, *Acta Crystallogr.*, 1996, **C52**, 416.

64   A. G. Evdokimov and F. Frolow, *Acta Crystallogr.*, 1996, **C52**, 3218.

65   L. Eriksson, A. Pilotti, R. Stenutz and G. Widmalm, *Acta Crystallogr.*, 1996, **C52**, 2285.

66   Y. Ate, K. Harata, M. Fujiwara and K. Ohtu, *Langmiur*, 1996, **12**, 636 (*Chem. Abstr.*, 1996, **124**, 176 670).

67   H. Reinke, D. Rentsch and R. Meitschen, *Carbohydr. Res.*, 1996, **281**, 293.

68   P. J. Cox, O. A. Armishaw and J. L. Wardell, *J. Chem. Res. (S)*, 1996, 140; (M) 815.

69   T. Popek, J. Mazurek and T. Lis, *Acta Crystallogr.*, 1996, **C52**, 1558.

70   C. Fernandez-Castaño and C. Foces-Foces, *Acta Crystallogr.*, 1996, **C52**, 1586.

71   A. Linden, C. K. Lee and K. F. Siew, *Acta Crystallogr.*, 1996, **C52**, 1797.

72   A.-M. Lebuis, D.-S. Lee and A. S. Perlin, *Aust. J. Chem.*, 1996, **49**, 299.

73   M. Lechtenberg, A. Nahrstedt and F. R. Fronczek, *Phytochemistry*, 1996, **41**, 779.

74   Z. Galdecki, A. Fruzinski, O. Achmatowicz and B. Szechner, *Pol. J. Chem.*, 1995, **69**, 269 (*Chem. Abstr.*, 1996, **124**, 9 149).

75   W. Yinqui, J. M. Waters and L. F. Blackwell, *J. Chem. Soc., Perkin Trans. 1*, 1996, 1449.

76   C. J. Bailey, D. C. Craig, C. T. Grainger, V. J. James and J. D. Stevens, *Carbohydr. Res.*, 1996, **284**, 265.

77   G. Foerster, R. O. de la Cruz, G. Bendas and P. Nahn, *Prog. Colloid Polym. Sci.*, 1995, **98**, 201 (*Chem. Abstr.*, 1996, **124**, 9 183).

78   D. K. Watt, D. J. Brasch, D. S. Larsen, L. D. Melton and J. Simpson, *Carbohydr. Res.*, 1996, **285**, 1.

79   P. W. K. Loo and J. R. Rubin, *Tetrahedron*, 1996, **52**, 3857.

80   B. Kamenar, N. K. Hulita, I. Vikovic, G. Kobrehel and G. Lazarevski, *Acta Crystallogr.*, 1996, **C52**, 2566.

81   F. Yvelin, M.-Y. Zhang, J.-M. Mallet, F. Robert, Y. Jeannin and P. Sinaÿ, *Carbohydr. Lett.*, 1996, **1**, 475.

82   S. D. Hulleman, W. Helbert and H. Chanzy, *Int. J. Biol. Macromol.*, 1996, **18**, 115 (*Chem. Abstr.*, 1996, **124**, 232 925).

83   R. Dutzler, Y. F. Wang, P. J. Rizkallah and J. P. Rosenbusch, *Structure*, 1996, **4**, 127.

84   K. Harata, Y. Nagano, H. Ikeda, T. Ikeda, A. Ueno and F. Toda, *J. Chem. Soc., Chem. Commun.*, 1996, 2347.

85   D. Mentzafos, A. Terzis, A. W. Coleman and C. de Rango, *Carbohydr. Res.*, 1996, **282**, 125.

86   B. D. Blasio, S. Galdiero, M. Saviano, C. Pedone, E. Benedetti, E. Rizzarelli and S. Pedotti, *Carbohydr. Res.*, 1996, **282**, 41.

87   T. Steiner and W. Saenger, *Carbohydr. Res.*, 1996, **282**, 53.

88   T. Steiner, F. Hirayama and W. Saenger, *Carbohydr. Res.*, 1996, **296**, 69.

89   J. E. Anderson, *J. Org. Chem.*, 1996, **61**, 3511.

90   H. Yoshii, T. Furuta, K. Kawasaki and H. Hirano, *Ogo Toshita Kagaka*, 1995, **42**, 243 (*Chem. Abstr.*, 1996, **124**, 30 165).

91   M. Selkti, H. P. Lopez, J. Navaza, F. Villani and C. de Rango, *Supramol. Chem.*, 1995, **5**, 225 (*Chem. Abstr.*, 1996, **124**, 117 745).

92   J. Matulic-Adamic, L. Beigelman, S. Portmann, M. Egli and N. Usman, *J. Org. Chem.*, 1996, **61**, 3909.

93   P. Verlhac, C. Leteux, L. Toupet and A. Veyrières, *Carbohydr. Res.*, 1996, **291**, 11.

94   T. Hosoya, Y. Ohasi, T. Matsumoto and K. Suzuki, *Tetrahedron Lett.*, 1996, **37**, 663.

95   R. Csuk, M. Schaade and C. Krieger, *Tetrahedron*, 1996, **52**, 6397.

96   W. Zheng, J. A. D. Matte, J.-P. Wu, J. J.-W. Duan, L. R. Cook, H. Oinuma and Y. Kishi, *J. Am. Chem. Soc.*, 1996, **118**, 7946.

97   A. Ernst and A. Vasella, *Helv. Chim. Acta*, 1996, **79**, 1279.

98   W. J. Ferguson, S. Parson and R. M. Paton, *Acta Crystallogr.*, 1996, **C52**, 3067.

99   A.-L. Wang, S.-J. Lu, H.-X. Fu, H.-Q. Wang and X.-Y. Huang, *Carbohydr. Res.*, 1996, **281**, 301.

100   A. Wang, S. Lu, X. Fu, H. Wang, L. Huang and X. Huang, *Jiegou Huaxue*, 1996, **15**, 284 (*Chem. Abstr.*, 1996, **125**, 143 143j).

101   G. J. Gainsford, P. C. Tyler and R. H. Furneaux, *Acta Crystallogr.*, 1996, **C52**, 1274.

102   R. D. Clercq, R. Herdewijn and L. V. Meervelt, *Acta Crystallogr.*, 1996, **C52**, 1213.

103   P. Dokurno, A. Wisniewski and J. Blazejowski, *Pol. J. Chem.*, 1995, **69**, 1273 (*Chem. Abstr.*, 1996, **124**, 87 593w).

104   W. H. Ojala and W. B. Gleason, *Acta Crystallogr.*, 1996, **C52**, 3188.

105   R. Anulewicz, I. Wawer, B. Piekarska-Bartoszewicz and A. Temeriusz, *Carbohydr. Res.*, 1996, **281**, 1.

106   D. Gomez de Anderez, H. Gil, M. Helliwell and J. M. Segrada, *Acta Crystallogr.*, 1996, **C52**, 252.

107   T. W. Brandstetter, C. de la Fuente, Y.-L. Kim, R. I. Cooper and D. J. Watkin, *Tetrahedron*, 1996, **52**, 10711.

108   V. V. Mossine, C. L. Barnes, G. V. Glinksy and M. S. Feather, *Carbohydr. Res.*, 1996, **284**, 11.

109   J. G. Fernandez-Bolanos, S. Garcia, J. Fernandez-Bolanos and M. J. Dianez, *Carbohydr. Res.*, 1996, **282**, 137.

110   V. V. Mossine, C. L. Barnes, G. V. Glinksy and M. S. Feather, *Carbohydr. Lett.*, 1995, **1**, 355.

111   P. W. Smith, S. L. Sollis, P. D. Howes, P. C. Cherry, K. N. Cobley, H. Taylor, A. R. Whittington, J. Scicinsky, R. C. Bethell, N. Taylor, T. Skarzynski, A. Cleasby, O. Singh, A. Wonacott, J. Varghese and P. Colman, *Bioorg. Med. Chem. Lett.*, 1996, **6**, 2931.

112   J. Cobo, M. Melguizo, A. Sanchez and M. Nogueras, *Acta Crystallogr.*, 1996, **C52**, 148.

113   J. N. Low, G. Ferguson, J. Cobo, M. Melguizo, M. Nogueras and A. Sanchez, *Acta Crystallogr.*, 1996, **C52**, 145.

114   A. T. Ung, S. G. Pyne, B. W. Skelton and A. H. White, *Tetrahedron*, 1996, **52**, 14069.

115   P. Merino, F. Junquera, F. L. Merchan and T. Tejero, *Acta Crystallogr.*, 1996, **C52**, 3197.

116   M. T. Lakin, N. Mouhous-Riou, C. Lorin, P. Rollin, J. Kroon and S. Pérez, *Carbohydr. Res.*, 1996, **290**, 125.

117   M. Cudic, B. Kojic-Prodic, V. Milinkovic, J. Horvat, S. Horvat, M. Elfosson and J. Kihlberg, *Carbohydr. Res.*, 1996, **287**, 1.

118   Z. Urbanczyk-Lipowska, K. Sunwinska, D. Mostowicz and M. Chmielewski, *J. Chem. Crystallogr.*, 1995, **25**, 693 (*Chem. Abstr.*, 1996, **124**, 202 824).

119   J. T. Link, S. Raghavan, M. Gallant, S. J. Danishefsky, T. C. Chou and L. M. Ballas, *J. Am. Chem. Soc.*, 1996, **118**, 2825.

120   T. W. Brandstetter, C. de la Fuente, Y.-L. Kim, L. N. Johnson, S. Crook, P. M. de Q. Lilley, D. J. Watkin, K. E. Tsitsanou, S. E. Zographos, E. D. Chrysina, N. G. Oikonomakos and G. W. J. Fleet, *Tetrahedron*, 1996, **52**, 10721.

121   K. Shinozaki, K. Mizuno and Y. Masaki, *Chem. Pharm. Bull.*, 1996, **44**, 927.

122   F. Morís-Varas, X.-H. Qian and C.-H. Wong, *J. Am. Chem. Soc.*, 1996, **118**, 7647.

123   R. Hoos, J. Huixin, A. Vasella and P. Weiss, *Helv. Chim. Acta*, 1996, **79**, 1757.

124   T. D. Heightman, M. Locatelli and A. Vasella, *Helv. Chim. Acta*, 1996, **79**, 2191.

125   D. Marek, A. Wadouachi, R. Uzan, D. Beaupere, G. Nowogrocki and G. Laplace, *Tetrahedron Lett.*, 1996, **37**, 49.

126   Y. Tsuda, Y. Sato, K. Kanemitsu, S. Hosoi, K. Shibayama, K. Nakao and Y. Ishikawa, *Chem. Pharm. Bull.*, 1996, **44**, 1465.

127   O. Varela, P. A. Zunszain, D. O. Cicero, R. F. Baggio, D. R. Vega and M. T. Garland, *Carbohydr. Res.*, 1996, **280**, 187.

128   J.-Y. L. Questel, N. Mouhous-Riou and S. Pérez, *Carbohydr. Res.*, 1996, **284**, 35.

129   I. A. Litvinov, V. A. Nauniov, L. I. Gurarii and E. T. Mukmenev, *Izv. Akad. Nauk, Ser. Khim.*, 1993, 406 (*Chem. Abstr.*, 1996, **124**, 87 503).

130   D.-X. Liem, I. D. Jenkins, B. W. Skelton and A. H. White, *Aust. J. Chem.*, 1996, **49**, 371.

131   M. A. Brown, P. J. Cox, R. A. Howe, O. A. Melvin and J. L. Wardell, *J. Chem. Soc., Perkin Trans. 1*, 1996, 809.

132   M. A. Brown, P. J. Cox, O. A. Melvin and J. L. Wardell, *Main Group Met. Chem.*, 1995, **18**, 175 (*Chem. Abstr.*, 1996, **124**, 117 700).

133   R. W. Tjerkstra, M. L. Verdonk and J. Kroon, *Carbohydr. Res.*, 1996, **285**, 151.

134   B. V. Yang and M. A. Massa, *J. Org. Chem.*, 1996, **61**, 5149.

135   S. Ianelli and M. Nardelli, *Acta Crystallogr.*, 1996, **B52**, 2853.

136   M. D. R. Hernandez-Medel, O. Lopez-Marquez, R. Santillan and A. Trigos, *Phtochemistry*, 1996, **43**, 279.

137   R. Lai and S. Martin, *Tetrahedron Asymm.*, 1996, **7**, 2783.

138   T. Hudlicky, K. A. Abboud, J. Bolonick, R. Mauryn, M. L. Stanton and A. J. Thorpe, *Chem. Commun.*, 1996, 1717.

139   M. W. Wieczorek, J. Blaszczyk and B. W. Król, *Acta Crystallogr.*, 1996, **C52**, 1193.

140   C. Gack and P. Klüfers, *Acta Crystallogr.*, 1996, **C52**, 2972.

141   P. Barbaso, A. Currao, J. Herrmann and R. Nesper, *Organometallics*, 1996, **15**, 1879.

142   H. C. Freeman, D. A. Langs, C. E. Nockolds and Y. L. Oh, *Aust. J. Chem.*, 1996, **49**, 413.

143   V. Graingeot, C. Brigando, B. Faure and D. Benlian, *Acta Crystallogr.*, 1996, **C52**, 3229.

144   I. D. Spiers, C. H. Schwalbe and S. Freeman, *Acta Crystallogr.*, 1996, **C52**, 2575.

145   S.-K. Chung, Y.-T. Chang, D. Whang and K. Kim, *Carbohydr. Res*, 1996, **295**, 1.

146   I. D. Spiers, C. J.Barker, S.-K. Chung, Y.-T. Chang, S. Freeman, J. M. Gardiner, P. H. Hirst, P. A. Lambert, R. H. Mitchell, D. R. Poyner, C. H. Schwalbe, A. W. Smith and K. R. H. Solomons, *Carbohydr. Res.*, 1996, **282**, 81.

147   D. Lampe, C. Liu, M. F. Mahon and B. V. L. Potter, *J. Chem. Soc., Perkin Trans. 1*, 1996, 1717.

148   Y. Suguwara and H. Urabe, *Mol. Cryst. Liq. Cryst. Sci. Technol., Sect. A*, 1996, **277**, 615 (*Chem. Abstr.*, 1996, **125**, 58 957h).

149   C. G. Suresh, K. Sakthivel and T. Pathak, *Acta Crystallogr.*, 1996, **C52**, 1776.

150   F. H. Hausheer, N. D. Jones, P. Seetharamulu, J. B. Deeter, L. W. Hertel and J. S. Kroin, *Comput. Chem.*, 1996, **20**, 459 (*Chem. Abstr.*, 1996, **125**, 301 452).

151   P. Klüfers and P. Mayer, *Acta Crystallogr.*, 1996, **C52**, 2970.

152   A. K. Das, A. Mukhopadhyay, N. Das, S. K. Mazumdar and V. Bertolasi, *Acta Crystallogr.*, 1996, **C52**, 2615.

153   S. D. Soni, T. Srikrishnan and J. L. Alderfer, *Nucleosides Nucleotides*, 1996, **15**, 1945.

154   S. Neidle, J. P. Chattopadhyaya, N. Hossain and A. Papchikhin, *Acta Crystallogr.*, 1996, **C52**, 3173.

155   P. Singh, M. Zottola, S. Huang, B. Ramsayshaw and L. G. Pederson, *Acta Crystallogr.*, 1996, **C52**, 693.

156   Q. Gao, A. Sood, B. Ramsayshaw and L. D. Williams, *Acta Crystallogr.*, 1996, **C52**, 1823.

157   I. Basnak, M. Sun, T. A. Hamor, N. Spencer and R. T. Walker, *Nucleosides Nucleotides*, 1996, **15**, 1275.

158   F. Seela, M. Zulauf, H. Rosemeyer and H. Reuter, *J. Chem. Soc., Perkin Trans. 2*, 1996, 2373.

159   J. Murfurt, E. Stulz, H. U. Trafelet, A. Zingg, C. Leumann, M. Hazenkamp, R. Judd, S. Schenker, G. Strouse, T. R. Ward, M. Förtsch, J. Hauser and H.-B. Burgi, *Acta Crystallogr.*, 1996, **C52**, 713.

160   V. M. Kolb, P. C. Colloton, P. D. Robinson, H. G. Lutfi and C. Y. Meyers, *Acta Crystallogr.*, 1996, **C52**, 1781.

161   P. D. Robinson, C. Y. Meyers, V. M. Kolb and P. C. Colloton, *Acta Crystallogr.*, 1996, **C52**, 1215.

162   F. Freeman, R. A. Scheuerman and J. W. Ziller, *Acta Crystallogr.*, 1996, **C52**, 2006.

163   S. Neidle, D. C. Capaldi, C. B. Reese and P. D. Roselt, *Acta Crystallogr.*, 1996, **C52**, 2332.

164   G. Biswas, T. Chandra, N. Garg, D. S. Bhakuni, A. Pramanik, K. Avasthi and P. R. Maulik, *Acta Crystallogr.*, 1996, **C52**, 2563.

165   A. Kittaka, N. Yamada, H. Tanaka, K. T. Nakamura and T. Miyasaka, *Nucleosides Nucleotides*, 1996, **15**, 1447.

166   A. Kittaka, Y. Tsubaki, H. Tanaka, K. T. Nakamura and T. Miyasaka, *Nucleosides Nucleotides*, 1996, **15**, 97.

167   J. Farràs, M. del Mar Lleó, J. Vilarrasa, S. Castillón, M. Matheu, X. Solans and
      M. Font-Bardia, *Tetrahedron Lett.*, 1996, **37**, 901.
168   R. Wolin, D. Wang, J. Kelly, A. Afonso, L. James, P. Kirschmeier and A. T.
      McPhail, *Bioorg. Med. Chem. Lett.*, 1996, **6**, 195.
169   K. Sakhthivel, T. Pathak and C. G. Suresh, *Tetrahedron*, 1996, **52**, 1767.
170   A. P. Higson, G. K. Scott, D. J. Earnshaw, A. D. Baxter, R. A. Taylor and
      R. Cosstick, *Tetrahedron*, 1996, **52**, 1027.
171   K. Haraguchi, H. Tanaka, S. Saito, S. Kinoshima, M. Hosoe, K. Kanmuri,
      K. Yamaguchi and T. Miyasaka, *Tetrahedron*, 1996, **52**, 9469.
172   M. Sun, I. Basnak, T. A. Hamor and R. T. Walker, *Acta Crystallogr.*, 1996, **C52**, 2556.
173   L. Pickering and V. Nair, *Nucleosides Nucleotides*, 1996, **15**, 1751.
174   K. Hisa, A. Hittaka, H. Tanaka, K. Yamaguchi and T. Miyasaki, *Nucleosides
      Nucleotides*, 1996, **15**, 85.
175   A. Marx, P. Erdmann, M. Senn, S. Körner, T. Jungo, M. Petretta, P. Imwinkelried,
      A. Dussy, K. J. Kulicke, L. Macko, M. Zehnder and B. Giese, *Helv. Chim. Acta*,
      1996, **79**, 1980.
176   N. Hossain, N. Blaton, O. Peeters, J. Rozenski and P. A. Herdewijn, *Tetrahedron*,
      1996, **52**, 5563.
177   L. Santana, M. Teijeira, E. Uriate, C. Teran, V. Casellato and R. Graziani, *Nucleo-
      sides Nucleotides*, 1996, **15**, 1179.
178   S. N. Mikhailov, N. Blaton, J. Rozenski, J. Balzarini, E. D. Clercq and B. Herde-
      wijn, *Nucleosides Nucleotides*, 1996, **15**, 867.
179   M. A. Siddiqui, J. H. Ford, C. George and V. E. Marquez, *Nucleosides Nucleotides*,
      1996, **15**, 235.
180   G. S. Padiyar and T. P. Deshadri, *Nucleosides Nucleotides*, 1996, **15**, 857.
181   M. J. Thompson, A. Mekhalfia, D. L. Jakeman, S. E. V. Phillips, K. Phillips,
      J. Porter and G. M. Blackburn, *Chem. Commun.*, 1996, 791.
182   D. J. Cooke, J. R. Lu, E. M. Lee, R. K. Thomas, A. R. Pitt, E. A. Simister and
      J. Penfold, *J. Phys. Chem.*, 1996, **100**, 10298.
183   O. Haba, K. Yokota and T. Kakuchi, *Chirality*, 1996, **7**, 193.
184   J. Barek, R. Hrncir, J. R. Moreira and J. Jima, *Collect. Czech. Chem. Commun.*,
      1996, **61**, 333 (*Chem. Abstr.*, 1996, **125**, 11 237h).
185   E. R. Arndt and E. S. Stevens, *Biopolymers*, 1996, **38**, 567 (*Chem. Abstr.*, 1996, **125**,
      11 251h).
186   N. Zhao, P. Zhou, N. Berova and K. Nakanishi, *Chirality*, 1996, **7**, 636 (*Chem.
      Abstr.*, 1996, **124**, 261 516).
187   D. Rele, N. Zhao, K. Nakanishi and N. Berova, *Tetrahedron*, 1996, **52**, 2759.
188   K. Satsumabayashi, Y. Nishida and K. Tanemura, *Nippon Shika Daigaku Kiyo,
      Ippam Kyoiku-kei*, 1996, **25**, 43 (*Chem. Abstr.*, 1996, **125**, 301 364).
189   Y. Nishida and K. Satsumabayashi, *Nippon Shika Daigaku Kiyo, Ippam Kyoiku-kei*,
      1996, **25**, 55 (*Chem. Abstr.*, 1996, **125**, 276 352).
190   S. Arimori, M. Takeuchi and S. Shinkai, *Chem. Lett.*, 1996, 77.
191   T. D. James, H. Shinmori, M. Takeuchi and S. Shinkai, *Chem. Commun.*, 1996, 705.
192   T. D. James, K. R. A. S. Sandanayake and S. Shinkai, *Angew. Chem., Int. Ed. Engl.*,
      1996, **35**, 1910.
193   F. Seela and Y. Chen, *Chem. Commun.*, 1996, 2263.
194   J. Fan and A. C. Austen, *Thermochim. Acta*, 1996, **266**, 9 (*Chem. Abstr.*, 1996,
      124, 146 607).
195   R. J. Wilson, A. E. Beezer and J. C. Mitchell, *Thermochim. Acta*, 1996, **264**, 27
      (*Chem. Abstr.*, 1996, **124**, 146 683).

196   P. Perkkalainen, H. Halttunen and I. Pitkaenen, *Thermochim. Acta*, 1996, **269**, 351 (*Chem. Abstr.*, 1996, **124**, 202 820).

197   M. C. Ramos-Sanchez, M. T. Barrio-Arrendondo, A. I. D. Andres-Santos, J. Martin-Gil and F. J. Martin-Gil, *Thermochim. Acta*, 1996, **262**, 109 (*Chem. Abstr.*, 1996, **124**, 146 672).

198   A. Marini, V. Berbenni, G. Bruni, V. Massarotti, P. Mustareui and M. Villa, *J. Chem. Phys.*, 1995, **103**, 7532.

199   P. Germani, M. Bilal and C. de Brauer, *Thermochim. Acta*, 1995, **259**, 187 (*Chem. Abstr.*, 1996, **124**, 87 559).

200   V. T. Yilmaz, A. Karadag and H. Icbudak, *Thermochim. Acta*, 1995, **261**, 107 (*Chem. Abstr.*, 1996, **124**, 9 169).

201   E. Ultiz, *J. Therm. Anal.*, 1995, **45**, 767 (*Chem. Abstr.*, 1996, **124**, 117 856).

202   A. H. Otto, S. Schrader and G. Reinisch, *Starch/Staerke*, 1996, **48**, 29 (*Chem. Abstr.*, 1996, **124**, 290 009).

203   I. Shehatta, *React. Func. Polym.*, 1996, **28**, 183 (*Chem. Abstr.*, 1996, **124**, 290 034).

204   M. Jezowska-Bojczuk, S. Lamotte, H. Kozlowski and P. Decocic, *Spectrosc. Biol. Mol. Eur. Conf., 6th*, 1995, 451 (*Chem. Abstr.*, 1996, **124**, 146 664).

205   M. K. Shukla and P. C. Mishra, *J. Mol. Struct.*, 1996, **377**, 247 (*Chem. Abstr.*, 1996, **125**, 11 295a).

206   I. Tuzov, K. Craemer, S. N. Maganov and M.-H. Whangbo, *New J. Chem.*, 1996, **20**, 37 (*Chem. Abstr.*, 1996, **124**, 317 727).

207   I. Tuzov, K. Craemer, S. N. Magonov and M.-H. Whangbo, *New J. Chem.*, 1996, **20**, 23 (*Chem. Abstr.*, 1996, **124**, 317 726).

208   K. Craemer, S. Demharter, R. Muelhampt, H. Frey, S. N. Magonov, I. Tozov and M.-H. Whangbo, *New J. Chem.*, 1996, **20**, 5 (*Chem. Abstr.*, 1996, **124**, 317 708).

209   A. Sandana, J. P. Alarie and T. Vo-Dinh, *Talanta*, 1995, **42**, 1567 (*Chem. Abstr.*, 1996, **124**, 146 637).

210   T. L. Netzel, K. Nafisi, J. Headrick and B. E. Eaton, *J. Phys. Chem.*, 1995, **99**, 17948.

211   M. Suzuki, H. Tanaki, K. Tanaka, K. Narita, H. Fujiwara and H. Ohmari, *Carbohydr. Res.*, 1996, **288**, 75.

212   B. Jankiewicz, R. Soloniewicz and A. Socha, *Acta Pol. Pharm.*, 1995, **52**, 331 (*Chem. Abstr.*, 1996, **124**, 56 462).

213   T. Kimura, O. Shirata, A. Suzuki and Y. Ohtsu, *Kuromatogurafi*, 1996, **17**, 156 (*Chem. Abstr.*, 1996, **125**, 87 033j).

23
Separatory and Analytical Methods

The apparent relative decline in this chapter cannot be taken as evidence that separations and analyses are becoming less significant in carbohydrate chemistry. Rather the reverse seems more probable; many sophisticated methods are being used routinely, their availability is taken for granted and they are not identified as key features of papers.[*]

1 Chromatographic Methods

1.1 Gas-Liquid Chromatography – Acetylated glycosyl fluorides have been identified by this procedure,[1] and capillary gas chromatography has been employed in pyrolysis studies of disaccharides using a Curie-point pyrolyser.[2]

1.2 Thin-layer Chromatography – Sugar phosphates and nucleotides have been detected and quantified by densitometry on thin layer plates following hydrolysis with alkaline phosphatase and staining for phosphate with molybdic acid and Malachite green.[3] The effect of solvents on the relative stability of the complexes formed between sugars and alditols and lanthanide cations has been examined by thin-layer ligand exchange chromatography,[4] and the products of reaction between sugars and ethyl acetoacetate in the presence of zinc chloride have been monitored by TLC with detection using Liebermann-Burchard reagent.[5]

1.3 High-pressure Liquid Chromatography – A short review in Japanese has described recently developed methods for the separation and analysis of oligo- and polysaccharides. Compounds were separated using HPLC with graphitized carbon column and detection with pulsed amperometric methods. Structural analysis was effected by FAB-MS and ^{13}C NMR spectroscopies. Enzymically synthesized branched cyclodextrins were amongst the products described.[6] ^1H NMR spectroscopy and HPLC separations were used in the study of the hydrolysis of small cello-oligosaccharides with degree of polymerisation 4-6 by a cellobiohydrolase.[7] Disaccharide units of the glycosaminoglycans chondroiton sulfate, dermatan sulfate and hyaluronic acid were released in more than 90%

---

[*]Alas a further explanation can be given: papers in *J. Chromatogr.* were inadvertently overlooked this year. They will be abstracted for Vol. 31.

---

yield by solvolyses in DMSO containing 0.1% water and determined by HPLC separation on strong anion exchange resin columns.[8] Reverse phase HPLC methods were used in the determination of mono- and oligo-saccharides derivatized with *p*-aminobenzoic acid ethyl ester.[9] Various partially methylated 1,5-anhydroglucitol derivatives carrying ester groups formed during reductive cleavage of polymers have been separated by HPLC.[10]

A novel approach to the investigation of acyl migration rates of 1-*O*-acyl β-D-glucopyranuronates involved separating the isomers on reverse phase HPLC columns and passing the solutions containing the products into online NMR flow probes in a 600 MHz NMR spectrometer.[11] Separations of diastereoisomeric glycosides of terpene alcohols have been conducted using HPLC methods involving a cyclodextrin stationary phase.[12]

A novel and very sensitive procedure for determining the absolute configuration of sugars involving an HPLC procedure operating on a picamole scale involved making the derivatives **1** by use of a resolved acid which was coupled with a peracetylated glycosyl chloride.[13]

1

High pH anion exchange chromatography (HPAEC) has been applied to the separation of neutral *N*-linked oligosaccharides,[14] oligosaccharides derivatized by *p*-nitrobenzylhydroxylamine[15] and various D-mannose derivatives. In the last case the relationship between retention times and structure was examined.[16]

1.4 Column Chromatography – Activated charcoal columns were used in the separation of oligosaccharides derived from hydrolysis of konjac glucomannan.[17]

2 Electrophoresis

Electrophoretic studies on some quercetin glycosides have been reported.[18]

3 Other Analytical Methods

An electrochemical method involving electrocatalytic oxidation with TEMPO was suitable for the determination of glucose in aqueous solution.[19] The cyanine dye **2**, which carries two phenylboronic acid moieties displays increased fluorescence on binding to monosaccharides and provides a novel detection method.[20] Compound **3** also contains a phenylboronic acid residue and binds D-glucose 6-

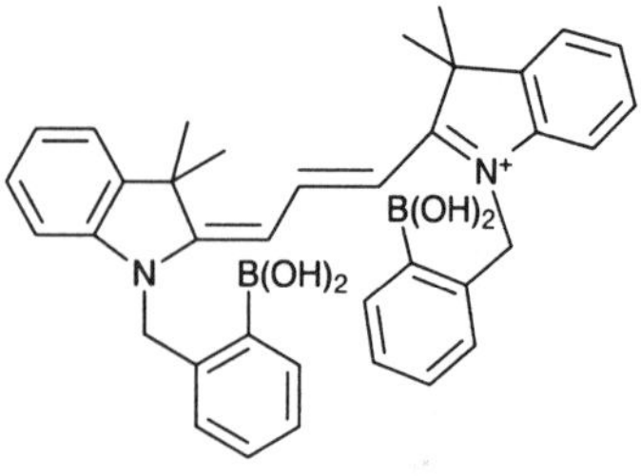

2

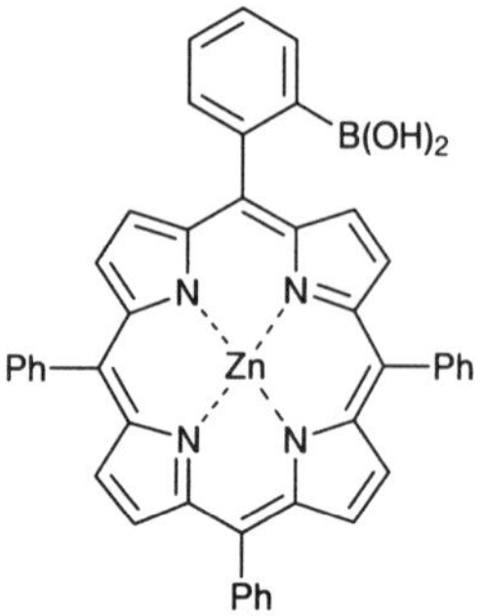

3

phosphate in a two-point manner, the boronic acid moiety complexing with the 1,2-diol and the zinc binding the phosphate ester. The reagent allows glucose 6-phosphate to be differentiated from glucose 1-phosphate by NMR which is otherwise difficult.[21] Sequence analysis of reducing oligosaccharides by use of 8-amino-2-naphthalenesulfonic acid and methyl chloroformate which, by way of a Schiff base, cleaves the reducing *N*-glycosidic bond unit.[22] Maltobionic acid can be determined by use of three specific enzymes: a hydrolase which gives D-glucose and D-gluconic acid, a kinase which converts the acid to the 6-phosphate, and a dehydrogenase which converts this phosphate to ribulose-5-phosphate. The reactions can be followed spectrophotometrically by the production of NADPH monitored at 340 nm.[23]

References

1   B. A. Bergamaschi and J. I. Hedges, *Carbohydr. Res.*, 1996, **280**, 345.

2   M. Wada, S. Fujishige, S. Uchino and N. Oguri, *Kobunshi Ronbunshu*, 1996, **53**, 201 (*Chem. Abstr.*, 1996, **124**, 343 864).

3   H. Ohyama, C. Matsubara and K. Takamura, *Chem. Pharm. Bull.*, 1996, **44**, 1252.

4   Y. Israëli, C. Lhermet, J.-P. Morel and N. Morel-Desrosiers, *Carbohydr. Res.*, 1996, **289**, 1.

5   R. Imura, *Kumamoto-kenritsu Daigaku Seikatsu Kagabuku Kiyo*, 1995, **1**, 23 (*Chem. Abstr.*, 1996, **124**, 176 658).

6　K. Koizumi, *Oyo Toshitsu Kagaku*, 1994, **41**, 465 (*Chem. Abstr.*, 1996, **124**, 87 496).

7　V. Harjunpaa, A. Teleman, A. Koivula, T. T. Teeri, O. Teleman and T. Drakenberg, *VTT Symp.*, 1996, **163**, 203 (*Chem. Abstr.*, 1996, **125**, 143 157).

8　G. Qiu, H. Toyoda, T. Toida, I. Koshiishi and T. Imanari, *Chem. Pharm. Bull.*, 1996, **44**, 1017.

9　H. Kwon and J. Kim, *Anal. Sci. Technol.*, 1995, **8**, 859 (*Chem. Abstr.*, 1996, **124**, 317 709).

10　C. K. Lee and E. J. Kim, *Carbohydr. Res.*, 1996, **280**, 59.

11　U. G. Sidelmann, S. H. Hansen, C. Gavaghan, H. A. J. Carless, R. D. Farrant, J. C. Lindon, I. D. Wilson and J. K. Nicholson, *Anal. Chem.*, 1996, **68**, 2564. (*Chem. Abstr.*, 1996, **125**, 87 053).

12　C. Salles, J. C. Jallageas and J. Crouzet, Riv. Ital. EPPOS, 1993, **4**, 90 (*Chem. Abstr.*, 1996, **124**, 56 440).

13　C. Bai, Y. Nishida, H. Ohrui and H. Meguro, *J. Carbohydr. Chem.*, 1996, **15**, 217.

14　J. M. McGuire, M. Douglas and K. D. Smile, *Carbohydr. Res.*, 1996, **292**, 1.

15　M. Pauly, W. S. York, R. Guillen, P. Albersheim and A. G. Darvill, *Carbohydr. Res.*, 1996, **282**, 1.

16　Z. Li, S. Mou, W. Liao and D. Lu, *Carbohydr. Res.*, 1996, **295**, 229.

17　G. Y. Xu, L. W. Cheng, Y. M. Xu, Y. H. Ma and P. J. Zhou, *Chim. Chem. Lett.*, 1996, **7**, 131 (*Chem. Abstr.*, 1996, **124**, 343 865).

18　N. Sulochana, P. Nagarajan and C. M. Kamalina, *J. Indian Chem. Soc.*, 1996, **73**, 145 (*Chem. Abstr.*, 1996, **125**, 114 963).

19　Y. Yamauchi, H. Maeda and H. Ohmori, *Chem. Pharm. Bull.*, 1996, **44**, 1021.

20　M. Takeuchi, T. Mizuno, H. Shinmori, M. Nakashima and S. Shinkai, *Tetrahedron*, 1996, **52**, 1195.

21　T. Imada, H. Kijima, M. Takeuchi and S. Shinkai, *Tetrahedron*, 1996, **52**, 2817.

22　S.-P. Hong, A. Sano and H. Nakamura, *Anal. Sci.*, 1996, **12**, 491 (*Chem. Abstr.*, 1996, **125**, 143 154).

23　Y. Shirokane, A. Arai, R. Uchida and M. Suzuki, *Carbohydr. Res.*, 1996, **288**, 127.

24
Synthesis of Enantiomerically Pure Non-carbohydrate Compounds

1   Carbocyclic Compounds

Intramolecular Horner-Wadsworth-Emmons (HWE) reactions have been used to prepare carbohydrate-annulated cyclopentenones and cyclohexenones. Thus, 3-C-allyl-3-deoxyglucose derivative **2** was prepared following ring opening of epoxide **1**, elaborated to cyclization substrates **3** or **5**, which then underwent HWE reactions to afford the fused bicyclic compounds **6** and **4**, respectively (Scheme 1). This work is conceptually very analogous to the earlier work of Jenkins and co-workers who prepared such fused cyclopentenones using aldol addition-eliminations in place of HWE methods for ring closure (Vol.29, p.359). Analogous chemistry from the D-*allo*-epoxide isomer of **1**, proceeding *via* the C2 allyl intermediate from axial epoxide opening, provides a route to the isomeric cyclopentenone and cyclohexenone.[1]

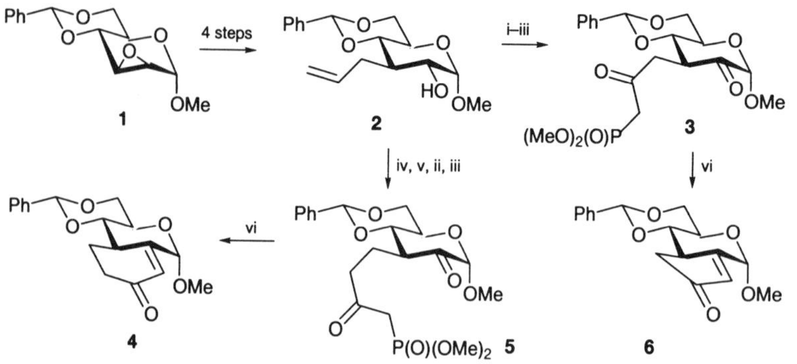

Reagents: i, OsO_4, NMNO; ii, $(MeO)_3P(O)CH_2Li$; iii Dess-Martin periodinane; iv, $BH_3.DMS$; v, PCC; vi, K_2CO_3, 18-crown-6, 60 °C.

Scheme 1

A number of radical cyclizations of carbohydrate-derived intermediates have been reported this year for synthesis of carbocycles. One interesting example of tandem radical cyclizations involves the first example of a product derived by a formal radical [2+2+2] cycloaddition. Thus, reaction of **7** using triphenyltin radical gave a mixture of **8**, **9** and formal [2+2+2] product **10**, while the bispropargyl analogue **11** gave a mixture of **12** and [2+2+2] product **13**.[2]

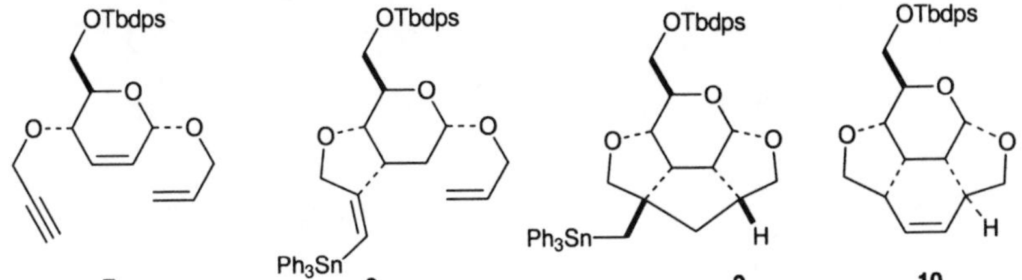

Two examples of radical cyclizations of open-chain carbohydrate derivatives have been reported. The first involves elaboration of 3-deoxyglucose derivative **14** to **15**, which, under radical deoxygenation conditions cyclizes to **16**, an intermediate for synthesis of prostaglandin IPF$_{2\alpha}$-I (Scheme 2).[3] An aminocyclitol precursor **18**, related to part of the glucosidase inhibitor trehazolin, has been prepared from D-mannose by radical atom transfer reactions of the oximino acetylene **17**. Isomeric analogues of the ultimate aminocyclopentitols such as **19** were also prepared. (Scheme 3). This methodology was also applied to a variety of other aminocyclopentitol analogues.[4]

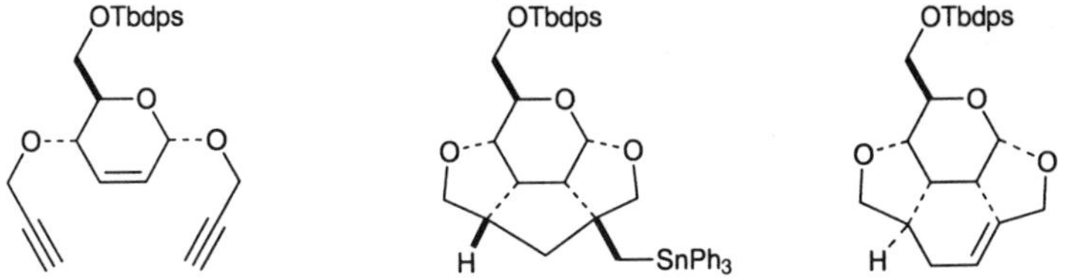

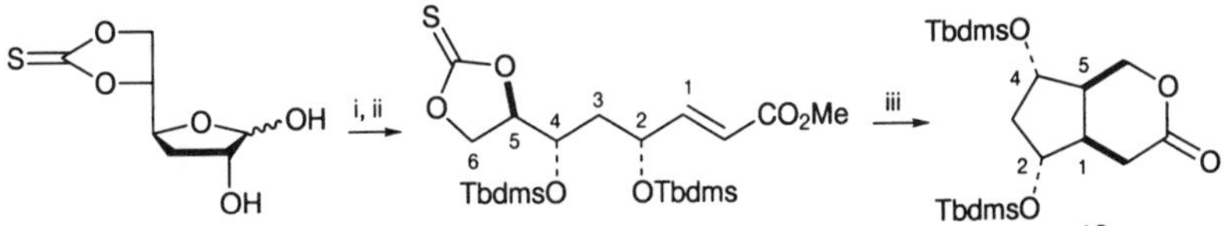

Reagents: i, Ph$_3$P=CHCO$_2$Me; ii, TbdmsCl, ImH; iii, Bu$_3$SnH, AIBN.

Scheme 2

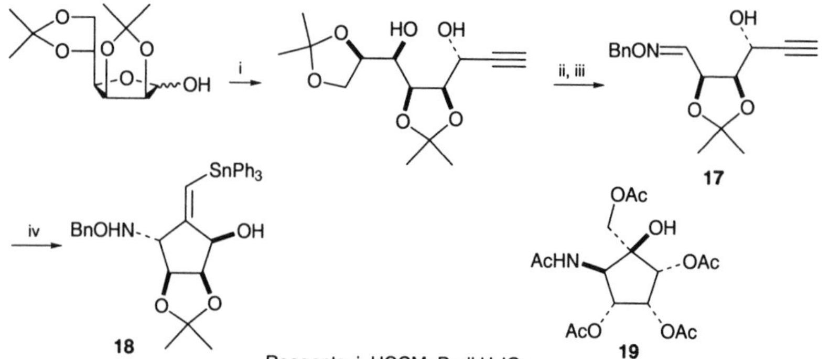

Reagents: i, HCCMgBr; ii,H$_5$IO$_6$;
iii, NH$_2$OBn; iv, Ph$_3$SnH, Et$_3$B.

Scheme 3

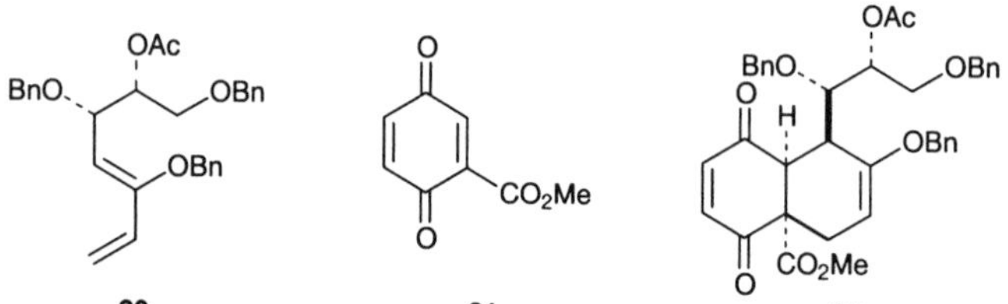

Two examples of electrocyclic reactions for synthesis of carbocycles were reported. Diels-Alder reaction of D-glucose-derived diene **20** with quinone **21** gave **22**, which epimerized on base catalysis to the *trans*-fused analogue.[5] Levoglucosenone-derived enediene **23**, was elaborated to triene **25** (*via* **24**), thermal [3+3] rearrangement giving cyclohexadiene **26**, then converted to **27**, an

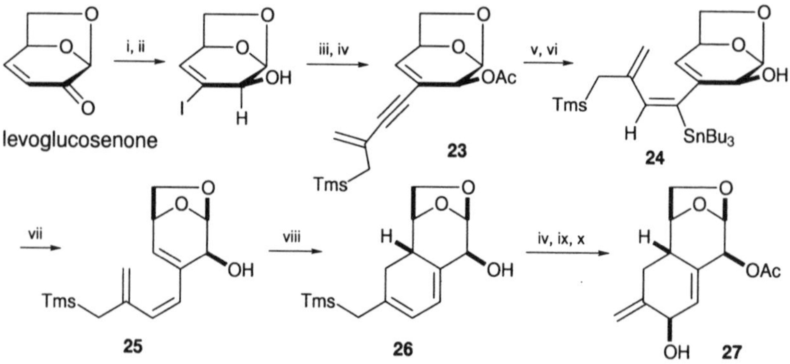

Reagents: i, I$_2$, Py; ii, NaBH$_4$, CeCl$_3$; iii, HCC-C(=CH$_2$)-CH$_2$Tms, Pd(OAc)$_2$, PPh$_3$, CuI, *n*-BuNH$_2$; iv, Ac$_2$O; v, Bu$_3$SnH, NiCl$_2$(PPh$_3$)$_2$; vi, K$_2$CO$_3$, MeOH, 30 min.; vii, K$_2$CO$_3$, MeOH, 24 h; viii,PhMe, D; ix, *m*-CPBA; x, 2N HCl.

Scheme 4

intermediate in a synthesis of tetrodotoxin (Scheme 4). Other similar cyclizations were described.[6]

2 Lactones

L-Erythrulose has been converted into lactone **28** and also epoxyalcohol **29** (as intermediates for natural product synthesis).[7]

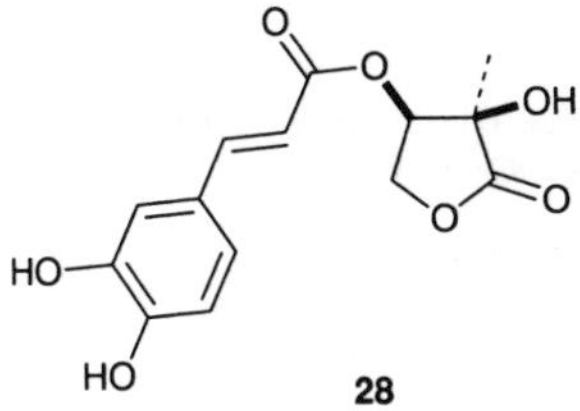

A formal total synthesis of canadensolide **33** has been achieved starting from L-arabinose-derived epoxide **30**, converted to lactone **31** by known methods (Vol. 26, p. 159). Conversion of **31** to known canadensolide intermediate **32** involved debenzylation then Wittig reaction (Scheme 5).[8]

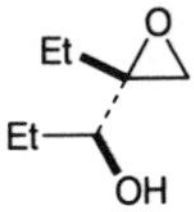

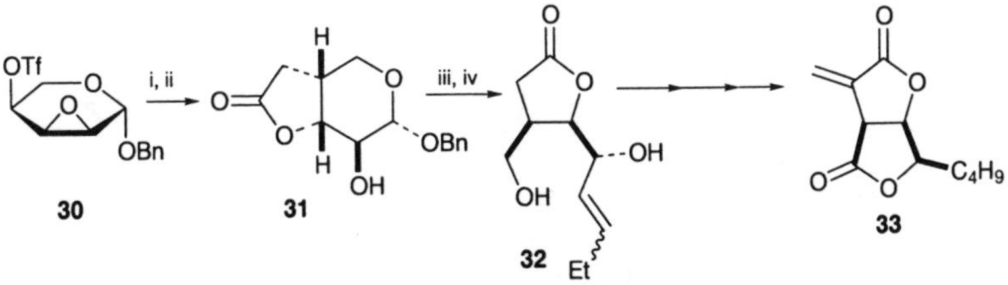

Reagents: i, LiCH$_2$CO$_2$Bu-*t*; ii, TFA; iii, H$_2$, Pd(OH)$_2$/C; iv, CH$_3$CH$_2$CH=PPh$_3$.

Scheme 5

The conformationally constrained diacyl glycerol analogue **38** was prepared by samarium diiodide mediated reaction of known ketone **34** with methyl acrylate to give **35**. Acid catalysed delactonization followed by periodate cleavage then gave **36**, which was elaborated, *via* **37**, to **38** (Scheme 6). Similar chemistry was employed to prepare **39** (and its *E*-isomer), the side chain being introduced by aldol reaction.[9]

The same group further extended this chemistry to prepare lactones **40** and **41**, and also used a modified approach towards other diacylglycerol (DAG) analogues **44** and **45**. This latter approach involved synthesis of the spirocyclic lactone intermediate **43** by addition of a Grignard maganesium alkoxide to **42** and the PCC oxidation (Scheme 7).[10]

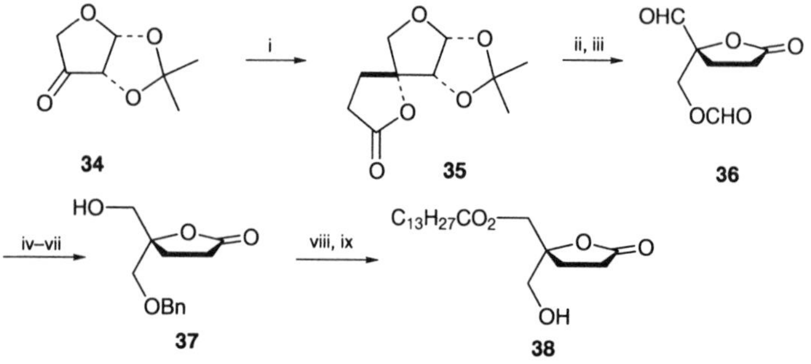

34   i →   **35**   ii, iii →   **36**

iv–vii →   **37**   viii, ix →   **38**

Reagents: i, SmI₂, CH₂=CHCO₂Me; ii, HCl, THF; iii, NaIO₄; iv, CH₂=PPh₃; v, BnBr, Ag₂C
vi, OsO₄, NaIO₄; vii, NaBH₄; viii, C₁₃H₂₇COCl; ix, BCl₃.

Scheme 6

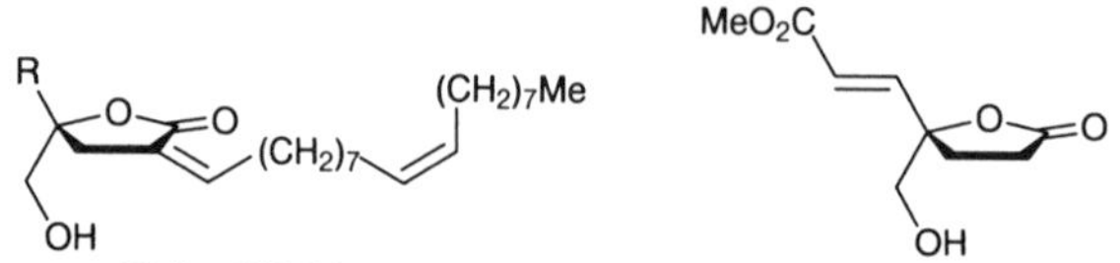

39 R = CH₂OAc
40 R = CH=CHCO₂Me

41

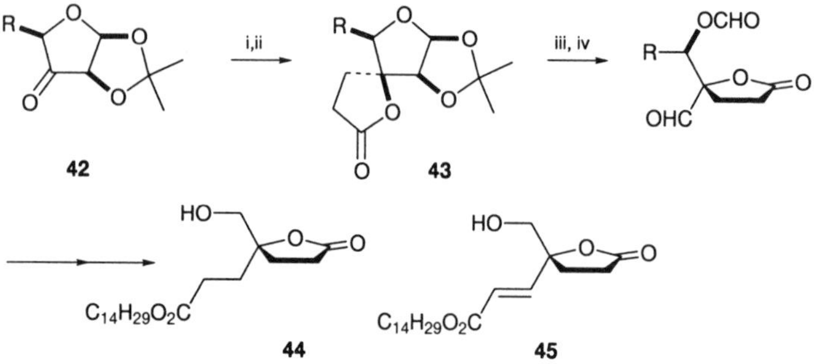

42   i,ii →   **43**   iii, iv →

→   **44**   **45**

Reagents: i, ClMgO(CH₂)₃MgCl; ii, PCC; iii, HCl; iv, NaIO₄.   R = CH₂OTbdps

Scheme 7

Other conformationally constrained diacylglycerol analogues were also prepared by Marquez's group. Thus, **46**, derived from tri-*O*-acetyl-L-glucal, was
converted to **47** through deacylation-lactone isomerization, primary alcohol
protection, acylation with the C₁₄ acid and final debenzylation. Additionally, L-
galactonolactone was elaborated to **48**, epimeric with **47**, the key steps being a
samarium diiodide mediated deoxygenation (Scheme 8). Further analogues **49**
and **50** were also prepared from **34** through a Wittig reaction, Ireland-Claisen
rearrangement sequence.[11]

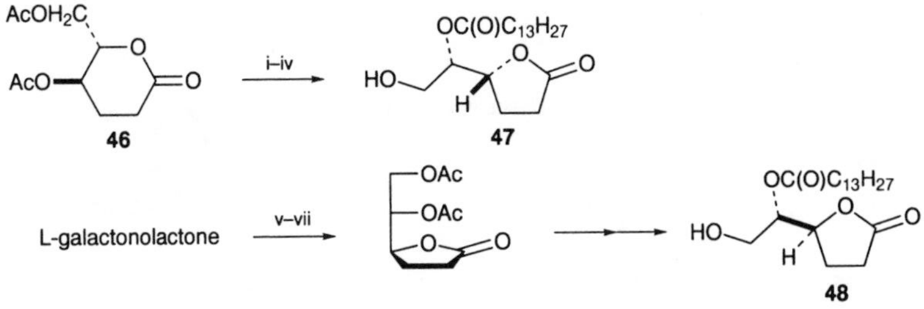

Reagents: i, HCl(aq.); ii, Bu$_2$SnO then BnBr, CsF; iii, C$_{13}$H$_{27}$COCl; iv, BCl$_3$, CH$_2$Cl$_2$, −78 °C; v, Ac$_2$O, Py; vi, H$_2$, Pd-C, Et$_3$N; vii, SmI$_2$, HMPA.

Scheme 8

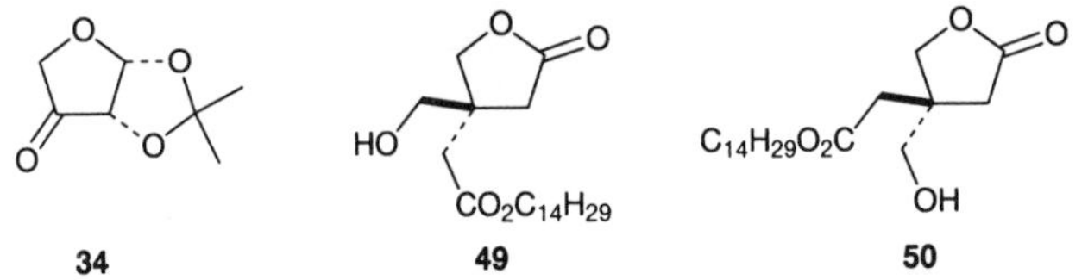

The monocerin analogue **54** was prepared in 5 steps from the known D-glucose derived enal **51**, key steps being the diastereoselective hydrogenation of **51**, cyclization of **52** through *C*-glycosidation and final benzylic oxidation of **53** (Scheme 9).[12]

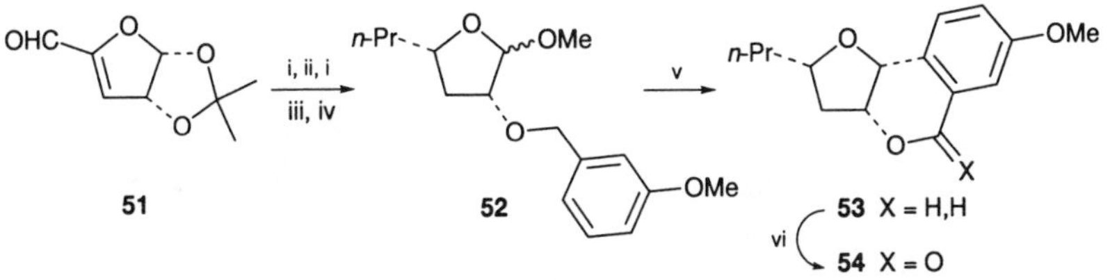

Reagents: i, H$_2$, Pd-C; ii, Ph$_3$P=CHMe; iii, MeOH, H$^+$;
iv, *m*-MeOC$_6$H$_4$CH$_2$Cl, NaH; v, SnCl$_4$, CH$_2$Cl$_2$; vi, CrO$_3$, Py.

Scheme 9

3   Macrolides, Macrocyclic Lactams and Their Constituent Segments

Several analogues of erythromycin containing a nitrogen atom in the macrocyclic ring – 'azalides' – have been reported. For example, **57**, was prepared from **56**, a degradation product of erythromycin, and **55** derived from D-xylose. The key steps in ring formation involved reductive amination and Mitsunobu lactonization.[13] An epimer of **57** was prepared analogously from a D-lyxose derivative.

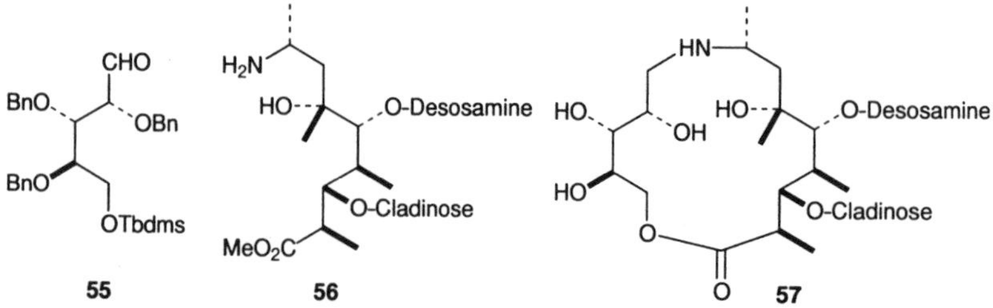

55   **56**   **57**

Compound **59**, the C_8-C_{19} part of the immunosuppressant agent rapamycin, was prepared from diethyl dithioacetal **58**.[14]

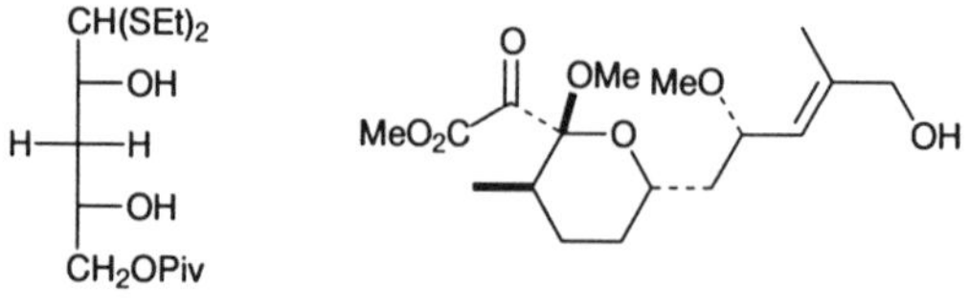

58   **59**

The dibranched glycoside **62** from which the C18-C23 subunit of macrolide lasonolide A (**63**) can be obtained was prepared in several steps from **60** (see Chapter 14 for synthesis) *via* epoxide **61**. Axial epoxide opening with LAH followed by oxidative alkene cleavage and reduction converts **61** to **62**.[15]

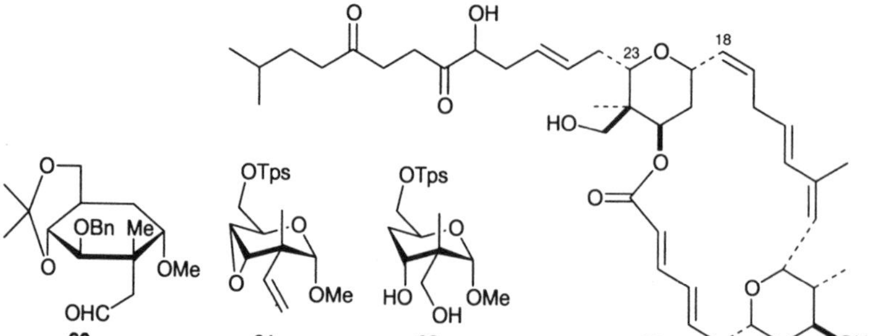

60   **61**   **62**   **63**

4   Other Oxygen Heterocycles, Including Polyether Ionophores

4.1   Five- to Ten-Membered O-Heterocycles – (-)-Allo-muscarine **66** was prepared in a number of steps from D-glucose, *via* the known **64**. Key steps are inversion of the C-3 hydroxyl group through benzoate displacement and deoxygenation of C-2 and C-5 *via* the reduction of cyclic sulfide **65** (Scheme 10).[16]

In the area of dioxolanes, enantiomeric chiral equivalents of glycolic acid **67** and **68** were prepared from D-mannitol *via* tosylates **69** and **70** (bis-acetalization

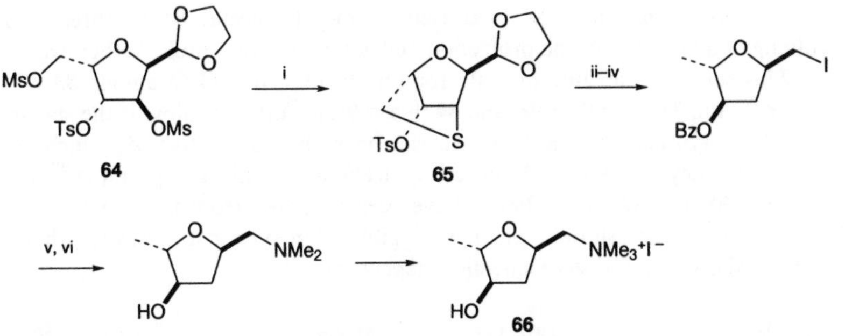

Reagents: i, NaHS; ii, Raney Ni; iii, KOBz, DMF; iv, deprotection then iodination; v, $NHMe_2$, EtOH; vi, NaOH, EtOH.

Scheme 10

of D-mannitol, lead tetraacetate cleavage, reduction and tosylation) which underwent elimination (with *t*-BuOK) and ozonolysis. The enolates of **67** and **68** can be diastereoselectively alkylated.[17]

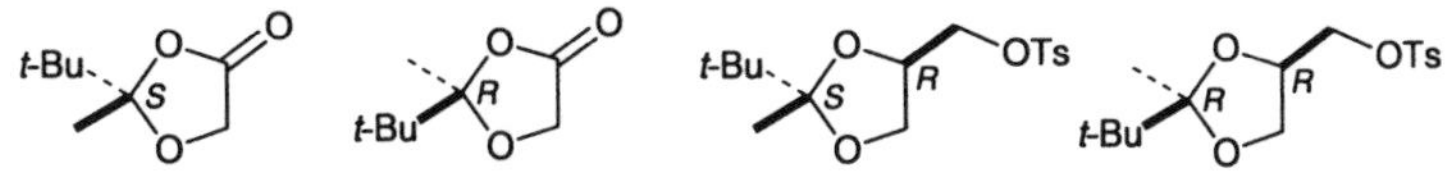

67 **68** **69** **70**

Reductive elimination of sugar tosyl hydrazones using sodium hexamethyldisilazide/LAH provides a route to various homochiral dioxolanes. Thus, D-xylose, D-arabinose and D-xylose derivatives **71-73** were converted to isopropylidene acetals **74-76**, while D-xylose derivatives **77** and **78** provided 6-membered acetals **79** and **80**, respectively.[18]

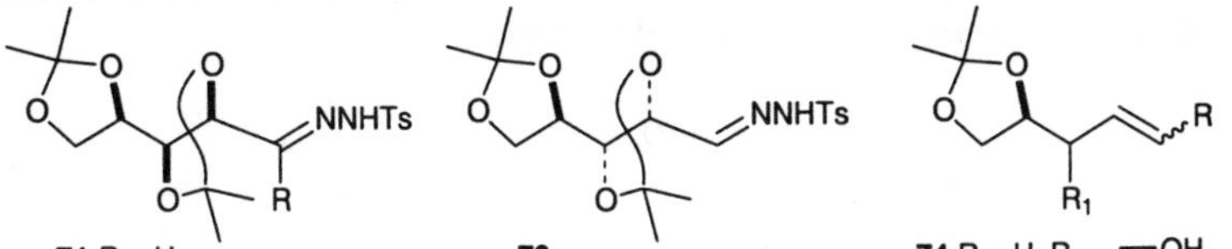

71 R = H
72 R = Me

73

74 R = H, R1 = —OH
75 R = H, R1 = ''''''OH
76 R = Me, R1 = —OH

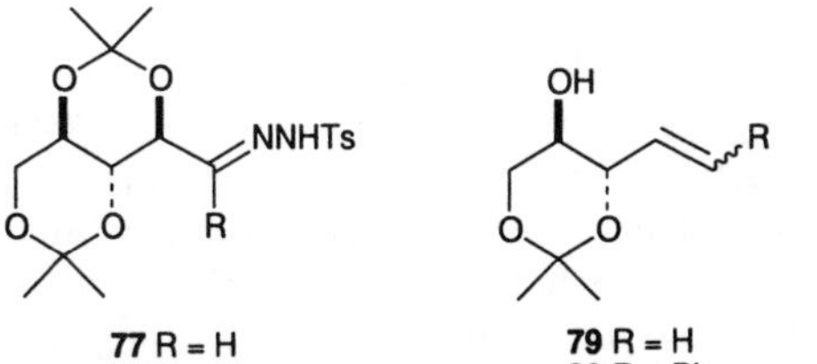

77 R = H
78 R = Ph

79 R = H
80 R = Ph

6-*O*-Pivaloyl-D-galactal was converted to bicyclic acetal **81** in three steps involving O-3,O-4 stannylene protection and activation, and radical ether cyclization. This served as an intermediate for furanosides **82** and furanoids **83** (*E*:*Z* 66% for R=Ph, 31% for R=*i*-Pr) and **84** (β:α = 9:1).[19] L-Erythrulose is the starting material for synthesis of the furan **86** *via* the known derivative **85**, which was treated with dicyanomethane followed by acetic anhydride and pyridine.[20] Two analogues, **88** and **89**, of LL-P880γ have been prepared from D-glucose-derived **87**. These analogues showed potency in gibberellin-synergistic assay with rice seedlings, showing a C1′ stereochemical dependence.[21]

| **81** | **82** R = Me, Oct, 4-penten | **83** R = Ph, *i*-Pr | **84** |

| **85** | **86** | **87** | **88** R = Et **89** R = Bu |

Palladium catalysed alkylation of 1,6-anhydro-2-chloro-2,3,4-trideoxy-β-D-*erythro*-hex-3-enopyranose (**90**) with a variety of activated 1,3-dicarbonyl nucleophiles provides mostly **91** (75-85%) with some of the minor regiosiomer **92**. One of these major products, **93**, was then elaborated to **94**, a well known intermediate in synthesis of thromboxane A₂, B₂ and analogues.[22]

| **90** | **91** | **92** | **93** | **94** |

Allylation of 3,4-diacetyl-D-arabinal or 3,4-diacetyl-D-xylal (with BF₃, CH₂=CHCH₂Tms) gives *C*-glycoside **95** in high yields (90-96%) and with high diastereoselectivity (29:1 to 36:1). This was then elaborated with stereochemical inversion at C-4, palladium-catalysed allylic substitution and Wacker oxidation to **96**, a precursors to pseudomonic acid **97**.[23] Two precursors of the spiro-fragment of tautomycin, **100** and **101**, were prepared from **98** and **99** respectively.[24]

95   96   97

98 X = H
99 X = OAc   100   101   102   103   104

Diisopropylidene-D-glucose is the starting material for a synthesis of **104**, a model for the side chain of terpentecin. The known intermediate **102** underwent Grignard addition and Barton deoxygenation followed by osmylation, primary OH tosylation and acetal deprotection to give **103**. Subsequent key steps were spiro-epoxide formation with sodium methoxide and secondary OH oxidation proceeding with temporary anomeric silylation (with TbdmsOTf) to give **104**.[25]

105 X = —OH
106 X = ---OH   107   108   109   110 X = —Me
111 X = ---Me   112 X = H
113 X = —Me
114 X = ---Me

Carbohydrate epoxyalcohols are converted to ketones by triflate formation, reaction with pyridine, sodium borohydride reduction, and finally acidification. For example, epoxide **105** or **106** are converted to **108** by reduction and then hydrolysis of the pyridinium salt **107**. Similarly, epoxyalcohols **109**, **110** and **111** are converted to **112**, **113** and **114** respectively.[26]

The 9-membered cyclic ethers **116** have been prepared *via* oxy-Cope rearrangement of **115**, ultimately derived from D-glucose.[27] The 10-membered bis-ether **118** was prepared *via* Bergman cyclization of the bis-acetylene **117** (R=2,3,4,6-tetra-*O*-benzyl-β-D-glucosyl). Solution conformational analysis was reported from NMR experiments. Dialkyne **117** was prepared by Pd-catalysed coupling of the β-D-glucosyl-1-alkyne and 1,2-diiodobenzene, and a mechanism was proposed for formation of **118** involving H radical abstraction from the 2-*O*-benzyl ethers by the Bergman diradical intermediate.[28]

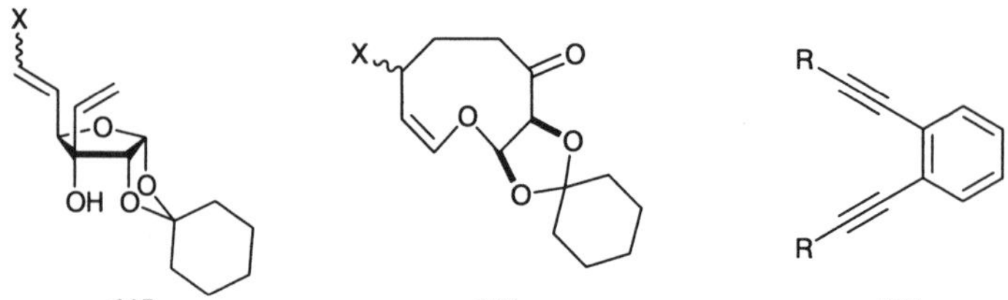

115   116   117

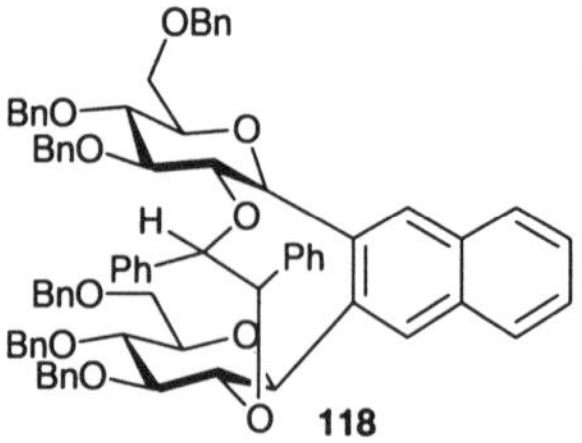

X = CHO, CN, CO$_2$Et, CO$_2$Me

118

A total synthesis of (+)-lanomycin (**121**) has been reported, key steps being asymmetric crotylation of the precursor aldehyde (using Roush's tartrate boronate) generating **119**, and then elaboration to the *C*-glycoside **120** (Scheme 11).[29]

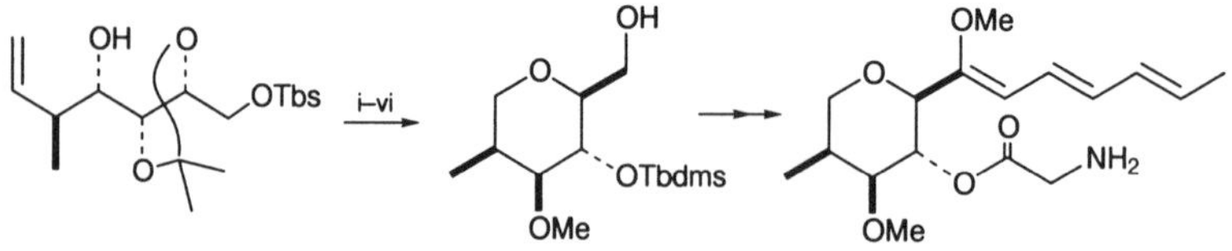

119   120   121

Reagents: i, NaH, MeI; ii, HCl, THF; iii, 2,4,6-Me$_3$C$_6$H$_2$SO$_2$Cl, Et$_3$N; iv, NaH, THF, then TbdmsCl; v, O$_3$, MeOH then NaBH$_4$; vi, BF$_3$.OEt$_2$.

Scheme 11

L-Arabinose is the starting material for a synthesis of *C*-glycoside **123**, corresponding to the C1-C8 fragment of the antifungal agent ambruticin. The key ring-forming step involves intramolecular reaction of the secondary OH freed in reaction with boron tribromide of the intermediate enol ether derived by Wittig reaction of **122** (Scheme 12).[30]

An analogue (**126**) of the immunosuppressant PA48153 (with a ring methoxy in place of the ethyl group of the natural product) has been prepared from D-galactose derivative **124**. Homologation was achieved by Wittig reaction, with subsequent reasonably diastereoselective hydroboration at C-6 and further

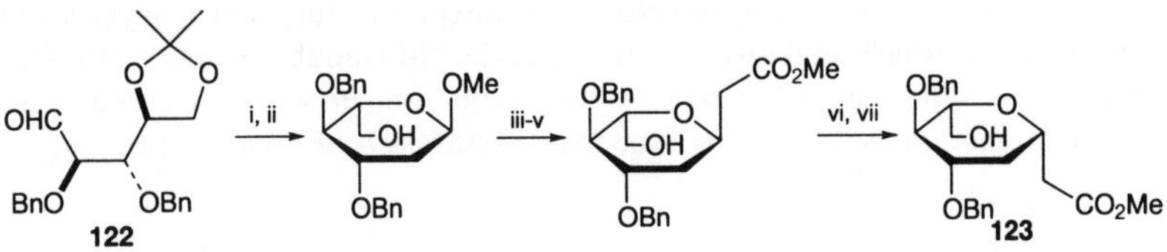

Reagents: i, LDA, MeOCH₂PPh₃Br; ii, BBr₃; iii, BF₃.OEt₂, allylTms; iv, O₃, MeOH, DMS; v, PDC, MeOH; vi, NaOMe, MeOH; vii, MeOH, HOBT, EDC.

Scheme 12

homologation followed by dideoxygenation (PhSiH₂/AIBN) of intermediate bis-xanthate **125**.[31]

1-*C*-Allyl-2,3,4,6-tetraacetyl-α-D-glucose has been converted into **130**, a degradation product of the zooxanthellatoxins (marine toxins) which established their absolute stereochemistry. Wacker oxidation of **127** allows subsequent ketalization (deacylation then ketal/acetal forming conditions) to provide intermediate **128**, which after homologation using an appropriate lithium acetylide then reduction provides **129**, which is then a substrate for using Sharpless AD (on the derived aldehyde) to introduce the side chain chiral centres and directly afford the side chain cyclic hemiacetal of **130**.[32]

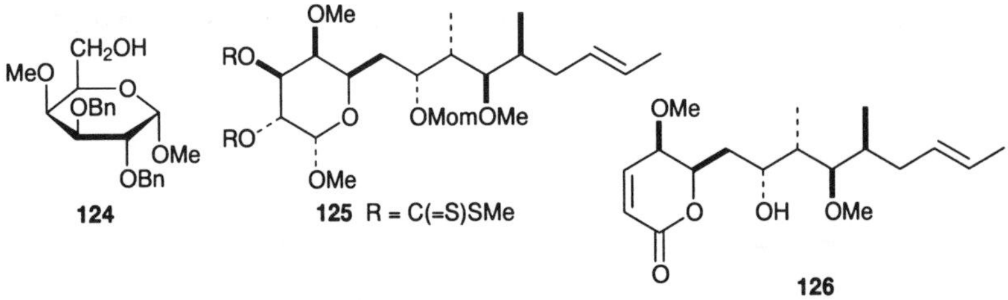

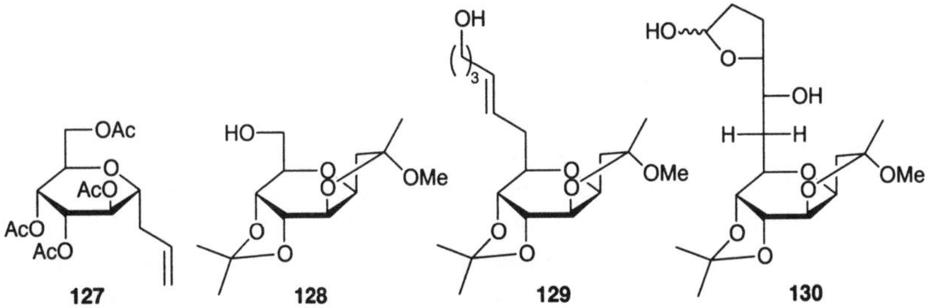

4.2 Polycyclic O-Heterocycles – The lactone **132** was prepared from L-xylose-derived lactone **131** by bromoacylation, Wittig-type cyclization and epimerization.[33] This lactone was then converted to a mixture of furanoid *C*-glycals **133** and **134** (3:1) which underwent intramolecular Mitsunobu reaction to give furanofurans which proved separable. The major isomer was hydrogenated to give the furanofuran subunit **135** of the toxin erythroskyrine (Scheme 13).

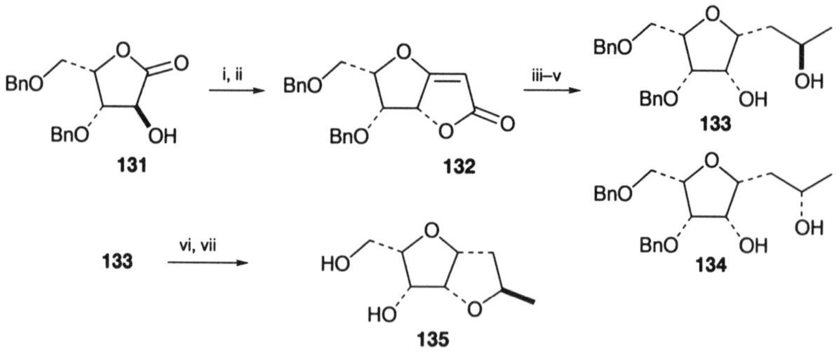

Reagents: i, BrCH₂CO₂Bn, Py; ii, PPh₃ then DBU; iii, H₂, Rh-Al₂O₃; iv, DIBAL; v, MeMgCl; vi, PPh₃, DEAD; vii, H₂, Pd-C.

Scheme 13

Deoxygenation of one of the separated mono-xanthates of 1,4:2,5-dianhydro-D-glucitol provides **136**, an intermediate used for synthesis of non-peptidyl HIV-1 protease inhibitors, such as **137**.[34]

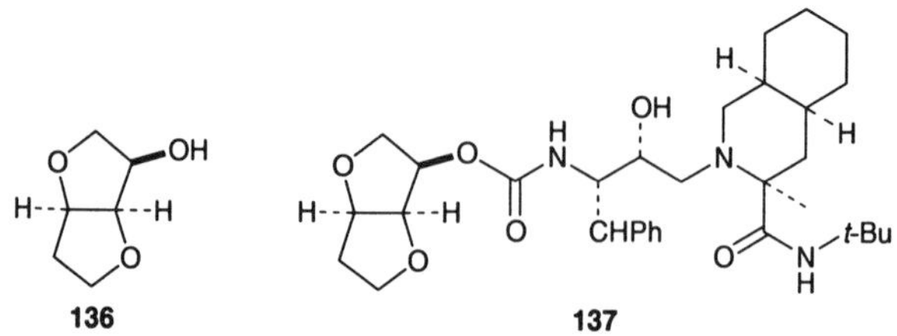

Base catalysed (catalytic DBU in trifluoroethanol) ring contractions of pyranoid *C*-glycosides provides a new route to some 2,5-dihydrofurans and tetrahydrofuro[3,2-*b*]furans. Thus, **138-140** were converted to **141-143**, though yields were generally modest (<50%). The mechanism proposed in the case of the bicyclic products, for example, involves retro-Michael ring opening, silyl migration (in the case of **138**) and conjugate addition recyclization onto the intermediate α,β,γ,δ-system, proton transfer and then final hydroxyl conjugate addition.[35]

Furan fused carbohydrates **144** and **145** have been prepared by Grignard additions to C-4 aldehyde-bearing hexopyranosides, followed by intramolecular displacement of the triflate derivatized alchol by the 6-OBn goup. Similar cyclization chemistry was used to convert **146** to **147**.[36]

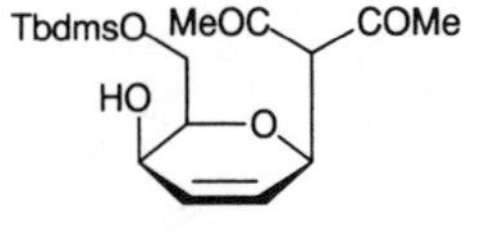

138 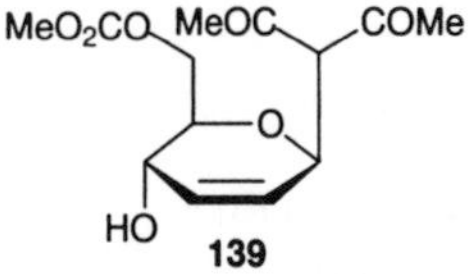 139 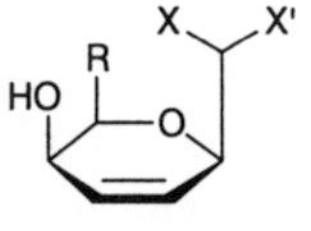 140

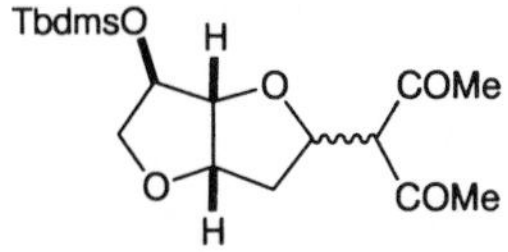

141 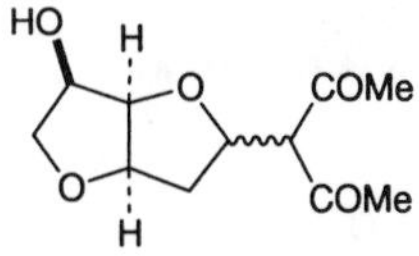 142 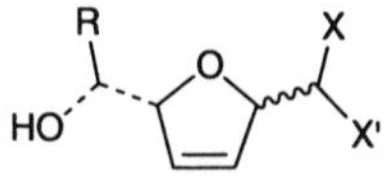 143

R = H, Me, CH₂OTbdms,
X = COMe, CO₂Me
X' = COMe or NO₂

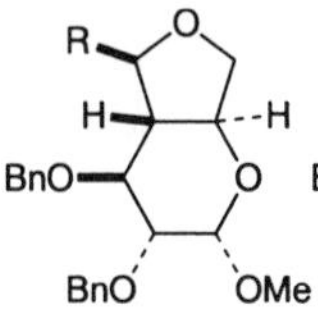

144 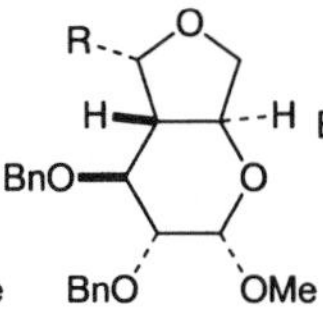 145 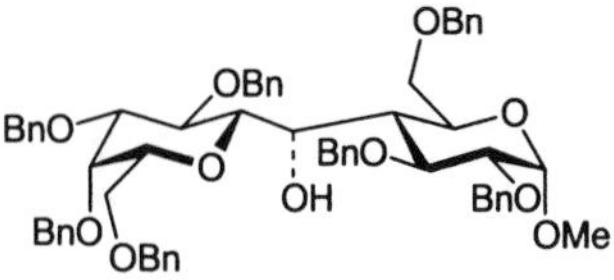 146 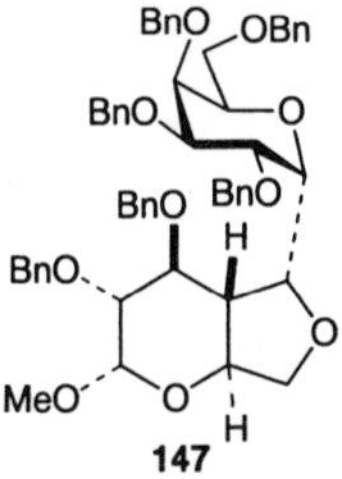 147

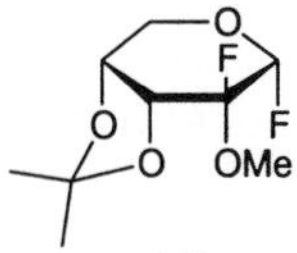

148 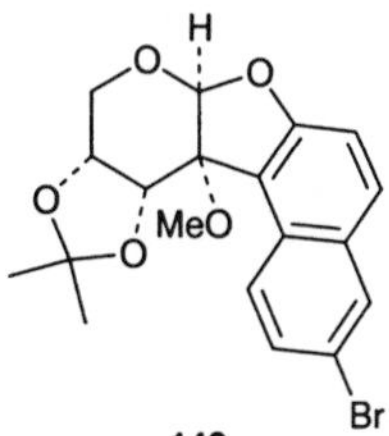 149 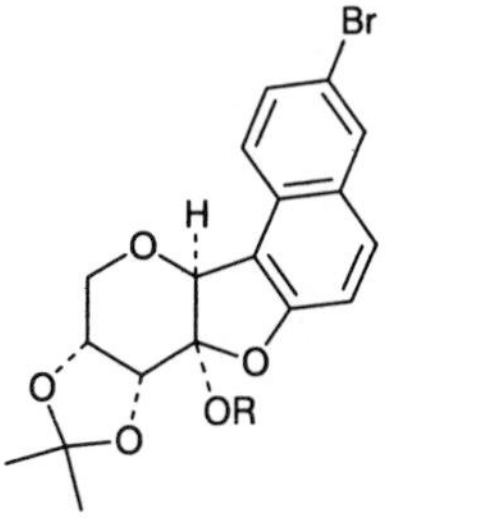 150 R = 6-Br-2-naphthyl

Reaction of 2-fluoroglycosyl fluoride **148** with 6-bromo-2-naphthol with Cp₂HfCl₂-AgOTf catalysis affords the fused carbohydrate-containing tetracycles **149** and **150** in 44% and 21% yields respectively. Similar chemistry was applied to phenol derivatives.[37]

Intramolecular Heck reactions (5-8% Pd(OAc)₂, PPh₃) of hex-2-enopyranosides bearing tethered aryl halides (**151**) or vinyl bromides (**152** and **153**) provide a route to a number of fused pyrans (**154**, **155** and **156** respectively). Gluco analogues of such enopyranoside, however, do not cyclize because they lack the necessary syn β-H for β-elimination in the Pd-alkyl intermediate.[38]

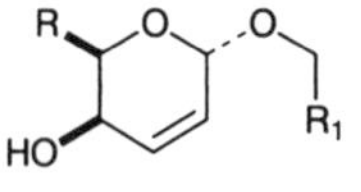

151 R₁ = 2-iodophenyl, R = Me, CH₂OH
152 R₁ = C(Br)=CH₂, R = CH₂OH
153 R₁ = 2-bromocyclopentenyl,
         2-bromocyclhexenyl, R = CH₂OH

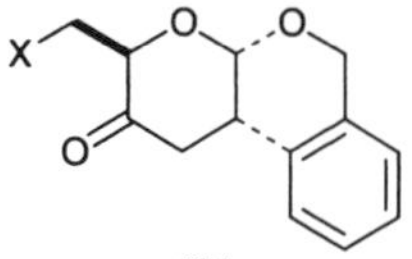

154
(X = H, OH)

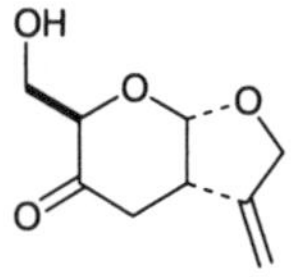

155

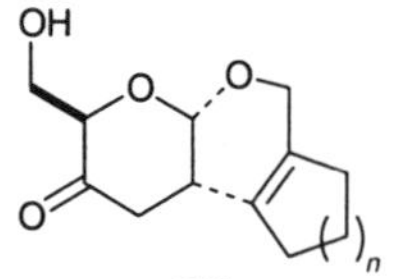

156

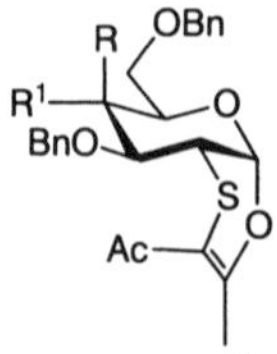

157 R = H, R¹ = OBn   **159**
158 R = OBn, R¹ = H

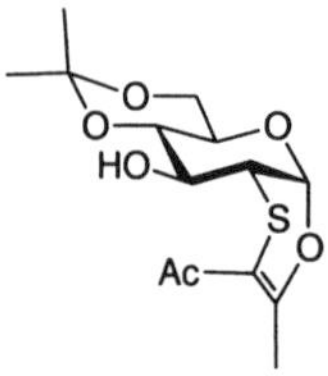

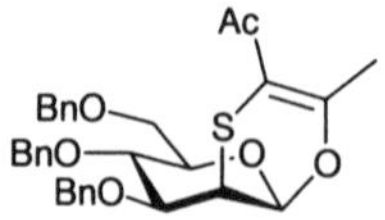

160

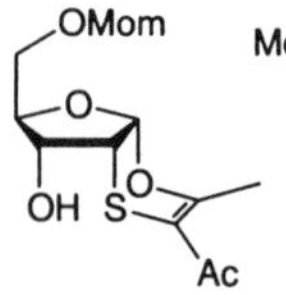

161

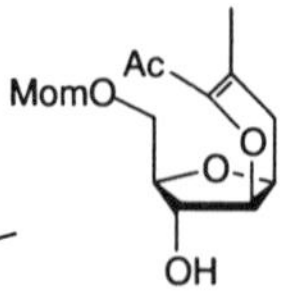

162

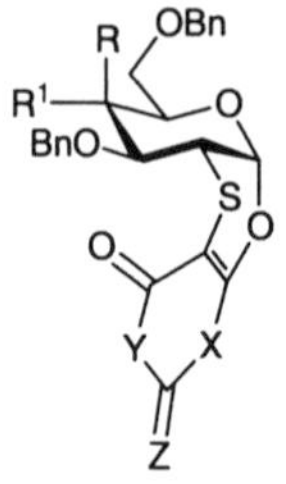

163 X = Y = NMe, Z = O
164 X = CH₂, Y = O, Z =

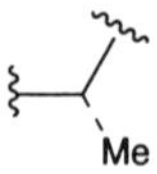

R = H, OBn   R¹ = OBn, H

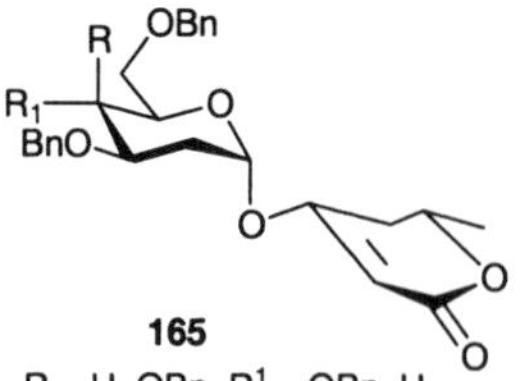

165
R = H, OBn   R¹ = OBn, H

A variety of glycals undergo hetero-Diels Alder reactions with pentan-2,4-dione-3-thione (and pyridine or lutidine catalysis) to provide a range of fused heterocycles **157-162**. Similarly, analogous cyclic hetero-dienes react with tri-*O*-benzyl-D-glucal or -galactal to give tricyclic products **163** and **164**. Raney-Ni treatment of the latter leads to the glycoside **165**. Thus, the overall Diels Alder-desulfurization process provides a stereocontrolled method for glycoside synthesis.[39]

Sinaÿ's group have used silicon tethering to facilitate radical addition reactions for construction of *C*-disaccharides. Thus, **166**, prepared by reacting the constituent monosaccharide derivatives with BuLi and dichlorodimethylsilane, undergoes radical reaction (AIBN) followed by cleavage of the silyl tether (TBAF) to provide *C*-disaccharide **167** (see also Chapter 3) which is related to the carbon skeleton of the herbicidins.[40]

166   **167**

Two syntheses of (-)-hongconin **170** have been reported. Both rely on the conjugate addition of the anion generated from **168** to either levoglucosenone[41] or to **169**.[42] In the former case, reductive opening of the anhydro sugar followed by reductive deoxygenation of the original sugar's hydroxymethyl group (by mesylation, iodination and treatement with Zn,Cu, AcOH) is required.

168   **169**   **170**

171 R = 4-MeOPh   **172**   **173**

A biomimetic synthesis of syringolide 2 **173**, a microbial elicitor, involves hydrogenation under acidic conditions of carbohydrate derivative **171** to give **172** which is directly converted to **173** by alumina in THF, the ease of this conversion suggesting that it is a possible model for biosynthetic construction of **173**.[43]

Intramolecular hetero-Diels Alder reaction of *in situ* generated *N*-benzyl nitrones derived from **174** and its C-3 epimer gives a mixture of **175-176** (50%, 18% and 32%) and of **177** and **179**, respectively.[44] Intramolecular hetero-Diels Alder reaction of nitrile oxides derived from **180** (*in situ* by treatment with NaOCl) gives **181** then converted to fused oxazepine **182**, related to the diterpenoid natural product, zoapatanol.[45]

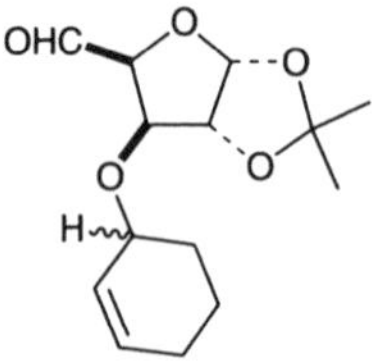

174 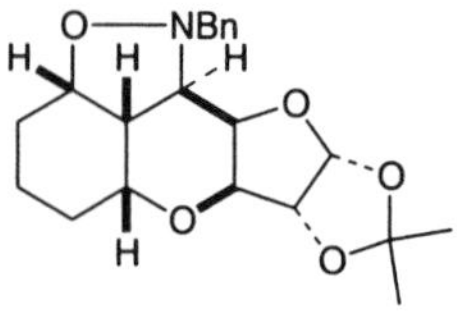   **175** 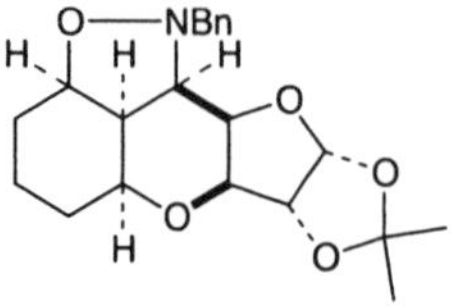   **176**

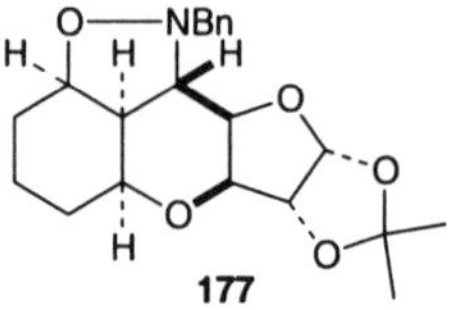

177 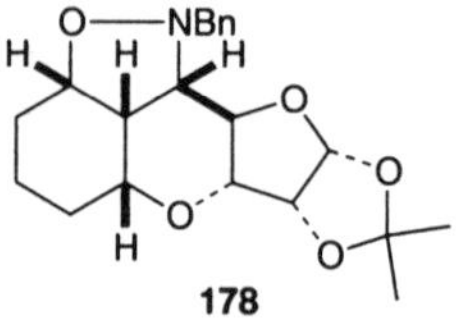   **178** 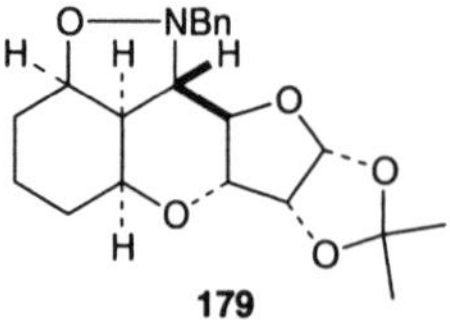   **179**

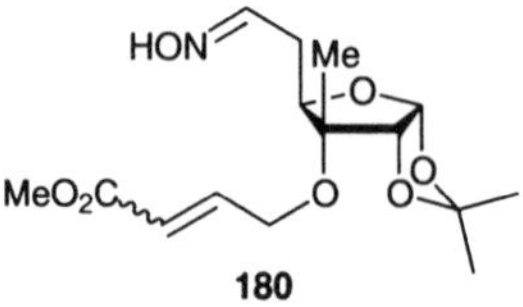

180 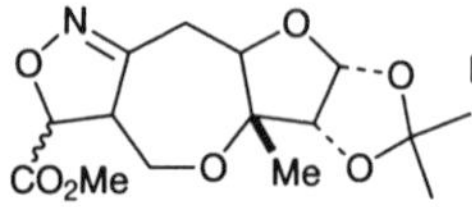   **181** 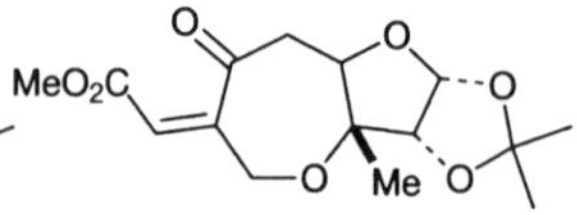   **182**

The bicyclic acetal **183** was prepared from D-glucose as a potential lipooxygenase inhibitor (Scheme 14).[46] Wittig coupling of the homologated deoxycarbohydrate derivatives **184** and **185** afforded **186**. Iodoetherification and concurrent glycoside ring opening using IDCP gave **187**; Wittig methylenation of the released aldehyde group, boron trifluoride catalysed etherification glycoside ring opening and a further Wittig methylenation provided bis-THF **188**. Similar chemistry with the alternate isomer from iodoetherification of **186** (i.e. addition of iodine to the other side of the alkene) gave **189**.[47]

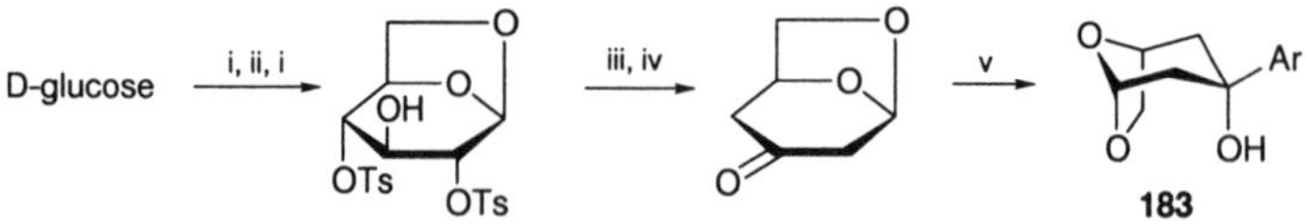

D-glucose $\xrightarrow{i, ii, i}$ $\xrightarrow{iii, iv}$ $\xrightarrow{v}$ **183**

Reagents: i, TsCl, Py; ii, DBU; iii, LiEt$_3$BH; iv, PCC; v, ArLi.

Scheme 14

The first total synthesis of staurosporine and of *ent*-staurosporine has been reported in full (see Vol. 29, p. 378). The glycal-derived epoxide **191** (from dimethyldioxirane epoxidation of the precursor glycal) was treated with the sodium salt of **190** and the product, **192**, was then elaborated using iodoamination as the the key cyclization step, radical deiodination then affording **193**, an intermediate for synthesis of *ent*-staurosporine.[48]

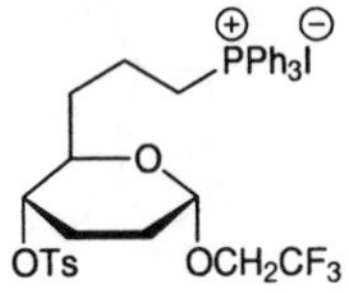

184 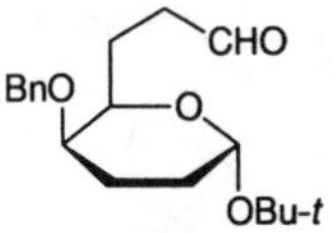 **185** 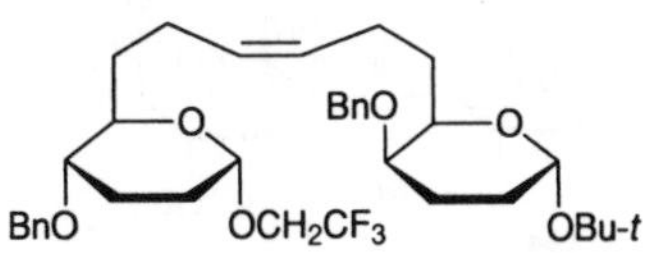 **186**

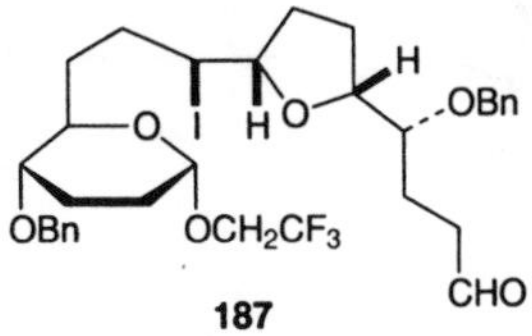

187

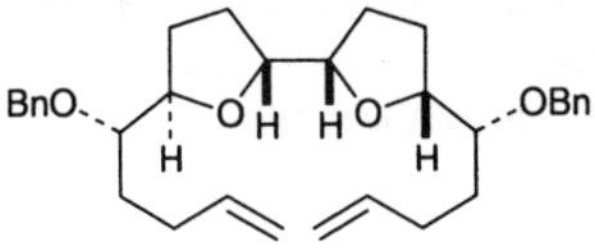

188 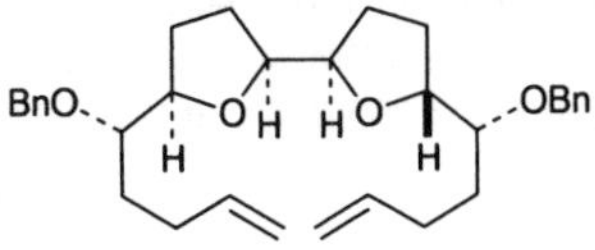 **189**

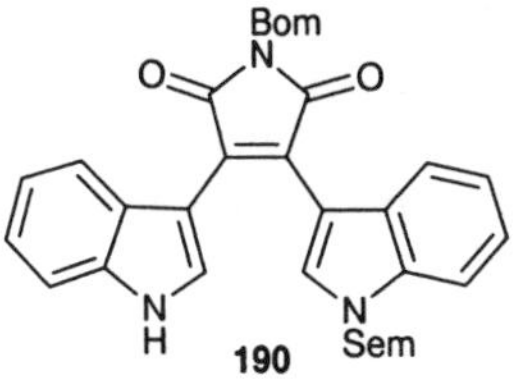

190 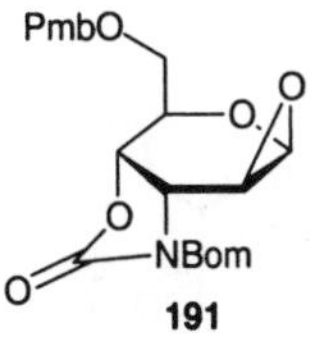 **191** 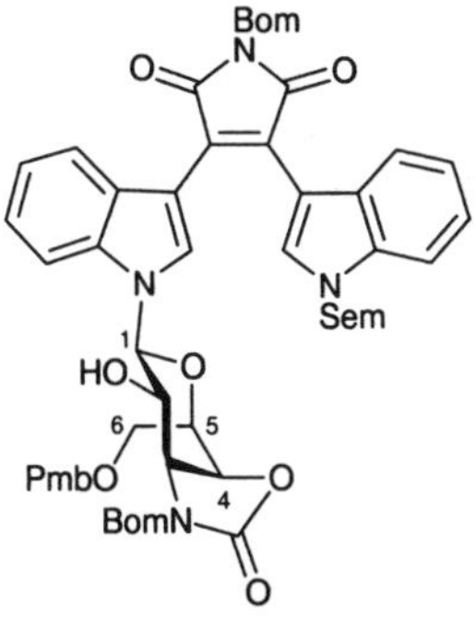 **192** 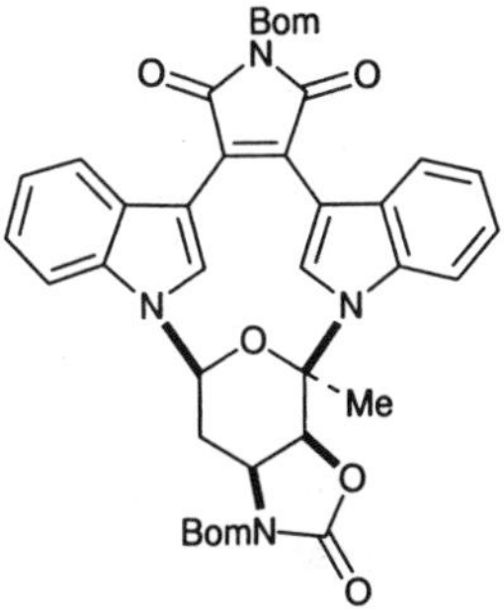 **193**

A review on the chemistry and biology of the zaragozic acids (squalestatins), including discussion of a number of approaches from sugars, has appeared this year [132 refs].[49] Heathcock's group have reported a complete total synthesis of zaragozic acid A *via* **195**, prepared from carbohydrate precursor **194** (Scheme 15).[50] The group had prepared **195** from the natural product and then reconverted it to the natural product to establish it as a viable relay compound for total synthesis.[51]

Complex polyether marine natural product toxins continue to attract synthetic attention, and this year has again seen several interesting papers in this area. The complete relative stereochemistry of maitotoxin has finally been resolved with the publication of a major paper by Kishi's group on the synthesis of various fragments of the natural product. D-Ribose was converted to **196**, containing the C10-C15 component (this synthesis utilized diastereoselective indium-mediated

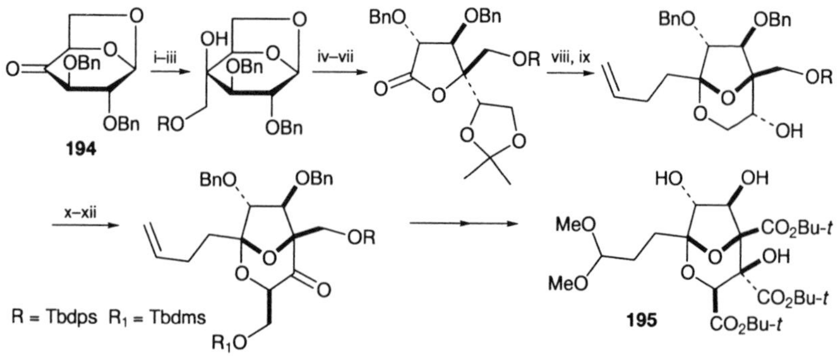

194

195

R = Tbdps R₁ = Tbdms

Reagents: i, Me₂(*i*-PrO)SiCH₂MgCl; ii, H₂O₂, MeOH, NaHCO₃, THF; iii, TbdpsCl, ImH, DMF; iv, TFA, Ac₂O; v, NaOMe, MeOH; vi, acetone, H⁺; vii, PDC; viii, CH₂=CHCH₂CH₂CeCl₂; ix, HCl, H₂O; x, DMSO, TFAA, Et₃N; xi, *t*-BuLi, CH₂O; xii, TbdmsCl, ImH, DMF.

Scheme 15

allylation and Roush asymmetric crotylboration reactions). All eight diastereomers about C12-C14 were prepared to allow identification by difference NMR of the natural product stereochemistry. A similar multiple diastereomer synthesis approach was applied for the preparation of **197** to identify the C35-C39 stereochemistry, **198** to identify the C63-C68 stereochemistry and **199** to identify the C134-C142 stereochemistry (diastereomer diversity centres are all asterisked; starting carbohydrate materials are indicated).[52]

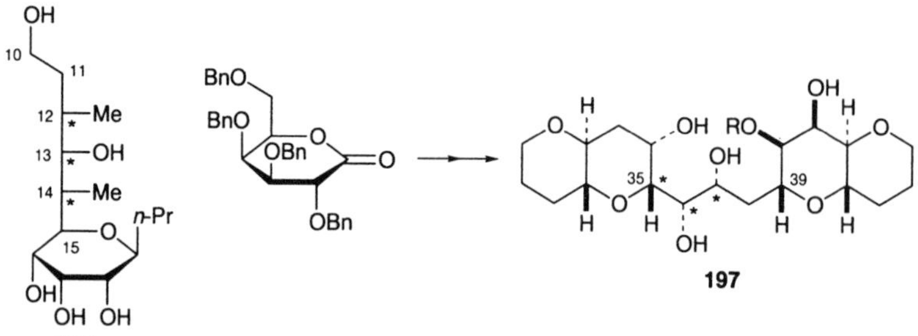

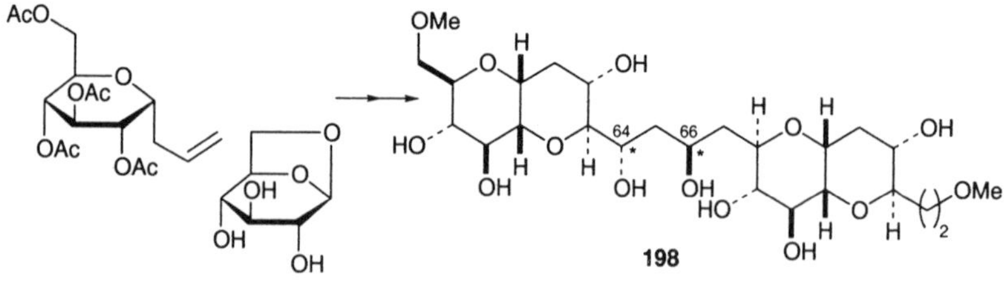

196

197

198

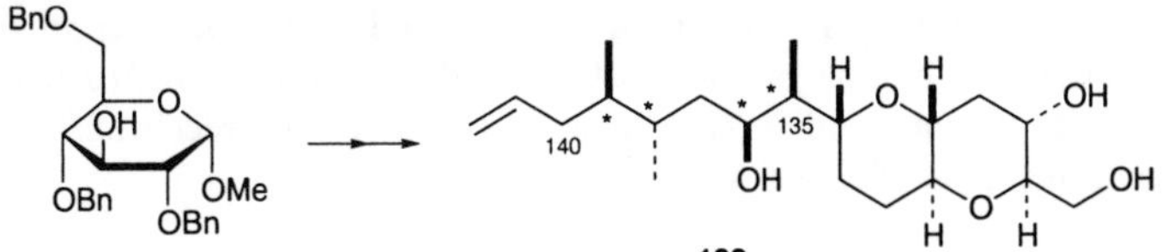

199

Ciguatoxin synthesis has been the subject of two carbohydrate-based approaches this year. D-Galactose was converted into the intermediate **201** *via* **200**, thence to **202** and on to **203**, the BC ring component of ciguatoxin. The enantiomer of **203** was also prepared.[53]

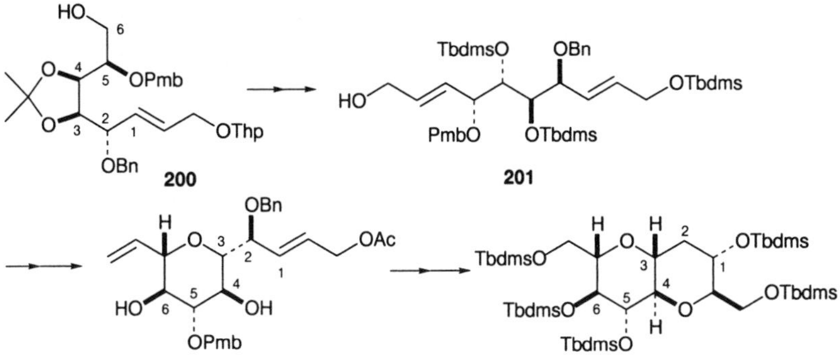

200   **201**

202   **203**

Another report focuses on developing methodology for fused medium ring ethers found in ciguatoxin using cobalt hexacarbonyl acetylene derivatives to activate adjacent carbocationic centres. Thus, **204** (ultimately derived from D-glucal) was converted to **205**, which on treatment with boron trifluoride undergoes cyclization of the sugar ring-derived hydroxyl group onto the propargylic cationic centre giving **206**. Reductive removal of the cobalt then reveals **207**. Analogous cyclization chemistry was demonstrated for the formation of 8- and 9-membered fused ethers of this type, all of which have relevance to ciguatoxin synthesis.[54]

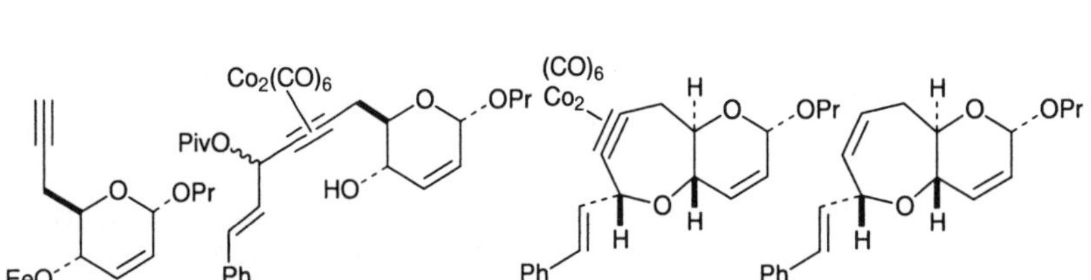

204   **205**   **206**   **207**

A synthesis of the C1-C13 segment of the halichondrins has been reported. Key steps involved the homologation of **208** to **209**, cyclization to **210** and then elaboration to the vinyl iodide **211** which has the C1-C13 stereochemistry of the halichondrins (P,P=cyclohexylidene).[55]

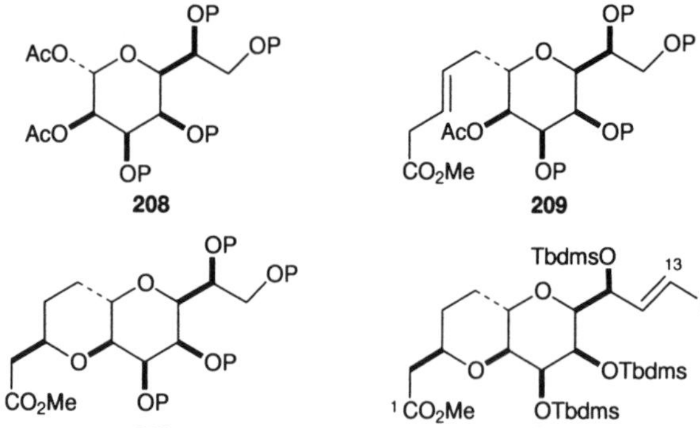

208

209

210

211

Nicolaou's group have developed copper(I)-promoted Stille coupling of glycal triflates, e.g. **212**, and C-1 stannylated glycals, e.g. **214**, for the synthesis of complex polyether marine toxin components. The methodology was applied to targets such as **213** (in which oxepines are coupled) and to maitotoxin component **215**. Couplings even for these complex products proceed in good yields.[56]

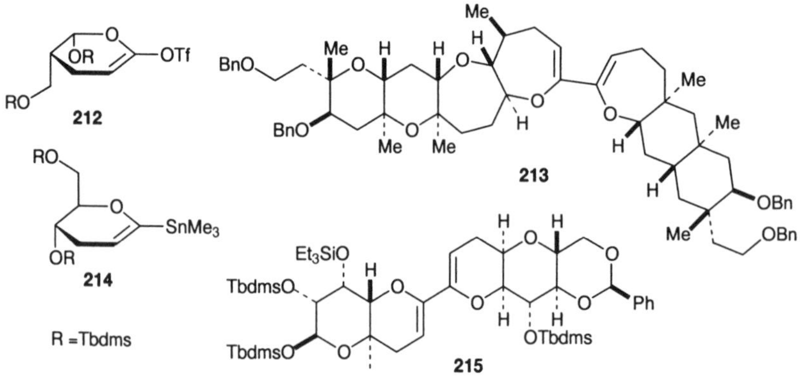

212

214

R =Tbdms

213

215

The C1-C9 and C16-C23 subunits of the polyether antibiotic lysocellin have been synthesized from D-glucose and D-mannitol respectively. (For preliminary report see Vol. 26, p. 312).[57, 58]

Fraser-Reid and co-workers have reported investigations into the reactivity of an intermediate towards tetrodotoxin. Thus, **216** did not undergo acetolysis, but the (armed) benzyl analogue **217** did, providing **218**. Compound **217** was elaborated to **219** by enolate *C*-alkylation when X=O.[59]

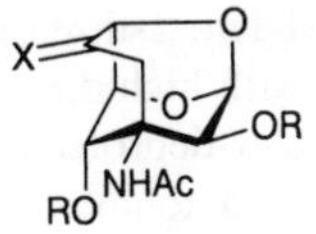

216 R = Ac
217 R = Bn
X = O, CH$_2$

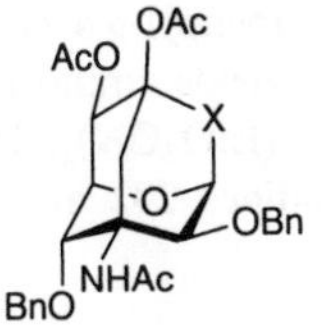

218

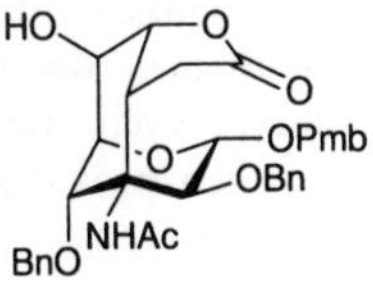

219

5   N- and S-Heterocycles

The aziridinium-fused furanoside **221** was prepared from D-ribose derivative **220**, and thence elaborated to acyclic aziridine **222** and compound **223**, which are both 3,4-disubstituted L-glutamates (Scheme 16).[60] [2+2] Cycloadditions of sugar vinyl ethers **224** and **225** to tosylcyanide followed by reduction with RedAl afforded the β-lactams **226** and **227**. These were elaborated to the tricyclic systems **228** and **229**.[61]

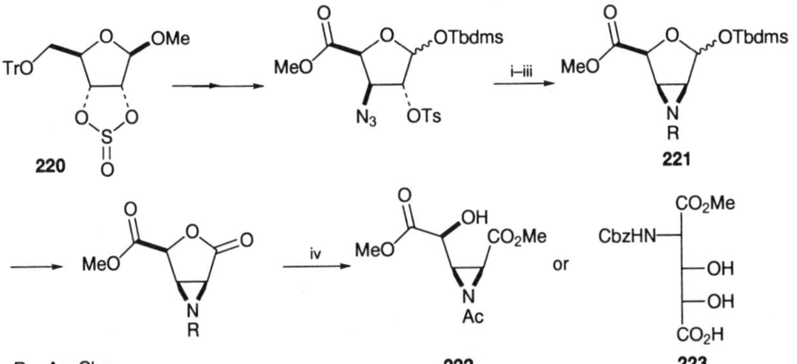

220

221

R = Ac, Cbz

222

223

Reagents: i, H$_2$, Pd-C; ii, Et$_3$N, DMF; iii, Ac$_2$O or CbzCl, Py; iv, MeOH, BF$_3$.Et$_2$O.

Scheme 16

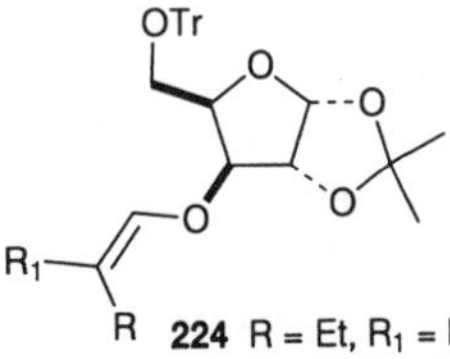

224 R = Et, R$_1$ = H
225 R = H, R$_1$ = Et

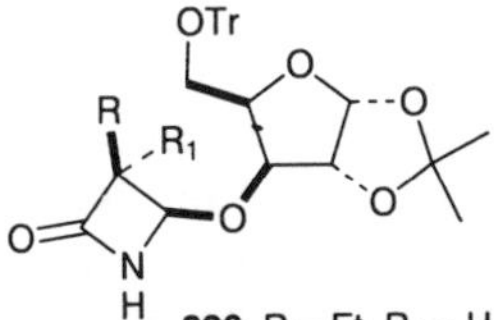

226 R = Et, R$_1$ = H
227 R = H, R$_1$ = Et

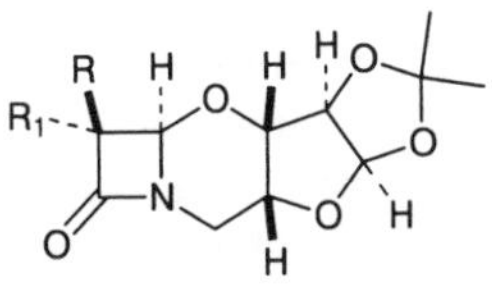

228 R = Et, R$_1$ = H
229 R = H, R$_1$ = Et

Addition of a range of organometallic reagents RM (MeLi, BuLi, Me$_3$Si-CH$_2$Li, PhLi, 2-C$_4$H$_3$-SLi, CH$_2$=CHMgBr, CH$_2$=CHCH$_2$MgBr, PhCH$_2$MgBr and *p*-anisylCH_2MgBr) to *N*-2-*O*-dibenzyl-3,4-*O*-isopropylidene-L-threose imine **230**, then subsequent deprotection, yields aminotriols **231** with high d.e. (>95:5). Mitsunobu-type cyclization [PPh_3, $(EtO_2CN)_2$, Py] and *O*,*N*-deprotection (H_2, $Pd(OH)_2/C$) then provides pyrrolidines **232** and **233**, related to the antibiotic anisomycin.[62]

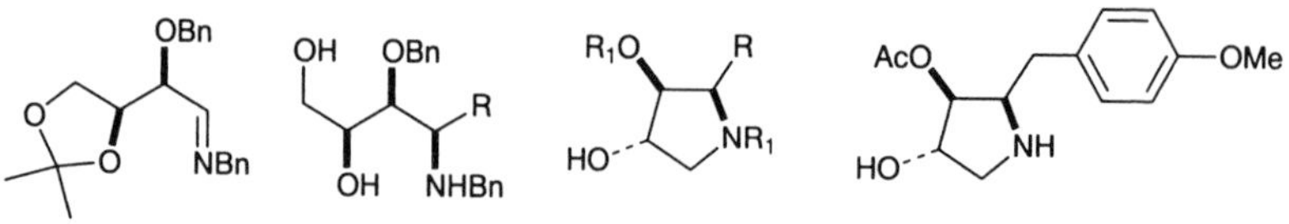

230 **231** **232** R_1 = Bn **233** R_1 = H Anisomycin

The pyrrolidine **235**, a (2*S*,3*S*)-dihydroxy-(4*S*)-amino acid moiety of the gastroprotective substance AI-77B, was prepared from the known alkene **234** in several steps. The same starting alkene also provided a lactam analogue of **235**, namely **236**, also a component of AI-77B (Scheme 17).[63]

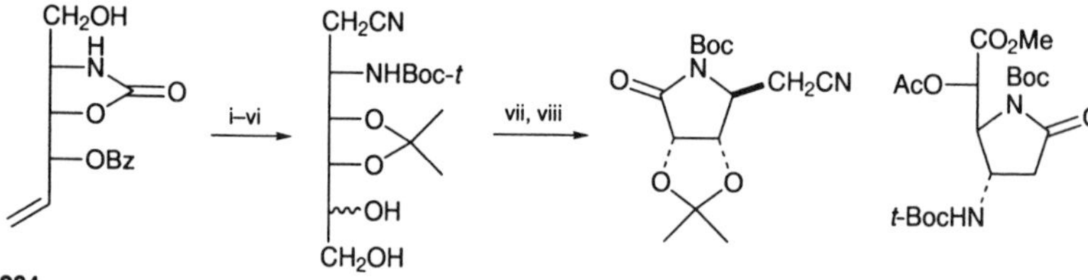

234 **235** **236**

Reagents: i, TsCl, Py; ii, NaCN, DMSO; iii, (t-BuO_2C)O; iv, K_2CO_3; v, $C(Me)_2(OMe)_2$; vi, OsO_4, NMNO; vii, $NaIO_4$; viii, CrO_3, Py.

Scheme 17

D-Xylose has been converted to (2*S*)-3-(indol-3-yl)propane-1,2-diol **237** by two different routes, one involving direct Fischer indolization of **238**.[64] The dibenzyl-dithioacetal **239** was elaborated to the fused triazoline **240** following reaction with MCPBA. Initial oxidation was followed by elimination of acetic acid allowing intramolecular 1,3-dipolar cycloaddition reaction to construct the triazole ring.[65] The bicyclic *N*,*S*-acetals **242** and **241** were prepared by reaction of the 2,3-*O*-isopropylidene-D-ribofuranose with 2-aminoethane thiol followed by Mitsunobu reaction. These products are considered analogues of castanospermine and australine.[66]

The pyrrolidine natural product **245** has been prepared in a number of steps from D-glucose *via* **243** and **244**.[67] The antifungal antibiotic (+)-preussin **248** was prepared from the known intermediate **246** (Vol. 27, p. 210, Scheme 8) *via* selective deoxygenation giving **247**.[68]

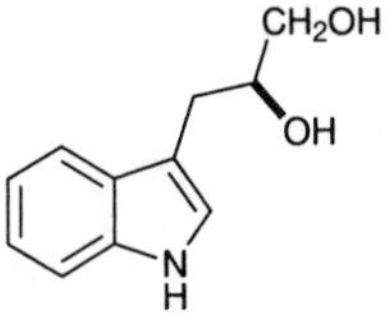

CH$_2$OH

OH

237

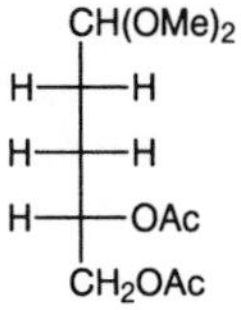

CH(OMe)$_2$
H——H
H——H
H——OAc
CH$_2$OAc

238

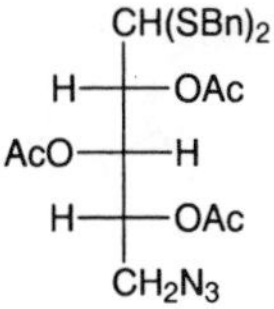

CH(SBn)$_2$
H——OAc
AcO——H
H——OAc
CH$_2$N$_3$

239

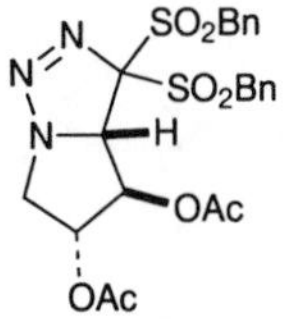

SO$_2$Bn
SO$_2$Bn
H
OAc
OAc

240

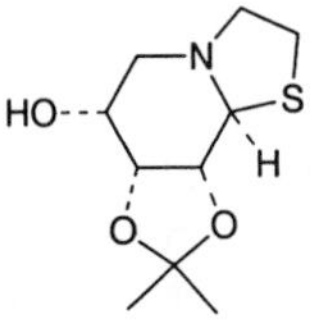

HO

241

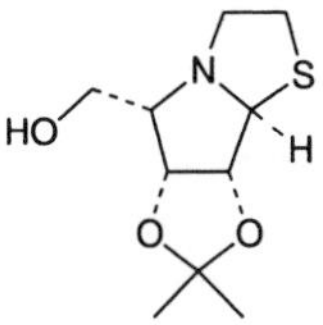

HO

242

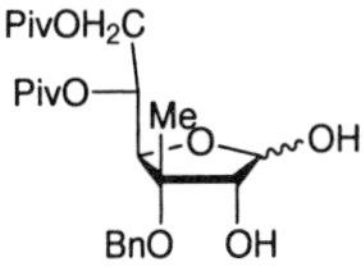

PivOH$_2$C
PivO
Me
O
OH
BnO   OH

243

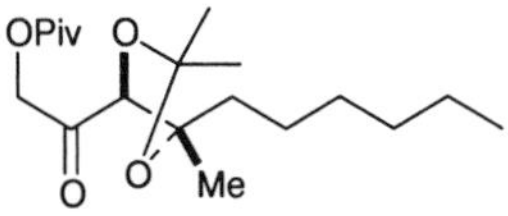

OPiv
O
Me

244

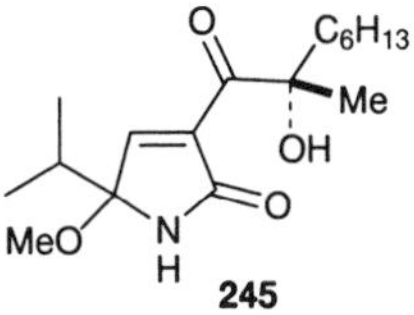

O   C$_6$H$_{13}$
Me
OH
O
MeO   N
H

245

R
O   N
Bn
X   OBn

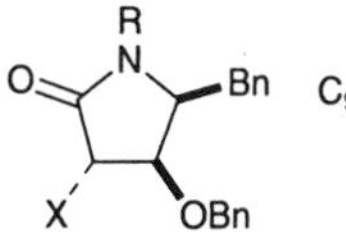

246 R = Mpm, X = OBn
247 R = Boc, X = H

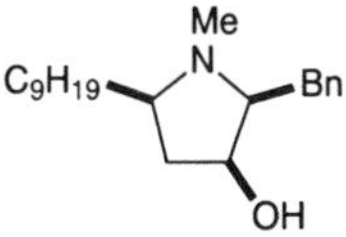

Me
C$_9$H$_{19}$ N   Bn
OH

248

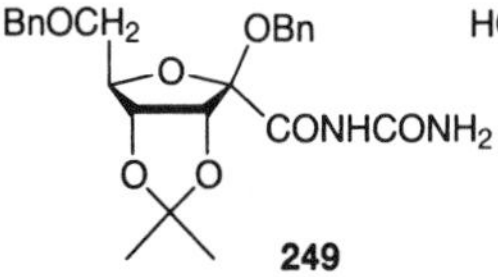

BnOCH$_2$ OBn
O
CONHCONH$_2$
O   O

249

HOCH$_2$ HN
O   O
NH
OH OH O

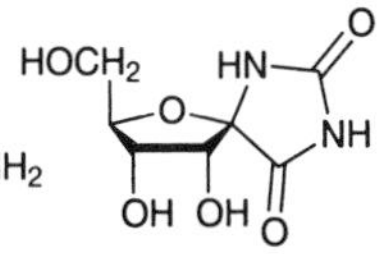

250

Full details of a synthesis of (+)-hydantocidin **250** and 5-*epi*-(+)-hydantocidin from D-fructose *via* **249** have been reported (see Vol. 27, p. 339, ref. 74 and p. 237, ref. 102 for preliminary reports).[69]

A series of *N*-substituted D-*arabino*-piperidinol phosphates, e.g. **253**, were prepared from the 1,2,3-inositol triphosphate **251**, by periodate cleavage yielding **252**, which was converted to **253** by aminolysis and NaBH$_4$ reduction. A wide range of alkyl and functionalized R groups were used.[70]

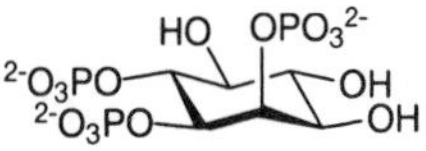

HO   OPO$_3{}^{2-}$
$^{2-}$O$_3$PO   OH
$^{2-}$O$_3$PO   OH

251

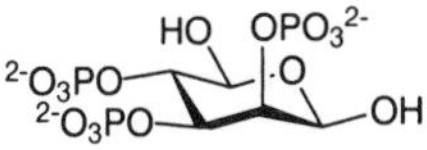

HO   OPO$_3{}^{2-}$
O
$^{2-}$O$_3$PO   OH
$^{2-}$O$_3$PO

252

OPO$_3{}^{2-}$
NR
$^{2-}$O$_3$PO
$^{2-}$O$_3$PO

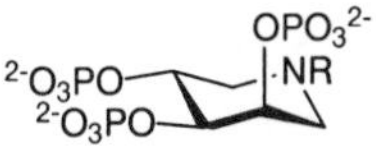

253

A multi-step synthesis of α-fucosidase inhibitor **255** (K$_i$=6.4 μM) from **254** was reported (Scheme 18). Neither yeast α-glucosidase or *E.coli* β-galactosidase are inhibited by **255** at concentrations <1 mM (See also Chapter18).[71]

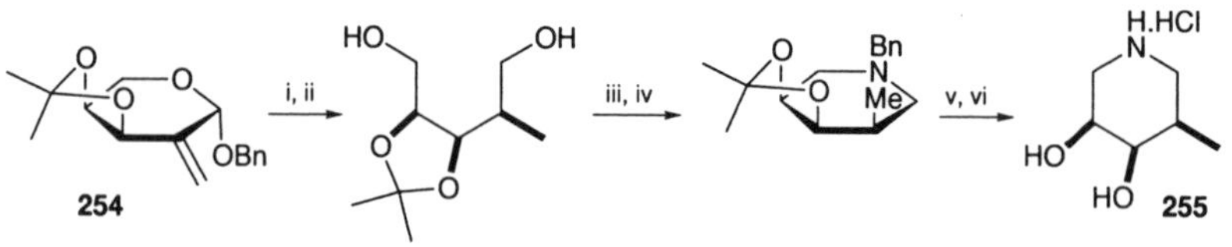

Reagents: i, H$_2$, Pd-C, NH$_3$; ii, Na, NH$_3$; iii, TsCl, Py; iv, BnNH$_2$; v, TFA, H$_2$O; vi, H$_2$, Pd-C, 1M HCl.

Scheme 18

This year has seen continued interest in synthesis of indolizidine alkaloids and related compounds from sugars. D-Arabinose derivative **256** was homologated by Wittig reaction, Mitsunobu-type azidation affording **257**. Subsequent epoxidation, and bicycle construction *via* double alkylation through epoxide opening and *N*-alkylation, followed by full debenzylation afforded diastereomers **258** and **259**.[72]

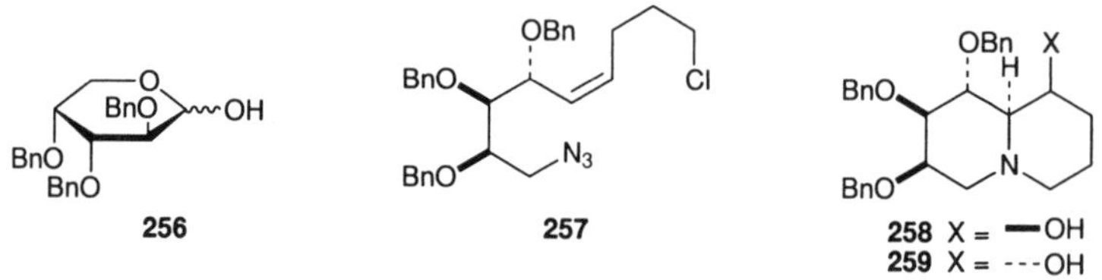

The same authors also reported a similar strategy (using epoxide opening) starting from **260**, proceeding *via* lactam **261** to ultimately provide **262** and **263** (Scheme 19). These are ring expanded analogues of the pyrrolizidine alkaloids alexine and australine respectively.[73] A further approach to such alkaloids has also been reported by the same workers. Lactone **264** was converted to **265** using Sharpless epoxidation, and thence, *via* lactam **266**, to (-)-swainsonine **267** (Scheme 20).[74]

A synthesis of castanospermine has been described commencing with the chiral lactam **268**. *N*-Allylation, selective primary benzyl-acetyl exchange, deacylation, oxidation and Wittig homologation gave **269**. Metathesis reaction gave the

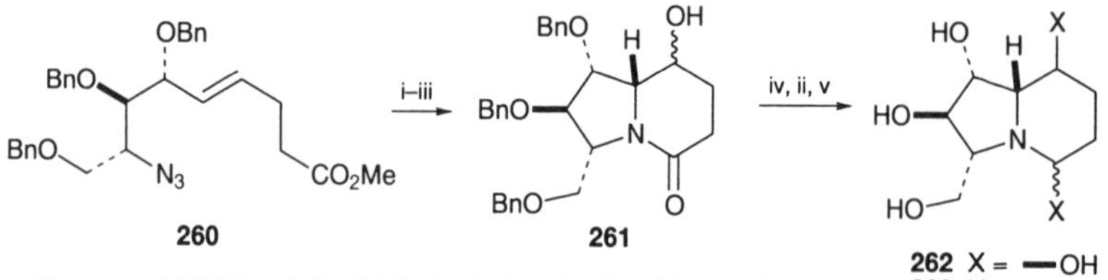

Reagents: i, MCPBA; ii, H$_2$, Pd-C; iii, NaOMe; iv, BH$_3$.SMe$_2$; v, OH$^-$ resin.

Scheme 19

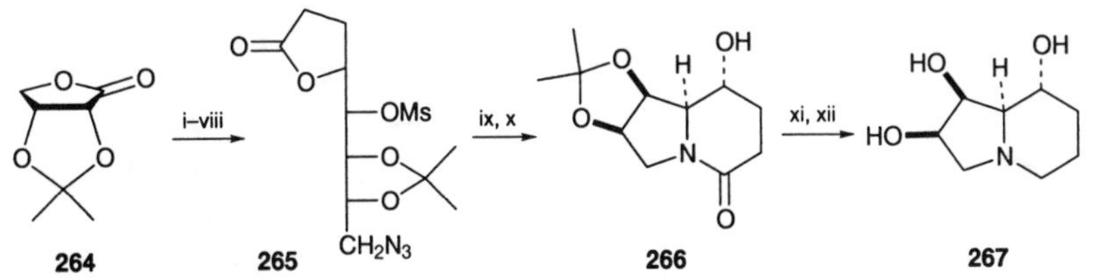

Reagents: i, DIBAL; ii, CH$_2$=CHMgBr; iii, TbdmsCl; iv, CH(OMe)$_3$, EtCO$_2$H, heat; v, Sharpless epoxidation; vi, Bu$_4$NF; vii, MsCl, Py; viii, NaN$_3$; ix, Pd(OH)$_2$, H$_2$; x, NaOMe; xi, BH$_3$.SMe$_2$; xii, HCl,

Scheme 20

bicyclic **270**, which underwent osmylation, conversion to the cyclic sulfate, reductive ring opening, lactam reduction and finally full deprotection to yield **271** (Scheme 21).[75]

Another synthesis of castanospermine commences with D-glucose derivative **272**, ultrasonic diastereoselective allylation and then benzylation giving **273**. This was then elaborated in five steps to diketone **274** and thence to **271** (Scheme 22).[76]

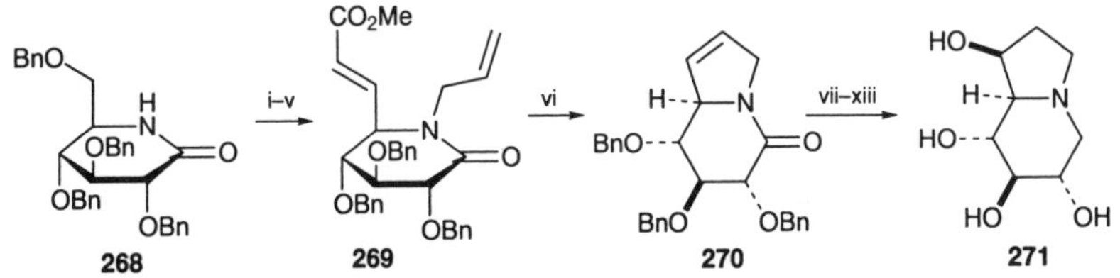

Reagents: i, CH$_2$=CHCH$_2$MgBr, KOH; ii, Ac$_2$O, FeCl$_3$; iii, NH$_3$, MeOH; iv, Dess-Martin periodinane; v, Ph$_3$P=CHCO$_2$Me; vi, Cl(PCy$_3$)$_2$Ru=CHCH=CPh$_2$; vii, OsO$_4$, NMNO; viii, SOCl$_2$; ix, NaIO$_4$; x, NaBH$_4$; xi, H$_2$SO$_4$; xii, BH$_3$.SMe$_2$; xiii, H$_2$, Pd-C.

Scheme 21

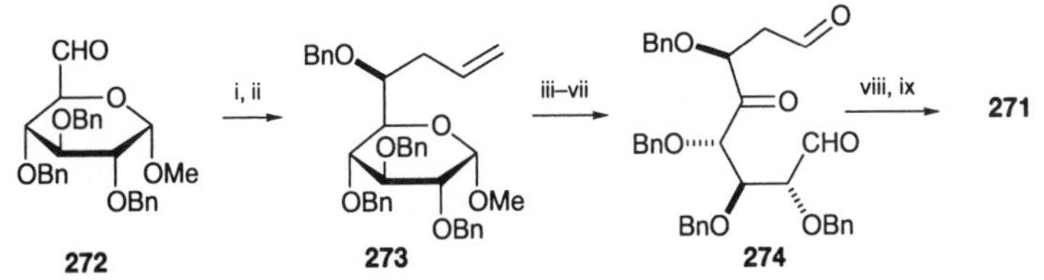

Reagents: i, CH$_2$=CHCH$_2$Br, Sn, ultrasound; ii, BnBr, NaOH; iii, IDCP; iv, Zn, EtOH; v, DMSO, (COCl)$_2$; vi, O$_3$, Ph$_3$P; vii, HCl; viii, NH$_4$CO$_3$, NaCN, BH$_3$; ix, HCO$_2$H Pd-C.

Scheme 22

A synthesis of (2S)-2-hydroxycastanospermine (**280**) and (2S)-2-hydroxy-6-*epi*-castanospermine (**277**) starts with chain extended sugar **275**.

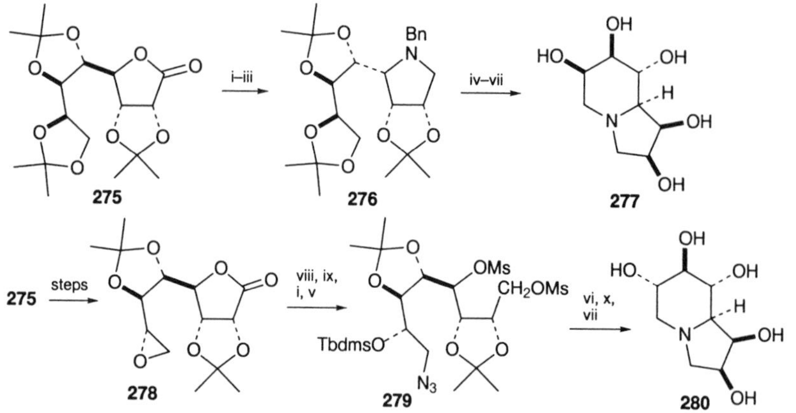

Reagents: i, LiBH₄; ii, MsCl; iii, BnNH₂; iv, TsOH, MeOH; v, MsCl, Py; vi, H₂, Pd-C; vii, TFA, H₂O; viii, NaN₃; ix, TbdmsOTf; x, TBAF.

Scheme 23

The former synthesis proceeds *via* epoxide **278** which was ring opened by azide, the resulting alcohol was protected and full reduction of the lactone and mesylation gave **279**. Reductive cyclization and deprotection then provided **280** (Scheme 23). The 6-*epi* analogue **277** was prepared by converting **275** to pyrrolidine **276**. Selective acetal removal and primary hydroxyl mesylation then allowed reductive cyclization followed by deprotection to give **277**. Similar chemistry was also used to convert **281**, closely related to **275**, to (2*R*)-2-hydroxy-6-*epi*-castanospermine **282**.[77]

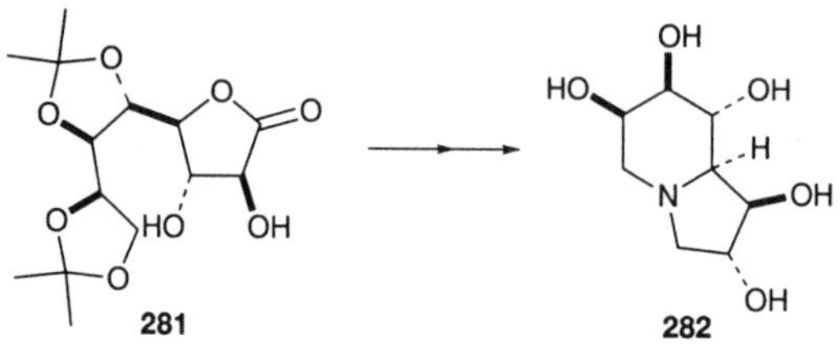

Radical deoxygenation was used for conversion of **283** to **287**, (7*R*)-7-hydroxy-L-swainsonine. The former was also elaborated to L-(+)-swainsonine **286** *via* **284** and **285** (Scheme 54), and also to the dehydrogenated analogue **288**. L-Swainsonine has K_i of 0.45 μM as an inhibitor of L-rhamnosidase (naringinase).[78]

The sulfonium analogue **292** of 8-*O*-methylswainsonine has been prepared from D-glucose derived **289**. Key steps involved cleavage of the benzylidene acetal to the 4-*O*-benzyl-6-bromide, and thioacetyl substitution of the 6-Br. Base then freed the 6-SH function which displaced the C-3 mesylate generating the *S*-heterocycle **290**. Benzylation, glycosidic hydrolysis, Wittig homologation and reduction then leads to thio sugar *C*-glycoside **291**, which was elaborated to **292** (Scheme 25).[79]

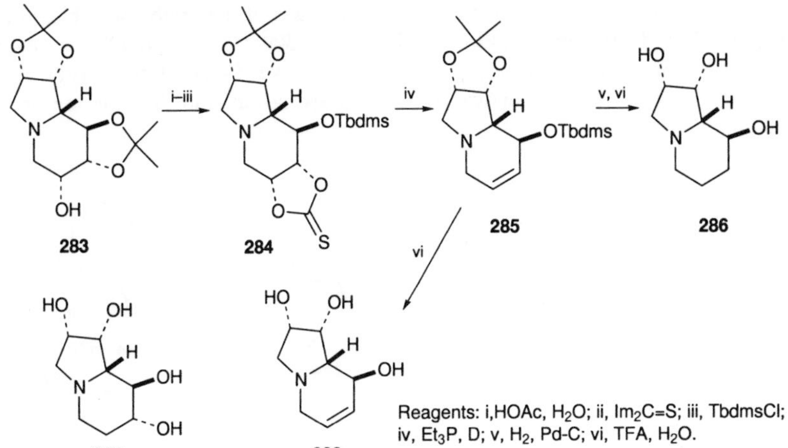

283 → **284** → **285** → **286**

287 **288**

Reagents: i, HOAc, H$_2$O; ii, Im$_2$C=S; iii, TbdmsCl;
iv, Et$_3$P, D; v, H$_2$, Pd-C; vi, TFA, H$_2$O.

Scheme 24

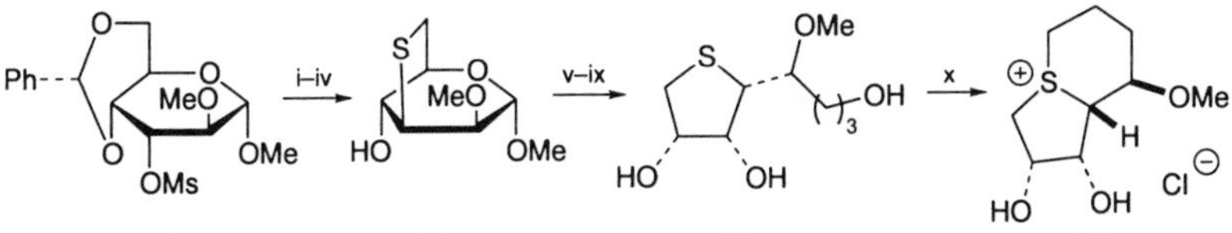

289 **290** **291** **292**

Reagents: i, NBS, CCl$_4$; ii, KSAc; iii, NaOMe, MeOH; iv, BnBr, NaH; v, TFA, H$_2$O;
vi, Ph$_3$P=CHCO$_2$Me; vii, H$_2$, Pd(OH)$_2$; viii, LAH; ix, Birch reduction; x, TsCl, Py.

Scheme 25

A number of spiroheterocyclics have been obtained through photochemical reactions of succinimide-containing carbohydrates. Thus, **293** generated [5.6] spirocycle **294** in 36-49% yield on photolysis (254 nm), and **295** gave a mixture of **296** (24%) and its spiro-centre epimer (12%).[80] A simpler monocyclic 7-membered *N*-heterocycle, (3*R*,4*R*)-3-amino-4-hydroxyazepine (**298**) has been prepared from **297** (itself prepared from D-isoascorbic acid).[81]

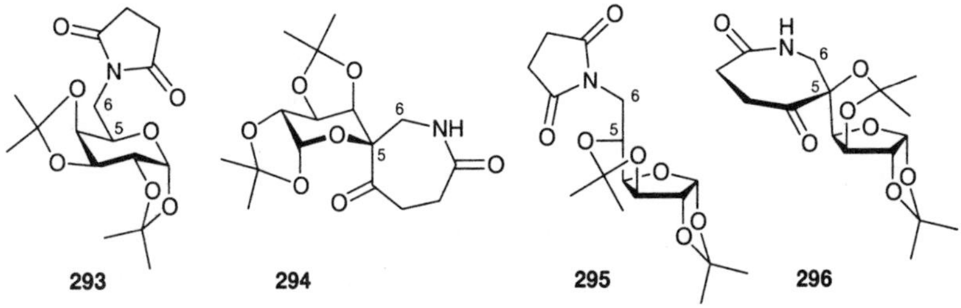

293 **294** **295** **296**

L-Mannitol was the starting material for synthesis of the orally available, non-peptidic HIV-1 protease inhibitory cyclic sulfone **300**. Compound **299** was made by known methods (Vol. 11, p. 97) and iterative *C*-benzylations (PhCHO/Al(OPr-*i*)$_3$ then NaBH$_4$/Ni(Acac)$_2$) then *O*-alkylation and *S*-oxidation gave **300**.[82]

297   **298**   **299**

300 R = Bn, *m*-NH$_2$Ph,

301 X = -NCS
302 X = NH$_2$ **303**   **304**   **305**   **306**

New carbohydrate mimetics based on cyclic thioureas have been prepared. Thus, isothiocyanate **301** or amine **302** (either D-*allo* or D-*gluco*) were converted to the thioureas **303**. Deprotection gave acyclic compounds **304**, and co-evaporation with water then IR-45 (OH⁻) ion exchange resin gave **305** (D-*allo* case) or **306** (D-*gluco*).[83]

An enantioselective synthesis of *R*-laudanosine (**309**) and of *R*-glaucine (**310**) has been achieved starting from L-gulonolactone **307**. Key steps involve acylation to give **308** and stereocontrolled dehydrative cyclization followed by periodate oxidation of the residual acyclic component.[84]

307   **308**   **309**   **310**

6   Acyclic Compounds

The acyclic acid **315**, a component of the anti-tumour calyculins, and lactone **314**, have been prepared from D-glucosamine *via* **311**, which was converted either to **312** or **313**, intermediates for **315** and **314** respectively (Scheme 26).[85] The same group have reported a very similar strategy to **315** also from D-glucosamine in a separate paper providing further details (see section 5).[63]

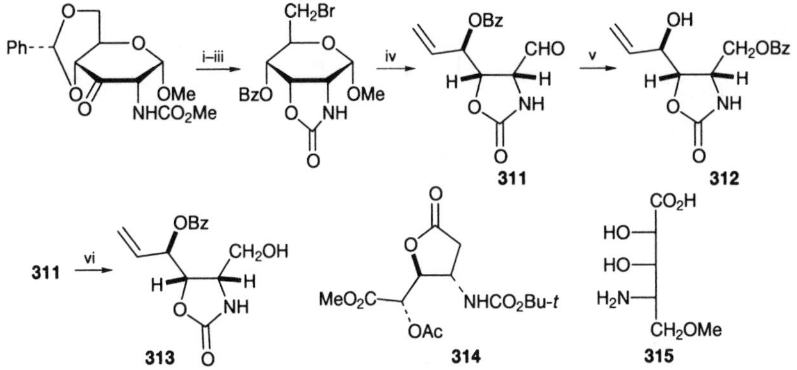

Reagents: i, NaBH₄; ii, NaOMe, MeOH; iii, NBS, BaCO₃, CCl₄; iv, Zn, EtOH; v, NaBH₄, 12 hr, r.t.; vi, NaBH₄, 20 min, 0 °C.

Scheme 26

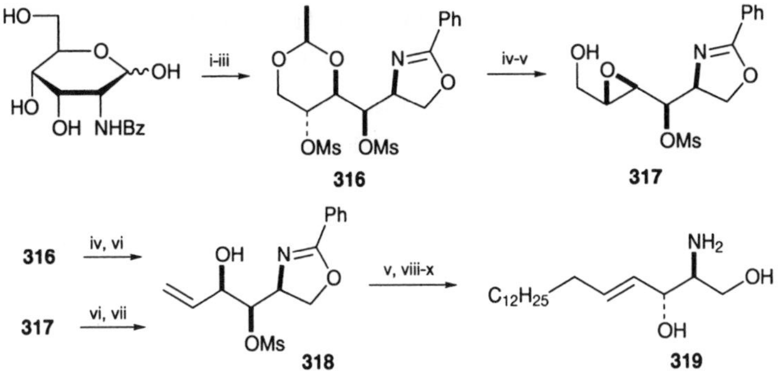

Reagents: i, (MeCHO)₃, H⁺; ii, NaBH₄, *i*-PrOH; iii, MsCl, H₂O, Et₃N; iv, TiCl₄, PhSH; v, K₂CO₃, MeOH; vi, PPh₃, I₂, Py; vii, BuLi (3 eq.); viii, C₁₂H₂₅MgBr, CuCN; ix, HCl; x, NaOH.

Scheme 27

This year has again seen several synthetic routes to sphingosines and analogues. Oxazolidinone **312** was converted by the same workers to D-*erythro*-sphingosine **319** in six steps[86] and *N*-benzoyl glucosamine has also been elaborated to this compound. Key intermediates were either the mono-mesylated diol **318** or the epoxyalcohol **317**.[87] The route *via* **318** avoiding an epoxide intermediate was shorter, but only proved applicable on scales of 200 mg or less (Scheme 27).

Sphingosine **319** has also been prepared from D-xylose derivative **320**, and also then converted to ceramide **324**. Wittig homologation and mesylation of **320** gave **321**, which underwent S_N2' substitution to give intermediate **322**, elaborated by standard means to **323** and thence **324** (Scheme 28).[88]

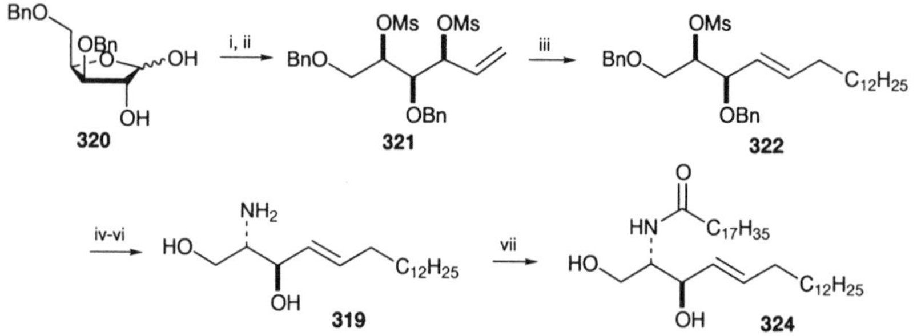

Reagents: i, Ph$_3$PCH$_3$Br, BuLi; ii, MsCl, DMAP, Et$_3$N; iii, C$_{12}$H$_{25}$MgBr, CuCN; iv, NaN$_3$, DMF; v, LAH; vi, Na, NH$_3$; vii, Stearic acid *N*-succinimidyl ester.

Scheme 28

D-Glucosamine was the starting material in a synthesis of (3*S*,4*S*)-statine **328**, and also of the analogue with the isopropyl group of statine replaced by a phenyl group. Thus, glucosamine derivative **325** underwent oxidative cleavage, reduction, iodination and elimination to give **326**. Treatment with base generated oxazolidinone **327**, then elaborated to **328** (Scheme 29).[89]

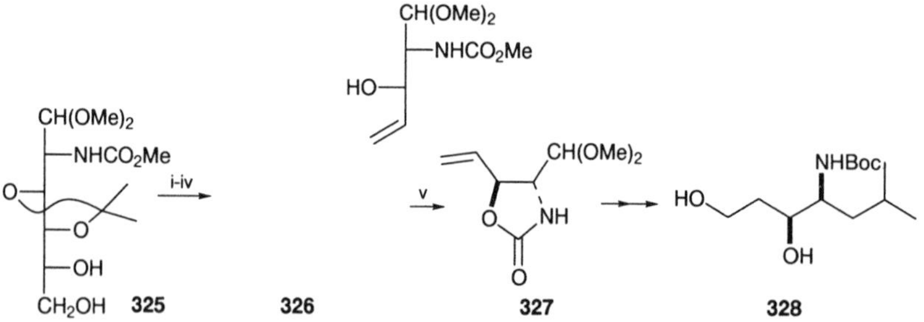

Reagents: i, NaIO$_4$; ii, NaBH$_4$; iii, I$_2$, PPh$_3$, ImH; vi, Zn, *i*-PrOH; v, NaOMe.

Scheme 29

Aminotriol derivative **330** was prepared from the D-lxyose-derived **329**. Amide formation and detritylation allowed for glycosylation with the tetra-*O*-benzyl-α-D-glucosyl bromide, final hydrogenation then giving **331**, which has immunosuppressant and anti-tumour properties (Scheme 30).[90]

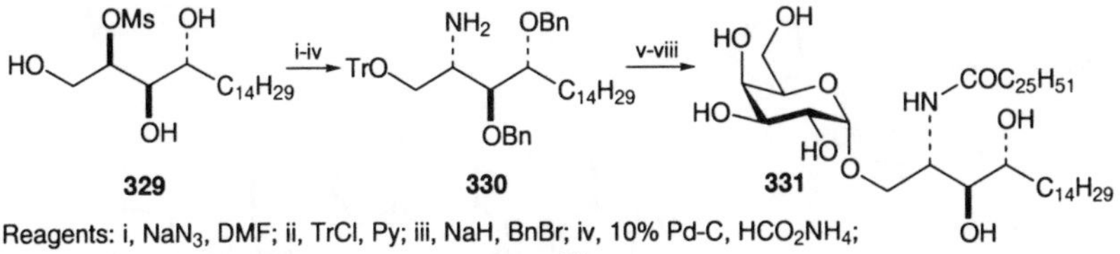

Scheme 30

Reagents: i, NaN$_3$, DMF; ii, TrCl, Py; iii, NaH, BnBr; iv, 10% Pd-C, HCO$_2$NH$_4$;

v, C$_{25}$H$_{51}$CO$_2$H, WSC-HCl; vi, 10% HCl; vii, [structure] n-hex$_4$Br; viii, Pd(OH)$_2$, 4-Me cyclohexene, D

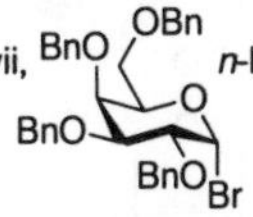

D-Mannitol has been converted into leukotriene LTB$_4$ methyl ester *via* the known intermediate **332**. Acetal hydrolysis with 1,2-ethanedithiol and TiCl$_4$-AsPh$_3$, and cleavage of the symmetrical diol then sequential 'double' formyl olefination using an arsenic ylide gives **333**, which was then elaborated in seven steps to **334** (LTB$_4$) (Scheme 31).[91]

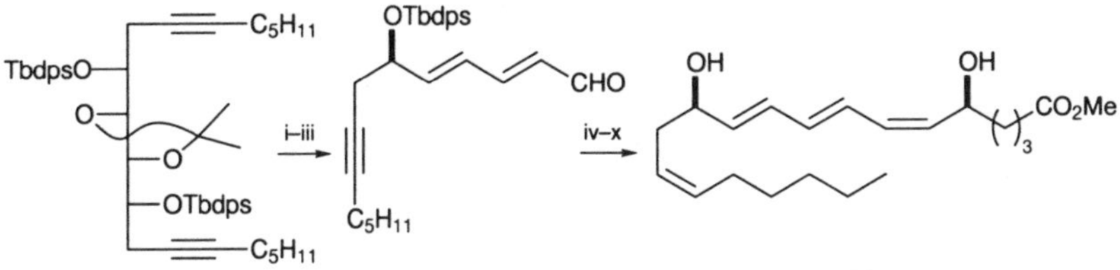

Reagents: i, HSCH$_2$CH$_2$SH, TiCl$_4$-AsPh$_3$; ii, Pb(OAc)$_4$; iii, 2.5 eq. [AsPh$_3$CH$_2$CHO]$^+$Br$^-$, K$_2$CO$_3$; iv, NaBH$_4$-CeCl$_3$;

v, CBr$_4$, PPh$_3$; vi, PPh$_3$; vii, BuLi, OHC [structure] CO$_2$Me ; viii, H$_2$, Pd-CaCO$_3$; ix, TBAF; x, K$_2$CO$_3$, MeOH.

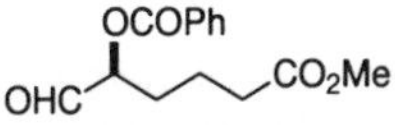

Scheme 31

The previously reported alkene **335** (from L-glucose, vol. 28, p. 373) has now been elaborated to 12-*epi*-prostaglandin F$_{2\alpha}$.[92] The C35-C43 subunit of dolastatin G (**336**) has been prepared from levoglucosenone.[93]

Tri-*O*-acetyl-D-glucal (*via* 2,3-dideoxyglycoside **337**) is the starting point for a synthesis of **339**, a hydroxyethylene dipeptide isostere related to an HIV-1 protease inhibitor.[94] The ring opening of aziridine **338** is a key step (Scheme 32).

Bestatin **343**, a potent inhibitor of leukotriene A4 hydrolase, has been prepared from **340**. The peptide link is formed by reaction of the product of reduction of **341**, giving peptide linked **342** (R=Leu-OBn). Deprotections and substitution of the homo benzylic OH by amino (*via* the azide) gives bestatin (Scheme 33).[95]

D-Threitol derivative **344** has been converted into various chain lengthened targets bearing acid functions. Specifically, **344** was converted to **346**, and to **345**. Both syntheses utilized Wittig strategies. Products **346** or **345** were then converted to trapoxin-containing affinity agents, or trapoxin B and its ^{3}H labelled analogue, respectively (Scheme 34).[96]

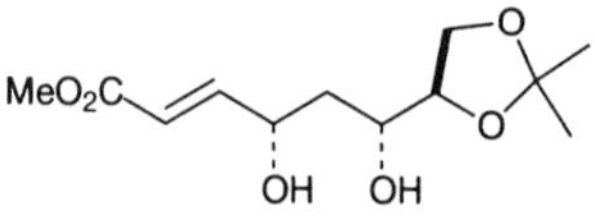

335

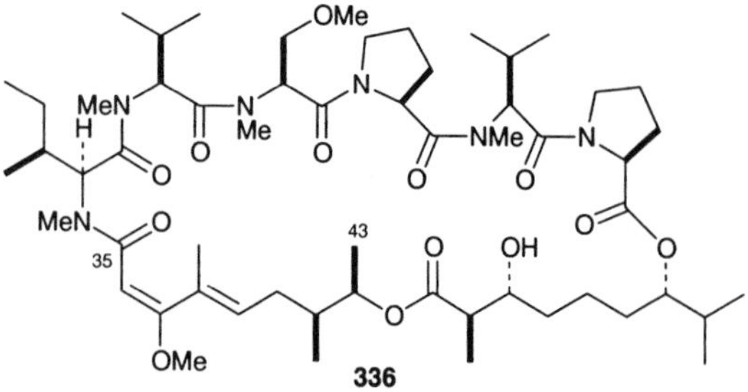

336

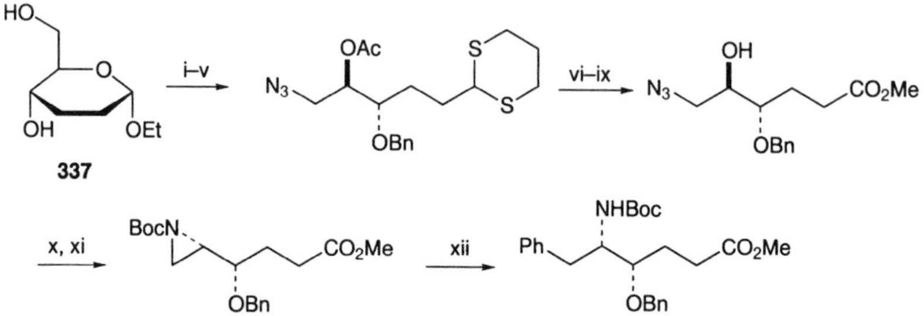

337

338   **339**

Reagents: i, TsCl, Py; ii, NaN$_3$, DMF; iii, NaH, BnBr; iv, HS(CH$_2$)$_3$SH, BF$_3$.OEt$_2$; v, Ac$_2$O, Py, DMAP
vi, HgO (red), BF$_3$.OEt$_2$, THF-H$_2$O; vii, PDC, DMF; viii, CH$_2$N$_2$; ix, K$_2$CO$_3$, MeOH;
x, PPh$_3$, DEAD, C$_6$H$_6$; xi, Boc$_2$O; xii, PhMgBr, CuBr, Me$_2$S.

Scheme 32

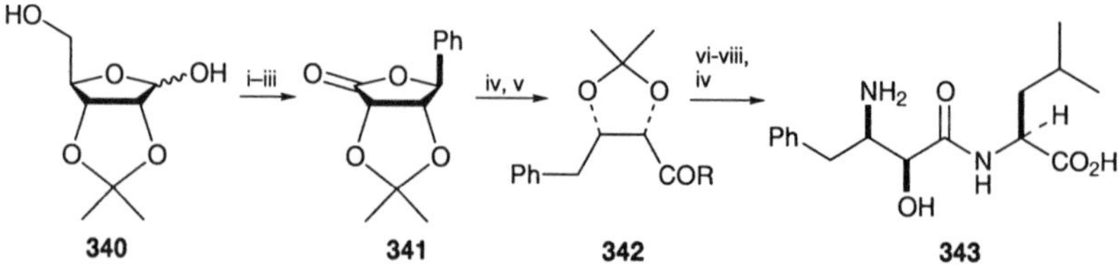

340   **341**   **342**   **343**

Reagents: i, PhMgBr; ii, NaIO$_4$; iii, Jones oxidation; iv, H$_2$, Pd-C; v, LeuOBn, Et$_3$N,
1-ethyl-3-(3-dimethylaminopropyl)-carbodiimide.HCl; vi, TFAA; vii, 1-Me,2-F-Py.PTS, Et$_3$N;
viii, NaN$_3$, HMPA.

Scheme 33

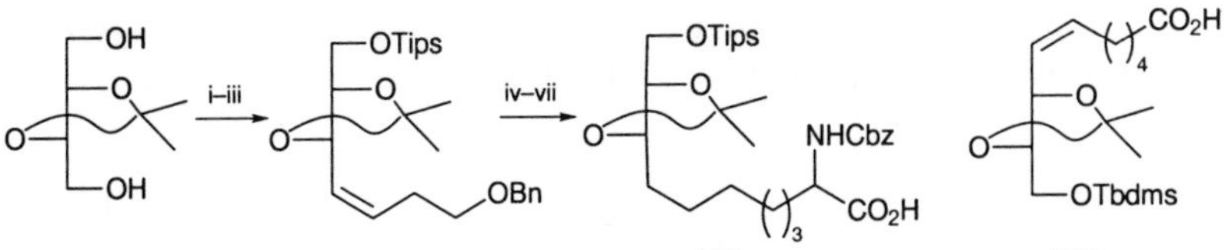

344   **345**   **346**

Reagents: i, TipsCl, NaH; ii, Swern; iii, PPh₃=CHCH₂CH₂OBn; iv, H₂, Pd-C; v, PPh₃, CBr₄;

vi, Mg, Et₂O; vii, CuBr, DMS.

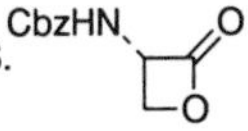

Scheme 34

Two routes to bisubstrate analogues **349**, **354** and **355** targeted at HMGR (3-hydroxy-3-methylglutaryl coenzyme A reductase) have been reported. Levoglucosan derivative **347** was elaborated to **348** and thence to **349**. The synthesis of **354** or **355** commenced with **350**, conversion to the Weinreb amide *via* intermediate acid, temporary silylation, reaction with the anion (on methyl) of **356** and final desilylation giving **351**. Diastereoselective keto alcohol reduction to **352** or **353** was achieved using alternative reductants to give predominantly the new alcohol with either *S* or *R* configurations predominantly. Standard manipulations converted these to **354** or **355** (Scheme 35).[97]

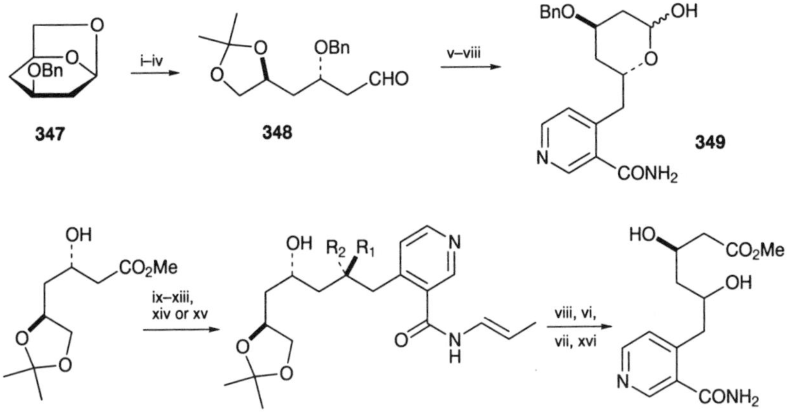

347   **348**   **349**

350   **351** R₁,R₂= C=O   **354** R₁=OH, R₂=H
            352 R₁=OH, R₂=H   **355** R₁=H, R₂=OH
            353 R₁=H, R₂=OH

Reagents: i, H⁺, H₂O; ii, NaBH₄, EtOH; iii, Me₂CO, TsOH; iv, Swern; v, s-BuLi,
vi, MeOH,H⁺; vii, NaIO₄; viii, NH₃, MeOH; ix, NaOH, MeOH;
x, MeNHOMe, BOP; xi, TmsCl; xii, s-BuLi, **356**; xiii, TBAF; xiv, Me₄NHB(OAc)₃;
xv, Et₂BOMe, air then NaBH₄; xvi, Ag₂O, MeOH.

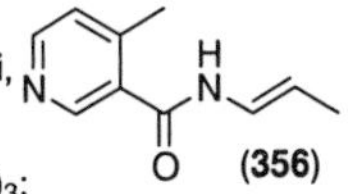

(356)

Scheme 35

The amidoenones **357** and **358** (see Chapter 16) were converted in several steps to 4,5,6-trihydroxynorleucine **360** and 5-hydroxynorvaline **359** respectively.[98]

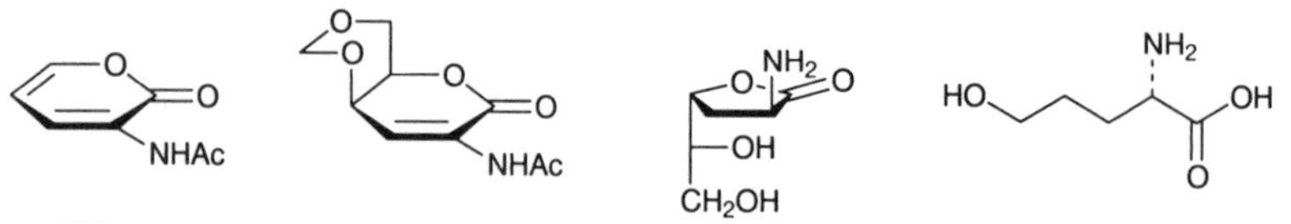

The *N*-TFA methyl ester derivative of a novel β-amino acid component **364** of the tetrapeptide toxin nodularin has been synthesized from D-glucose *via* **361**, **362** and **363** (Scheme 36).[99]

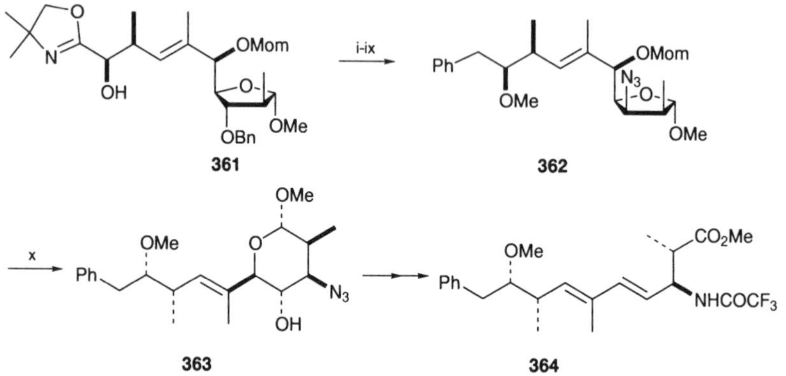

Reagents: i, TFA, H$_2$O; ii, LiAlH$_4$; iii, TsCl, Py; iv, NaOH, MeOH; v, PhMgBr; vi, KH, MeI; vii, H$_2$, Pd(OH)$_2$; viii, Tf$_2$O, Py; ix, LiN$_3$; x, HCl, MeOH.

Scheme 36

The enantiomer **366** of fungal metabolite YM-47522 has been prepared (thereby characterizing the natural product) in 13 steps from **365**. The key alkene stereochemistries were installed by substitution of a terminal iodide with *E*-1,2-ditributylstannylethene, conversion to the vinyl iodide and PdCl$_2$(MeCN)$_2$-catalysed coupling with another organotin reagent (Scheme 37).[100]

The D-*ribo* analogue and the analogue with the alternative quaternary centre configuration of the immunomodulator conagen A (**368**) have been prepared from **367**.[101] The C18-C25 portion **371** of bafilomycin A$_1$ has been prepared from **369** *via* furanose **370** (Scheme 38).[102]

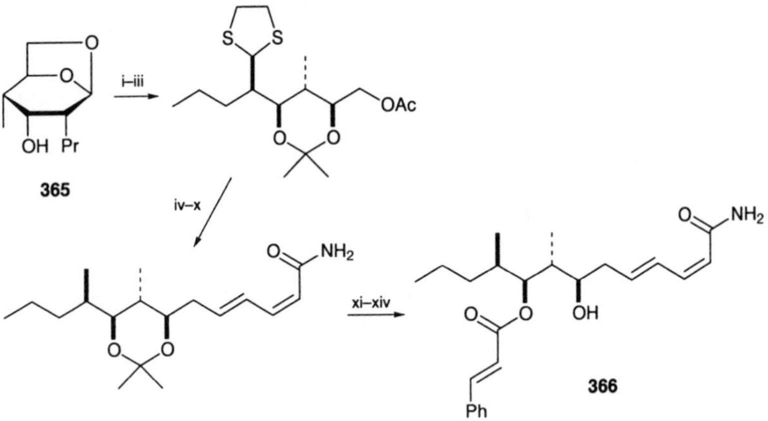

Reagents: i, HSCH₂CH₂SH, BF₃.OEt₂; ii, Ac₂O, Py; iii, Me₂CO, Me₂C(OMe)₂, H⁺; iv, Ra Ni;
v, K₂CO₃, MeOH; vi, TsCl, Py; vii, LiI; viii, Me₂CuLi.LiCN, (E)-Bu₃SnCH=CHSnBu₃; ix, NIS;
x, (Z)-Bu₃SnCH=CHCONH₂, PdCl₂(MeCN)₂; xi, AcOH, H₂O; xii, Et₃SiCl, Py;
xiii, PhCH=CHCOCl, DMAP; xiv, HF, H₂O, MeCN.

Scheme 37

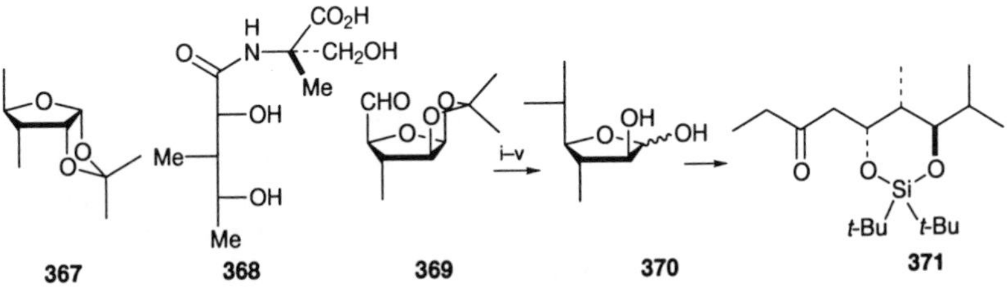

Reagents: i, MeMgI; ii, PCC; iii, Ph₃P=CH₂; iv, H₂, Ni; v, H⁺.

Scheme 38

7 Carbohydrates as Chiral Auxiliaries, Reagents and Catalysts

This year has again seen a considerable number of uses of carbohydrate
derivatives as either removable auxiliaries or as chiral ligands/catalysts for
asymmetric reactions.

Marshall and Elliot have prepared the chiral building block **375** from glucose-
derived **372**. This chemistry involved conjugate addition of stannyl cuprate to
372, trapping of the intermediate enolate as its Tbdms ether to afford diastereomer **373** as the major isomer. Stannylallylation of undec-2-yneal with **373** then
introduced the second chiral centre of **374**, silylation of the new alcohol and
oxidative cleavage providing **375**.[103]

372   373   374   375

376   377   378   379

380   381   382

Cycloaddition reactions have again provided several examples. A review on synthesis of azasugars using hetero-Diels Alder reactions of nitrosodienophiles includes coverage of examples of such dienophiles bearing carbohydrate auxiliaries [55 refs].[104] An example of the type of cycloaddition chemistry discussed in this review has been used in a synthesis of 1,6-dideoxynojirimycin (382) and its D-*allo*- and D-*gulo*-analogues 379 and 381 respectively. All syntheses utilized the hetero-Diels Alder reaction of D-mannose derived 376 with oxime ether 377 followed by *N*-protection and stereoselective osmylation to give 378. Hydrogenation of 378 gave 1,6-dideoxy-D-*allo*-nojirimycin 379, whilst derivatization of 378 as its cyclic sulfate 380 allowed alternative elaboration to a mixture of 1,6-dideoxynojirimycin 382 and 1,6-dideoxy-D-*gulo*-nojirimycin 381.[105]

The acrylate 383, derived from 1,4:3,6-dianhydro-D-glucitol, undergoes Diels Alder reactions with cyclopentadiene giving >99% *endo* products, with a 92% selectivity in favour of the *S-endo* product 384. The analogous reaction of the 1,4:3,6-dianhydro-D-mannitol-derived ester, catalysed by ethylaluminium dichloride, also gives >99% *endo* products, but with a 90% diastereoselectivity in favour of the *R-endo* product.[106]

An interesting example of a chiral aryl chromium complex enhancing diastereoselectivity in carbohydrate auxiliary-influenced acrylate Diels Alder reactions has been described by Shing and co-workers. Thus, for example, diethylaluminium chloride catalyzed reactions of complex 385 with 2-methylbutadiene

383   **384**   **385**

proceeded with about 90% d.e., whereas without the Cr complex formation the d.e. was only around 56%. Simlar enhancements were seen in reactions with other dienes.[107]

Stoodley and co-workers have reported another example (recent volumes in this series) of using anomerically-linked tetra-O-acetyl-D-glucose as an auxiliary, in this case for diastereoselective hydrogenation of α,β-unsaturated enol ethers. 1-Formyl-2,3,4,6-tetra-O-acetyl-β-D-glucopyranose (R*OCHO, where R*=glucosyl auxiliary) underwent Wittig reaction to give enol ethers **386** and **387** (n=1) with the appropriate Wittig reagents. Alternatively, sodium enolates were added to tetra-O-acetyl-β-D-glucopyranosyl bromide giving **389** or **387/388** (n=0, 1). Hydrogenation of all these enol ethers causes modest *Re* face addition to give **390-392**; fractional crystallization allowed isolation of a single isomer in most cases.[108] Glycosidic cleavage could then be effected by treatment with methanolic hydrogen chloride, for example providing access to (αS)-α-hydroxymethyl-γ-butyrolactone **393** in 96% e.e *via* **387** (n=0, X=O).

386   **387** X = O   **389**   **390** R = $CO_2Me(Et)$
          388 X = CH_2 **391** R = COMe

392   **393**

Two reports of different diastereoselective cyclopropanations have appeared this year. The first approach involved using carbohydrate enol ethers **394** and **395**, combined with Cu(OAc)$_2$ or Cu(OTf)$_2$ and either of two chiral ligands. In the case of **394**, Cu(OTf)$_2$ and ligand **399** gave 74% yield and >97:3 *trans:cis* ratio of **396**, but with 60% d.e. for the predominant (*trans*) isomer (**395** gave **397**). Using **398** as ligand along with Cu(OAc)$_2$, **394** gave a poorer *trans:cis* ratio of 81:19, but with >95% d.e. for the *cis* isomer (and only 4% d.e. for the *trans*). Reaction of **395** using **399** as ligand and Cu(OTf)$_2$ catalysis gave 88:12 *trans:cis*, 14% d.e. for the *cis* and 65% d.e. for the *trans* isomers.[109]

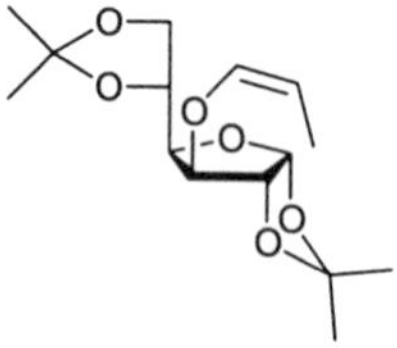

394 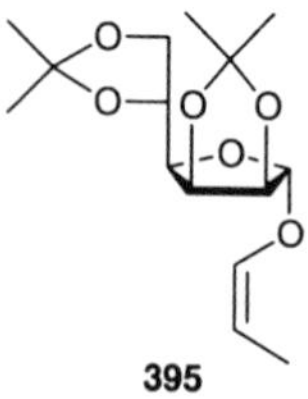   **395** 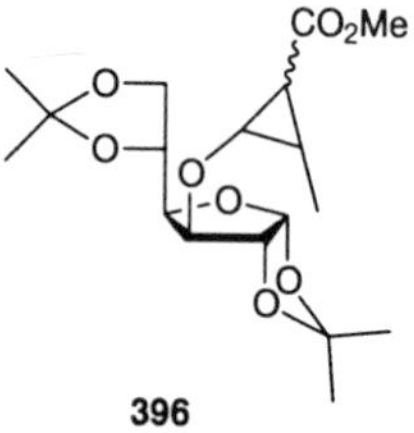   **396** 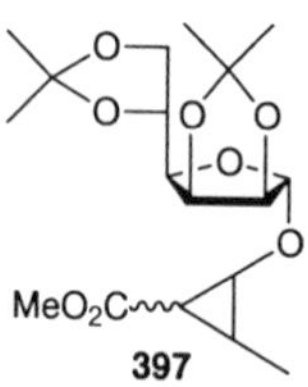   **397**

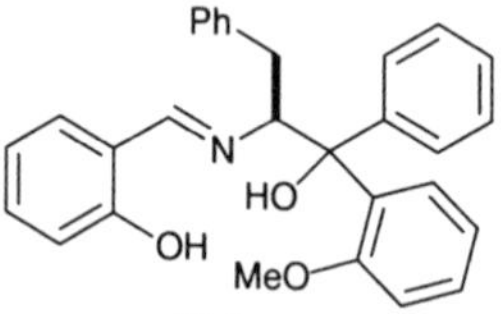

398 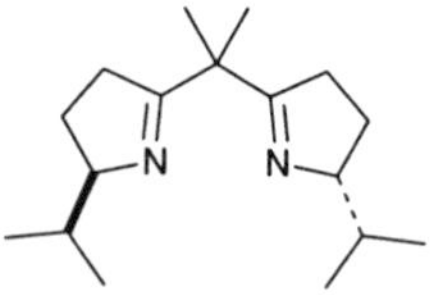   **399**

The glucose-derived allylic glycosides **400** and **401** (from *R*-perillyl and *S*-perillyl alcohols respectively) undergo cyclopropanations with chloroiodomethane and diethyl zinc to give **402** and **403** (R=sugar) with >97% and >99% d.e. respectively, both reactions proceeding in good yield. This work is very similar to that of Charette's group reported previously, Vol. 25, p. 335.[110]

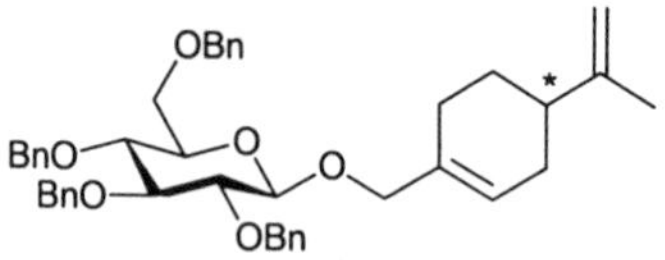

400 * = R
401 * = S

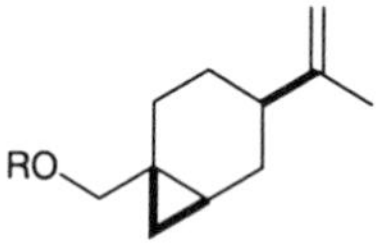

402

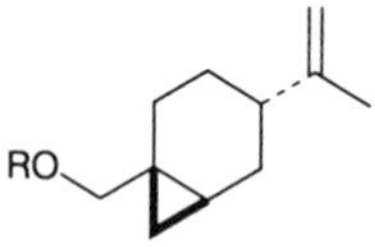

403

Carbohydrate auxiliaries have been used for diastereocontrol in Darzens reactions. Thus, **404** reacted with *p*-anisaldehyde and sodium hydride to give **405** and its diastereomer in 65:35 ratio. The major isomer **405** was further elaborated *via* α-hydroxy acid **406** to *ent*-diltiazen **407**, the enantiomer of a known calcium channel blocker. 6-Chloroacetyl-1,2:3,4-diisopropylidene-D-galactose also reacted as did **404** in a Darzens reaction, generating a single isomeric glycidic ester.[111]

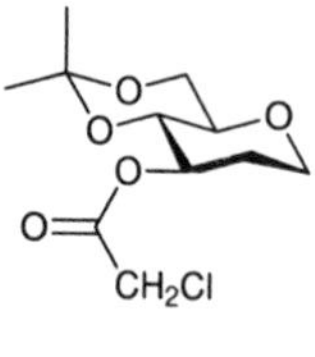

404 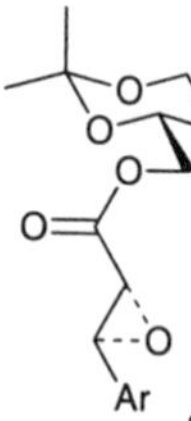   **405** 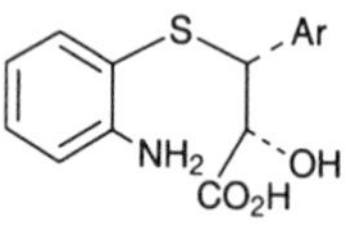   **406** 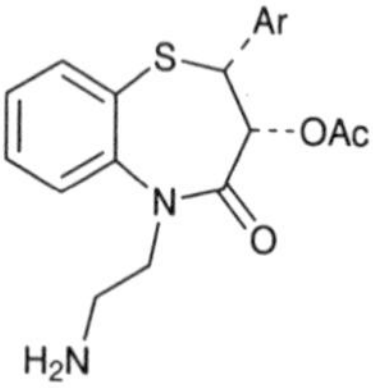   **407**

Several examples of use of carbohydrates as auxiliaries to control enolate alkylation diastereoselectivity have appeared. The lithium enolate of **408** (R=H) was benzylated to give esters with the BnCH(Me)CO$_2$ group having the *R* configuration in up to 92% d.e. When R=Tms, the diastereomer of *R*-chirality in the side chain is obtained in 42% d.e., but when HMPA was added this selectivity switches to give predominantly the *S* configuration at the new centre in 91% d.e.[112] Benzylation of lithium enolates derived from diisopropylidene-β-D-fructo-pyanose-1-esters **409** gave modest yields (34-56%) and poor diastereoselectivities (*S*:*R* of 3:2 and 7:3 when R=Me and piperonyl respectively).[113] The D-xylose-derived oxazolidinones **410** were evaluated as auxiliaries for enolate alkylations of amide derivatives **411**. A wide range of analogues were evaluated using a range of electrophiles. With aromatic electrophiles the *R* product predominated with 5:1 to 17:1 d.s. and yields of around 60%. In all other cases, however, yields were lower and selectivities variable in both magnitude and direction.[114] Another example of a bicyclic carbohydrate-derived oxazolidinone auxiliary is **412** (from D-glucose). Methylation of the lithium enolate, or naked enolate generated using the commercially available P$_4$ base, gave only monomethylation (R=Me), while ethylation and allylation gave mixtures of mono- and dialkylated products, the former however formed with high diastereoselectivity in the same sense as methylation.[115]

408

409

410 R = H
411 R = C(O)CH$_2$R$_1$
X = C(Me)$_2$ or CHPh

412 R = R$_1$ = H

Aldol like reactions using (*N*-diphenylmethylene) glycinates **413** and **414** (see Chap. 7) under phase transfer conditions give low diastereoselectivity in all cases.[116] Dicyclohexylidene-D-glucose reacts with Ph(Me)(O)PCl to give a mixture of **415** and **416** in 95:5 ratio using triethylamine as base, and 30:70 ratio using pyridine. Reactions with Grignard reagents then liberate the new chiral phosphine oxides **417** and **418**.[117] Another interesting example using carbo-hydrate chirality to establish phosphorus chirality is the diastereoselective synth-esis of *R*$_p$ phosphorothioate dinucleotide **421**. Reaction of **419** with 6'-*O*-Tbdms thymine followed by 3'-*O*-Tbdms thymine gave **420** in a 6:1 diastereomeric ratio. Thionation and cleavage of the auxiliary (with retention at P) afforded the dinucleotide **421** (see also Chapter 20).[118]

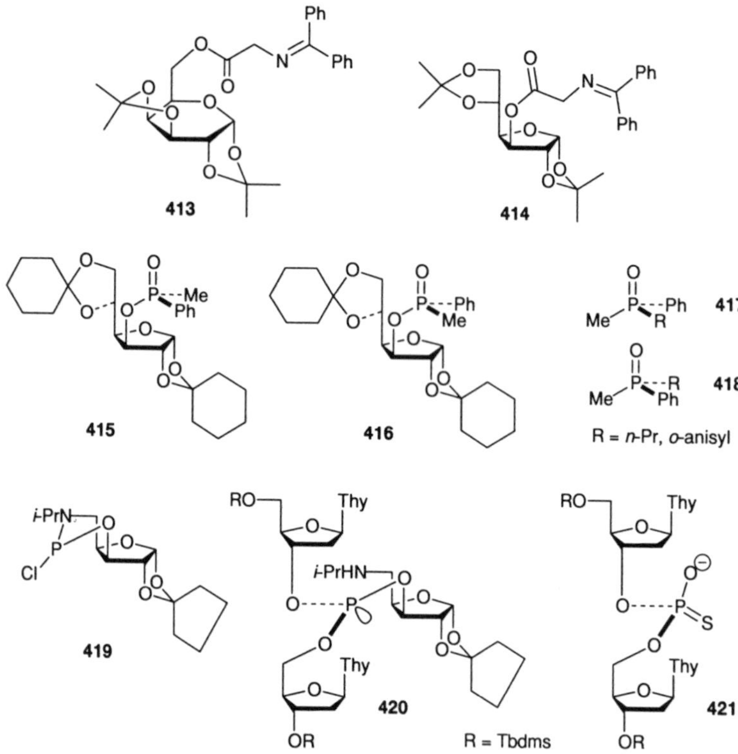

413	**414**

415	**416**	$Me\overset{O}{\underset{R}{\overset{\|}{P}}}{-}{-}Ph$ **417**

$Me\overset{O}{\underset{Ph}{\overset{\|}{P}}}{-}{-}R$ **418**

R = *n*-Pr, *o*-anisyl

419	**420**	**421**

R = Tbdms

A series of lactones **423** (R$_1$=Me, Ph and n=1,2) have been prepared with 30-99% e.e. using D-glucose-based anhydro sugars as chiral auxiliaries, the key stereochemical step being diastereoselective ketone reduction of **422** (Scheme 39).[119]

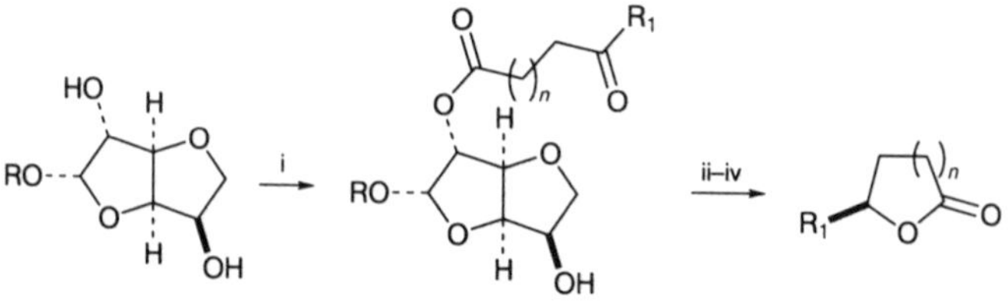

R = Me, Bn	**422**	**423**

Reagents: i, R$_1$C(O)CH$_2$(CH$_2$)$_n$COCl, Py; ii, ZnCl$_2$, NaBH$_4$; iii, LiOH; iv, HCl.

Scheme 39

The C$_2$-symmetric homochiral pyrrolidine **424** has been prepared in several steps from L-mannono-1,4-lactone *via* L-mannitol (for the enantiomer of **424** see Vol.25, Chapter 14, ref. 108).[120] Another interesting C$_2$-symmetric ligand the bipyridyl compound **426** was prepared from diacetone-D-glucose through oxidation and then diastereoselective addition of 2-bromo-6-lithiopyridine to give **425** followed by nickel dichloride/Zn-catalysed coupling.[121]

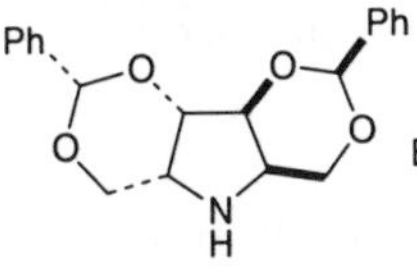

424 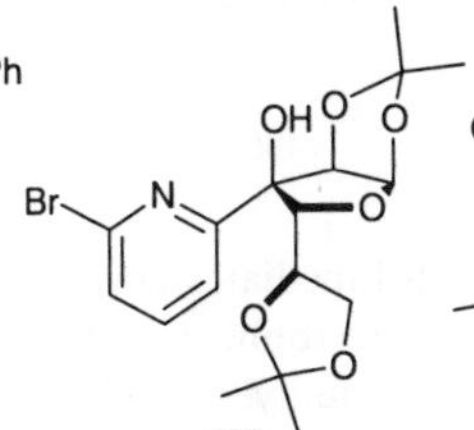   **425** 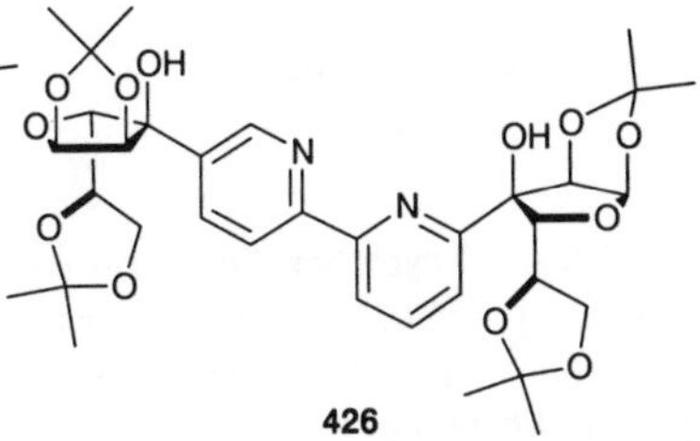   **426**

N-(β-Mercaptoethyl)pyrrolidine **427**, which is also C$_2$ symmetric, was effective for catalysis of diethylzinc addition to aldehydes,[122] as was the D-ribonolactone-derived hydroxypyrrolidine **428**.[123]

The amino alcohols **429** and **430** (prepared from diacetone-D-glucose, the amino group introduced by displacing a C-6 tosyl, and **430** prepared by deoxygenation of C-3) catalysed (at 5 mol%) diethylzinc addition to several aldehydes but with e.e.s of only 9-46% (*S* products predominated).[124]

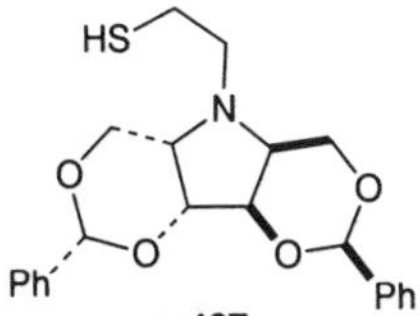

427 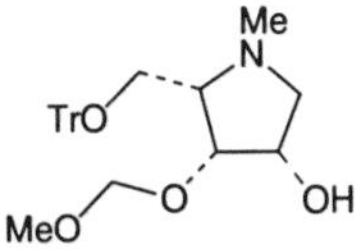   **428** 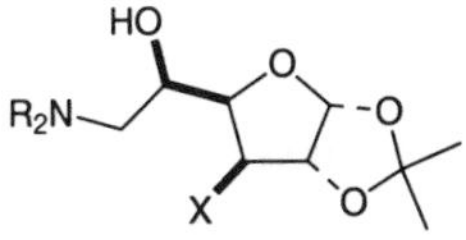   **429** X = OMe
430 X = H

Bisphosphinite carbohydrate derivatives have already seen very successful application to asymmetric hydrogenations. Evaluation of **431** as a ligand for Pd-catalysed asymmetric allylation with sodium dimethyl malonate gave **432** in 16-39% e.e. (*R* selectivity) with electron rich R groups (3,5-dimethyl, 3,5-di-*t*-butyl-4-methyl- and 3,5-ditrimethylsilylphenyl) and 17-77% e.e. (but *S* selectivity) with electron poor R groups (3,5-ditrifluoromethyl-, 3,5-difluoro-, 4-trifluoromethyl and 4-fluorophenyl). This electronic catalyst tuning could have valuable impact if e.e.s can be further optimized.[125]

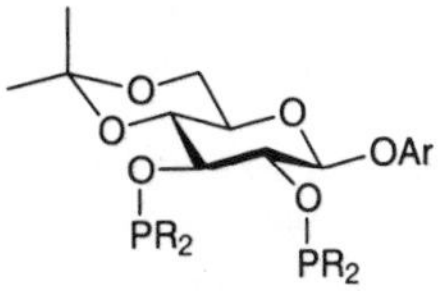

431 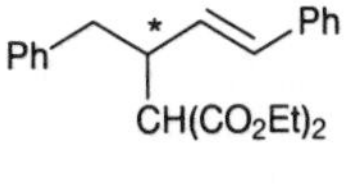   **432** 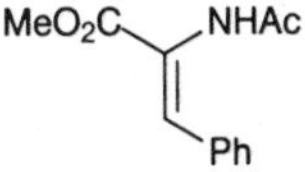   **433**

434   **435**

$R_1 = (CH_2)_2CO_2Me$
$R_2 = CH_2CO_2Me$
$R_3 = (CH_2)_2CO_2H$
$R_4 = CH_2CO_2H$

1,2:5,6-Di-*O*-cyclohexylidene-D-mannitol mediates asymmetric McMurry coupling (2 eq. $TiCl_3/Mg$, -70°C to r.t.) of acetophenone to give a 5.7:1 *dl:meso* selectivity.[126] Tetrafluoroborate salts of methyl 4,6-*O*-benzylidene-bis-2,3-(*O*-diphenylphosphine)-D-glycopyranoside Rh(COD) complexes catalyse asymmetric hydrogenations of **433**. The enantioselectivity (giving *S* configurations predominantly) decreases with the number of axial groups in the catalyst, for example β-D-*gluco* gives 90% e.e., falling to 60% e.e. for the corresponding α-D-*galacto* analogue.[127]

An example of a carbohydrate auxiliary being used for chromatographic enantioseparation is the derivatization of **434** with 3,4-*O*-benzylidene-D-ribonic acid δ-lactone. These diastereomeric products could then be separated by HPLC. The enantiopure **434** (containing a chiral quaternary centre) was used to prepare **435**, believed to be involved in the biosynthesis of porphyrins.[128]

References

1   M. Pipelier, M. S. Ermolenko, A. Zampella, A. Olesker and G. Lukacs, *Synlett*, 1996, 24.

2   J. Marco-Contelles, *J. Chem. Soc., Chem. Commun.*, 1996, 2629.

3   M. Adiyaman, J. A. Lawson, S.-W. Hwang and S. P. Khanapure, *Tetrahedron Lett.*, 1996, **37**.

4   J. Marco-Contelles, C. Destabel, P. Gallego, J.-L. Chiara and M. Bernabé, *J. Org. Chem.*, 1996, **61**, 1354.

5   S. Raul, R. Roy, S. N. Suryawanash and D. S. Bhakuni, *Tetrahedron Lett.*, 1996, **37**, 4055.

6   M. Bamba, T. Nishikawa and M. Isobe, *Tetrahedron Lett.*, 1996, **37**, 8199.

7   J. A. Marco, M. Carda, F. Gonzalez, S. Rodriguez, J. Murga and E. Falomir, *An. Quim.*, 1996, **91**, 103 (*Chem. Abstr.*, 1996, **125**, 196 189).

8   Y. Al-Abed, N. Naz, D. Mootoo and W. Voelter, *Tetrahedron Lett.*, 1996, **37**, 8641.

9   J. Lee, S. Wang, G. W. A. Milne, R. Sharma, N. E. Lewin, P. M. Plumberg and V. E. Marquez, *J. Med. Chem.*, 1996, **39**, 29.

10 J. Lee, R. Sharma, S. Wang, G. W. A. Milne, N. E. Lewin, Z. Szallasi, P. M. Plumberg, C. George and V. E. Marquez, *J. Med. Chem.*, 1996, **39**, 36.

11 J. Lee, N. E. Lewin, P. Acs, P. M. Plumberg and V. E. Marquez, *J. Med. Chem.*, 1996, **39**, 4912.

12 K. Mallareddy and S. P. Rao, *Tetrahedron*, 1996, **52**, 8535.

13 S. J. Waddell, J. M. Eckert and T. A. Blizzard, *Heterocycles*, 1996, **43**, 2325.

14 J. D. White and S. C. Jeffrey, *J. Org. Chem.*, 1996, **61**.

15 M. K. Gurjar, P. Kumar and B. V. Rao, *Tetrahedron Lett.*, 1996, **37**, 8617.

16 V. Popsavin, O. Beric, M. Popsavin, J. Csanádi and D. Milkovic, *Carbohydr. Res.*, 1996, **288**, 241.

17 P. Renaud and S. Abazi, *Helv. Chim. Acta*, 1996, **79**, 1696.

18 S. Chandrasekhar, S. Mohapatra and M. Takhi, *Synlett*, 1996, 759.

19 V. Jaouen, A. Jégon and A. Veyrières, *Synlett*, 1996, 1218.

20 J. L. Marco and G. Martn, *Tetrahedron: Asymm.*, 1996, **7**, 2191.

21 M. Kirihata, M. Ohe, I. Ichimoto and Y. Kimura, *Biosci. Biotech. Biochem.*, 1996, **60**, 677.

22 M. Matsumoto, H. Ishikawa and T. Ozawa, *Synlett*, 1996, 366.

23 A. Balog, M. S. Yu and D. P. Curran, *Synth. Commun.*, 1996, **26**, 935.

24 Y. Jiang and M. Isobe, *Tetrahedron*, 1996, **52**, 2877.

25 K. Takao, Y. Aiba, H. Ito and S. Kobayashi, *Chem. Lett.*, 1996, 931.

26 Y. Al-Abed, N. Naz, M. Khan and W. Voelter, *Angew. Chem., Int. Ed. Engl.*, 1996, **35**, 523.

27 A. V. R. L. Sudha and M. Nagarajan, *Chem. Commun.*, 1996, 1359.

28 J. Xu, A. Egger, B. Bernet and A. Vasella, *Helv. Chim. Acta*, 1996, **79**, 2004.

29 S.-H. Khang and C.-M. Kim, *Synlett*, 1996, 515.

30 L. Liu and W. A. Donaldson, *Synlett*, 1996, 103.

31 K. Yasui, Y. Tamara, T. Nakatani, I. Horibe, K. Kawada, K. Koizuni, R. Suzuki and M. Ohtani, *J. Antibiotics*, 1996, **49**, 173.

32 H. Nakamura, K. Fujimaki and A. Murai, *Tetrahedron Lett.*, 1996, **37**, 3153.

33 R. Rossin, P. R. Jones, P. J. Murphy and W. R. Worsley, *J. Chem. Soc., Perkin Trans. 1*, 1996, 1323.

34 A. K. Ghosh, J. F. Mincaid, D. E. Waters, Y. Chen, N. C. Chaudhuri, W. J. Thompson, C. Culberson, P. M. D. Fitzgerald, H.-Y. Lee, S. P. McKee, P. M. Munson, T. T. Duong, P. L. Darke, J. A. Zugay, W. A. Schlief, M. G. Axel, J. Lin and J. R. Huff, *J. Med. Chem.*, 1996, **39**, 3278.

35 A. Tenaglia and J.-Y. L. Brazidec, *Chem. Commun.*, 1996, 1663.

36 A. T. Khan, H. Dietrich and R. R. Schmidt, *Synlett*, 1996, 131.

37 M. I. Matheu, R. Echarri, C. Domènech and S. Castillón, *Tetrahedron*, 1996, **52**, 7797.

38 A. Tenaglia and F. Karl, *Synlett*, 1996, 327.

39 C. Capozzi, A. Dios, R. W. Frank, A. Geer, C. Marzabadi, S. Menichetti, C. Nativi and M. Tamarez, *Angew. Chem., Int. Ed. Engl.*, 1996, **35**, 777.

40 A. J. Fairbanks, E. Perrin and P. Sinaÿ, *Synlett*, 1996, 679.

41 J. S. Swenton, J. N. Freskos, P. Dalidowicz and M. L. Kerns, *J. Org. Chem.*, 1996, **61**, 459.

42 P. P. Deshpande, K. N. Price and D. C. Baker, *J. Org. Chem.*, 1996, **61**, 455.

43 J. P. Henschke and R. W. Rickards, *Tetrahedron Lett.*, 1996, **37**, 3557.

44 A. Bhattaharjee, A. Bhattaharjya and A. Patra, *Tetrahedron Lett.*, 1996, **37**, 7635.

45 T. K. M. Shing, C.-H. Wong and T. Yip, *Tetrahedron: Asymm.*, 1996, **7**, 1323.

46 D. Delorme, Y. Ducharme, C. Brideau, C.-C. Chan, N. Chauret, S. Desmarais, D. Dubé, J.-P. Falguegret, R. Fortin, J. Guay, P. Hamel, T. R. Jones, C. Lepine, C. Li, M. McAuliffe, C. S. McFarlane, D. A. Nicoll-Griffith, D. Riendeau, J. A. Yergey and Y. Girard, *J. Med. Chem.*, 1996, **39**, 3951.

47 Z. Ruan, P. Wilson and D. R. Mootoo, *Tetrahedron Lett.*, 1996, **37**, 3619.

48   J. T. Link, S. Raghavan, M. Gallant, S. J. Danishefsky, T. C. Chou and L. M. Ballas, *J. Am. Chem. Soc.*, 1996, **118**, 2825.

49   A. Nadin and K. C. Nicolaou, *Angew. Chem., Int. Ed. Engl.*, 1996, **35**, 1623.

50   S. Caron, D. Stoermer, A. K. Mapp and C. H. Heathcock, *J. Org. Chem.*, 1996, **61**, 9126.

51   D. Stoermer, S. Caron and C. H. Heathcock, *J. Org. Chem.*, 1996, **61**, 9115.

52   W. Zheng, J. A. DeMatte, J.-P. Wu, J. J.-W. Duan, L. R. Cook, H. Oinuma and Y. Kishi, *J. Am. Chem. Soc.*, 1996, **118**, 7946.

53   T. Oka, K. Fujiwara and A. Murai, *Tetrahedron*, 1996, **52**, 12091.

54   M.Isobe, S. Hosokawa and K. Kira, *Chem. Lett.*, 1996, 473.

55   D. P.Stamos and Y. Kishi, *Tetrahedron Lett.*, 1996, **37**, 8643.

56   K. C. Nicolaou, M. Sato, N. D. Miller, J. L. Gunzer, J. Renaud and E. Untersteller, *Angew. Chem., Int. Ed. Engl.*, 1996, **35**, 889.

57   K. Horita, T. Inoue, K. Tanaka and O.Yonemitsu, *Tetrahedron*, 1996, **52**, 531.

58   K. Horita, K. Tanaka, T. Inoue and O.Yonemitsu, *Tetrahedron*, 1996, **52**, 551.

59   C. S. Burgey, R. Vollerthun and B. Fraser-Reid, *J. Org. Chem.*, 1996, **61**, 1609.

60   P. Dauban, A. Chiaroni, C. Riche and R. H. Dodd, *J. Org. Chem.*, 1996, **61**, 2488.

61   B. Furman, Z. Kaluza and M. Chmielewski, *Tetrahedron*, 1996, **52**, 6019.

62   U. Vieth, O. Schwardt and V. Jäger, *Synlett*, 1996, 1181.

63   K. Shinozaki, K. Mizuno, H. Wakamatsu and Y. Masaki, *Chem. Pharm. Bull.*, 1996, **44**, 1823.

64   S. Lajsic, G. Cetkovic, M. Popsavin and D. Miljkovic, *Collect. Czech. Chem. Commun.*, 1996, **61**, 298 (*Chem. Abstr.*, 1996, **125**, 33 988f).

65   P. Norris, D. Horton and D. E. Giridhar, *Tetrahedron Lett.*, 1996, **37**, 3739.

66   D. Marek, A. Wadouachi, R. Uzan, D. Beaupere, G. Nowogrocki and G. Laplace, *Tetrahedron Lett.*, 1996, **37**, 49.

67   R. Shiraki, A. Sumino, K.-i. Tadano and S. Ogawa, *J. Org. Chem.*, 1996, **61**, 2845.

68   H. Yoda, H. Yamazaki and K. Takabe, *Tetrahedron: Asymm.*, 1996, **7**, 373.

69   N. Nakajima, M. Matsumoto, M. Kirihara, M. Hashimoto, T. Katoh and S. Terashima, *Tetrahedron*, 1996, **52**, 1177.

70   M. Malmberg and N. Rehnberg, *Synlett*, 1996, 361.

71   A. Hansen, T. M. Tagmose and M. Bols, *Chem. Commun.*, 1996, 2649.

72   W. H. Pearson and E. J. Hembre, *J. Org. Chem.*, 1996, **61**, 5537.

73   W. H. Pearson and E. J. Hembre, *J. Org. Chem.*, 1996, **61**, 5546.

74   W. H. Pearson and E. J. Hembre, *J. Org. Chem.*, 1996, **61**, 7217.

75   H. S. Overkleeft and U. K. Pandit, *Tetrahedron Lett.*, 1996, **37**, 547.

76   H. Zhao and D. R. Mootoo, *J. Org. Chem.*, 1996, **61**, 6762.

77   A. A. Bell, L. Pickering, A. A. Watson, R. J. Nash, R. C. Griffiths, M. G. Jones and G. W. J. Fleet, *Tetrahedron Lett.*, 1996, **37**, 8561.

78   B. Davis, A. A. Bell, R. J. Nash, A. A. Watson, R. C. Griffiths, M. G. Jones, C. Smith and G. W. J. Fleet, *Tetrahedron Lett.*, 1996, **37**, 8565.

79   I. Izquierdo, M. T. Plaza and F. Aragón, *Tetrahedron: Asymm.*, 1996, **7**, 2567.

80   C. E. Sowa, M. Stark, T. Hiedelberg and J. Thiem, *Synlett*, 1996, 227.

81   A. Tuah, M. Sanière, Y. L. Merrer and J.-C. Depazay, *Tetrahedron: Asymm.*, 1996, **7**, 2901.

82   C. V. Kim, L. R. McGee, S. H. Krawczyk, E. Harwood, Y. Harada, S. Swaminathan, N. Bischofberger, M.-S. Chen, J.-M. Sherrington, S.-F. Xiong, L. Griffin, K. C. Cundy, A. Lee, B. Yu, S. Gulnik and J. W. Erickson, *J. Med. Chem.*, 1996, **39**, 3431.

83   J. L. J. Blanco, C. O. Mellet, J. Fuentes and J. M. García, *Chem. Commun.*, 1996, 2077.

84   Z. Czarnocki, J. B. Mieczkowski and M. Ziolkowski, *Tetrahedron: Asymm.*, 1996, **7**, 2711.
85   K. Shinozaki, K. Mizuno and Y. Masaki, *Heterocycles*, 1996, **43**, 11.
86   K. Shinozaki, K. Mizuno and Y. Masaki, *Chem. Pharm. Bull.*, 1996, **44**, 927.
87   T. Murakami and M. Hato, *J. Chem. Soc., Perkin Trans. 1*, 1996, 823.
88   Y.-L. Li and Y.-L. Wu, *Liebigs Ann. Chem.*, 1996, 2079.
89   K. Shinozaki, K. Mizuno, H. Oda and Y. Masaki, *Bull. Chem. Soc. Jpn.*, 1996, **69**, 1737.
90   M. Morata, E. Sawa, K. Yamaji, T. Sakai, T. Natori, Y. Koezuko, H. Fukushima and K. Akimoto, *Biosci. Biotech. Biochem.*, 1996, **60**, 288.
91   Z.-H. Peng, Y.-L. Li, W.-L. Wu, C.-X. Liu and Y.-L. Wu, *J. Chem. Soc., Perkin Trans 1*, 1996, 1057.
92   S.-W. Hwang, M. Adiyaman, S. P. Khanapure and J. Rokach, *Tetrahedron Lett.*, 1996, **37**.
93   T. Mutou, T. Kondo, T. Shibata, M. Ojika, H. Kigoshi and K. Yamada, *Tetrahedron Lett.*, 1996, **37**, 7299.
94   T. K. Chakraborty and K. K. Gangakhedkar, *Synth. Commun.*, 1996, **26**, 2045.
95   K. Koseki, T. Ebata and H. Matsushita, *Biosci. Biotech. Biochem.*, 1996, **60**, 534.
96   J. Taunton, J. L. Collins and S. L. Schreiber, *J. Am. Chem. Soc.*, 1996, **118**, 10412.
97   C. Taillefumier, D. de Fornel and Y. Lapleur, *Bioorg. Med. Chem. Lett.*, 1996, **6**, 615.
98   A. P. Nin, R. M. de Lederkremer and O. Varela, *Tetrahedron*, 1996, **52**, 12911.
99   N. Sin and J. Kallmerten, *Tetrahedron Lett.*, 1996, **37**, 5645.
100   M. S. Ermolenko, *Tetrahedron Lett.*, 1996, **37**, 6711.
101   A. Kovács-Kulyassa, P. Herczegh and F. J. Sztaricskai, *Tetrahedron Lett.*, 1996, **37**, 2499.
102   K. Toskima, T. Jyojima, H. Yamaguchi, H. Murase, T. Yoshida and M. Nakata, *Tetrahedron Lett.*, 1996, **37**, 1069.
103   J. A. Marshall and L. M. Elliott, *J. Org. Chem.*, 1996, **61**, 4611.
104   J. Streith and A. Defoin, *Synlett*, 1996, 189.
105   A. Defoin, H. Sarazin and J. Streith, *Helv. Chim. Acta*, 1996, **79**, 560.
106   A. Loupy and D. Monteux, *Tetrahedron Lett.*, 1996, **37**, 7023.
107   T. K. M. Shing, H.-F. Chou and I. H. F. Chung, *Tetrahedron Lett.*, 1996, **37**, 3713.
108   D. A. Larsen, A. Schofield, R. J. Stoodley and P. D. Tiffin, *J. Chem. Soc., Perkin Trans. 1*, 1996, 2487.
109   R. Schumacher and H.-U. Reissig, *Synlett*, 1996, 1721.
110   A. B. Charette, H. Juteau, H. Lebel and D. Deschênes, *Tetrahedron Lett.*, 1996, **37**, 7925.
111   A. Nangia, P. B. Rao and N. N. L. Madhavi, *J. Chem. Res.*, 1996, (S) 312; (M) 1716.
112   M. Kishida, T. Eguchi and K. Kakinuma, *Tetrahedron Lett.*, 1996, **37**, 2061.
113   P. R. R. Costa, V. F. Ferreira, K. G. Alencar, H. C. L. Filho, C. M. Ferreira and S. Pinheiro, *J. Carbohydr. Chem.*, 1996, **15**, 691.
114   P. Köll and A. Lützen, *Tetrahedron: Asymm.*, 1996, **7**, 637.
115   M. N. Keynes, M. A. Earle, M. Sudharshan and P. G. Hultin, *Tetrahedron*, 1996, **52**, 8685.
116   G. Salladié, J.-F. SaintClair, M. Philippe, D. Semeria and J. Maignon, *Tetrahedron: Asymm.*, 1996, **7**, 2359.
117   A. Benabra, A. Alcudia, N. Khiar, I. Fernández and F. Alcudia, *Tetrahedron: Asymm.*, 1996, **7**, 3353.
118   Y. Jin, G. Biancotto and G. Just, *Tetrahedron Lett.*, 1996, **37**, 973.

119   V. Nair and J. Prabhakaran, *J. Chem. Soc, Perkin Trans 1*, 1996, 593.
120   B. Giese and S. N. Müller, *Tetrahedron: Asymm.*, 1996, **7**, 1261.
121   M. A. Peterson and N. K. Dallay, *Synth. Commun.*, 1996, **26**, 2223.
122   Y. Masaki, Y. Satoh, T. Makihara and M. Shi, *Chem. Pharm. Bull.*, 1996, **44**, 454.
123   M. Ikota and H. Inaba, *Chem. Pharm. Bull.*, 1996, **44**, 587.
124   B. T. Cho and N. Kim, *Synth. Commun.*, 1996, **26**, 855.
125   N. Nomura, Y. C. Mermet-Bouvier and T. V. RajanBabu, *Synlett*, 1996, 745.
126   N. Balu, S. K. Nayak and A. Banerji, *J. Am. Chem. Soc.*, 1996, **118**, 5932.
127   R. Selke, M. Ohff and A. Riepe, *Tetrahedron*, 1996, **52**, 15079.
128   M. A. Cassidy, N. Crockett, F. J. Leeper and A. R. Battersby, *J. Chem. Soc., Perkin Trans. 1*, 1996, 2079.

Author Index

In this index the number in parenthesis is the Chapter number of the citation and this is followed by the reference number or numbers of the relevant citations within that Chapter.

401

Murray, W.M. (9) 55
Murthy, P.P.N. (21) 42
Mussini, P.R. (3) 58
Mustareui, P. (22) 198
Mutou, T. (24) 93
Mutter, M. (3) 99
Myers, B. (1) 2; (18) 97

Nadin, A. (24) 49
Nafisi, K. (22) 210
Nagae, H. (4) 102
Nagai, M. (20) 2, 3
Nagai, Y. (3) 178, 226; (10) 44;
    (16) 47; (20) 299; (22) 39
Nagamura, Y. (3) 211
Naganawa, H. (19) 25, 66
Nagano, Y. (22) 84
Nagarajan, M. (5) 40, 41; (8) 20,
    21; (11) 12; (12) 8; (18) 29;
    (24) 27
Nagarajan, N. (10) 38
Nagarajan, P. (3) 116; (23) 18
Nagase, S. (3) 211
Nagashima, M. (19) 86
Nagatsu, A. (7) 39
Nagy, J.O. (4) 63; (10) 11, 12
Nahn, P. (22) 77
Nahrstedt, A. (22) 73
Nair, V. (20) 159, 301, 302, 305;
    (22) 173; (24) 119
Naito, S. (3) 128
Nakae, T. (3) 154
Nakahara, Y. (3) 148; (4) 115,
    142, 143, 171; (7) 19
Nakai, K. (8) 4; (18) 11; (19) 22
Nakai, T. (3) 292
Nakai, Y. (3) 250, 258; (4) 135;
    (6) 31
Nakajima, M. (4) 7; (19) 72
Nakajima, N. (10) 17; (19) 67, 68;
    (24) 69
Nakajima, S. (18) 165; (19) 86
Nakamura, A. (20) 66
Nakamura, H. (9) 29; (13) 23; (23)
    22; (24) 32
Nakamura, K. (4) 109; (5) 47; (13)
    32
Nakamura, K.T. (20) 37, 38; (22)
    165, 166
Nakamura, M. (3) 62
Nakamura, S. (3) 62, 128
Nakamura, Y. (17) 39
Nakane, H. (19) 89
Nakanishi, H. (4) 175; (22) 52
Nakanishi, K. (22) 186, 187

Nakanishi, T. (19) 89
Nakano, Y. (11) 25
Nakao, K. (3) 184; (11) 27; (16)
    19, 20; (22) 126
Nakashima, H. (4) 124; (5) 49; (7)
    78
Nakashima, M. (23) 20
Nakashizuka, M. (4) 102
Nakata, M. (3) 284; (13) 16; (19)
    31; (24) 102
Nakatani, T. (13) 27; (24) 31
Nakatsubo, F. (4) 79, 161, 178; (7)
    44, 49
Nakatsuto, F. (5) 50; (15) 6
Nakayama, M. (22) 28
Nakayama, T. (3) 105
Nakazawa, K. (9) 19; (19) 75
Nakmura, Y. (21) 31
Namane, A. (20) 232, 251
Namazi, H. (7) 20
Nangia, A. (24) 111
Narasaka, K. (7) 87
Nardelli, M. (22) 135
Narimatsu, H. (10) 43
Narita, K. (22) 211
Nash, R.J. (3) 236; (6) 7; (8) 28;
    (9) 4; (10) 19-21, 62; (16) 23,
    51; (18) 54, 71; (24) 77, 78
Nashed, M.A. (6) 8; (7) 16
Nasr, A.Z. (10) 57, 58; (18) 47
Natarjan, S. (22) 61
Nativi, C. (3) 66; (11) 9; (24) 39
Natori, T. (24) 90
Naumov, V.A. (18) 16; (22) 129
Naundorf, A. (15) 5
Naurinsh, Y. (20) 63
Navarre, N. (20) 303
Navaza, J. (22) 91
Naven, T.J.P. (22) 43, 45
Nawrot, B. (20) 247
Nayak, S.K. (24) 126
Naz, N. (14) 29; (24) 8, 26
Nazarov, A.A. (7) 62
Nechaev, A. (20) 238
Neelima, N. (20) 54
Neenan, J.P. (20) 333
Neidle, S. (22) 154, 163
Neidlein, U. (21) 66
Neilsen, C. (20) 11
Nelles, G. (4) 188
Nepogod'ev, S.A. (3) 70; (4) 129,
    183
Nesper, R. (3) 269; (22) 141
Neszmelyi, A. (5) 53
Netzel, T.L. (22) 210
Neumann, K.W. (4) 99

Neuß, O. (14) 2; (17) 29
Newton, M.G. (8) 7; (20) 87
Ng, C.J. (2) 16; (6) 6
Nga, N. (18) 112
Nguedack, J.-F. (14) 28
Nicas, T.I. (19) 81
Nichols, C.J. (19) 61
Nicholson, J.K. (23) 11
Nicolaou, K.C. (4) 74; (19) 83;
    (24) 49, 56
Nicoll-Griffith, D.A. (24) 46
Nicolosi, G. (18) 124, 128
Nicotra, F. (3) 290; (13) 37; (17) 3
Niddam, V. (20) 310
Niedzielski, C. (7) 71
Nieger, M. (14) 2; (17) 29
Nielsen, C. (20) 74, 91, 107
Niemela, K. (16) 11
Nifant'ev, E.E. (7) 62
Nifant'ev, N.E. (3) 141; (4) 53, 68,
    108; (21) 73
Niklasson, G. (18) 18
Nikolaev, A.V. (4) 122
Nikolaeva, T.G. (20) 331
Nikonov, G.K. (7) 9
Nillroth, U. (18) 18
Nilsson, K.G. (3) 170
Nilsson, U. (3) 228; (11) 39
Nin, A.P. (16) 14; (24) 98
Ning, J. (4) 91
Ninomiya, S. (9) 19; (19) 75
Nishi, N. (4) 201; (6) 12
Nishida, M. (20) 52
Nishida, Y. (3) 167; (21) 4, 5; (22)
    188, 189; (23) 13
Nishihara, S. (10) 43
Nishijima, M. (10) 10; (16) 27
Nishikatani, M. (22) 32
Nishikawa, T. (24) 6
Nishikitani, M. (3) 220; (21) 51
Nishimura, S. (3) 88, 163
Nishimura, T. (3) 154; (4) 79, 161
Nishimura, Y. (9) 76, 83; (18) 66;
    (19) 14, 71, 73, 74
Nishino, H. (7) 39
Nishio, T. (3) 241; (12) 9
Nishioka, I. (7) 47
Nishiyama, S. (19) 56, 64
Nishizawa, M. (7) 17
Nishizono, N. (20) 133, 321
Nissl, J. (2) 73
Niyazymbetov, M.E. (14) 41
Nockolds, C.E. (22) 142
Noda, M. (3) 313; (19) 89
Noda, N. (7) 68
Noe, C.R. (20) 276